THE
FIRE OF HEPHAESTUS

RETHINKING SCIENCE, RETHINKING NATURE

KARL ROGERS

Copyright © 2016, 2020 Karl Rogers

First and Second Editions

Published by Trébol Press

Los Angeles, California

www.trebolpress.com

All rights reserved. No part of this publication may be reproduced or transmitted in any form or by any means, electronic or mechanical, including photocopying, recording, or any other information storage and retrieval system, without permission in writing from the publisher.

info@trebolpress.com

ISBN-10: 0-9903607-5-X

ISBN-13: 978-0-9903607-5-9

Printed in the United States of America on recycled paper.

For My Grandfather

M.C. Davey

CONTENTS

1 Introduction: Moving Beyond the Traditional Philosophy of Science   1

2 Empiricism and Realism   43

3 Origins of Mechanical Realism   115

4 The Technological Framework   173

5 Scientific Experience and Observations   225

6 The Technological Society   271

7 Back to Nature   367

References   423

"What we observe is not Nature itself, but Nature exposed to our method of questioning."

--Werner Heisenberg (1971: 58)

INTRODUCTION:

BEYOND THE TRADITIONAL PHILOSOPHY OF SCIENCE

Science has brought wonderful new inventions into the world. Explanatory theories such as the atomic theory of the Periodic Table or Darwin's theory of evolution through natural selection have provided us with general explanations of all the phenomena under their rubric, such as how all the chemical elements can be comprised of atoms, which are in turn comprised of permutations and combinations of a simple set of three subatomic particles (protons, neutrons, and electrons), or how the wide variety of species of life on the planet Earth can be explained as natural variations selected through reproductive fitness and adaptability to their environment. Einstein's General Relativity not only explains gravitation and the perihelion of the planet Mercury, but also predicts a vast range of phenomena, such as gravity lensing, black holes, and time dilation in atomic clocks undergoing different rates of acceleration. The Standard Model of Elementary Particles reduces all matter to a set of fundamental particles called fermions and all radiation to a set of fundamental quanta called bosons, each associated with specific kinds of fundamental force (electroweak, strong nuclear, gravitation). These kinds of theory, when taken together as a whole, provide us with a comprehensive and general scientific worldview. It allows us to explain the variation and structure of the world in terms of fundamental elements and the basic principles or rules by which those elements are combined to form the totality of all that exists. The scientific worldview explains the wide variety of life on Earth and the structure of the Universe, and thereby tempts us to believe that all natural phenomena can be explained in terms of a single theory based on a simple set of elements and rules for their combination. Such a Grand Theory or Theory of Everything would be the culmination of scientific endeavors for over 400 years, beginning with the Scientific Revolution of the sixteenth and seventeenth centuries. In terms of grand ideas of cosmic order and unity under a simple set of axioms or logical principles, especially mathematical principles and laws, science finds its precursors in the ancient philosophies of Greece and Egypt, Babylon and Sumeria, and India. These were the ideas that gestated throughout Europe, North Africa, and the Middle East during the so-called Medieval or Middle-Ages, and were rediscovered afresh during the Renaissance.

To make myself clear from the outset, I do not doubt that science brings new powers, knowledge, and experiences, all of which have transformed the world and the human condition within it. My line of questioning is not whether science provides us with a comprehensive worldview or general explanation of the world. I do not doubt that science tells us something about the world and provides us with the ability to use that understanding to change the world and our material conditions. My line of questioning is one of asking: how does science do this? How is science possible? What are its historical and cultural conditions? What are its intellectual origins? Within what contexts are its discoveries made and its theories tested? How does science encounter the world? How is science done in practice? It is my assumption that we cannot hope to understand science, its methods and their results, unless we know what science is and how it is possible. This involves acknowledging that science is done by human beings, living and acting from within societies, with their own culture and history, and this insight shows us the conditions and contingencies of scientific theory and knowledge. It involves acknowledging that the existence of science, it methods and results, depend on how certain human beings think about the world, organize their actions within it, and interpret its results. This insight can be a bitter pill to swallow if one is looking for science to provide one with certain and universal knowledge of an asocial, ahistorical, and non-human nature. The bitterness of this pill, however, is itself merely the product of categories of thinking that have been imposed on us by the traditional philosophers of science. One must either be an objectivist or a relativist about science, according to these categories. To acknowledge the historical and cultural contingency of science and the scientific worldview is often dismissed as "relativism" by the traditional philosophers of science, and by doing so allows them to ignore all the problems and lines of thinking that such an acknowledgement brings with it. It seems that even to question the human contingencies upon which objectivity depends is to transgress categorical boundaries and limits. Yet it is on such a journey of transgression that I invite the reader to join me.

Who are the traditional philosophers of science? The traditional philosophers of science include Bertrand Russell, Alfred North Whitehead, Carl Hempel, Otto Neurath, Rudolf Carnap, and Karl Popper, and their followers.[1] These philosophers all assumed without question that physics, chemistry, and biology are natural and empirical sciences, and the theories produced by these sciences are representations (descriptions and approximations) of "the Laws of Nature." These "laws" are assumed to exist independently of science. The theoretical representations of these laws revealed through scientific methods are tested against experience via experiment. For the traditional philosophers, philosophy is the handmaiden of science and her job is to clarify and organize the concepts and methods of science into a methodology—the *logos* (logic, explanation, account, or first principles) of the scientific method. They tell us that science is theoretical, logical, rational, abstract, and based on reason and experience. Science is the intellectual organization and application of methods to produce theories and test them against experience via experiments. Experimental apparatus and instruments are merely the technological means to provide "the facts" to determine empirically (via measurement and observation) which theory "fits the facts" better than the others. The traditional philosophers of science all agree that science is successful at doing this and achieves progress in empirical accuracy and predictive power. They all agree that the representations provided by scientists are "closer to the truth" or "more logically consistent" about Nature than past representations. They all agree that science has led to a growth in knowledge.

Disagreements between the traditional philosophers of science are concerned with the character of the scientific method by which representations are rationally ordered and justified by the empirical data—the sum total of measurements and observations—or shown to be inaccurate and in need of correction. Here the traditional philosophers divide themselves into two categories: empiricists and realists. The empiricists tend to focus their philosophical efforts on the task of deciding which series of logical operations best characterize the general (or universal) principles of the scientific method of deciding between theories. The realists tend to emphasize the importance of causal explanations to the intelligibility of the scientific method. For the empiricist, the scientific method is

---

[1] Hempel (1966); Russell (1962); Whitehead (1962, 1997); Carnap (1995); Neurath & Carnap (1955); Popper (2002, 1975, 1974). Contemporary advocates of the traditional philosophy of science include Bhaskar (1975, 1986, 1989); Harré & Madden (1977); Harré (1986); Wolpert (1992); Maxwell (1998); Psillos (1999); Norris (2000); Ladyman & Ross (2009); Chakravartty (2010); French (2014).

primarily a logical process of justification of the choice between theories on the basis of "the facts," whereas, for the realist, it is primarily a logical process of discovery and the role of theory is to explain "the facts." For the empiricist, science largely operates in *the context of justification*, whereas, for the realist, science largely operates in *the context of discovery*.

What have the traditional philosophers got wrong? First of all, the traditional philosophers have ignored the anthropological dimension to science. They have ignored that science is an activity conducted by human beings. As a result, they have ignored the being for whom scientific knowledge constitutes knowledge, for whom it is possible to have an experience and communicate it in a learned and public language, and without whom science would not exist at all. They have ignored the being to whom knowledge and experience happen and are meaningful. They have ignored the fact that there are intelligibility requirements for knowledge and experience, and, therefore, they have failed to understand that meaning and truth are conditioned and motivated by human choices and interests, culture and tradition, conventions and customs, and leaned ways of thinking about the world. They have ignored how scientific knowledge and experience emerge from and are situated within a history and culture, and it is this history and culture that gives science its meaning, value, and truth. Science is a way of thinking and acting in the world that is practiced by human beings for a variety of different reasons, and science is the product of human agency. This subjects it to social and psychological influences that motivate and shape how science is understood and done in practice. The scientific method is itself contingent and conditioned. What qualifies as scientific knowledge and experience, measurement and observation, theory and its adequate test are all contingent on the outcomes of the application of standards and norms that are decided and established through human agency in the context of the culture within which that human agency occurs and is given its meaning as a means of exploring and understanding the world. If we hope to understand science, we need to examine and understand the contingencies and conditions that inform and constrain the methods and results of science, and this cannot be done by dismissing such lines of inquiry as "relativism" and paying them no further attention.

Science must be communicated between human beings if it is to be shared, tested, and corroborated. This requires the existence of a shared language and other shared representational practices, such as drawing diagrams or using mathematical equations, or, if they do not already exist, they must be developed from already existing communicative and representational practices. Human agency and culture are involved at all stages of language-use and representational practices, including those used during private reflection, and their further development to deal with novel phenomena. Language and representations are causal in making observations and measurements possible in the first place, alongside the possibility of their successful communication, but when mediated by shared communicative and representational practices, the medium itself imposes cultural limits and possibilities on what can be communicated and understood, as well as providing the intellectual resources required to communicate and understand anything, and these cultural influences inform, constrain, and select between the possibilities of communication and representation. How we understand the world and communicate that understanding are always mediated by our culture. If we hope to understand science, we need to understand how specific communicative and representational practices mediate (inform, constrain, and select) thinking, action, and experience to produce observations and measurements, knowledge and theory. We need to understand how human beings came to understand the possibilities and limits of science as a means of knowing the natural world.

Science is researched, communicated, taught, tested, and developed between human beings to satisfy human purposes. Its methods, theories, and results are produced, disseminated, and reproduced via the production and reproduction of representational and material practices, which includes the guidelines of how to perform and interpret them properly, and, consequently, science should be treated in the same way as all cultural practices, including all the practical arts and crafts, music, sculpture, literature and poetry, agriculture and animal husbandry, horticulture, medicine, religion, navigation and cartography, philosophy, etc. We can understand science as a cultural practice only once we recognize that science is a product of human agency and situated within relations between humans, whose thought and language mediated by shared communicative and representational practices, as well as having shared beliefs about the world and how to act within it, and thereby informed and constrained by shared standards and norms, all with a history of development and refinement, controversy and struggles, choices and roads not taken. This recognition moves our thinking about science beyond the abstractions of the traditional philosophy of science, and it opens our reflections on science to include studies and insights into science as a human activity, wherein the sum total of methods, theories, and results are a specific set and arrangement of organized representational, interpretation, and material practices, with their own history of invention and development; they are made meaningful in relation to a historically developed set of cultural norms, standards, and values.

This affords us the possibility of studying and understanding science and the culture from which it came in the same way as we would study and understand any human activity, regardless of whether this behavior occurred in a so-called primitive or civilized culture, whether it is a religion or any traditional practices and interpretations. Unless we 'go native' and lose our critical distance—and simply believe what scientists say about science and the meaning of their own behavior—we should not treat science differently from any organized human activity which involves the use of representational and material practices to give meaning and significance to the results of the activity. This usage is learned through shared language, community membership, education and other forms of cultural dissemination. It is this shared background of practices and interpretations that allows human beings to share the same perspective or worldview, share resources and cooperate towards the innovation and refinement of the same methods to solve shared problems and satisfy the same needs; it also allows us to talk meaningfully of practitioners sharing the same culture when we observe the same behavior given the same interpretation by the practitioners themselves.

In order to tell this story about science, we must explain how science came to be and how it spread. We need to place the kinds of human activity that produces and reproduces science in the historical and cultural contexts of its practitioners developing new representational and material practices to reveal and explore new phenomena in new ways, and communicate these practices to other people by demonstrating how to make them meaningful to others in the same way as they are to the practitioners; thereby initiating others into these new representational and material practices and allowing their reproduction and further dissemination by others to others. This is achieved by understanding how the instructions or guidelines about how to reproduce specific human practices—what we might term as the scientific method—are transmitted by teaching others how to reproduce the same representational and material practices, and interpret them using the same language and cultural terms. Through education and other kinds of cultural dissemination, theoretical descriptions and explanations of how the world works in terms are taught and conveyed in forms that are meaningful to those human beings versed and skilled in how to interpret and articulate those descriptions and explanations in the terms of a shared understanding or worldview, and it is this shared understanding of the world and how to live in it that constitutes culture.

Importantly, in Western culture, the traditional philosophers have ignored the technological nature of science. Science has produced the means of production by which human beings satisfy material needs and change material conditions. It brings forth new powers into the world and changes the world and culture from which it came, and, thereby, makes new possibilities and problems. It is developed and given its cultural value because science innovates new techniques, instruments, and machines as its products and resources, and also provides us with new representations of the world by which these products and resources are interpreted and given meaning. These changes and innovations opens an expanding horizon of future trajectories of innovative solutions, and new problems and unforeseen consequences. The changes and new problems they bring, which require further changes and solutions, constitute the horizon of research and development. Science makes new experiences and discoveries possible, which brings the possibility of new powers and challenges, thereafter changing both science and the culture from which it came. Within scientific culture, the scientific method can be understood as the ongoing project of using science to further advance and integrate the human organization of representational and material practices, alongside the development of unified theoretical explanations of the world, thereby moving towards an increasingly universal and comprehensive conception of human agency within the world, as well as improve the efficiency and productivity of human labor and administrative methods.

Scientific knowledge and truth presuppose (and require) a particular mode of engagement with beings, namely scientific activity, wherein mathematical representations inform and constrain theoretical concerns and representations by revealing and excluding aspects of experience, and this cannot be separated from the being, namely human beings, for whom mathematics and scientific theories are meaningful and (approximately) true. Once we admit human agency into the philosophy of science, this involves admitting language, culture, and history too, as well as all the factors that influence human agency and life, into which we are born, learn to speak (and read and write) from others (our family, schools, other children) in a language that pre-exists each of us, and thereby gains meanings and truths from others, which of course we claim as our own through the very processes by which we encounter and learn them, just as others did before us. All our efforts to describe or explain our experiences of non-human nature are made in human terms and involve cultural and historical influences that have shaped the very language, including mathematics, within which those human experiences of the non-human were made meaningful and truthful. Without some Archimedean point on which a human being can stand outside of the Earth—outside history and culture, as well as outside Nature—the scientific objectification of Nature cannot be sustained as anything other than a cultural and historical prejudice in favor of the meaning and truth of mathematics, based on little more than a naturalistic faith in human culture, language, reason, and perception as *somehow* corresponding to objective reality.

Cultural relativism and Science

Cultural relativism has been at the forefront of anthropology since it was pioneered in America in the early twentieth century by the German anthropologist Franz Boas. For Boas, culture generates and shapes human interactions and behavioral norms, and he used this concept of culture to establish and legitimate bringing together the disciplines of archaeology, linguistics, and physical anthropology as sub-disciplines of cultural anthropology. In his 1911 book *The Mind of Primitive Man*, Boas rejected the traditional (nineteenth century) anthropological classification scheme that categorized peoples who practice writing as "civilized" and peoples who do not practice wring as "primitive." He not only rejected this division of humanity into these two categories of people, but Boas also rejected the 'order of rank' between different kinds of society that it entailed. Instead, he argued that cultural plurality is a fundamental characteristic of the human species, and the biological, linguistic, and ethnographic traits of any group of human beings is an autonomous product of cultural changes and their historical development within specific geophysical surroundings.

Boas' early studies of Baffin Island Inuits, and later his wider studies of Northwestern Pacific peoples, confirmed his belief in "the psychic unity of mankind," as he termed it, which idealized the assumption that no culture was inherently superior to any other with respect to intellectual capability. All other things being equal, different peoples around the world had the same level of intellectual ability but exercised it differently. Different cultures had different ways of figuring out how to satisfy the same needs. There is not any *a priori* reason to assume that behavior in one culture is better than behavior in another, if both ways of behavior are capable of satisfying the same need. Thus the cultural differences between peoples are more the result of historical accidents rather than any inherent biological differences (i.e. racial differences) between different groups of people. There is no impartial reason to believe that nineteenth century Europeans were more advanced, or intellectually superior, than Baffin Island Inuits.

Influenced by his early studies in physical geography and the problems of perception that arise when trying to make quantitative measurements of the colors reflected off different kinds of water, Boas became critical of the idea of objective and value-neutral science, especially in the field of anthropology wherein human beings from one culture attempt to study human beings from another culture. He opposed the dominant notions of social evolution that had come out of nineteenth and early twentieth century interpretations of Charles Darwin's *The Descent of Man* (1871) and *Origin of the Species* (1859). Boas was a staunch critic of biological theories of race, as well as biological determinism in general. He was particularly concerned about the application of biological theories of evolution to anthropology and how we understand the origin and fundamental properties of human nature. According to Boas, these interpretations were not only based on misrepresentations of Darwin's ideas, but also resulted in what Boas termed as pseudoscience, which had been used by European anthropologists to justify ideological assertions of white racial supremacy and the manifest destiny of Western civilization in general. These pseudoscientific interpretations had been particularly damaging for anthropology, having been promoted for ideological purposes rather than scientific ones.

Throughout his whole life, Boas tirelessly continued to oppose racism and notions of cultural superiority and institutionalized notions of white supremacy, and, in his later life, he explicitly opposed Nazi Germany and their race-based pseudoscience. Boas' approach not only called for respecting the cultures of so-called "primitive" peoples, but focused on how individuals learn, negotiate, and navigate the representational and material practices of their culture, thereby showing how so-called "primitive" people were not some still existing version of the prehistoric ancestors of Europeans, but were independent and alternative cultures in their own right. Instead, Boas dedicated his intellectual efforts to challenging and discrediting biological determinism in anthropology. For Boas, cultural anthropology is not concerned with the biological inheritance of individual anatomical, physiological, or mental characteristics, but is instead concerned with the diversity of these traits among peoples found in different geographical areas and social classes. It was the task of anthropology to investigate the differentiation of "the psychic unity of humanity" and reveal the historical sequence of events that led to these differences—and the plurality of human cultures around the world, by showing how culture was the dominant factor that shaped human behavior. In order to understand how any particular culture is acquired and disseminated, and how culture mediates human behavior, beliefs, and representational and material practices, Boas emphasized the importance of looking at human agency and the interactions between human beings. This is especially important when attempting to understand how human beings learn and use languages.

Boas advocated *cultural relativism* and emphasized interactions between different cultural groups (i.e. migration) as the primary causal factor during cultural change and innovation. (Today, we would also include trade alongside migration as being primary factors involved in cultural dissemination and change throughout human history.) Although convinced that migration was the dominant causal factor in cultural change, Boas did not deny that (Darwinian) evolution played a role in shaping human bodies and needs, but he did deny that it was the dominant determinant in the development and differentiation of linguistics, perception, meaning, knowledge, and the social organization of symbolic and material culture. He proposed a fluid, plastic, and dynamic idea of culture, which required anthropology to be practiced scientifically on the basis of ethnographic fieldwork rather than asserting the superiority of the culture of the observer over the culture of the observed. Ethnographical field research, when combined with historical and archaeological evidence of migration, showed that cultural boundaries are permeable and overlapping, as human beings from different cultures meet and interact. By comparing different skull sizes and shapes between Inuits and Europeans, Boas showed that not only was there a considerable degree of overlap between the spread in measurements between these two groups of peoples, but that diet and general health during childhood were the dominant factors that governed skull size and shape, rather than race or inherent biological differences between the two peoples. These measurements and observations supported Boas' belief in the dominance of ethnographical factors, such as geographical and sociological conditions, which Boas termed "the influence of the surroundings" (what we would term today as "environmental conditions"), over how society and human behavior changes, and this put historical and contextual analysis, alongside studies of physical geography and archaeology, at the forefront of Boa's ground-breaking ethnographic methodology for the new discipline of cultural anthropology.

Today, the cultural anthropology of science is a field in its own right. Since Thomas Kuhn's *The Structure of Scientific Revolutions* (1962), there has been a growing awareness of the historical and cultural dimensions of natural science, and the possibility of studying and explaining changes in representational and material practices in natural science as the cultural and historical products of human agency within a community of practitioners. Kuhn criticized the traditional philosophy of science for failing to recognize the historical dimension of scientific theory and practice, and how this historical dimension conditions and generates changes in scientific culture and thinking. He explained that changes in theory and the scientific worldview do not necessarily correspond to the outcomes of logical tests of scientific theory in relation to observations and measurements, but, instead, are the products of consensus within the scientific community. Natural science and the scientific worldview are the products of human relations within a specific culture with its own historical development. How science is taught, published, and researched within the scientific community and its institutions (such as schools, colleges, universities, research groups, prestigious and peer-reviewed journals, textbook editors and publishers, funding bodies and committees, Noble Prize committees and awards, and popular science magazines, television, and other media) gives science its cultural and normative value as a method of inquiry into Nature. Kuhn described science as going through periods of what he termed as *normal science*, during which basic concepts and fundamental representations, alongside conventions and standards of measurement, knowledge and methodology, and intellectual values and norms constitute a *paradigm*. The paradigm is shared by the scientific community acting as a community of practitioners. The task of normal science is to apply the paradigm to problem-solving activities within all areas that are determined from within the paradigm to be scientific problems, which are the problems that can be solved from within the paradigm. The line of demarcation between science and non-science, and what constitutes "good" or "bad" science, are determined within the paradigm through consensus between practitioners, and therefore what constitutes a valid line of inquiry and what does not.

While the word 'paradigm' finds its etymological stem in the fifteenth century, from Latin and Greek, meaning "to show side by side," which is largely compatible with Kuhn's general usage of the term, Kuhn's notion of paradigm has been criticized for its vagueness. Margaret Masterman famously counted 21 different uses of the word 'paradigm' in the first edition of Kuhn's *Structure of Scientific Revolutions*.[2] Kuhn's notion of 'paradigm' has also been criticized for its circularity, given that Kuhn originally defined 'paradigm' as being the collection of values, standards, assumptions, and representational practices shared by the same scientific community, and he defined a 'scientific community' as being a collection of practitioners sharing the same paradigm. Kuhn conceded this and took considerable pains to clarify the definition of the term 'paradigm' in postscripts to subsequent editions.[3] He also modified the term 'paradigm' in his 1977 work *The Essential Tension* to include material practices and the technical aspects of scientific research, and termed it as a *disciplinary matrix*. Putting aside the criticism of the ambiguity and circularity of Kuhn's original definition of the term 'paradigm,' and, for now, also putting aside problems that arise from his incommensurability thesis, his basic insight about the cultural and historical nature of science still stands. Kuhn's insight tends to run contrary to the epistemological individualism (or 'naturalism') dominant in the traditional philosophy of science, as he emphasizes the cultural and historical dimensions involved in the development and acceptance of new theories. Rather than arriving at a new scientific theory in an intuitive flash of individual genius, as many of the traditional philosophers have presumed, and its truth accepted or rejected by the wider community on the basis of its self-evident empirical success or failure, the situation is more of a complex human struggle taking a generation or more, involving many people and institutions, and the struggle itself often transforms the form and content of theories, as well as the criteria under which "the facts of experience" are determined, weighed, interpreted, accepted or rejected. It is not necessarily the case that the accepted scientific theory is the best explanation of the experimental results, but rather is the theory that has survived the cultural and historical processes involved in establishing consensus about the demonstration of the successes or failures of any particular theory.

How scientific theories and methods are selected and tested are conditioned and determined by the scientific community and paradigm within which the sciences are categorized as empirical and natural sciences, their methods and standards are established as a mode of inquiry into natural phenomena, and the scientific worldview is secured as

---

[2] Masterman (1965)
[3] Kuhn (1970)

corresponding to a world that supposedly exists independently of scientific inquiry. The paradigm produces and reproduces preconceptions about the object-area of any scientific inquiry, whether it even exists, and how to study it. Anything outside of the paradigm cannot be considered to be part of science. Not only is the object-area of scientific disciplines determined by the current paradigm, but, within specific disciplines, the paradigm also conditions decisions about which problems in the discipline are considered to be fundamental problems and how they should be approached and solved, and which scientists, experiments, and theories are considered to be *exemplars* used to teach a new generation of students how the discipline should be practiced if it is be accepted as being scientific. Today, disciplines such as physics, chemistry, and biology are paradigmatically categorized under "natural sciences," while psychology, sociology, and anthropology are paradigmatically categorized under "human sciences," and parapsychology, astrology, and homeopathy are paradigmatically categorized under "pseudosciences." The epistemological standards and norms of science are produced and reproduced through institutions, such as universities, journals, and scientific societies, and thereby shows that science is determined in relation to other human beings and their beliefs, values, and norms. What qualifies as "a scientific observation," "explanatory success," "a testable theory," and "the scientific method"—what qualifies as "science" itself—are culturally determined through consensus using concepts and representations that are historically contingent and change with time. Thus, what qualifies as "the empirical" is contingent on the cultural and historical factors, which includes values and other supposedly non-scientific factors that condition and determine the content of the paradigm.

Kuhn described periods of change within science as *scientific revolutions*. These historical changes occur as shifts between paradigms, and thereby a radical change in the total set of standards of knowledge, methodologies, basic concepts, fundamental representations, intellectual values and norms, and pedagogical methods to teach new generations of scientists how to be scientists. Paradigm shifts occur periodically to deal with crises in the old paradigm. These crises occur when a set of anomalies and contradictions cannot be explained in the terms of the old paradigm, which also cannot solve a growing number of fundamental problems, and these simply cannot be ignored or dismissed any longer. Within the scientific community, there is a growing awareness that the old paradigm is fundamentally flawed in its assumptions, cannot progress yet remains incomplete, and a tipping point has been reached. A new paradigm arises when a new generation of scientists invent new concepts, ideas, and fundamental representations to explain and resolve the contradictions and anomalies of the old paradigm, and thereby solve the fundamental problems it cannot solve, and thereby create a new way of thinking about science and how the natural world can be explored and understood.

When the concepts and definitions of the new paradigm do not share common meanings with those of the old paradigm, the two paradigms are said to be *incommensurable* with each other—lacking common measures—and therefore do not share common referents or objects of theory. 'Matter' and 'position' did not mean the same thing in Aristotle's physics as they did in Galileo's physics; nor did the two physics share the same standards of observation and hypothesis testing. Some concepts, such as "impetus" in the Aristotelian physics, do not have any corresponding concept in the new physics. The Aristotelian emphasis on understanding the purpose or *telos* of a natural phenomenon, as a guiding explanatory principle in how it comes into being, is an idea that has no place in the Galilean physics. The basic concepts, definitions, and fundamental representations of Galileo's new science were radically different from those of the Aristotelian philosophy, leading to a different set of epistemological rules, principles, and standards. Even though the Copernican heliocentric system and Ptolemaic earth-centric system have about the same level of predictive power (or empirical accuracy), the older ideas of the quadrant and retrograde motion simply do not make sense or have any use in the new system, and it was the geometrical simplicity and coherence of the Copernican system, along with its appeal to the Neo-Platonism of the Renaissance, that inspired Galileo (Kuhn, 1957). It is therefore quite impossible to compare the two physics, without simply siding with one paradigm over the other, as they are simply different ways of thinking about the world and are referring to different things. Thus the 'paradigm shift' can only be understood in wider cultural context in relation to clash between a traditional way of thinking and a new generation of thinkers.

Galileo's new physics not only arose in response to the failures and dead ends of Aristotelian physics, but as a reaction against the dominance of the scholastic tradition of Medieval Aristotelianism and the Catholic Church over university learning and natural philosophy. The basic concepts, definitions, and fundamental representations of Galileo's new science were refined and developed into the Newton's laws of motion and universal gravitation, and further refined over the next two hundred years to become what is now known as Classical Physics. This paradigm provided the fundamental representations and mathematical methods to develop the classical theories of mechanics, thermodynamics, acoustics, ballistics, electromagnetism, gravity, and matter, culminating in the Laws of Thermodynamics, Maxwell's Laws of Electromagnetism, the wave theory of light, and the atomic theory of matter and radiation, within an absolutely determined universe of matter and radiation behaving in accordance with immutable mathematical laws in absolute space and time. After the Scientific Revolution of the sixteenth and seventeenth century, the natural philosophers and physicists of the eighteenth century refined and developed the methods of empirical science—what Kuhn termed as the normal science phase—which by the end of the nineteenth century had become a profession and positivistic enterprise, promising to explain everything in terms of a simple set of mathematical laws, from which all possible events and interactions could be derived and tested against observation and measurement. Yet, by the turn of the twentieth century, the inability of the Michelson-Morley interferometer to measure the Earth's aether drift and the inability of Classical Physics to explain this null-result. Along with Classical Physics' inability to explain the spectrum of blackbody radiators, the photoelectric effect, and the contradiction between Newtonian mechanics and Maxwell's equations, the Michelson-Morley null result had resulted in a crisis that was only resolved when a new generation of physicists were able to free themselves from the old concepts of absolute space and time, and the physical determinism and continuity they entailed, and develop and explore new theories such as Einstein's Theory of Relativity, Lorentz transformations, Quantum Theory, the Uncertainty Principle, Wave-Particle Duality, Bose-Einstein and Fermi statistics, the Big Bang Theory, and the Standard Model of the subatomic theory of matter and radiation. It is this new way of looking at how to solve the scientific problems of the day and resolve the contradictions and anomalies of the old paradigm that constituted what Kuhn termed as the revolutionary science phase.

Terms like "mass" or "charge" simply do not mean the same thing in the new paradigm of Quantum-Relativistic Physics as they did in the old paradigm of Classical Physics, and other terms, such as "spin," do not have any correlate in the old paradigm at all. The new paradigm involves a whole new way of approaching scientific practice, explanation, and representation that would not make any sense in the terms of the old paradigm. This is why a 'paradigm shift' is so disturbing and disruptive to the older generation of scientists. It is not simply the case that the new physics was able to explain observations that the old physics could not. The new physics offered a radical new scientific worldview and conceptual framework that not only had different interpretations of how to engage in scientific activity, but it also offered a new way of thinking about the meaning of observing and measuring natural phenomena. It transformed how natural science was understood. This led Kuhn to make what is arguably the most controversial point of his whole thesis: that is impossible to compare observations made from within either of the two paradigms, without siding in favor of one paradigm over the other, and therefore it is impossible to say that science progresses or better describes the facts of experience, given that "the facts of experience" are dependent on which paradigm one is operating under when the facts of experience were determined, collected, or discovered. The traditional philosophers have been unaware that what constitutes "a fact" or "an explanation" of a fact is itself contingent on the cultural and historical factors that determined whether it is established through consensus as a fact or an explanation of it, and whether it can be assumed to be obvious or self-evident, given already existing ideas and assumptions. At all stages, and at every level, theories and the facts of experience remain contingent and mediated by an already-existing culture, which is transmitted and embodied in human behavior, and, at no stage, do human beings encounter any objective or context-independent natural world.

Over the last fifty years, the history and philosophy of science (HPS) has provided us with a great deal of detailed research into the historical and cultural contexts of novel experimentation and observation, and the innovation of representational and material practices in those contexts, and how the facts of experience are constructed.[4] As Kurt Danziger argues in his 1990 book *Constructing the Subject*, it has become fundamental to studies in HPS that scientific observation is understood as a human activity situated in historical and cultural contexts, and these reveal the theoretical concepts, representations and interpretations involved in making observations possible and intelligible as repeatable observations of otherwise invisible and unknowable theoretical entities. In *Measurement and the Making of Meaning* (1990), HPS scholar David Gooding also advanced this argument on the basis of his analysis of the laboratory notebooks of Michael Faraday and historical studies of nineteenth century experiments in electromagnetism and the development of "the field" as the fundamental representation of how electromagnetism works. By looking at historical case studies of how working scientists actually performed experiments, published their findings, dealt with controversy and criticism, and weaved their work back into the wider scientific community and culture from which it emerged, we can see how the historical and cultural contexts of building, performing, and interpreting experiments gave form and content to "the facts of experience" and how they were described, interpreted, and explained in theoretical terms. These historical studies of the fundamental importance of the development and refinement of representational and material practices to produce and reproduce new phenomena, which involves the development and demonstration of new skills and apparatus to other people, by communicating to others how to reproduce the experiment. These studies have shown us that theories and experiments are developed in relation to each other through the ongoing development of representational and material practices, and how the cognitive and logical comparison between theoretical expectations and observations occurs within historical and cultural contexts.

In his 1986 book *The Neglect of Experiment*, physicist and historian Allan Franklin argued that the traditional philosophy of science has neglected the role of experiment in the epistemology of physics. Franklin showed that paying attention to the history of experimental particle physics reveals how theory and experiment are interweaved by human agency during the actual construction, development, use, and reception of

---

[4] Hackmann (1978); Billig (1987); Suchman (1987); Gooding, Schafer, & Pinch (1989).

experiments, such as Millikan's oil-drop experiment to measure the charge to mass ratio of the electron, to produce results that are used by physicists to confirm (or refute) a theory or decide between competing theories. Franklin's interpretation of the history of experimental particle physics has been confirmed and further developed by physicist and historian Peter Galison in his books *How Experiments End* (1987) and *Image and Logic* (1997). Galison's work shows how the intellectual and material culture of high-energy physics must be examined as situated and developed within a community of practitioners, and shows how the cultural and historical contexts of the development and use of representational and material practices shape consensus regarding theory and the meaning of experimental results. Franklin and Galison confirm historian and philosopher Ian Hacking's thesis in his 1983 book *Representing and Intervening* that representational and material practices are interweaved and inform each other when making observations and measurements during the construction and performance of actual experiments. Rather than passively observing the results of experiments, and then comparing these with theoretical predictions, real scientists actively use theories to invent new forms of representation and use these to design, perform, and interpret experiments, measurements and observations, and communicate to the scientific community how to repeat these experiments, measurements and observations.

Experimentation is a theory-laden activity. It takes considerable education and training, effort and time, and cultural resources—intellectual and material resources—to translate the experiences acquired within the context of the experiment into observations and measurements. It takes further work to interpret these experiences as revealing "objective" context-independent facts, general theories, and universal laws, and this additional work also occurs in historical and cultural contexts, adapting and modifying available cultural resources—including linguistic and cognitive resources—in order to establish something new and weave this new product into an already existing scientific culture as a resource available to others. It requires successful communication with other people to reproduce scientific results and make them meaningful within scientific culture. Scientific thinking does not occur in isolation from the culture from which it emerges, and it does not follow any linear trajectory or "internal logic," and what constitutes science or the scientific method changes from time to time. Non-scientific conditions shape human relations, beliefs, behavior, and decisions, and these contingent modes of human agency generate and transform the nature of the scientific method along with the knowledge and the representations of reality it produces and reproduces.

What Kuhn's *Structure of Scientific Revolutions* did for HPS and advancing the cultural anthropologist's story about how natural science occurs in historical and cultural contexts, which motivate and shape human behavior—changing how human beings think about and experiment on natural phenomena—and explaining why theories and knowledge in the natural sciences are contingent and mutable, Michel Foucault's *The Order of Things* did for social theories of science in general, especially epistemologies of the human or social sciences, and how the categories of language and representation are social constructs.[5] In this book, Foucault explored "the will to truth" in operation throughout the history of Western civilization. He described this "will to truth" as an endless consequence of the human desire for certainty and power rather than a progressive development towards enlightenment. Foucault analyzed knowledge in Western civilization by performing *an archaeology of knowledge* and revealing *strata of epistemes* (standards or rules for acquiring and ordering knowledge) that have periodically undergone radical transformations to deal with anomalies and contradictions that occur in the dominant discourses of social power

---

[5] *The Order of Things* was first published in 1966 in French under the title *Les Mots et les Choses* (Paris: Gallimard) and the first English translation was published in 1970.

at different periods. By taking texts at different periods and looking for shared rules of organization, Foucault identified five archeological strata or *epistemes*: Medieval, Renaissance, Classical, Nineteenth Century, and the Present. Foucault defined an *episteme* as being the total set of relations that unite, at a given period, the discursive and textual practices that give rise to epistemological figures, methods, categories, and formalized systems of knowledge, such as sciences. Looking at the texts at the beginning of each period, Foucault uncovered that way that each *episteme* produced and reproduced a different classification system of knowledge and representations, which not only allowed distinctions between "truth" and "error" by providing a foundation for knowledge, but made the previous *episteme* nonsensical or flawed. Once we recognize this, we must also accept that there is no reason to presuppose that the present *episteme* will not be considered to be nonsensical or flawed in the future.

The medieval *episteme* formed a discourse in which the world and the Word of God were the one and the same, and the Church was its ordained interpreter and the diviner of its rules. The order of things was understood as the interconnected symbols of the Word of God, and it was the right and duty of the Church and its scholastic tradition to provide the systematic categorization and collection of the order of things and their proper interpretation. For the Church, this exegesis of the world was as much a political as a hermeneutic or theological one, given that the Church was both the executive and custodian of the authority to reveal this exegesis, thereby denying any epistemological basis for prioritizing experience over the authority of the Church. Only the Church could give the proper interpretation and meaning of experience in terms of God's plan and the human place within it, as revealed by the text of the world and the Bible—as interpreted and taught by the Church. The medieval *episteme* was made absolute by the Church's power. From the perspective of the present, the medieval *episteme* formed a strange and almost unthinkable way of looking at the world. It was a period focused entirely on suffering and pleasure in which an exegesis was to be performed to understand the human place within God's creation. The world was read like a book for signs and portents on how one should live one's life in accordance with the Word of God, who, after the fall of Adam and Eve from Eden into a world of suffering and toil, had benignly given us clues so we could know God's plan to redeem each human soul by teaching Man how to overcome Original Sin and find the key to the paradise of heaven by solving the riddle of the world. Knowledge was an unending chain of resemblances and similarities, which were connected together to determine the human place in the order of things, as a privileged place in God's creation. This provided a system of knowledge that allowed the medicinal properties of herbs and parts of animals, for example, to be known through the resemblance between the shape of a flower or animal part and the afflicted organ or appendage. Thus the mandagora root (or mandrake) was beneficial for fertility; the red lotus was good for the heart if taken in the morning when the flower blossomed; powdered rhino horn would act as an aphrodisiac because the horn was shaped like a phallus; and a poultice for bravery could be made from a lion's heart. A similar way of thinking can be still found in traditional Chinese medicine. These similarities were placed there by God to allow people to understand the order of things.

With the Renaissance, all this changed. The failure of the Church to maintain itself as custodian of the Word of God and the exegesis of the world was the result of its failure to maintain its absolute authority and power. This was more the result of Martin Luther than it was Galileo, but it opened a space that allowed new forms of discourse about the world. This new possibility resulted in a mutation of the rules of discourse that opened up a gap between words and things, and the philosophy of Rene Descartes became possible. Henceforth there were things on one side of the gap and different ways of describing them in words on the other. Language and thought became *representational*. The structures of thought, language, and the world became distinct from one another, and it became possible to have the idea that there was a discoverable way of bridging the gap—a correspondence—between representations of things and the things themselves. This was necessary for the subject-object distinction to become possible via the division of the world into subjective and objective properties, and for mathematics to become the language in which "the Book of Nature" was written—a *mathesis universalis*—that could bridge that gap. This change allowed the rise of a new *episteme* that categorized language as a system of representation—that represented things to the human mind—and science ordered language into a system of eternal and universal knowledge about the world, upon which the possibility of rational thinking was premised. Language henceforth became a grid that could be superimposed over reality to reveal its essential structure whereby essence was revealed in terms of proper names and categories. Natural history and science became possible—the Enlightenment became possible and found its culmination in the philosophy of Immanuel Kant.

The classical *episteme* began with the Enlightenment promise and exciting prospect of success in revealing the essential order of things by placing them under the correct categories of names. The Enlightenment promised complete and eternal knowledge, which would give Man certainty and power over Nature, predict and exploit Nature, and allow Man to know himself, be free, moral, and rational, and build and live in a civilized world, within which men could live as equals in accordance with their own individual reason and conscience. By the turn of the eighteen century, the Classical *episteme* had arisen on the possibility of ordering the whole Universe into categories and tables of objects and their proper relation with each other. The whole classical system of knowledge was organized around the task of naming and identifying the proper category and relations of things, thereby providing eternal and universal knowledge of the order of things. The study of Man in terms of customs, grammar, history, and the wealth of nations became possible, and, according to Foucault, the idea of Man arose from the classical *episteme*. Of course this does not imply that Foucault denied that men and women existed before the eighteenth century. What he meant was that the abstract category of Man, under which all men and women could be listed, as well as all the peoples of the Earth, was an invention of the eighteenth century. Nature was another category against which Man could be held as distinct from, and even in opposition. Man and Nature were only brought together under the category of Creation, which was itself placed under the category of the Creator: the God of Man and Nature. Aboriginal and indigenous peoples around the world could be henceforth subcategorized as "uncivilized" or "primitive," or even as "savages," closer to "animals" and "Nature" than "civilized man." It was this kind of categorization that was frequently used to justify the colonial exploitation of "primitive man" by "civilized man" and requiring the intervention of "civilized man" for their proper education and entry into "civilization." The relationship between Man and Nature, as epitomized by the Biblical fallen Adam and Daniel Defoe's *Robinson Crusoe*, became defined in terms of Man's use of reason, language, knowledge, and the practical arts to overcome and master Nature through ingenuity and tools.

The classical *episteme* and its nominalization and categorization of all things reached its limit. Akin to the Tower of Babel, it reached its dizzy heights, but fell short of its goal and came crashing down in the nineteenth century. An awareness of change, progress, variation, mutation, and differences came to dominate nineteenth century thought. Words became variable in what they could represent and taxonomies revealed their incapacity to accommodate certain borderline cases, anomalies, and "monsters." The classical *episteme* failed to provide accurate predictions and could not incorporate a concept of Time in its systems of representation and categorization. History was no longer understood as a category under which past events and persons were listed. Instead, History became capitalized and understood *as a force* that obeyed its own laws of development. Through the new awareness of Time and History, the ideas of self-organization, progression, and evolution came to dominate scientific thinking and discourse, and knowledge about the world could be ordered in terms of sequences and succession of events, where the task of science is to reveal the principles by which the world orders itself. Franz Bopp was able to discover the shared origins of the grammatical forms of Indo-European languages in Sanskrit and how the meaning of words had changed in time in relation to the development of a succession of languages, such as Greek, Latin, and Nordic languages, which shaped the development of European languages. Philology became possible. George Cuvier's idea of reducing animals and plants to self-organizing structures (coining the term "organisms") comprised of the temporal development and arrangement of organic cells became fundamental to how life was understood. The modern science of biology was born. David Ricardo was able to describe the wealth of nations in terms of economic laws of development that became the cornerstone of the science of economics, the notion of free trade, and globalization. Economics could be treated as a force that determines the actions of people, rather than the other way around, and hence people need to understand that force and apply it to national trade policies, if they hope to increase the wealth of their nation. The idea of economic forces as simultaneously creative and destructive became possible, and therefore economic forces could be described in the same terms as natural forces, as objective and transformative forces with their own logic of development. Hence, at the beginning of the nineteenth century, the classical systems of categorizing grammar, natural history, and wealth were replaced with philology, biology, and economics.

"The March of Progress" had been invented and "the Future" became something objective towards which people should strive for their own liberation. Once History became understood as a force (spirit or will), the philosophies of G.W.F. Hegel and Arthur Schopenhauer became possible, which formed the points of departure that led to Friedrich Nietzsche's idea of the *Übermensch* as the overcoming of Man at the end of History. History was a force of its own, which shaped and determined Man, who was destined to participate for better or worse in making "the Future" happen. Whether given optimistic or pessimistic conclusions in the nineteenth century philosophies and sciences, History became understood as an irreversible motion forward—an arrow of Time—which drove human thoughts and actions, rather than the other way around. Charles Darwin's *Origin of the Species* and *Descent of Man* became possible as theoretical descriptions of the relation between Man and Nature, wherein History became a special category of Natural History, itself the consequence of Time and the motion of matter in space, in all its possible variation. Karl Marx's historical dialecticism became possible and his definition of the essence of Man as one of producing the conditions of Man's existence by transforming Nature into the means to produce those conditions. Man's thoughts and actions were determined by his material conditions, and, through transforming Nature, Man makes new thoughts and actions possible, thereby making his own history as something distinct from and superimposed over his natural history as the biological species *Homo sapiens*. Even in those idealistic nineteenth century philosophies which had some space for radical creative freedom, such as Henri Bergson's *Time and Free Will* (1889), *Matter and Memory* (1896), or *Creative Evolution* (1907), an extrapolation of the present into the past was necessary for a proper orientation towards the future, and the psychic creation of reality should be viewed as a historical process or construction towards mental liberation and enlightenment rather than a steady-state or equilibrium.

In the present *episteme*, a new awareness has become possible. We are led back to the place where Friedrich Nietzsche asked "Who speaks?" and Stéphane Mallarmé answered "the Word." It has now become possible to reveal that constructive role of language in shaping human thought and action, especially those uses of language to know the world and the nature of things. This awareness is centered on language itself—as both the subject and object of language—and this has resulted in *a language anxiety* about the problems that arise when language becomes the means to reveal itself. Knowledge takes the form of either formulations or interpretations of how language is able to reveal itself, and how text and discourse are produced, disseminated, and reproduced. Reflexivity and irony become intellectual routes to circumvent the inevitable paradoxes that arise when language is used to reveal language. It is within this *episteme* that Foucault's discourse became possible, along with postmodern and poststructuralist thought. Jacque Derrida's *Of Grammatology* became possible. We are led back to this place not by science but by language itself. It signals the disappearance of Man and allowed Foucault to diagnose a crisis in Western culture (including science and logic) due to the impossibility of achieving the kind of transcendental knowledge that philosophy has identified itself with since Plato and Aristotle. Concerned with discovering the origin and essence of things, and thereby discovering universal and eternal truths, this crisis in Western civilization has resulted from the anthropocentric idea that all questions about being must be questions about human being and the possibility of "knowing ourselves," while cloaking such questions in the poetry of some transcendental realm in which truth resides that is somehow accessible to special kinds of discourse and narrative. Thus the social and natural sciences are caught up with this crisis because their contradictory commitments to objectivity and anthropocentrism cannot reveal the being for which knowledge is intelligible as knowledge, without undermining its project of revealing the truth which is the truth regardless of whether it is known or not. This returns us to back to Socrates and problem of knowing how human beings can have knowledge at all. Accordingly, in the present *episteme*, science (including mathematics) are a special kind of literature or narrative that has gained its authority through structures and patterns of language acquisition and usage, without having any concrete referent or corresponding to anything outside of itself. All language use becomes something metaphorical or poetical—a *catachresis* that always misses its mark—all the while being taken seriously as literal with its objective referents. "Man" is nothing more than the product of the classical *episteme*, and is disappearing into becoming the vehicle for language—the place where language happens and from which it flows—and nothing else. It is language that tells us what we are and what we can say. It is this line of thinking that lead us back to that place from which Ludwig Wittgenstein finished his *Tractatus*: "Whereof one cannot speak, thereof one must remain silent."

It is from within the present *episteme* that the sociological critics of the traditional philosophy of science has arisen. While the sociology of science dates back to the work of Robert King Merton in the 1930s[6], which in many respects remained within the nineteenth century *episteme*, since Foucault and Kuhn a so-called strong program in the sociology of scientific knowledge (SSK) has arisen to study science as a sociological phenomenon, wherein the scientific production and reproductions of representations is the object of study.[7] The roles of social consensus, authority, and power in the definition, content, ordering, and dissemination of scientific knowledge and the associated representation of reality have been discussed and analyzed at great length. The nature of

---

[6] Merton (1938, 1968, 1978)
[7] Greenberg (1967); Bloor (1976); Latour & Woolgar (1979); Knorr-Cetina (1981); Collins (1985); Latour (1987); Pickering (1987).

the scientific method, knowledge, and the representation of reality have been studied as socially constructed phenomena.[8]

Traditional philosophers of science have been heavily criticized by the sociologists of science for their neglect to attend to the social nature of scientific practices and experiments.[9] The ways scientists reproduce findings and deal with the obstacles to communicating the techniques and knowledge needed to replicate new results have been examined and discussed at length in science studies over the last half century. By paying close attention to the social and material practices within the real-time processes of scientific work, it has become possible to study the interaction between investigative, observational, and communicative techniques with the content of what is experienced or disclosed as being a natural phenomenon, mechanism, or law at work within the laboratory. Historical and contemporary scientists have been studied at work in order to analyze in detail how they have chosen their materials, built, organized, and operated experiments, and produced representations, facts, and knowledge. The question of how procedures for the evaluation and dissemination of scientific expertise and knowledge have become institutionalized has become central to the way that real experiments have become understood within the sociology of science. The social importance of material practices to the construction and communication of facts, concepts, techniques, has not only become central to the way that real experiments are understood, but these studies bring the social and historical conditions of both scientific discourse and technological practices into question. It takes considerable effort and time to translate the experiences acquired within the local context of the experimenters' work into context-independent facts, general theories, hypotheses, or universal laws, which are then available to other scientists.

In SSK, scientific knowledge is determined by social conditions and factors, and all scientific theories, whether determined to be true or false, are conditioned by human relations and interests, and the structures of language itself, with all its specialized formations and interpretations. The anthropocentric core and contingency of objective knowledge is revealed. As with HPS, in SSK there has been considerable focus upon the way that scientists negotiate the dissemination of their work from the local context of the laboratory to the wider public domain of shared representations, results, and facts. Scientists have been studied at work and how they have chosen their materials, built, organized, and operated experiments, and produced representations, facts, and knowledge has been analyzed at length. HPS and SSK critics of the traditional philosophers of science have also examined that way that the rules by which the representational and material practices have been ordered, related, and analyzed have changed and become accepted as part of "the scientific method." By paying close attention to the ways that representational and material practices are innovated and ordered within the real-time processes of scientific work, the interaction between investigative, observational, and communicative techniques with the content of what is experienced or disclosed have also been studied in detail. The tactics that scientists use to

---

[8] Merton (1970), Mathias (1972), Feibleman (1982), Easlea (1983), Harding (1986), Ince (1986), Marcus (1986), Billig (1987), Bijker (1987), Yearly (1988), Danziger (1990), Haraway (1991), Kaufman (1993), Aronowitz (1996), Galison and Stump (1996), Galison (1997), Stiegler (1998), and Gordo-Lopez and Parker (1999).

[9] Latour and Woolgar (1979), Ihde (1979), Knorr-Cetina (1981), Levidow and Young (1981), Heelan (1983), Collins (1985), Franklin (1986), Galison (1987), Latour (1987), Suchman (1987), Gooding, Pinch, and Schaffer (1989), Gooding (1990), Ihde (1991), Pickering (1992), Crease(1993).

reproduce findings and deal with the obstacles to communicating the techniques and knowledge needed to replicate new results have also been examined at length.

From HPS and SSK, a new generation of Science and Technology Studies (STS) has grown to examine science and technology comprehensively in terms of social relations, economics, politics, and power.[10] In relation to both historical and sociological researches, STS has explored in detail the determining role of social consensus, context, authority, and power in the definition, content, ordering, and dissemination of scientific knowledge and the scientific worldview, which mirrors and reinforces the cultural norms and prejudices of the society within which science emerges. According to STS, it is only as a result of the social conditions and organization of technology that science could become culturally situated as a special kind of activity that is capable of knowing and explaining otherwise unknowable and inexplicable causes of the structures of the world. Hence, according to STS, technology is itself a social construct that can be explained in terms of social relations, conditions, and norms. As part of the process of undermining the traditional philosophy of science, STS proponents attack the foundations upon which "science" has gained authority and power, and, as a consequence, they tend to be concerned with the social organization, legitimization, authority, and justification of "the facts" and how their discovery is reported, disseminated, and received. They look at how science has become weaved into the fabric of society, and how this in turn has shaped how we understand what "the scientific method" is.

By examining the institutionalization of procedures for the evaluation and dissemination of expertise and knowledge, the importance of the public reception of representational and material practices for the construction and communication of facts and theories has become central to the way that real experiments are understood within HPS, SSK, and STS. These studies bring the social and historical conditions of both representational and material practices into question, and thereby challenge their objectivity and correspondence to structures within Nature, understood as an underlying material world. Some STS proponents have gone even further and attempted to place "Nature" into historical and social contexts in order to show that scientific knowledge, observations, practices, and values are completely contingent constructions and instruments of authority and power, reflecting only cultural norms and prejudices, and there is no such thing as "the scientific method" nor any rational means to demarcate science from non-science.[11] Focusing on the social construction of "non-human agency" in science, these proponents of STS argue that objectivity and Nature itself are social constructs and nothing else. In the light of this insight, we should be critical of the traditional distinctions between nature/society, human/non-human, fact/fiction, subject/object, and truth/power, given that the traditional philosophers of science have ignored the social conditions and factors that allow such distinctions to be disseminated and accepted intelligible and "self-evident" assumptions about the world. Even if we remain uncomfortable with the idea of Nature as a social construct, we still have to address the critiques and problematization of objectivity and value-neutrality raised by proponents of HPS, SSK, and STS. The traditional philosophers have ignored the linguistic and conceptual practices that allow such distinctions and dualities to be postulated and articulated as meaningful and necessary, and this reveals how the traditional philosophers are immersed in a culture within which such distinctions are ready-made and available as intellectual resources and linguistic artifacts.

---

[10] Foucault (1980); Easlea (1983); Hughes (1983); Harding (1986); Ince (1986); Haraway (1989, 1991); Galison and Stump (1996)
[11] Haraway (1989); Latour (1991); Fuller (1993)

Taking the insights of HPS, SSK, and STS in account, we can see that science cannot be properly understood without understanding the human relations and interests that produced it as a cultural and historical phenomenon. It cannot be properly understood without understanding the society within which it is emergent and how it gains its cultural status as a special rational activity. In this respect, proponents of HPS, SSK, and STS have taken seriously Nietzsche's, James's, Heidegger's, Dewey's, and Wittgenstein's insight that truth is a value.[12] As a result of their uncritical acceptance of the subject/object distinction, traditional philosophers of science tend to consider values to be "subjective" or "merely psychological," and, therefore, values have no correspondence to the objective or natural world. They assert a clear distinction between the "pure" science concerned only with truth and the "applied" science of value in the wider world. According to this traditional view, how people in wider society value the products of scientific work is not the responsibility of scientists and has nothing to do with the form and content of science, and how science is weaved into society does not change its meaning because science is "self-evidently" value-free by definition. The proponents of STS and SSK (and to a lesser extent HPS) have taken considerable pains to show example after example of actual scientific work where this is evidently not the case, and they have critically examined the "objectivity" or "neutrality" of science in relation to the political and economic values of the wider society. They have also taken considerable pains to demonstrate how the very character of "science" is transformed by the demands of the wider world in which science is emergent and integrated, and, consequently, means different things to different people at different times, very much dependent on the contexts in which human discourse and practices are declared to be "scientific." This is true of all science, and not just ideologically-driven cases of science or pseudo-science, such as phrenology in nineteenth century Britain, Lysenkoism in the Soviet Union, or 'the Aryan science' of Nazi Germany. The fundamental insight shared by proponents of HPS, SSK, and STS is that we need to look at the totality of scientific discourse and practice—how science is done, reported, disseminated, and received by working scientists—and how choices and directions of scientific research are emergent from and weaved back into the culture in which science-as-practiced is situated and developed, and given meaning. It may well turn out, if we were to apply the abstract definitions of "the scientific method" asserted by various traditional philosophers of science to real scientific research, that working scientists do not practice "science" at all. This in itself should be deeply troubling to the traditional philosophers of science. However, traditional philosophers of science continue to reject (or ignore) the insights of HPS, SSK, and STS, and, when they are aware of them at all, they dismiss them as "relativism," and no further discussion is required.

The Neglect of Technology

Detailed discussion of the traditional philosophers of science must wait until the next chapter, but one thing they all have in common is that they tend to ignore technology completely— being nothing more than a means to an end—or place it under the category of "applied science" and forget about it. They have unanimously neglected or rejected the very idea that technology might have any substantive role or agency in shaping the development and content of theories within experimental sciences such as physics, chemistry, and biology. Experiments simply provide us with "empirical data" to verify or falsify theories by testing their correspondence to predictions and observations.

---

[12] Rorty (1981)

Traditional philosophers of science have paid no attention whatsoever to how experiments are actually designed, constructed, and performed as tests of theory—after all, that is the job of scientists—and they have ignored the technological activities involved in the production and implementation of fundamental representations of material practices, which are termed as "natural mechanisms" e.g. friction or electromagnetic induction, required to design and build experiments in the first place as a means of disclosing those mechanisms and testing theoretical representations and predictions of machine performances. The outcomes of experiments (the machine performances reported as observations and measurements)—the "data"—are simply "the facts" against which the empirical adequacy of theories are to be judged. These "facts" are loosely considered to be synonymous with "experience."

It is often supposed by the traditional philosophers of science that theory precedes and anticipates experiment in the form of hypotheses, conjectures, or predictions, and experiments are simply designed to test them. The traditional philosophers of science presuppose that science provides us with a rational understanding of Nature and modern technologies embody that understanding as "applied science." This "received wisdom" is that technology is the logical consequence of the application of scientific knowledge and rational thought to human problems in the material world. Technology is supposedly only a means to satisfy human purposes and has no role in shaping scientific knowledge, the conception of rationality, or human intentionality. The construction of theories is supposedly a purely intellectual affair for which the technology of experimentation does not have any constitutive scientific role. The experimental apparatus and methodology are supposed to be ontologically and epistemologically neutral. Technology is something that, at most, is a matter for applied ethics. This view presupposes that rational thought and logic transcend the material world and that the primary relationships between the human mind and the world are those of cognition, manipulation, and control. It is an anthropocentric view of the non-human. It presumes that technology enhances and extends the powers of the human mind and senses without changing or directing either.

In the traditional view of the Scientific Revolution of the sixteenth and seventeenth centuries, the rise of modern technology is taken to be derivative from the mathematical sciences and modern technology is taken to be "applied science."[13] However, the proponents of this view have not provided us with a satisfactory account of how this "application" occurred. It was for this reason that Alfred North Whitehead considered the way that highly abstract mathematically formulated theories have been "effectively applied to practical affairs" to be the "paradox" of modern science.[14] Ernst Nagel argued that the relations between modern science and technology are not as obvious and clear as the traditional philosophy of science has assumed.[15] He maintained the traditional view that modern technology is applied science but was aware that the character of "application" is an ambiguous one. How did the mathematical sciences lead to modern technology? How were mathematical and technological practices both conceptually connected? How was this conceptual connection justified?

Hence I argue that the traditional realist and empiricist philosophies of science are—*at best*—only intelligible in terms of an internal rationale, which gives realism and

---

[13] As well as the traditional philosophers of science mentioned above, Strong (1966), Westfall (1971), Koyré (1992), and Wolpert (1992) are all examples of philosophers of science for whom this derivative relationship between modern science and technology was self-evident.
[14] Whitehead (1997: 32)
[15] Nagel (1961)

empiricism cultural and personal meaning, and—*at worst*—they are completely irrational projections over the natural world. The traditional philosophers of science are simply conforming to a historically and culturally embedded paradigm, and reproducing and reinforcing its "self-evidence" through assertion and repetition. If we are to understand how this paradigm was possible, and how it relates representational and material practices to each other, we need to examine the mechanical realist foundation of natural science and the technological activities of experimentation, and look closely at how this has shaped and interweaved the theoretical and empirical aspects of science. We need to examine how the natural sciences, and both realist and empiricist interpretation of their results, were conceptually possible. We need to examine how the conceptual relation between science and technology was historically and culturally established and underwrites the work of scientists (and traditional philosophers of science), while also being concealed by the cultural normalization of mechanical realism within epistemology, thereby allowing a technical form of knowledge—and the technical form of rationality it presupposes—to conceal the metaphysics upon which it is founded and assert itself as the knowledge of how the Universe—the Grand Machine—works.

Hence, according to the traditional philosophers of science, scientific instruments such as telescopes and microscopes simply increase our perceptual ability to see what is "out there," say planets or bacteria. The use of instruments such as X-ray scanners, electron microscopes, and Geiger counters have allowed scientists to understand perceived phenomena in the visible world, such as materials and their properties, in terms of otherwise invisible entities, such as molecules, atoms, and radiation. Accordingly, the practical value of such instruments clearly "proves" that science has made considerable advances and progresses, and nothing more needs to be said. Scientific instruments are means to an end—and nothing more. Technology has no further relevance to the philosophy of science—and, once scientists have finished with them, these instruments and detectors are only objects for historians of science to catalog and study. Yet, contrary to the traditional philosophers, I argue that technological innovation has not only made new observations and experiments possible—expanding the horizon of science—but it has also transformed our experience and conception of reality. Once we understand how technology has a substantive and transformative role in how scientists understand the natural phenomena they are investigating, we can see that technology changes how we understand both the methods and objects of scientific research. Using a microscope or a Geiger counter does not merely involve seeing or detecting what is there. One must interpret the behavior of the instrument in terms of an understanding of how it works in order to understand what it reveals and also what we mean by "it" and "there."

Making an observation using scientific instruments requires using techniques to represent what one sees in terms of mechanisms that explain how the instrument works. These techniques are ordered into sets of procedures and operations within a technological framework—that has been learned by scientists in order to use the instrument and make the observation or measurement in question—and that framework transcends the context within which any particular experiment occurs. The methods and instruments of observation are connected via representations of mechanisms to theoretical abstractions of the phenomenon being observed or measured. For example, visual and mathematical representations of the mechanism of ionization connect how scientists understand how a Geiger counter works with their understanding of how radioactivity works, and these can be intellectually connected with theoretical abstractions of the interaction between matter and radiation, given in a mathematical form capable of providing measurable variables and constants, such as potential difference and charge, which changes of which can be interpreted as radiation passing through ionized gas.

These visual and mathematical representations substantively transform how scientists design, build, operate, and understand experiments, instruments, and detectors within ongoing scientific research into radioactivity, and thereby shape how scientists test hypotheses and theories about radiation and matter, and therefore places conditions on what qualifies as a testable hypothesis and scientific theory.

Scientists require explanatory representations of how some observation or measurement was made, if that observation or measurement is to be successfully communicated, repeated, and tested as something "empirical." This is also widely accepted by scientists as being a condition for objectivity, neutrality, and impartiality. However, the traditional philosophers have ignored how representations are utilized or tested in practice by working scientists, and, as a consequence, they have ignored how *testability-constraints* precondition the way scientists (and some philosophers) theorize about the natural world. Utilizable and testable representations can take many forms, such as textual or verbal accounts, drawings or diagrams, apparatus or instruments, photographs or graphs, simulations and models, analogies and demonstrations, and do not need only to take the form of mathematical theories, formulae, and logical deductions. However, in order to work as representations of anything, they must be given meaning in relation to fundamental conceptions of the natural phenomenon in question and how the scientific instrument works to reveal it. They must be quantifiable and manipulable via technology. Theories must provide *operational variables* to be testable. Profound changes in our ability to represent and explain natural phenomena involves the development of new conceptions of what it is possible to cognate, represent, manipulate, and control. The way that scientists describe and explain the performances of new experimental apparatus in terms of a substratum of invisible, physical structures and interactions, such as genes, molecules, atoms, electronic charges, waves, particles, forces, fields, laws, coupling constants, or whatever, does not passively arise from the scientist's experience of the flashing lights, graphical outputs, and changing read-outs of detectors, instruments, and machinery. These "observations" and "experiences" require fundamental conceptions, representations, and explanations of what the responses of the apparatus mean—made meaningful in relation to conceptual juxtapositions between machine performances and natural mechanisms—in order for them to be unified into an experience of the effects of invisible, physical entities, laws, and mechanisms that made the particular machine performances possible.

How does understanding technology help us understand the relation between theory and experience? Let's take a look at the example of Newtonian gravitation. We all have the experiences of objects falling when dropped, the sunrise and sunset, and seeing the moon move across the night sky. The act of categorizing these distinct experiences as manifestations of universal gravitation requires a non-empirical conceptualization of these phenomena. Otherwise there is no basis for considering these phenomena to be unified and explicable in the same terms. Novel scientific research involves the fundamental creation and extension of new concepts rather than collecting mere facts because a fact is only what it is in the light of a fundamental conception. The creation of new concepts is characteristic of Kuhn's notion of revolutionary science, and, as Kuhn pointed out, scientists are only capable of performing the work of normal science by presupposing and applying these concepts to experience as something given to experience. The mathematical practices of the new physics allowed mathematical reasoning to become an intuitive and shared means to discover the truth, thereby transforming the conception of "the empirical" in terms of Euclidean geometry and its projection over the phenomenon in question (with discrepancies and inaccuracies explained by invoking some other mechanisms e.g. friction). This remained the case for the experimental sciences and European mathematics until the nineteenth and twentieth centuries' construction of non-Euclidean geometries. The precepts of mechanical realism had allowed both experimental science and modern technology to be conceptually possible, but what became transformed by the subsequent development of non-Euclidean mathematics was the conception of "natural mechanism" itself. While the theoretical terms of classical and quantum thermodynamics (i.e. when describing and predicting the behavior of black body radiators) may well be incommensurable (as Kuhn argued), both "paradigms" of physics remain commensurable at a methodological level by sharing the same conception of "the empirical" and how to "test" theories. The results of the double-slit experiment may well utilize radically incommensurable conceptions between classical and quantum physics in their explanations of its results, but the experiment itself is secured by mechanical realism as a means of disclosure of the way Nature behaves.

This insight allows us to see that "paradigm shifts" or scientific revolutions (as Kuhn termed them) maintained continuity through mechanical realism, which underpinned the post-sixteenth century's conception of technological powers and natural phenomena as having the same unitary origin and manifest according to the same principles or laws, regardless of how those mechanisms were represented and theoretically understood, thereby allowing nineteenth century classical physics to have methodological and metaphysical commensurability with twentieth century quantum physics, and remain as strata within the same technological framework and its ongoing development of its mathematical projections. The reduction of the phenomenological lived-world of human experience to the mechanical worldview required the famous distinction between primary and secondary qualities: an *a priori* distinction between those properties supposedly possessed by material bodies and those effects supposedly due to the interaction between material bodies with human sense organs and minds. This distinction itself was a transdiction made to account for the fact that human experience is not of a mechanical world and also provided the possibility of a mechanical account of human perception, thereby establishing the starting point of the mathematical projection of the mechanical world view over the human body and the science of physiology. This fundamental transformation in the conception of the human body established the scientific legitimacy of dissection and paved the way for the transformation of natural history into modern biology, and psychology into neurology and computer science.

The purpose of the art of experimentation is to explore its own possibilities by making them happen. It is metaphysical in the sense that it aims to equate and unify these interventions and material practices under an operational principle derived from non-empirical conceptions of science, technology, and natural phenomena. By showing how these metaphysical precepts are related to modern technology, through concepts of "mechanism" and "law," it is my intention to show how experimental science has been successful in the discovery of novel phenomena and achievement of predictive accuracy, while maintaining a critical distance from the scientific realists' application of the concepts of "natural mechanism" and "natural law" to natural phenomena. Hence it my intention to show that the existence and progress of experimental science does not support or refute metaphysics that lies at its epistemological heart, and, consequently, the technological successes of science cannot be used by scientific realists to support their claims about the ontology of the world. However, far from supporting a positivistic interpretation of physics, it is questionable whether a natural science based upon a metaphysical project of experimentation using machines and instruments is an empirical science at all!

Technology has radically transformed human agency and the experience of the natural world. It has transformed the relation between society and the natural world in irreversible and unpredictable ways to construct an artificial world—a technological society—wherein human beings hope to be liberated from toil and suffering, and have new experiences and powers. To the extent that we define ourselves in terms of what we do, our technological activities, at least in part, define who we are. They structure how we participate in society, shaping the horizon of possibilities towards which we aspire, providing that we conform to the discipline they demand. The Ancient Greek poet Sophocles (497-406 BCE) warned us that we will become increasingly dependent on the artificial world that we construct to liberate us from our physical limitations and protect us from "the evils" of the natural world.[16] Sophocles lamented the tragedy of the human condition—existing in a cosmic order that is indifferent to human suffering—for which *techne* (artifice) is the means to escape suffering by conquering Nature through agriculture, medicine, mechanics, and architecture. *Techne* allowed human beings to live comfortably in cities, develop politics and the other arts, but, as Sophocles warned us, ultimately, humanity will become dominated by its dependence upon artifice, which perpetually creates and destroys, as it drives human beings to innovate an artificial world that is beyond human control.

Technological innovation not only allows the discovery of new phenomena, such as black holes and viruses, but also allows for the production of "exotic" phenomena, such as superfluidity in Ultra-Low Temperature Physics or matter-antimatter collisions in High Energy Physics, which could not exist outside of the experimental apparatus, supposedly due to interfering complexities of the natural world. Thus technological innovation allows scientists to explore "natural mechanisms" that cannot occur naturally, due to the postulated interference from other "natural mechanisms" or the inability to achieve a specific set of conditions under which such mechanisms would occur naturally, except through artificial means. This consideration not only problematizes the traditional metaphysical assumptions about "the natural," but it also problematizes the traditional conception of "the empirical" as well. This is not only peculiar to subatomic or quantum physics, molecular chemistry, or genetics. It applies to macroscopic science too. We all have experiences of objects falling to the ground when dropped and staying on the ground when left there, but there is nothing passively immediate in these experiences that allows us to unify these experiences as the consequences of mass, inertia, and gravitation, let alone being able to equate and unify these experiences with the movement of the moon and sun across the sky, the shape of the Milky Way, or the metric of a space-time continuum. We also would not be able to understand and have experiences of the effects of electricity and magnetism in terms of electromagnetic fields, charges, currents, and voltages, without fundamental conceptions by which we could equate and unify our experiences of the behavior of otherwise distinct entities within the world. There is nothing innate in our ability to conceptualize lightning as an electrical phenomenon akin to the behavior of an electrical spark between two charged plates. We have learned how to see these connections. Similarly in biology, there is nothing immediate in the observation of the traits of peas that leads us to understand them in terms of inherited characteristics, let alone in terms of genes and DNA. New experiences and discoveries in science require new conceptions of experience and observations; thereby allowing otherwise separate phenomena to be brought together under the same general causes or principles, which can be tested through technological activity. In order to understand how discovery and justification are made possible and brought together through experimentation, we need to understand how new conceptions of phenomena are made

---

[16] Sophocles (1993)

and related to representational and material practices in order to make new experiences possible and meaningful in terms of natural mechanisms and technological activity.

Insights into how technology mediates scientific experience and theory allows us to move beyond the traditional philosophy of science, and help us explain how natural science is possible without simply assuming that explanatory success and discovery are "self-event" confirmations of the correspondence between the scientific worldview and the objective structures of the natural world. Yet, according to the traditional philosophers of science, by appealing to abstracts such as "simplicity" or "correspondence," the "explanatory power" of natural science is apparently a "self-evident" confirmation of the success of natural science, and it needs no further explanation of how it is even possible for human beings to learn how to explain ordinary experiences in scientific terms, thereby translating experience in terms of unexperienced entities and mechanisms. Yet if we pay close attention to ordinary experience, we encounter very little that immediately translates into the language of scientific theories and explanations. How is it possible for human beings to explain lightning in terms of electromagnetism? How are we able to describe our experience of witnessing lightning in terms of abstracts such as charge, electrical resistance and conductivity, and potential difference? It is not immediate to ordinary experience that lightning is akin to the electrical spark between two charged plates of metal. It is an analogy. How did we learn how to make these kinds of analogy meaningful as theoretical explanations of ordinary experiences of natural phenomena when those theoretical terms are taken from contexts of explaining machine performances within different and artificial contexts?

It is only when a certain level of technological development made corresponding observations and experiences possible—for example, eighteenth and nineteenth century efforts to map out the contours of an electromagnetic field by using an electric circuit and a magnetic needle as a probe—that those experiences obtained their autonomy as being based on "sensory data" or "the facts" and the technological framework (and its history) that made them possible disappeared from view. Thereafter, the electromagnetic field and electromagnetism were understood as being objectively "there," and it could be "observed" as something "real," which could explain natural phenomena like lightning. It was only after this work was done and successfully weaved into culture could ordinary experience be described meaningfully in scientific terms, and the causes of lightning explained in terms of electrical charge and potential difference. Scientific activity became a matter of first working out how to represent and explain "it" in the laboratory, and later using "it" to explain a natural phenomenon outside the laboratory, while simultaneously developing "it' as a mechanism to develop new scientific instruments and detectors, which could be used in scientific research in any laboratory; thereby further confirming "its" independent reality as a "natural mechanism" in terms of its instrumentality and utility in ongoing technological activity outside of the initial context of discovery. It is this confirmation that allows the experienced outcomes of technological activity to become represented as something objective and real, independent of technological activity and human experience. It was only after this level of historical and cultural abstraction had been achieved that the traditional philosopher of science (from the vantage point of his armchair) is able to declare that scientific instruments and detectors are nothing more than the neutral means to test "theory" against "experience," and then argue about the logic of those tests—abstracted into a set of logical propositions and operators—and how theory can be refuted or corroborated on the basis of those tests.

Thus the traditional philosophers can argue about the results of the famous double-slit experiment and their meaning for Heisenberg's Uncertainty Principle and the

Copenhagen Interpretation, for example, without troubling themselves about how the double-slit experiment was even possible as an experiment into the fundamental nature of matter and radiation. They do not trouble themselves with the question of how the fundamental nature of matter and radiation became a fundamental problem for science, or why scientific methods to explore this problem took the forms that they did. Scientists simply discovered it by following some "logic" of discovery and applying "the scientific method" to experience. We can see that the traditional philosophers of science are already considerably "downstream" (or have "gone native") when it comes to understanding the nature of science. Yet, once we move "upstream" and question the underlying assumptions of natural science, we can see that technological innovation has transformed the trajectories, methods, and form and content of research within science, and, when metaphysically underwritten, allowed new representations of natural phenomena to be theoretically developed on the basis of new machine performances, and thereby connected within an ongoing, historically constructed and incomplete technological framework. New machines modify the technological framework and not only produce new phenomena, data, and facts, but also provide new techniques of representation and explanation in terms of newly discovered or postulated mechanisms, which are not only used to refine the framework but to also open up a new horizon of research. Technological innovation disrupts the framework and places experimentation in the context of discovery, as experimenters learn how to re-stabilize the framework and confirm the independent existence of whatever mechanism they are exploring through technological activity. These new mechanisms bring new challenges and possibilities—new research programs and unexpected consequences—thereby allowing both the technological framework and scientific worldview to be refined or revised in terms of new techniques and representations, which forms the foundation upon which new models and theories can be formulated and tested, and in the process corroborate itself as the means by which natural phenomena can and should be explored.

It is for this reason that, from the outset, the argument of this book departs from the traditional philosophy of science rather than rehash their ideas and attempt to argue against them from within their own paradigm. Such an effort would be quixotic and almost certainly be doomed to failure. Instead, I shall move beyond the traditional philosophy of science and offer an alternative philosophy of science and technology that takes both into account, as culturally conditioned and mediated human activities that need to be understood in historical and social contexts, yet does not reduce Nature to a social construct. Instead of rejecting the traditional assertions of scientific knowledge, objectivity, and rationality, I shall endeavor to explain them and show how to reveal their historical origin and development within society, as cultural norms and assumptions about human behavior and its significance, and how these norms and assumptions shape scientific experience, representational and material practices, and the scientific worldview. In this way, I shall not only analyze science in the context of justification, thereby explaining empiricism, but I shall also analyze it in the context of discovery and explain how scientific discovery is possible, without conceding scientific realism.

When traditional philosophers of science discuss experimentation at all, they do so only "in the light of theory," and they presume that the representational and material practices used in real experimental work are quite simply the neutral (value-free) and man-made techniques used by scientists to disclose "the facts of experience" against which the predictions of theories are to be compared. The traditional philosophers of science have accepted this division of labor between science and philosophy, and "the facts" are whatever scientists ultimately present as "the facts." Having left it to scientists to provide "the facts," the philosophers of science focus on the nature of science and its methodology. When they leave it to scientists to operate in the realm of "the empirical," traditional philosophers of science assume that there is a clear distinction between theory and experience, and ignore how the activities of theorizing and experimenting are interweaved when done in practice by working scientists. Without even realizing it, traditional philosophers of science share a particular conception of "the empirical" in their thinking about the so-called natural sciences. This conception is accepted as a "self-evident truth" and not questioned at all because the traditional thinking about science has been preconditioned and made possible as the result of a hidden metaphysical conception of Nature.

This might seem to the reader to be a strange thing to say, given that the traditional philosophers of science have told us that metaphysics plays no part in empirical science. Yet, if by the term "metaphysical" we mean that which must necessarily be true for "the physical" to be conceived of in the ways that we do, which is something that, as thinking beings capable of reasoning and imagining, we project over experience in order to give it meaning and speculate about it in particular ways. This involves an intellectual decision about the essential nature of the phenomena of experience and the correct way to think about them. As such, this decision does not come from experience, but instead involves the application of *a priori* concepts to experience to interpret it and give it meaning, and therefore this decision conditions how we think about phenomena in advance of our experience of them. It conditions our thinking by requiring a categorical division of our experience of phenomena into two categories of characteristics: those that belong to the phenomenon—which we term as its properties—and those that do not. When those characteristics under the category of not-properties are discarded from accounts of experience in order to give an empirical description of the phenomenon, thereby allowing us to recognize, learn, and communicate the facts about the properties of the phenomenon, this necessarily involves a reduction of experience in accordance with whatever rule of selection we apply to experience. "The empirical" is a subcategory of experience that requires the application of a rule of selection determined in accordance with a metaphysical concept. When the traditional philosophers of science talk of empirical science as being opposed to metaphysics, they do so by ignoring the metaphysical concepts required for there to be empirical science at all.

What are these metaphysical concepts? Where do they come from? How were the rules of selection derived from these concepts and applied to experience to give us scientific experience? In *On the Metaphysics of Experimental Physics* (2005), I termed the metaphysical conception of "the physical" in operation in physics as *mechanical realism*, defined it and explored its implications, explained how it was necessary for the emergence and establishment of physics in the sixteenth century as an empirical and natural science, and explained how it had become hidden behind a particular conception of "the empirical" which presupposes it as a set of operational precepts in experimentation. Mechanical realism was a novel conceptualization of machines that emerged out the fifteenth century Aristotelean studies in mathematics and mechanics at the University of Padua in Italy, which studied the six simple machines (the wheel, pulley, inclined plane, wedge, lever, and screw) from which all mechanical devices could be built. The Italian Aristotelians explained these simple mechanisms in terms of natural laws that could be described in terms of geometrical proofs, and led to the novel idea that the most efficient mechanism was a natural mechanism, which operated in accordance with natural law, and only machines that were built in accordance with natural law and the utilization of natural mechanisms would work in the real world. Alongside the mathematical poofs of the Italian Aristotelians and translations of the books of the ancient Greek mathematician and fabled inventor Archimedes, it was this idea that inspired Galileo Galilei to apply the study of mechanics to describe and explain the general motion of bodies in space. Galileo further refined mechanical realism by reducing the concept of a natural mechanism to be only those mechanisms that could be demonstrated via geometrical proof to be at work in machines, and therefore the construction of a machine was an essential step in the methodological demonstration of the existence of a natural mechanism and explanation of how it works. When described mathematically in terms of geometrical axioms, postulates, and proofs, the natural law that governed the operation of the mechanism was revealed. This was the fundamental intellectual step in the invention of the new physics through which experimentation and demonstration using machines became the method by which natural law could be disclosed to the experimenter, and when the fundamental mechanism at work was given mathematical form then it was shown to have been read from "the Book of Nature." This opened human imagination to the possibility of understanding the Universe and everything within it as a machine comprised of mechanisms operating on matter positioned in space and time. Ever since, the entirety of experimental physics has been exploring this possibility and testing its mathematical theories of fundamental mechanisms by building new machines as a demonstration of those mechanisms at work. This is as true today with the construction of machines like the Large Hadron Collider to demonstrate the existence of the Higgs Boson as it was for Galileo's construction of pendula to demonstrate that acceleration under gravity was constant and independent of the weight of a body.

What conception of "the physical" was presupposed by the classification of the mathematical and mechanistic sciences of experimental physics, chemistry, and biology as natural sciences? What made it possible to conceive of experimentation as a part of natural science that could reveal "the physical" and explore it through technological activity? What are the limits and justification of mathematical demonstration in contrast to a demand for a science based upon the facts of experience? How can we understand the relation between mathematics and experience in the context of experimentation? How does mathematics disclose the causes of experience to scientists? How does it allow the observer to grasp the essence of natural phenomena and their objective reality and truth? If experiments occur in artificial contexts, how is it possible to have scientific experiences of the realization and exercise of transfactual natural mechanisms and laws? And how did the answers to these questions become "self-evident"? How was this possible? How were the mathematical and mathematics connected in the new sciences? How does this provide science with objectivity and grant it access to an underlying realty? How are the natural sciences possible? What does "the empirical" mean in the natural sciences? Why is it taken for granted as "self-evident"? In *On the Metaphysics of Experimental Physics*, I answered these questions by explaining why mechanical realism was necessary for the so-called "hard sciences" such as physics, chemistry, and biology to become categorized by scientists and the traditional philosophers of science—as well as by the lay public in general—under the category of natural science, and the paradigmatic conception of "the empirical" at work in these sciences became largely taken for granted as "self-evident." Science has its historical origin in the metaphysics of the sixteenth century that changed how Nature was understood and it has made scientific experience and observation possible. This has changed how these have been constructed and refined in relation to a technological framework of machines, techniques, instruments, and mathematical methods, which is itself abstracted into a scientific worldview that is projected over the natural world and tested in relation to ongoing technological activity. This has profound implications for how Western civilization has been developed as a technological society since the Renaissance and Scientific Revolution, accelerated and intensified throughout the Enlightenment and Industrial Revolution, and how this society consumes, appropriates and transforms, and replaces the natural world with itself. An implicit societal faith in the technological society underwrites science and the traditional philosophy of science, and shapes the future of humanity and the future of life on Earth.

Once we understand mechanical realism and its implications, we can see some important ways we can rethink science, society, and our relation with Nature, with the hope that we can evolve our consciousness sufficiently to learn as a species how to better adapt to the ecology of the Earth—before we destroy it and ourselves along with it—and learn how to share this planet with all the beings that inhabit it. Before we can embark on this journey, it is important to clarify the philosophical and historical underpinnings of commonplace viewpoints regarding science and the natural world. The assumptions and presumptions taken for granted by many (if not most) scientists, philosophers, and the public in general have been produced and reproduced in natural philosophy and the philosophy of science since the seventeenth century. They condition, inform, and constrain how we think about science and the natural world, and have become an obstacle to rethinking and deepening our understanding of Nature and developing a more ecological existence on Earth. Hence it is essential that we move beyond the traditional philosophy of science and rethink the natural sciences.

In *Modern Science and the Capriciousness of Nature* (2006), I explained how both religion and science were responses to mankind's primordial fear of uncertainty and horror in the face of unpredictable and uncontrollable Nature. As well as providing human beings with great bounty, a home, inspiration and beauty, and the means to satisfy our needs, Nature resists and thwarts human intentions, and is responsible for the vulnerability, ageing, and mortality of the human body, leading to injury, decrepitude, dementia, cancer, and death. Nature also subjects humans to natural disasters, earthquakes, volcanos, tidal waves, diseases, droughts, pestilence, parasites, and predators. Nature also brings miscarriages and still-births, death during pregnancy or childbirth, or from infections afterwards, and a high infant mortality rate. And when not somehow drowning, starving, dehydrating, crushing, squashing, burning, rotting, eating, poisoning, or otherwise harming us, Nature looms above us with the possibility of a lightning strike or even a meteor crashing down on us. While the religious response is one of awe and dread, demanding supplication and sacrifice—or considering struggle and toil to be the consequence of a fall from a state of grace and communion with the divine—the scientific response has been the promise of new powers over Nature and the liberation of human beings from our natural conditions through technology. From the outset, science turned knowledge of Nature against Nature—as the test and confirmation of science's promise—and was implicated in a cultural project that grew out of the Scientific Revolution to construct an artificial utopia—a technological society—as a more predictable and controllable substitute for the natural world. However, this technological society tends towards becoming a totalitarian society that threatens to turn upon humanity; to transform human beings into a more predictable and controllable, and thereby more efficient and productive being whose identity is synonymous with his or her function in society. The rationality and achievability of this project becomes open to question once we examine its metaphysical foundations and the destructive form of political economy that has dominated the construction of the technological society and its tendency towards constructing a totalitarian society.

It is this fundamental concern with the substantive impact of technology on human freedom and the human condition that has been a recurring theme in the critiques of mass society and modern technology offered by many twentieth century philosophers and social theorists, including Max Weber, Martin Heidegger, José Ortega Y Gasset, Hans Jonas, Hannah Arendt, Lewis Mumford, Eric Fromm, Hans-Georg Gadamer, Walter

Benjamin, Jacque Ellul, and Langdon Winner. Carl Mitcham termed this as "the humanist tradition" of the philosophy of technology.[17] The advocates of "the humanist tradition" show how the trajectories and structures technological development inform and constrain human agency, intentionality, and experience. These theorists of technology rejected the assumption that technology is a neutral means, developed in accordance with either natural or historical laws, and also rejected the commonplace notion that human intentionality is independent from the means at our disposal. Technological innovation causes social changes and adaptations, as well as creating new possibilities for further exploration and development, thereby shaping human values and goals, but technological development is itself directed in accordance with human choices, attitudes, and expectations. Critical Theorists, including Theodor Adorno, Max Horkheimer, and Herbert Marcuse, developed and refined the application of Marxism and Freudian psychoanalysis to "mass society" and expressed deep concerns with the modern tendency towards the construction of a totalitarian society. They were concerned with the substantive role that positivistic science and technocracy have in constructing a technological society, wherein human thinking is reduced to instrumental reason and calculation. Critical of both industrial capitalism and state-socialism, these theorists were deeply concerned with the modern tendency towards irrationalism (barbarism) and also conformity to technical systems of production and patterns of consumption. Inspired by "the humanist tradition" and Critical Theory, contemporary social theorists, including Jürgen Habermas, Andrew Feenberg, and Richard Sclove, alongside other social theorists, including Thomas Hughes, Andrew Light, Donna Haraway, and Albert Borgmann, have developed substantive and critical theories of technology to help us critically reflect on the substantive impact of technology on human freedom and experience, as existential and sociological phenomena, taking seriously the structural interrelations between politics and technology, and critically examining how technological innovation effects the conditions for the rational development of advanced industrial society.

Perhaps Max Weber was right when he argued that in an advanced industrial society political decisions about the directions of science and its funding are constrained and directed by technocratic systems of calculation, assessment, and control.[18] A technocratic system establishes a technocratic elite that determine the framework and conditions that shape how political decisions about scientific research and technological innovation are made and the further development of society is planned. Although Weber considered science to be value-neutral, in the sense that it cannot tell us what to do or how we should live, he argued that the technocratic development of modern science has increasingly limited our freedom to pursue alternative values and courses of action.[19] Technical judgments are conditioned and contingent modes of ordering and organizing alethic modalities (our possibilities, limits, actualities, and contingencies) that endure from one generation to the next. Politics is no longer the art of the possible. Political judgments about the best course of action about the directions of scientific research presuppose technical judgments about what is possible and how best to achieve it. These judgments take the form of technically rational judgments made in terms of the alethic modalities of currently available technologies, the anticipated horizon of the technological development of society, and the new problems and needs created by technology. Human thinking and planning have become scientifically and technologically mediated and bounded at every level as the result of the culmination of historical choices, the current refinement of the scientific world view, and the actual technologies available to human beings. The identification of "the possible" has now become the jurisdiction of

---

[17] Mitcham (1994)
[18] Weber (1958; 1978)
[19] Weber (1972)

technocrats and technical experts, which, as Jacque Ellul also argued, has subordinated "the political" and "the epistemological" to "the technical" (Ellul, 1972). The technological society perpetually develops new means to unconsidered ends or goals, especially within the sciences and the production of new means of production and bringing new powers into the world, and this shows how technical rationality can dominate, inform, and constrain truth and knowledge, and suppress critical reasoning to such an extent that technological activity becomes an irrational end-in-itself.

In *Participatory Democracy, Science and Technology* (2008), I discussed why the democratization of the technological society is necessary to counter its totalitarian tendency and to maximize the creative potential of society required to better adapt to changes in the world. I argued that the practical value of participatory democracy rests on political and economic localization and self-sufficiency, wherein the people who make the decisions (regarding how best to solve shared problems and live together in communities) should be the same people who suffer the consequences of their decisions, thereby maximizing the societal capacity to learn from each other how to make better decisions, and optimizing pluralism and cultural diversity within an open-ended, changing, and complex world. The practical value of participatory democracy resides in its evolutionary value for the survival of the human species. Participatory democracies are built out of communities that are capable of maximizing inclusion and adaptation at the core of human decision-making capacity, thereby maximizing group lateral thinking and creativity, and generating the conditions for an open society, with its ideal form of being a free and enlightened society. When combined with our growing technological ability to communicate with each other, and learn the destructive ecological consequences of our technological transformation of the world, our own shared concern with survival and living a good life—even when we do not agree about the character of a good life—requires the democratic transformation of the technological society into *an ecotechnological society*. I shall discuss this further in chapter seven below.

Below, in chapters two, three, and four, I shall discuss how the traditional philosophical neglect of the substantive role of technology in shaping both "the empirical" and "the theoretical" has concealed the underlying metaphysical conception of Nature that underwrites both aspects of science. The natural sciences were only possible because of this underlying, historically situated and culturally embedded mechanical realist conception of the natural world and technology that allows both to be understood in the same terms as each other. Mechanical realism has been culturally embedded in Europe since the sixteenth century, and, as I shall show in chapter three, arose out of Aristotelian interpretations of mechanics and alchemy, which framed the point of departure for Galileo and the new sciences of physics and chemistry. Once we understand the metaphysical origin of the natural sciences and their historical foundation in mechanics and alchemy, which promised to provide great powers to mankind, we can understand how the traditional philosophy of science—with its realist and empiricist schools of thought—became possible through the European cultural acceptance of this metaphysical conception. Mechanical realism forms the foundation for the scientific worldview, and is empirically grounded on the technological innovations attributed to science and its application. The implicit acceptance of the identification of experience with the outcomes of the technological activities of experimentation was concealed by its own "self-evidence" as a culturally meaningful belief. Mechanical realism not only remains unquestioned within traditional realist and empiricist philosophies of science, but most traditional philosophers of science remain blissfully unaware that any such conception exists, even though the rationality of their intellectual labors depends on it.

If we wish to understand natural science and the scientific worldview, it is important to reflect upon how the world is investigated, represented, and experienced by working scientists, while questioning how the scientific knowledge of natural processes is possible on the basis of the use of artificial experimental apparatus, procedures, instruments, and techniques. We need to understand how the epistemology and ontology of modern science have been metaphysically unified into a scientific worldview and historically related in the context of the ongoing development of representational and material practices. We need to understand the technological framework of science and how it relates to the scientific worldview. How does experimentation achieve knowledge about the world? What kind of knowledge, if any, do scientists achieve? And, if scientists do achieve a kind of knowledge, which parts of the world do they achieve knowledge about? If experimentation is an artificial process—an art—and the objects of scientific thought are artifacts, then what do we mean by artifice? How does theory relate to "the artificial"? How does "the artificial" relate to "the natural"?

My argument is that the relation between "the artificial" and "the natural" is the outcome of activities that occur within a historically situated and culturally embedded technological framework of interventions, interpretations, expectations, possibilities, techniques, methods, and purposes. It is made meaningful in terms of the scientific worldview—interpreted as an experience of something that exists independently of whether it is actually experienced or not—in relation to sets of fundamental representations of natural mechanisms postulated to explain how machines work. How is this done? Understanding how scientists use technologies to innovate and understand new experiences allows us to understand the developmental trajectory of "the empirical" and how it substantively transforms how "the physical" is conceived. This requires that we reflect upon and analyze the process by which experimental interventions, visualizations, and models are located and interrelated within the scientific worldview and connected to each other within a technological framework. Rather than simply accept the categorical and positivistic duality between "theory" and "experience," as adopted by traditional philosophers of science—a duality inherited from the ancient soul-body duality—we need to uncover and explore the historical and social conditions involved in establishing and developing the epistemic relations between explanation and technological activity that give "observations" meaning as experiences of events or phenomena outside the context in which the experiences occurred. In this way, Galileo could present his experiments on the movement of cannon balls down inclined guttering in a room somewhere in northern Italy as demonstrating fundamental mechanisms of motion that are supposedly real everywhere, regardless of whether Galileo actually experimented on them. In order to understand scientific experience and how it relates to a transcendental realm of natural law, we need to take a close look at the context of those experiences, how they are conceived and represented, and what kind of knowledge arises from them.

How do scientists use and describe their instruments and detectors to explore and explain the world? By examining this question, we can discover how scientists relate their experiences to the theoretical disclosure of an underlying reality that is considered to be causal to the machine performances of the instruments and detectors, thereby providing scientific experience of natural mechanisms at work. This examination requires that we look at science as practiced by scientists and identify the presuppositions about reality that make both scientific experiences and their causal explanations possible. In this way, we can address how science is a creative activity in which scientific observation is theory-laden, as Paul Feyerabend (1975) used the term, but is also informed and constrained by prior scientific work, which does not permit an 'anything goes' ethos, and, instead, imposes discipline and a template over what constitutes permissible or legitimate scientific experience. How is scientific experience possible? How are scientists able to connect their experiences and interpretation of the artificial context of the experiment and apparatus with the wider natural world outside the laboratory? In order to understand how this connection was conceptually possible, in chapter three I shall examine the history of the origins of the experimental natural sciences, such as physics, biology, and chemistry, and see how these became viewed as the "hard" and "empirical" sciences. This historical origin explains how new and fundamental representations of Nature arose from the early sixteenth century Italian understanding of the Aristotelian science of mechanics, which allowed the mathematical laws describing the motion of the six simple machines (the wedge, lever, inclined plane, balance, wheel, and screw) to be represented as natural laws. These representations were projected over the phenomena of motion, as geometrical demonstrations, and provided the conceptual foundation of the new science of experimental physics. The sixteenth century understanding of methods of heating and distillation of substances, developed through alchemy, allowed new and fundamental representations of natural processes in terms of artificial process, as if there were no difference between the two kinds of process, except that the latter required human intervention and the former did not, which provided the conceptual foundation of the new sciences of chemistry, biology, and geology. The source of heat or pressure, for example, became irrelevant, and it became possible to represent natural change in terms of abstract variables and constants, which, in turn, could be used in the design, construction, and operation of novel machines, such as steam engines.

Later, through the innovation and development of new kinds of machines, new fields of physics, such as thermodynamics and electrodynamics, and later atomic and quantum physics, became possible as means to explore a stratified material world in terms of deeper fundamental mechanisms and laws. It was only through the projection of these new and fundamental representations over natural motion and processes, which could be represented and demonstrated in terms of the performance of machines within the closed and isolated confines of the laboratory experiment, while also being conceived as the outcome of natural laws and mechanisms that transcend and pre-exist the context of the experiment and demonstration. Machines became represented and understood as the realization and exercise of natural mechanisms in the juxtaposition and arrangement of materials, methods, and techniques within the technological framework of the experiment as a demonstration of natural mechanisms at work. In this way, the machine become conceptually transformed from being a metaphor for the natural phenomenon to being a literal means of disclosing underlying natural mechanisms and laws. This was a metaphysical and conceptual prerequisite for the mechanical worldview and scientific realism. Hence we can see how and why Francis Bacon and Galileo Galilei considered "the practical arts" as being the means to discover and test "the laws of Nature" through representational and material practices in the context of ongoing experimentation and the theoretical explanation of machine performances.

Scientists claim that they use technology for a very specific purpose of building and using machines to discover "the truth" about "the fundamental principles of Nature" or "the Laws of Nature." They do this by working out how these machines work in terms of fundamental "natural mechanisms." Scientists claim to be able to use their understanding of "natural mechanisms" to aid them in this ongoing process of innovating new machines to explore "deeper levels" of the material substratum, and, by doing so, scientists claim to increase their understanding of how invisible "natural mechanisms" work. Technological innovation is represented as the process through which it becomes possible to discover strata of more fundamental mechanisms using new generations of machines. It is this process that is both "self-correcting" and "self-corroborating" through the ongoing refinement and development of the technological framework itself. As well as building and using machines to acquire money, fame, glory, and for its own sake as a pleasure, scientists also pursue their art to learn truths and satisfy their curiosity about change and permanence in the natural world. How do scientists use machines to discover "natural mechanisms," "test" theories about "natural laws," and reveal "the truth" about "reality"? How do scientists understand the machines they use to do this? How does "the artificial" become a mode of disclosure of "the natural"—that which supposedly exists independently from and prior to "the artificial"? To what extent is the machine autonomous or controlled? How is Nature implicated in machine performances and new powers they bring into the world? To what extent is the machine operational in the realm of science and the way that scientists understand the world? Given that scientists turn to machines as an arbiter of fact, as non-human neutral instruments of measurement, then the technological framework mediates the construction of the judgment of what constitutes knowledge. How is this done? To what extent do machines create the facts of measurement and the reality that they disclose? To what extent is scientific experience technologically mediated or determined?

Of course the traditional philosophers have no time for such questions. Traditional philosophers of science have ignored how the technological framework of the experiment is historically and socially situated to present it as a possible and meaningful experiment. They have ignored how both experiments and observations have been constructed out of already existing representations and explanations of how the instruments and apparatus involved work in relation to theoretical expectations about the phenomenon under investigation, and, subsequently, how its results are related to an already existing cultural field of scientific knowledge, theories, and experimental research in order to design experiments in the first place. Consequently, the traditional philosophers of science are largely oblivious to the fact that scientific instruments and detectors did not fall from the sky ready-made with an instruction book, so they remain indifferent to the fact that the instructions about how to operate and interpret scientific instruments and detectors were all invented by human beings, alongside the invention of the instruments and detectors. All scientific instruments and detectors, and the instructions on how to use and interpret them, were innovated as a result of a history of socially situated labor processes, controversies, trials and errors, failures and dead ends, and protracted intellectual efforts. The goals and research programs within which these instruments and detectors were invented were emergent from a cultural background of interpretations, expectations, challenges, demands, and previous goals and research, and limited by the available technologies and resources, as well as the interpretations and expectations of other scientists and whomever is paying for the research. This cultural background from which all new goals and research emerge is embodied in scientific practices and experience through education and training, in publications and journals, in text books, in conferences, and in the research universities and laboratories around the world.

It is this cultural background that forms the technological framework of ongoing scientific activity wherein representational and material practices are related to each other—structured and organized—to gain their epistemological significance as a means to explore an otherwise invisible and unknowable objective reality via a metaphysical underpinning of the technological framework from within which science emerges. Traditional philosophers of science have presumed that technology enhances and extends the powers of the mind and senses without changing or directing either, and, they assume that the construction of theories is a purely intellectual affair for which the technology of experimentation does not have any constitutive role. However, as I shall argue in chapters four and five, new technologies have not only made new observations and experiences possible, but they have changed how science has developed as a series of disciplines and fields of study, by changing the object-sphere of those disciplines and fields. This has changed how scientists represent and explain natural phenomena by not only producing new techniques of representations and means of explanation, but fundamentally transforming the forms of representation and concepts of explanation into terms accessible to technological activity as a mode of exploration and intervention, and reducing the object-spheres of study to the machine performances of the instruments and detectors used to study them. Technology has fundamentally transformed human experience.

In chapters five and six, I shall describe how the history of science reveals *a dialectics of technology* that not only transforms how scientists (and traditional philosophers) understand the natural world in terms of natural mechanisms, but mediates—informs and constrains—what can count as scientific theory, experiment, or observation by mediating them in technological terms. Scientific instruments and apparatus allow one to see what is "out there" or "otherwise invisible" because a dialectical relation between how scientists already understand the phenomenon under investigation and how they interpret how the instrument or apparatus works is already situated within the technological framework. This dialectical relation occurs throughout the whole complex process of innovating the detector or instrument and making it available as a scientific detector or instrument, and weaving that into the technological framework to make the detector or instrument available to further scientific research. Not only does this explain how scientists labor under a presupposed relation between science and the natural world, without being conscious of that relation, but it also explain how the human understanding of the possibilities and limitations of further technological innovation determine what is measurable and observable, thereby informing and constraining the form and content of scientific theory and the scientific worldview, while simultaneously conveying upon technological activity a sense of discovery and objectivity through encountering limitations and resistances to ongoing technological innovation. Not only are specific stages of the development of this framework required before specific "experiences" (or "observations") can be possible and meaningful as the consequence of the exercise of real mechanisms, but those novel developments transform how past experiences or observations are subsequently understood and reinterpreted in the light of new experiences or observations, constructed in terms of the current "state of the art" of the technological activities of measurement and experimentation, and interpreted in the current refinements of the scientific worldview. Given that only that which can be disclosed via measurement and experiment can be included within scientific results *as something scientific*, the trajectories and outcomes of the dialectics of technology determine the form and content of science itself. This shows how theorizing is mediated—constrained and informed—by technological activities, as well as tested by them, to such an extent that the horizon of "the theoretical" is itself filtered, reduced, and shaped by anticipations of the future possibilities of "the measurable," and therefore "the real."

Novel experiments involve a dialectical relation between the theoretical and the instrumental, wherein novel techniques of making representations of the phenomenon under investigation are interweaved with novel techniques of intervening; changing the behavior of the phenomenon via techniques transform both in order to map out the contours of interventions and responses in terms of explanations of how the instrument or apparatus interacts with the phenomenon. This is far from being the straightforward and philosophically uninteresting process that the traditional philosophers of science have hitherto assumed it to be, when they have even been aware of it at all. The traditional notion that experimentation simply makes comparisons between theoretical predictions and observations is based on an abstract and false view of how experiments are actually designed, built, and performed. It ignores the extent that experimentation is an interpretive process based on complex material and representational practices that are understood against a background of already developed theories, techniques, instruments and apparatus, and thereby dialectical in how it is corrected and transformed through its own activity, and thereby changing the possibilities of future activity.

Once we pay attention to the role of the dialectics of technology within experimental natural science, especially in relation to transforming the criteria of testability and measurability, we can see how technological innovation transforms the criteria under which any theory is considered to be part of science, and, furthermore, how the established use of any scientific instrument or apparatus determines what qualifies as legitimate or illegitimate, meaningful or meaningless, promising or redundant scientific research. Technological innovation changes how scientists understand the natural world and the nature of science itself by making new research, observations, and representations possible; it also brings with it new challenges to achieve the anticipated possibilities of future innovation and research. This expanding horizon transforms what constitutes "science," thereby making possible a *dialectical relation between the background and foreground of the technological framework and the scientific worldview, wherein both are transformed in relation to the ongoing development and refinement of the other to resolve contradictions and anomalies within the framework itself, and it is this dialectical relation that gives science its objectivity as being referent to something outside of the human subject.* I shall discuss this important point—which is tantamount to a heresy within the traditional philosophy of science—and its implications throughout this book.

Hence the history of modern science is just as much a history of the innovation of new machines, instruments, and techniques, as it is a history of ideas, theories, and discoveries. Not only do new technologies create new observational possibilities, representations, and conceptions of the criteria for the possibility of possessing knowledge of laws and mechanisms that explain machine performances, but they transform how scientists understand and conceive those natural laws and mechanisms in quantified terms accessible to measurement and manipulation. Technological innovation transforms how scientists understand the nature of laws and mechanisms by reducing the theoretical understanding of how new detectors and instruments work to observe and measure natural phenomena to a set of quantified representations of machine performances that can be tested through ongoing technological activity. The artifice developed in designing, building, using, and interpreting novel technological instruments to make new forms of intervention in the behavior of natural phenomena is a particularly important feature of the novel dimension of scientific discovery. The metaphysical connection between the ongoing development of the technological framework of scientific research and the refinement of the scientific worldview provides mechanisms with "objectivity" within the experiment, which itself can be theoretically explained in terms of the exercise of natural mechanisms (in accordance with natural law) due to human interventions in the material substratum. This not only explains how scientific discovery is possible, as well as explaining how science explores an objective realm outside of the human subject, but this metaphysical conception is primarily necessary for a technological activity like experimentation to be understood as part of natural science at all.

The dialectics of technology and its metaphysical interpretation underwrites the traditional assumption that experimental science is autonomous as means of exploring the natural world. If we hope to understand how this fundamental conceptual "leap" between "the artificial" and "the natural" was possible, we need to reflect upon and analyze how scientific practices and experiences have been historically emergent, juxtaposed, and unified within a technological framework that has been interpreted metaphysically as a mode of disclosure of natural mechanisms and laws. We need to examine the ways that the scientific worldview has been developed, understood, and communicated in technological terms from its outset in the sixteenth century, and its epistemic success has been continually corroborated by the ongoing success of technological activity in bringing new powers into the world and constructing a technological society that is premised on the promise that it will be an improvement over the natural world. In this way, we can better understand the nature of "reality" and "experience" disclosed by scientific activity, and subject that understanding to philosophical scrutiny and question.

Once the autonomy of the natural and empirical sciences has been culturally established, which it had been by the eighteenth century, the underlying metaphysics that made it possible to conceive a set of technological activities as "a natural science" simply disappeared from view. It achieved the hallowed status of being "self-evident." Once this happened, scientific instruments and detectors became understood and represented as nothing more than the neutral means to make observations and measurements. Through positivism, the technological framework of the natural sciences became understood and represented as the neutral means to discover the facts of experience, to test hypotheses and conjectures, and refine the scientific worldview, and the constructive work of experimentation could be ignored by the traditional philosophers of science. Contrary to the traditional view, I accept that technology drives social change, which includes changes in knowledge and the rules by which it is obtained and given meaning.[20] Society also shapes technology.[21] We simply cannot separate scientific knowledge from technology, and the society and culture from which both emerge already intertwined.

We need to examine how experimenters experience and describe the things that they use to investigate the natural world. By examining how scientists experience the objects of scientific research, we can examine how they experience the way that these objects are presented as a mode of disclosure of a postulated invisible world of underlying reality. This involves examining the tacit presuppositions that are required to make those experiences and practices possible. Unmasking the presuppositions that are implicit in the history and trajectory of the theoretical and experimental practices of working scientists shows that there is frequently a gap between the practitioners' interpretation of their own practices and how those practices appear to one who does not share those presuppositions. Working practitioners unreflectively utilize tacit knowledge in their interpretations and actions. As Michael Polanyi pointed out, over sixty years ago, tacit knowledge is non-verbal and implicit to practices and is consequently not open to criticism by the practitioners because they are unaware of it.[22] The reality that scientists claim to discover is a product of human agency that cannot be separated from the social and historical contexts of material practices and interpretations from which they sprang. It is the historical and cultural conditioning of the activities and practices of the experimenter, embodied through training, that readily permit a naturalization of the use and interpretation of instruments in the laboratory. It should come as little surprise that

---

[20] White (1962), Jonas (1974, 1984), Winner (1977), Ihde (1983), Habermas (1987), Mitcham (1994).
[21] Gadamer (1966), Marcuse (1966), Horkheimer (1974), Baudrillard (1981), Lyotard (1991), Adorno (1994).
[22] Polanyi (1958)

most scientists are realists in the laboratory. As Robert Crease argued, *when experiments are inherently a series of performances*, interpretations are central to the act of producing measurements and observations from material events, whether or not scientists are aware of them.[23] These interpretations are only possible because scientists have learned a historically and culturally situated meaning which allows the judgments made by working experimenters in the real-time processes of laboratory experiments to provide meaning to their experiences and practices.

Traditional academic philosophy has also been heavily criticized for its neglect of the importance of technology to the human condition and the production of knowledge.[24] Critics of traditional academic philosophy have often placed so much emphasis on the importance of technology to human existence and knowledge that an "alternative tradition" has become fashionable, within which the experimental "natural" sciences, such as physics, chemistry, and genetics, are seen as forms of "applied technology."[25] Many historians of science and technology have also argued that technology led the sixteenth century "scientific revolution" and that modern science is "applied technology" to a lesser or greater degree.[26] It has become widely accepted that modern science is *technoscience* that can only be understood in relation to its uses within the culture in which it is emergent.[27] However, the characterization of a "natural science" such as physics as "applied technology" only reverses the problem of how the "application" occurred. How was technology "applied" in such a way as to become the "natural science" of physics? In order to understand experimental physics as "applied technology," we need to address how mathematics, machines, and natural phenomena were related. What understanding of technology do we need in order to understand the technological basis of the experimental natural sciences? We need to analyze modern science at a "deeper" level than merely pointing out that the use of mathematics and technology have been central to the experimental natural sciences since their origin in the sixteenth century. In order to understand the conceptual possibility of experimental science, we need to historically trace back its origin that permitted its current manifestation. What presuppositions about both natural phenomena and technology permitted the use of the latter to understand the former? As I shall argue in the following two chapters, this is a question of the metaphysics that underlies the whole intellectual legitimacy of the technological disclosure of natural mechanisms. Mechanical realism provided the operational precepts of experimental science and made the epistemological use of technology to disclose natural mechanisms and laws conceptually possible. How was this metaphysics possible? When did it occur? As I shall argue in chapter three, this metaphysics occurred during the fifteenth and sixteenth centuries and made the Scientific Revolution, the experimental sciences, and modern technology all conceptually possible. The precepts of mechanical realism allowed the mathematical description of the motions of the six simple machines (the wedge, the lever, the balance, the inclined plane, the screw, and the wheel) to be taken as descriptions of the fundamental natural motions. I shall discuss these precepts in detail in the chapter two. We need to understand how the reification of mathematics, as something objectively, eternally, and universally true, in the context of the Renaissance developments of the Medieval science of mechanics, allowed experimental physics to be metaphysically operational as a technological mode of disclosure of natural mechanisms,

---

[23] Crease (1993)
[24] Cohen (1955), Arendt (1958), Marx and Engels (1967), Durbin (1972), Winner (1977), Durbin (1978), Ihde (1979, 1983), Levidow and Young (1981), Feibleman (1982), Jonas (1984), Bijker (1987), Haraway (1991), and Stiegler (1998).
[25] Jonas (1974), Heelan (1983), Ihde (1991), and Mitcham (1994).
[26] White (1962), Merton (1970), Mathias (1972), Bennett (1986), Yearly (1988), Kaufman, (1993), Long (1997).
[27] Marcus (1986), Aronowitz (1996), and Gordo-Lopez and Parker (1999).

and for technology to be a consequence of the utilization of natural mechanisms in material practices. This runs deeper than a renaissance of Platonism. It provided the sixteenth and seventeenth centuries with both a methodological and an ontological foundation for the mechanical and experimental natural philosophies of Galileo, Descartes, Bacon, Gassendi, Newton, Boyle, Hobbes, *et al.* In chapters four and five, I shall describe how this methodological and ontological foundation was central to the methodology, intelligibility, and subsequent researches of experimental science, and in chapters six and seven, its implications for how we understand the relation between science and society, and the natural world shall be discussed.

So, what does it matter what philosophers of science think? After all, it is often quipped that science needs philosophers of science like birds need ornithologists. Yet, as Henryk Skolimowski pointed out,

"We are in a period of ferment and turmoil, in which we have to challenge the limits of the analytical and empiricist comprehension of the world as we must work out a new conceptual and philosophical framework in which the multitude of new social, ethical, ecological, epistemological, and ontological problems can be accommodated and fruitfully tackled. The need for a new philosophical framework is felt by nearly everybody. It would be a lamentable if professional philosophers were among the last to recognize this." Skolimowski (1976: 298)

Perhaps we need some common ground. As I have argued elsewhere, it is helpful to take a look at Plato's cave analogy.[28]

In the seventh book of Plato's *The Republic*, Socrates presented the cave analogy. We are asked to imagine people living since childhood in a cave deep underground. Their necks and legs are tied to fix them in place and prevent them from turning around. A fire burning far above and behind them provides light and they can only see the cave wall directly in front of them. Between the fire and the prisoners is a walled path along which puppeteers carry puppets of people and other animals made out of stone, wood, and every material. Some of these puppeteers are talking, while others are silent. The tethered people are only able to see the shadows of the puppets projected by the light of the fire. The echoes of the voices of the unseen puppeteers bounce off the cave wall and seem to come from shadows themselves. The prisoners think that the names they use to describe what they see applies to the shadows passing in front of them and that the shadows are truth and reality. Perhaps they would grant honors and praise to those who were sharpest at identifying the shadows as they passed by and were best at remembering the order of the procession. Perhaps they would consider those who devised some method for predicting the sequence of the shadows as being their wisest and most knowledgeable.

Socrates asked us to consider what would happen if one of these people were to be freed of his bonds and cured of his ignorance. Imagine that this prisoner was compelled to stand and turn his head, walk, and look up towards the light of the fire. Imagine him dazzled and pained by the light, unable to see the puppets whose shadows he had seen before, unable to make sense out of the voices that surround him. If he were told that his previous reality was comprised of only shadows and echoes and that he was now turned towards and closer to truth and reality, what do you think that his response would be? If he were shown each puppet in turn and ordered to tell us what it is, this poor man would no doubt be at a complete loss and would more than likely cling to his

---
[28] Rogers (2005; 2015)

familiar, previous truth and reality. If he were forced to look at the bright and burning light of the fire then no doubt his immediate inclination would be to flee back into the comfort of the gloom and shadows. Imagine that he was prevented from doing this and dragged, kicking and screaming, up the rough steep path, past the puppeteers, past the burning fire, and out of the cave. Imagine that he was dragged up to the surface by force and hurled into the sunlight. Imagine him blinded by the light of the Sun, unable to see a single thing that is now said to be true and real. He is stood, with his hands shielding his eyes, squinting at the painful and bright light that fills his vision.

Of course, some considerable time would be needed before he could see the world outside the cave. At first he would only be able to see most clearly at night. He would be able to see the moon and the stars, as well as the shadows and silhouettes of things in the moonlight. During the next day, he would be able to see shadows cast by the sunlight, then reflections cast into water, and only later would he be able to make out things directly. Finally he would be able to glimpse the Sun itself. This man might readily infer, as Socrates suggested he might, that the Sun governs everything in the visible world and is the cause of all the things that he has seen. Socrates then considered what the man would think if he remembered the cave, the prisoners, and the nature of truth and reality there. Would not the man think himself wiser now and pity the prisoners below? Would he not think himself in a happier condition than even the most honored and praised of the prisoners? Would he not rather suffer anything than return to the cave, share their opinions, and live like that again?

Socrates used the cave analogy to elucidate the relationship between the visible and the intelligible realms. The visible realm was to be likened to the cave. The journey to the surface was akin to the soul's journey to the intelligible realm and the study of the Sun was likened to the difficulty in coming to see and know the form of the good and rational life. Such knowledge could only be achieved with considerable difficulty. Socrates asked us to imagine what would happen if the man decided to return and help his former fellow prisoners. Imagine him returning into the darkness. His vision would now be dim in comparison with the prisoners. Would they not claim that his journey up to the surface had ruined his eyesight? Would they not ridicule him and the very idea that it was good to even try to travel to the surface? If he tried to free them then would they not resist or even try to kill him?

For Socrates, it was quite unsurprising that those who have grasped the form of the good and rational life should have great difficulty and be open to ridicule when they are compelled to contend in the courts about justice and laws with those who have never seen justice itself. How could we expect any society to be just when its rulers do not understand or care what justice is? In discussing the education of the ideal city's rulers, the philosopher-kings or guardians, Socrates proposed that number, calculation, and geometry (used within every craft and science, including physical training, poetry, music, and astronomy) are the subjects that have the greatest power to awaken intelligence in children and are also useful for the art of war (enabling the rulers to be adept at commanding the city's defense). For Socrates, the study of the science of mathematics was a pedagogical good because it provided the intellectual rigor, which he considered a prerequisite for turning the soul's intellect to the study of the form of truth. Mathematics does not convey such knowledge, but it sharpens the intellect and prepares it to grasp the forms of justice, beauty, and goodness through the dialectics of reasoning and argument, which was necessary for the purpose of understanding the good and rational life. Education should not be considered to be merely a process of instruction, putting knowledge into souls that lack it, like putting sight into blind eyes, but instead should be considered to be the process of turning the whole soul to see what is already within it. It should be the craft of redirecting the sight of the soul to bring out the knowledge of the good and rational life *that is already there*. Socrates termed the educational process, which brought people from the shadows into the light, to be the ascent to true philosophy.

What has this to do with the traditional philosophers of science? I ask you to imagine the philosophers of science to be the tethered prisoners, gazing at the wall of shadows, listening to the echoes that seem to come from the wall itself. Behind them are the working scientists, the shadow puppeteers, who use the representations they produce, the shadows on cave wall, to adjust their material practices, making and manipulating the shadow puppets, to develop and refine their knowledge and theoretical explanations of how the light of the fire interacts with the puppets to project shadows onto the cave wall. The puppeteers experiment with the form and materials of the puppets, to learn the inner principles of how to make shadows on the wall, and use this knowledge to learn about the fire. Meanwhile the traditional philosophers of science are sat with their backs to the shadow puppeteers, waiting for new shadows to be projected upon the cave wall, while the unseen puppeteers work with all kinds of materials to provide representations for the philosophers' scrutiny. These representations are taken to be the primitive experiences against which the predictions of competing scientific theories can be tested against observation—they are accurate predictions of the procession or they are not—and the fact that they can be thus tested gives the traditional philosophers of science confidence that they must, somehow, correspond to reality and truth. The procession of the representations—the shadows on the wall—becomes the objects of thought for the traditional philosophers of science, and they can maintain an absolute distinction between theory and experience, as they keep their backs turned towards the experimenters and ignore how representations are actually produced and used in experiments.

The philosophers argue and debate, using all the rhetorical and analytical skills their discipline provides, about which epistemology provides the best means to decide between theories on the basis of the representations and facts provided by the experimenters. They argue and disagree about whether any knowledge of a causal reality behind the representations is possible, while they agree upon the progress that theories have in matching predictions with the representations. Indeed, they consider the task of the philosophy of science to be that of devising some method for predicting the sequence of representations and relating them to competing theories. They consider those among themselves that are talented at working towards this task as being their wisest and most knowledgeable. They leave it to historians of science to remember the sequence of the procession of representations and theories, but the history of science has nothing more to teach them.

As critics of the traditional philosophy of science, we can turn around and tell the prisoners that the procession of shadows on the cave wall are produced by shadow puppeteers, while the discussion about the shadows is conducted by prisoners who are only seeing shadows projected on a wall. We can point out to the prisoners that they are indeed prisoners experiencing only shadows and echoes produced by the shadow puppeteers. We can pay close attention to how the shadow puppeteers practice and develop the art of shadow puppetry, and how the shadows have been interpreted by the prisoners from one generation to the next, and how these interpretations are historically and socially conditioned. We can also sit in the entrance of the cave and watch how the prisoners' interpretations of the shadows has reflected the norms and prejudices of a society with its own history. We can point out how the choice of puppets used by the shadow puppeteers and the order of their procession also reflects and reinforces the expectations, prejudices, fears, and desires of the tethered prisoners. We can argue that scientific work forms disparate, heterogeneous, fragmentary cultures of loosely associated sources of authority, technologies, research programs, and personalities, and science is a human activity (or social construct) and its interpretations reflect the values and norms of the society in which it occurs. We can challenge the traditional philosophers' attempts to demarcate science as a special rational activity that discovers the facts about the natural world and explains them. The prisoners will retort that we are "relativists" and they will continue as they did before.

So, why not leave the prisoners tied in the cave? Why should we care what they believe? The problem is that the traditional philosophers of science reinforce an impoverished view of Nature that justifies treating the natural world entirely in terms of its instrumentality—its use as an amalgam of natural resources and forces—and as a spectacle placed before them. The traditional philosophy of science amounts to propaganda in favor of the construction of an artificial world—a technological society—that replaces the natural world. This has resulted in an anthropocentric viewpoint that simultaneously places human beings in the center of this artificial world while also enslaving us to it. Meanwhile we are witnessing an ecological crisis wherein deforestation, overpopulation, agriculture, mining, industry, pollution, over-consumption, desertification, etc., not only threaten human health and well-being, but are also quite devastating for the other species we share this planet with. Mechanical realism has justified the technologization of Nature as something at our disposal while simultaneously technology degrades Nature by only viewing it in terms of its instrumental value. Any other view of Nature is dismissed as romanticism or irrationalism, and proponents of heretical views are soon marginalized and ignored. Hence, despite its seeming irrelevance, the traditional philosophy of science has become an obstacle to understanding the nature of science and technology, and how they have provided the means to replace Nature with an ersatz substitute that promises to build a better world. It is timely that we confront this impoverished view of Nature and show how it is implicated in the ongoing destruction of the ecology of the Earth. If we hope to overcome this ecological crisis, which some have termed the *Anthropocene* (or the sixth mass extinction), we need to move beyond the traditional philosophy of science and rethink both science and Nature.

2

REALISM AND EMPIRICISM

Why do we think of natural science as an inherently realist enterprise to discover and explain natural phenomena in terms of the "natural mechanisms" and "natural laws" that most scientists and some philosophers presuppose to pre-exist the attempts to discover them? Realists believe that there is an objective world that exists independently of all and any attempts to know it, and that the objects of experience exist independently of whether they are experienced or not. Most people are realists, at least about some things. Even so-called primitive peoples, with all their fears of evil spirits and demons, believe that they can step upon an unseen snake when walking through the forest, which is why it is a good idea to look out for them. To learn how to live in the world, to grow food, as well as learn all kinds of practical skills, all in good measure involve a degree of objectivity—paying attention to how the world really is, rather than how we so happen to think about it. Scientific realists believe that science provides a means to access and know that objective world by testing theories and hypotheses about it through observation and experiment. Scientific realism should be understood as a psychological motivation—an *internal rationale*—for experimental science that makes theorization and experimentation both meaningful as means to provide intelligible explanations of natural phenomena, given in terms of natural mechanisms governed by natural laws. It is a cultural expression of the psychological and social satisfaction of curiosity about the world by seeking to explain why the world is the way it is, which is a motivation shared by human beings from all cultures, expressed in all manner of different ways.

By understanding science as a human activity, we can see how the degree that people find their curiosity about the world satisfied through science, which makes it meaningful to them as a truth-seeking activity; we can see how the realist motivation is normative within both theory and experiment because this motive are presupposed by the activities involved in the methods and techniques of measurement and observation that provide the facts by which theories are tested and explored. How is this presupposition possible and how does it relate to the objective world? The objectivity of realism itself—as an internal rationale—is itself only conceptually possible and secured in thought because of an underlying *operational metaphysics*. This metaphysics is required to establish and explain how scientific knowledge is itself possible—to show how a specific epistemological interpretation of the interconnections between the technological activities of the experiment and mathematical modeling of the experiment have become transformed into the realist interpretation of experience in terms of the objective world. This metaphysics is also necessary for the scientific worldview to be related to truth via a concept of the natural world as something that transcends and exists independently from the context of any experimental or theoretical activities, but can be known through those activities. This metaphysics is required for the interrelated activities of theorizing and experimentation to be conceived as being means to discover both the outcomes and their causes within the contexts of performing those activities and interpreting them, and further representing and abstracting these outcomes and their causes in terms of natural laws and mechanisms that exist independently from theory and experiment. Without this operational metaphysics, the scientific realist interpretation of the results of experimentation would be completely irrational and the activities of experimental science would be meaningless as a cultural practice.

Mechanical realism provides a unifying conception of "the physical" and "the empirical" that underwrites (psychologically, culturally, and historically) the foundational philosophical principles and assumptions justifying the epistemological categorization of experimental science as a natural science.[29] It also is needed to secure a connection between knowledge about Nature and the mathematical modelling of machine performances as the means to gain knowledge of a natural world that exists independently of human beings. Without mechanical realism, neither "the facts of experience" (observations and the data produced during experimentation) nor the representations of "the causes of experience" produced during theorization would be meaningful in the particular contexts of performing any experiment or formulating a theory to explain it. Without this metaphysical conception of the objective world and the human ability to know it, scientists could not claim to have any knowledge about any natural structures, mechanisms, and laws that supposedly transcend the contexts of experimentation and theorization. They would only be able to talk about particular experiences within particular contexts, and make no explanation of them whatsoever. Without mechanical realism, there would be no reason whatsoever to categorize the referents of experiment and theory as "natural." There could be no intelligible correspondence to anything in Nature, which transcends the particular contexts of the experiment or its theoretical explanation, in relation to the particularities of the artificial contexts of scientific practices in the laboratory or observatory.

Without metaphysical presuppositions about the objects of its activities, science would have nothing to do with the natural world. It would remain subject to an instrumentalist, pragmatist, or constructionist interpretation of "the facts," bound up with human interests and the practicalities of life, and made only in terms of the experiment or theory's instrumental, practical, or social value in that context of development and application, and nothing more. The physical sciences would be branches of cybernetics and engineering, dealing with information and material problem-solving, and unintelligible as means to discover anything outside of their contexts of application and their value for the further development of technology and the infrastructure of human society. Any scientific realist would be horrified at that prospect. Yet, it is incumbent on the scientific realist to explain what they must presume about the world as a condition for being able to theorize and have knowledge about it. If the idea that natural mechanisms and laws could be discovered and revealed through technological activities is itself a cultural product that presupposes mechanical realism, the whole rationality of science is at stake unless that metaphysics is made explicit and presented clearly as a set of precepts for experimentation to be a means of gaining knowledge about the causes of natural phenomena. Without it, the claim that experimental science is a natural science would be completely unintelligible and irrational, and we would have great difficulty in explaining why science exists as a cultural activity.

Once we recognize that mechanical realism underwrites and explains the epistemology and methodology of modern science and technology, and has been implicitly presupposed to interpret the results of experimental physics since Galileo to the present day, we can understand why it cannot be replaced without making experimental science meaningless. We can also recognize that it is a methodological requirement for experimentation using machines to operate in the context of justification, by testing the empirical adequacy of theories, and in the context of discovery by finding

---

[29] Rogers (2005)

new phenomena. I shall describe in the next chapter how this metaphysics originated before the Scientific Revolution and made it possible. It has endured through the paradigm shifts from mechanical physics to quantum mechanical physics, and from Newton's Euclidean physics to Einstein's Lorentzian-relativistic physics. Mechanical realism underwrites all experimental science regardless of whether experimenters are building pendula to demonstrate constant acceleration due to gravity or whether they are building a particle accelerator to reveal the creation of matter. Mechanical realism provides the overarching conceptual foundation for the whole technological framework of experimental science, even though the content and particularities of that framework are historically conditioned and contingent on changing cultural influences and human relations. Its historical emergence was itself the emergence of a cultural differentiation—a new subculture—that established the new sciences as routes to new knowledge, experiences, and powers through technological activity.

Mechanical realism is premised upon the following set of operational precepts that provide all scientific practices with principles of action and their interpretation:

(i) Natural phenomena and technology can be explained using the same concepts, terms, methods, and categories;
(ii) Both natural phenomena and machines come into being by realizing the same causes; there is an unchanging and universal cause for every effect (or set of effects)—this does not preclude the possibility of complex multi-causal phenomena and machines;
(iii) Causes are related to their effects via fundamental mechanisms, which make both natural phenomena and machine performances possible, and govern their operation and results in an unchanging and universal manner; the realization and exercise of any mechanism is governed by a natural law, expressed in terms of mathematical relations; consequently, the performance of any machine is governed by natural law, as modelled through mathematics;
(iv) the distinction between natural phenomena and machine performances is one of a distinction between the conditions and context of the realization of a mechanism; machine performances are interpreted as being the consequence of the artificial arrangement of causes and their effects exercised in the material responses of technology; whereby machine performances require human interventions to realize this arrangement of mechanisms in the world, natural phenomena are interpreted as the arrangement of mechanisms that do not require human interventions to be realized in the world;
(v) The mathematical descriptions of the motions of mechanical devices and machines, which are mathematical descriptions of natural law, are presupposed to be symmetrical—isotropic (independent of spatial orientation), universal (independent of spatial location), and eternal (independent of time)—and, therefore, the changes and motions they describe are descriptions of repeatable and predictable mechanisms.

These precepts provided experimentation with a heuristic and epistemological framework for the conceptual establishment of an intelligible methodology to explore natural causes through technological activity. The acceptance of these precepts is a condition for the interpretation of "the facts of experience" discovered through experimental science to be interpreted as facts about natural phenomena, which is a precondition for the fundamental idea that mathematical theories can describe and explain mechanistic phenomena in terms of natural laws and mechanisms. These precepts are the basis for connecting and interrelating theory and practice in the experimental sciences by allowing observations made within the artificial context of the experiment to be represented and understood as observations of something that would happen independently of the experiment and its conditions and contingencies.

As an operational metaphysics, mechanical realism is distinct from *speculative metaphysics* (e.g. Kepler's harmonic theory of planetary motion, Gassendi's seventeenth century atomic theory of matter, Newtonian absolute space and time, Kelvin's idea of the aether as a thermal and electromagnetic medium permeating all space, or Everett's many-world interpretation of quantum mechanics). When scientists and philosophers of science talk of metaphysics, they usually are talking about speculative metaphysics and not operational metaphysics (for examples from quantum physics cf. Cohen, ed. 1997). Metaphysical speculations are proposed in order to make particular experiments and observations intelligible in terms of particular conceptions and interpretations of Nature, which themselves cannot be verified or falsified by means of experiment and observation. These speculations can be widely accepted as intelligible and (tentatively, pragmatically) adopted as if they were objectively established within the scientific paradigm, even though they change as that paradigm changes, with all the historical conditions for those changes. The speculative metaphysical idea of the existence of an aether as a medium for electromagnetic vibrations permeating all space, for example, was widely held by physicists, chemists, and natural philosophers throughout the nineteenth century. Yet it had not been observed and there was no experimental evidence for its existence. The very notion of vibrations of electromagnetic radiation in empty space was so unintelligible to nineteenth century physicists, chemists, and natural philosophers that they had to conceive of space permeated by some kind of medium. Otherwise their theories of waves made no sense. Yet after the famous 1887 Michelson-Morley null result and Albert Einstein's 1905 Special Theory of Relativity, the aether was considered as having no physical meaning, and unnecessary to explain the propagation of light, so it was considered irrelevant and obsolete, and a figment of the human imagination. It was replaced by space-time metrics—space and time are purely understood in terms of their measurement and the velocity of light—and the aether was soon forgotten, except by historians and some philosophers of science.[30] Cultural criteria for the acceptance of speculative metaphysical ideas are themselves replaceable and contingent within the historical development of experimental science—thereby differentiating science as a subculture, which itself is produced and reproduced through increasingly specialized human activities and relations, and new interpretations of those activities and relations give rise to new hypotheses and theories. These speculations not only allow human imagination to inspire theoretical and experimental activity, especially when modelling or performing 'thought-experiments', but they also allow for the selection between assumptions that can be accepted as plausible from those that cannot, and thereby shape the trajectories of scientific imagination and thinking.

All experiments use technologies as the means to explore the natural world. Experimental science is premised upon the legitimization and justification of techniques and the interpretation of machine performances. As a matter of epistemological principle, anything that cannot be disclosed via publicly accepted communicable and repeatable techniques of measurement and observation are excluded from being included as an object for replication, testing, or further research, and hence is excluded from "empirical science." All scientific results and data are emergent from and situated within a technological framework of machines, instruments, tools, and apparatus, each juxtaposed and related with sets of fundamental representations, techniques, and methods. This framework is constructed, refined, and modified by ongoing scientific activity and is made meaningful in relation to its theoretical interpretation as a means to observe, measure,

---

[30] Svenson (1972); Whittaker (1989)

and explore the causes of natural phenomena. For example, regardless of any personal experience or anecdotes, ghosts (and other so-called paranormal phenomena) cannot be included as legitimate objects for scientific research (i.e. parapsychology) without the invention and use of technological devices of measurement and observation (e.g. cameras, thermometers, motion detectors, and electromagnetic sensors). Of course, the use of such technology does not guarantee that ghosts exist—or can be interacted with or known in any way—but is a precondition for the search for ghosts to be a part of science, irrespective of whether that search is successful, and for any theories about ghosts to be explored and tested for their empirical adequacy. The same is true of any phenomenon, before it (and its effects) can be considered to be part of "the physical world."

Once we recognize how mechanical realism has reduced "the physical" to the operation of kinds of mechanisms and their effects on the motion of kinds of matter, we can see how, by pre-specifying the boundaries of "the physical" in terms of the means to explore it, the structures and content of scientific experience and theory are mediated and reduced by technology, which is historically and culturally conditioned and contingent upon the actualities of material and representational practices, and the values and ideals to which they are directed. Obviously the falsifiability of any theoretical prediction is dependent on the availability of technologies of observation and measurement, but what is less obvious is that the methods of making and interpreting observations and measurements depend on representations and interpretations of how the technologies of observation and measurement work. This understanding both preconditions and further shapes how "the physical" is both explored and theoretically understood, and thereby preconditions and constrains how "the empirical" is identified and selected, refined and further researched. Due to the empirical *testability constraints* on theory to provide (or derive) measureable quantities (operational variables) and observable results (predictions), which are empirically adequate, theory is also mediated by technology by being limited to forms amenable to mechanization (i.e. ratios, quantities, and operators). The "objects" of theoretical activity are technologically mediated—selected and shaped—in order to be accessible and intelligible objects for isolation, quantification, manipulation, control, observation and measurement. This allows theories to be "tested" via experiments and for experiments to be "modeled" by theory, and provides a dynamic technological trajectory of innovation and research, which, when successful, is taken to corroborate theory and disclose the objective world. In order to understand how technology mediates "the physical" and how it is explored and theoretically understood, and how "the empirical" is identified, selected, and related to "the physical," we need to look at how the representational and material practices of scientific activity are produced and related through experimentation, and fed-back into scientific activity to refine and develop it further, and thereby 'self-correct' and 'justify' science as an epistemological pursuit of knowledge of how "the physical" came to be the way it is.

As an operational metaphysics, mechanical realism determines what can be proposed as a speculative metaphysics. Only phenomena that can be situated within the technological framework of ongoing scientific activity can be considered as having any objective reality outside of human imagination and cultural representation. The meaning and necessity of an aether only arose because of realism about waves alongside the unthinkability of waves in a vacuum. The meaning of waves to nineteenth century physicists itself arose because of the usefulness of waves as a representational technique to model and explain a wide range of phenomena. The analogy between the waves in a tank of water and the waves proposed by electromagnetic theory broke down when physicists asked what is the analogous 'stuff' to the water, and hence the speculative metaphysical idea of the aether was developed and refined to secure the analogy presupposed by the nineteenth century wave theory of light.

The aim of theorizing is not just a matter of achieving empirical accuracy or predictive success. It is also a matter of using technical causal accounts to explain how to design, build, operate, and interpret machines in the production and explanation of novel phenomena of experience. From its outset in the sixteenth century, experimental science has limited itself to using mathematical theories to describe the law-like behavior of machine performances, and were only able to extend themselves to describe natural phenomena by reducing said phenomena to sets and sequences of machine performances. Scientists "test" the "empirical adequacy" of mathematical theories against measurements produced using instruments designed by using mechanistic causal accounts, and project these over natural phenomena and explain them using the same mechanistic causal accounts. As such, "the scientific enterprise" is implicitly a realist attempt to explain measurements and interventions in terms of causal mechanisms, and, hence, is a metaphysical, technological, and inherently realist enterprise pursued to discover the ontology of the causes of phenomena, and it cannot be considered to be based upon an empiricist epistemology at all.

However, realists cannot escape the empiricists' criticism that their commitment to realism is based on an internal rationale that has no rational justification. Realists have also ignored the extent that scientific explanations and descriptions are always technologically mediated and at no stage does scientific activity encounter an unmediated objective phenomenon. Realists also have assumed that they understand the "empirical success" of science, but they have simply taken the results of scientific work as givens and neglected to understand the technological framework from within which such work emerges. Realists are unable to provide any rational foundation for their commitment to realism, other than pointing out that scientific activity would be unintelligible without a commitment to realism; they cannot situate their own cultural and psychological beliefs within the ongoing project of developing the technological framework of scientific activity because their failure to recognize and understand the metaphysics of mechanical realism has led them to ignore the way that both "the empirical" and "the physical" have been metaphysically determined and technologically mediated from the outset. Whereas we should accept that scientific realism, at some level, always involves a naïve acceptance (or intuitional psychologism) of how scientists explain and describe new phenomena, in the terms given by scientists, realists cannot explain how experimental science began as a cultural activity in the first place, except by appealing to the "self-evident success" of science. Yet, the "self-evidence" of mechanical realism was culturally established long before its "self-evident success." It has become so widely and deeply accepted within the epistemology and ontology of the experimental sciences (which includes physics, chemistry, and biology) that it has ceased to be a part of metaphysics at all and has become an inarticulate and habitual (and irrational) set of cultural beliefs, values, and presuppositions. In this sense, mechanical realism can be treated as a psychological and anthropological phenomenon. Via education and the acceptance of scientific authority, mechanical realism has been culturally inherited and embodied in scientific discourse, experience, and practice. It has become the tacit and implicit foundation for all scientific paradigms. It remains concealed as a "self-evident truth" behind paradigmatic discourse and practices until its metaphysical stepping-stones have been disclosed. These stepping-stones are the precepts of mechanical realism.

Ian Hacking (1983) famously problematized the traditional philosophical distinction between experiment and theory. How are representational and material practices produced and related during real scientific experiments? Hacking characterized experimentation as making a series of interventions, and he examined how scientists use representations to plan and order these interventions. Hacking used the writings of Francis Bacon to show how the idea of experiment as intervention was fundamental to how it was understood from the outset. He also argued that theories and representations intervene as part of the process of making observations and measurements during experiments. Scientists use representations to make sense of what they are observing and also use representations when planning how to make an observation. Experimentation cannot be understood as simply a process of comparing representations with observations because representations are required in practice to make the interventions necessary for observations in the first instance. Representational and material practices are interrelated by making visual representations of "otherwise invisible entities" (e.g. "genes," "molecules," "valence bonds," "atoms," "radiation," "subatomic particles," "space-time curvatures," "superstrings," and "electromagnetic fields") and relating those representations to the interventions required to observe and measure those "otherwise invisible entities" during experiments. In other words, one cannot observe "electrons" without some representations of how "electrons" behave in specific contexts (e.g. "in the presence of an electromagnetic field," which is itself visualized analogously with a water wave). Visual representations are not only necessary to make the observation intelligible as an observation of an "otherwise invisible entity," but also are used as interventions to guide the processes by which experiments are performed, thereby preconditioning and selecting the assumptions underlying the methodology by which those observations and measurements are made. They make scientific experience possible. Furthermore, visual representations require interpretation—through the further use of theoretical or fundamental representations—to give theoretical meaning to material practices when making observations and measurements to compare with theory and thereby test it. They make corroboration and falsification possible and science can only proceed because of them.

Hacking is not alone in thinking this way. Many historically-minded philosophers of science, including Thomas Kuhn, Paul Feyerabend, and David Gooding, have shown that the boundary between observation and representation is ambiguous in the context of real-time experiment. This boundary changes with new instrumentations, concepts, and representational techniques. The innovation of new instruments has led to the growth of "the observable realm" and even Karl Popper acknowledged that "observations are made in the light of theories."[31] Philosophers of science must not only accept that the act of making an observation is guided by theory, but they must also accept that it is an essential condition for the intelligibility of experimentation that it is so guided. Not only are new experiences and theories made possible through the innovation of instruments such as the electron microscope and the microwave telescope, but representations are required to build the instruments in the first instance and make sense of those experiences, and they must be related and secured against how other instruments and machines are represented and understood.

Gooding provided us with a clear and detailed historical account of how the nineteenth century physicist Michael Faraday invented and used visualizations within the ongoing work of planning, constructing, and communicating his early experiments in electromagnetism.[32] According to Gooding, Faraday's work shows that the empiricist reduction of the ontology knowable through experimental science to only observable entities is extremely problematical because of the complex representations used to make meaningful relations between theory and practice that occur when novel instruments

---

[31] Popper (1975: 106-111, 423)
[32] Gooding (1990)

mediate novel observations in novel contexts, and require new techniques of representational and material practice for others to be able to replicate those experiments and observations. The prediction of possible novel phenomena is an important aspect of theoretical work, but it is not simply a matter of passively seeing what is there because one must also interpret what one sees in order to make the observation in the first place. Gooding acknowledged that Kuhn's (and Feyerabend's) point about how observations are "theory laden" was well made, but he argued (in agreement with Hacking) that observations are made in experiments through a series of interventions using pre-theoretical pictures, so many observations in novel experiments are not even formally "theory laden" but are made using tentative visual representations. Gooding termed these tentative visual representations as *construals* and used Faraday's construal of the electromagnetic field in terms of lines of force as an example. Experimenters make a construal of a novel phenomenon before making a theoretical interpretation of it, and use construals to show others how to make the same observation and to guide further interventions. The indirectness of perceptual experience when using such construals is a serious problem for empiricism because the intellectual processes involved in making novel observations by using construals are not based on empirical facts and/or logical propositions. These purpose of construals is to reveal an underlying mechanism that can be used to guide and explain material practices and their results to other people, thereby providing an object for replication, further research, and formal theorizing.

Of course, as Gooding and Hacking both understood, the growth of "the observable realm" through instruments is an important feature of scientific realist interpretations of those instruments. If the same entities can be observed with independent instruments then there is a good reason to believe that they exist independently of those instruments. For this reason, Hacking emphasized the importance of cross-checking and considered himself to be an *instrumental realist*. Thus, the instrumentality or usefulness of any theory in successfully predicting different experiments and observations using different kinds of instrument in different contexts supposedly provides a good reason to believe that the theoretical entities invoked by that theory exist outside of the context of particular experiments and their theoretical interpretation. If the belief in the objective and independent existence of theoretical entities (e.g. electrons or electromagnetic fields) is central to the intelligibility of experimental physics as a pursuit, and successfully used in the design of different experiments, all experimental scientists must adopt a realist interpretation of their practices. This point is highly problematical—if not fatal—to empiricism. Scientists use theories to discover and explain the elements of the world that they believe exists independently of human experience. For the scientific realist, the rationality of this intention is epistemologically justified *a posteriori* by scientific successes in realizing it. If novel observations and powers are made possible because of a new instrument then it would seem that there are good reasons to believe that the theoretical causal mechanisms utilized in the conception and design of that instrument must exist, even if any particular representation of it is open to revision and refinement. This belief is further corroborated if independent instruments, designed in different ways, all allow the observation of the same theoretically predicted phenomenon. Understanding the instrumentality of causal mechanisms in the construction of experiments in relation to other experiments is crucial for an adequate understanding of the epistemological relation between theory and experience in all experimental sciences.

However, once we take the operational metaphysics of mechanical realism into account then we can go further than Hacking and Gooding did. Theoretical objects are conceptually related to technological objects by using a concept of natural mechanism as the link between natural causes and effects. This is as true as it is for the use of computers as the explanatory trope in the sciences of psychology, neurobiology, and the study of human and artificial intelligence, as it is for the use of mechanical devices as the explanatory trope in physics and the study of matter and motion.[33] This shared metaphysical conception of "the physical" and "the empirical" allows scientific experience to be understood as the shared (public) experience of the consequences of the realization and exercise of (otherwise unrealized, unexercised, and unexperienced) natural mechanisms in the artificial contexts of the experiment that are governed by natural laws that transcend those contexts of realization. Only then can the juxtapositions and relations between representational and material practices be conceptually understood as disclosures and discoveries of those natural mechanisms at work in machine performances. Theories can then be presented as the mathematical abstraction of those technologically mediated disclosures, thereby connecting those theoretical representations with material practices; the derivation of machine performances from theory provides the predictions available for "empirical test" through further technological activity. This permits the progressive and internal development of both theory and technology in terms that relate to each other, thereby satisfying both testability constraints and providing science with instrumental or practical value, while allowing that development to be metaphysically and culturally represented as the ongoing (approximate and tentative) discovery of how Nature works—and interweave that representation into the scientific worldview of Nature as a grand and complex machine, thereby confirming "the empirical success" of science.

The machine performances of the apparatus are taken to be its natural responses to human interventions, when constrained within the context of the experiment, and theories of how those responses occur can be understood as having an objective referent in the natural world outside of those contexts, via the notion of mechanisms operating in accordance with mathematic laws. The approximate and tentative character of its results and their theoretical explanation are explicable in terms of the incompleteness of the technological framework within which representational and material practices are emergent, situated, and related with each other. This situation is made even more problematical for the empiricist when the initial innovation and development of a new instrument, such as a solar neutrino detector or electron microscope, is based upon a theory that utilizes unobservable mechanisms, such as neutrino oscillations or electronic quantum tunneling, and unobservable entities, such as neutrinos and potential barriers in its design, construction, and operations of the detector.

However, before discussing scientific realism further, I need to examine some of the most serious problems with empiricism (and its offspring: positivism and instrumentalism). Is empirical adequacy adequate? Below, I shall explain why it is not. In the following section, I shall argue that one of the most serious problems for empiricism is that the idea of the empirical adequacy of a theory does not provide a rational foundation upon which we can explain how experiments are designed, constructed, and operated as tests of that theory in the first place, and, furthermore, empiricism does not adequately explain how theory and observations are related through representational and material practices when designing, building, and performing real experiments. Empiricism requires a clear and distinct boundary between theory and observation that is not apparent in real scientific experiments, and, as I shall show below, largely engages with an abstract and imaginary version of science that would be meaningless if applied to scientific practice and experience. As a consequence, empiricism fails to provide any

---

[33] Dupuy (2000); Rogers (2005)

intelligible rationale for the cultural activity of experimental science.

The basic problem for empiricism is that the shift in the boundary between theoretically possible observations and "the observable realm" is navigated using instruments that are imagined, designed, built, and operated through the implementation of theories, interpretations, and representations based upon postulated entities and mechanisms that cannot be directly observed. We do not have any direct perceptual acquaintance with "objects" such as electrons, genes, molecules, or electromagnetic waves, etc., at any time or in any context. We infer their existence on the basis of the use of theoretical interpretations and representations of the outputs and performance of instruments and machines. Pointing at a tree and saying "tree" is not the same type of reference-act as pointing at a test-tube full of goopy liquid and saying "DNA," pointing at a photograph of a slightly curved line and saying "proton," or pointing at the change in the position of a magnetic needle next to an electric wire and saying "electromagnetic field." The construction and operation of instruments such as electron microscopes, x-ray scanners, neutrino detectors, Geiger counters, microwave telescopes, etc., is at best highly problematic for the empiricist, and at worst is meaningless and unscientific. This is fatal for a philosophy of science that aims at making scientific activity intelligible and meaningful.

The aim of theorizing is not just a matter of achieving empirical accuracy or predictive success. It is also a matter of using technical causal accounts to explain how to design, build, operate, and interpret machines in the production and explanation of novel phenomena of experience. From its outset in the sixteenth century, experimental science has limited itself to using mathematical theories to describe the law-like behavior of machine performances, and experimenters were only able to extend mathematics to describe natural phenomena by reducing said phenomena to sets and sequences of machine performances. Scientists "test" the "empirical adequacy" of mathematical theories against measurements produced using instruments designed by using mechanistic causal accounts, and project these over natural phenomena and explain them using the same mechanistic causal accounts. As such, "the scientific enterprise" is implicitly a realist attempt to explain measurements and interventions in terms of causal mechanisms, and, hence, is a metaphysical, technological, and an inherently realist enterprise pursued to discover the ontology of the causes of phenomena and it cannot be considered to be based upon an empiricist epistemology at all. Experimental science is an inherently metaphysical and cultural pursuit, and mechanical realism provides scientists with an internal rationale and dynamic explanatory power within contexts of discovery and justification, which gives scientific practices in the laboratory cultural meaning as a means to discover natural mechanisms and laws that exist outside the context of the laboratory, and also provides a means to test theoretical representations of these mechanisms and laws. As such, any adequate philosophy of science must transcend and opposes empiricism, positivism, and instrumentalism, none of which adequately capture the meaning of scientific activity. However, this raises the problem of how we can understand and explain scientific activity and culture if we do not share the realist internal rationale of scientists (and some philosophers of science). How can we question the "empirical success" of science and challenge the realist commitments of its proponents? How can we understand the subculture of science, without "going native"? Before answering these questions, I first need to explain the meaning and importance of the terms "empirical success" and "empirical adequacy," before showing why these terms are insufficient for any philosophy of science that aims to be a philosophy *for* science rather than one that is solely critical of science.

The inadequacy of empirical adequacy

Our experience of the world contains phenomena such as blue skies on clear summertime days. How would a scientist explain such a phenomenon? How would a physicist, for example, explain the experience of seeing the color blue to someone who had been blind since birth? Let us assume that the blind person is conversant in the language of modern physics. The physicist could explain the eye in terms of an optical device. S/he could explain how electromagnetic waves of a particular wavelength radiate from the sun, are refracted and scattered by particles in the atmosphere, are focused onto the retina by the lens of the eye, stimulate the rods and the cones of the retina, and are transformed into electromagnetic pulses in the optic nerves. S/he could then explain how these electromagnetic pulses travel through the optic nerves, travel through a network of nerves leading to the brain, generate electrochemical process in the brain's network of neurons, and are finally processed by the brain as the color blue. Let us assume that whatever theory or model the physicist utters is empirically adequate to the extent that its derivative resultant would be that the sky on a clear day (quantified in terms of humidity, pressure, and temperature) would have the color blue (quantified in terms of wavelength or frequency). Let us also assume that the blind person perfectly understands the physicist's explanation of how the eye and brain processes differences in the wavelength or frequency of light. But does the blind person know, on the basis of this explanation, what an experience of the color blue is?

There is one essential characteristic of the color blue that is missing from the physicist's explanation of blue light. That essential characteristic is the quality of the color of the clear daytime sky, *the blueness*, which is immediately experienced by all people who are able to see it. The blind person will not, from the physicist's explanation, have any idea whatsoever of what the experience of seeing the color blue is, nor that there is even the possibility of experiencing blueness as the result of understanding the physicist's explanation. The blind person would have no more idea of the experience of looking at the sky and experiencing blueness, as a result of the physicist's explanation, than s/he did at the outset. The physicist could talk of atoms, electrons, photons, matter, ions, radiation, wavelengths, refraction, the spectrum, prisms, electromagnetism, oscillations, coupling-constants, resonance, or whatever, but would be unable to introduce the quality of blueness into her/his explanation. Of course, the physicist could attempt to explain away the blueness of the sky on a clear day as a subjective illusion or epiphenomenon, but any arguments against the fact of the blueness of the clear daytime sky require assumptions and premises that would be more suspect than the indubitable experience of the blueness of the sky on a clear, summertime day. Blueness remains *surplus* to the physicist's explanations of the color blue. It lies outside the language of physics and remains undeniably *residua* to any attempt to explain it away.

This is true of all the qualia that are characteristic of the lived-world of human experience. The Humean empiricist is quite right to differentiate experience from scientific explanations of experience. The world of our experience is irreducible to scientific explanations; consequently, there is not any necessary connection between any particular scientific explanation and our experience of the world. We can explain the same experience in any number of different ways, and the sciences are inherently reductive. The philosopher Maurice Merleau-Ponty made this point about perception and the science of psychology in his criticisms of scientific accounts of perception.[34] Martin Heidegger made this point with regard to attempts to equate the lived spatiality and temporality of experience to any scientific conception of space and time.[35] The historian and philosopher of science J.W. Dunne made this point about all so-called empirical descriptions provide by science.[36] Yet, for the empiricist, this distinction between explanations and descriptions does not problematize how we understand science. As long as s/he can maintain a clear distinction between explanation and description within actual scientific work, the empiricist continues to claim that science aims at gaining knowledge by collecting and organizing experience. If descriptions and explanations of experience are independent from one another then we cannot claim to have knowledge of causes based upon knowledge of the facts of experience, claims the empiricist. Explanations may well have psychological and cultural value but they do not provide us with knowledge. Only experience does that. The empiricists' point is not that human experience is an existential condition for the existence of the world, which would be a form of solipsism at a species level; rather, the empiricist's point is that our knowledge about the world is limited to what we know based on our experience of the world. Experience is epistemologically primary, not ontologically primary. The only scientific knowledge we can attain, according to the empiricist, is the knowledge of patterns (sequences and conjunctions) of events, or relations between variables, within our experience of the world. We only encounter appearances, experienced as repetition and change, associations and correlations, regularities and constant conjunctions, and therefore an awareness of this epistemological limit allows us to guard against our imagination or prejudices getting the better of us, and substituting fictions for knowledge.

We can know that when there is lightning there is also thunder, and thunder follows lightning, but when it comes to explaining why this pattern occurs, we have to use reason, which has articulation constraints (language both structures and limits how we think) and has intelligibility constraints (we cannot think in terms that are unintelligible to us or incommunicable). Once we admit these constrains, we should also admit the necessity of psychological and cultural conditions on how we reason, and any "knowledge of causes" is riddled with psychological and cultural contingencies. These contingencies open us to the possibility of prejudice, preconceptions, and false assumptions in virtue of being contingent on human factors, such as being born into a world of pre-existing language and images, which we learn from others as we learn how to think, speak, read, write, and draw, and make sense of the world. We might explain thunder as caused by lightning, which is itself explained in terms of the potential difference caused by the ionization of clouds as ice particles rub against each other within the clouds, and the sound of the thunder is the air rushing back into the channel opened up in the air along the path of ionization of air between negative and positive charges. Or we might explain thunder as the laughter of the god of lightning. On the empiricist's account, the only reason why we prefer one explanation to the other is because of our psychological and cultural background, which provides us with the possibility of building our explanations on more basic elements (such as "ionization" or "laughter," and "negative and positive charges" or "the god of lightning") as if these were self-evident, and finding those explanations to be intelligible to ourselves and others within the same culture.

Empiricists and realists clearly disagree about the aims and outcomes of science. For the empiricist, the aim of science is to describe phenomena (in general terms, using specialized language) and provide the facts of experience, whereas for the realist the aim

---

[34] Merleau-Ponty (1999; 1987)
[35] Heidegger (1962)
[36] Dunne (1948: 11-17)

of science is to explain those facts of experience and disclose an objective reality. The realist claims that these explanations are part of knowledge, and phenomena such as lightning are explained by mechanisms and not gods. Of course, the onus is on the scientific realist to show that there are good reasons to believe that there is a connection between explanations of experience and the facts of experience, and s/he has to do more work than bang his or her fist on a table or kick a rock and declare it to be solid. Otherwise, realists fail to explain how theories and observations are connected to representational and material practices in the ongoing work of experimentation, and, as a result, how theory guides observation and makes it meaningful. The explanation of how theory guides observation and gives it meaning needs to be answered by a philosophy of science, if it is to adequately explain how science is done in practice, and how theoretical representations are related to the material practices performed during experiments.

My critique of empiricism starts from the observation that if we look at how science is actually done in practice—treating science as observable phenomena—we see that studies of the behavior of familiar objects, at most, constitute only starting points for experimental sciences. The majority of experimental work in experimental science involves the investigation of phenomena of which we have no direct experience whatsoever, but indirectly infer from the behavior of scientific instruments and apparatus. All changes in matter and energy, and their interchangeability—fundamental to the conception of "the physical" within the contemporary scientific worldview—are inferred from the behavior of scientific instruments and apparatus. For example, the experience of a change in the height or readout of a thermometer is not a direct sensory experience of heat. Of course, we can see the height or readout of a thermometer change as we feel the day become hotter or colder, but we need some education or instruction before we can make sense of thermometers as a device to measure temperature, and understand how the height of the mercury relates to measures of temperature and how that relates to heat. We were not born understanding Celsius and Fahrenheit. The relationship between changes in the readout of the thermometer and temperature is an interpretive one, requiring an explanation of how the thermometer works, and thereby admits theoretical explanation of the readout of the thermometer in terms of heat, the thermal properties of matter, and how these interrelate. Even a "simple" measurement of temperature using a thermometer implies an explanation why the instrument measures temperature, thereby capable of giving us a meaningful measurement of the heat of an object or environment, and therefore justifying the use of a thermometer to measure temperature. Without the explanation of how a thermometer works, the use of a thermometer to measure temperature would be irrational and meaningless. This is true for the relationship between all instruments and the phenomena they measure.

Once we recognize that measurement and observation require inferences, and inferences are not the same type of reference-act as perceptions, much of experimental activity transcends empiricism. The act of looking at the output of an electron microscope and seeing the internal structure of bacteria, or the act of looking at the output of a radio telescope and seeing a distant quasar, requires specialized interpretation and training to be able to perceive "what is there" in terms of a theoretical understanding of how the detector works. This is even apparent at the most basic level of interpretation of using a microscope to see small things and using a telescope to see things that are far away. Even understanding how a simple magnifying glass works requires inferences, representations, and theoretical interpretations. Any adequate philosophy of science must account for this, if it can make the use of instruments intelligible and meaningful within scientific activity.

This leads us to an important question: Do we need to be realists about causal mechanisms when the construction and use of instruments in experimental science is initiated and understood in terms of those causal mechanisms and experimental science is shown to technologically progress and achieve predictive accuracy by using those instruments? At first glance, it would seem that we do need to be realists (at least tentatively) in order to make sense of experimental science as a meaningful human activity. However, the situation is not quite that simple. Scientific theories are generalizations, idealizations, and abstractions, which focus on particular properties of phenomena under investigation that are modified in accordance with the particularities of use in context, especially when there is only partial regularity in that context. Empiricists are quite correct about that. As Nancy Cartwright argued in *How the Laws of Physics Lie* (1983), accuracy is really not a criterion in theoretical modeling. Assumptions and approximations are deliberately used to construct models of complex phenomena (such as weather patterns, for example) in simpler terms. Scientists *reduce* complex natural phenomena in order to facilitate manipulation and modeling within the context of the experiment (or use computer simulations to perform 'thought experiments') and theories are manipulated and modified—often on an *ad hoc* basis—in order to fit the facts. Thus representation is made less literal in order to model the world. Both theories and the objects to which they apply are constructed and then matched, in a piecemeal fashion, to the real situations to provide predictive accuracy, but rarely do theories and models match all the facts at once. Hence, Cartwright argued that the purpose of fundamental laws is not to describe reality but to simplify reality; theory describes the appearance of reality that "is far tidier and more readily regimented than reality itself."[37]

If scientists only produce simplifications of the appearance of reality, any success using these in practice is insufficient by itself to provide good reasons for a belief in the independent existence of the theoretical entities invoked within those descriptions. It remains logically possible that novel machines could be built on the basis of an instrumental, pragmatic, and piecemeal use of a collection of theories, representations, and models as heuristic guides, without these entities having any independent (non-instrumental) existence, and, consequently, the predictive success of any theory does not demonstrate the independent reality of its theoretical entities. Any machine can always be described and interpreted by using a different set of theoretical objects, and therefore its operations remain independent from any particular theoretical description, which remain perpetually open to refinement and revision. The attempts to produce a neat, consistent, unified, and abstract theoretical understanding of novel machines often retrospectively follow the innovation and development of that machine (as in the cases of the steam engine and the electromagnetic motor). Such theories can be replaced in the future with an alternative theory, and, consequently, as the empiricist would argue, empirically adequate theories are those theories that predict the behavior of phenomena, but this does not give us necessary and sufficient reasons to believe that the theoretical entities postulated in those empirically adequate theories really exist. It remains possible that either a more accurate theory with different theoretical entities will be invented as an alternative theory. The history of science has shown us that this happens, time and time again, and this historical contingency was fundamental to Kuhn's ideas of scientific revolution and paradigm-shift. The possibility of alternative theories seriously problematizes scientific realism by introducing reasonable doubt about the objective reality of contemporary theoretical causes and entities.

In her later work *A Dappled World* (1999), Cartwright argued that natural laws and the scientific theories that describe them are not universally applicable, but are true only *ceteris paribus* (all other things being equal), and the content of our best scientific theories

---

[37] Cartwright (1983: 62)

describes the *capacities* things have as a result of possessing specific features or properties. She argued that even the best descriptions of natural laws are highly limited in scope and apply only to very specific arrangements of capacities required to produce the regularities described by these laws. She termed these very specific arrangements as *nomological machines*. She defined a nomological machine to be an arrangement of objects and their properties that have stable capacities to produce empirical regularities that can be described by laws.[38] A nomological machine may be very simple, such as a rigid rod placed on a fulcrum that serves as a lever, or it may be very complicated, such as the Stanford Gravity Probe. The important feature of such machines, according to Cartwright, is that they possess capacities that generate regular, law-like behavior when the machines are set running under the right conditions. The right conditions include the fact that the machine is shielded from unwanted outside causal influences that would interfere with its operation.

Cartwright argued that capacities are more ontologically fundamental than empirical regularities because the existence of such regularities depends upon the existence of the capacities of a nomological machine. Regularities are hard to find because nomological machines operating in shielded conditions rarely occur naturally; it requires experimental intervention and control to shield a nomological machine in the right way for it to generate the appropriate regularities. Laws only hold so long as such arrangements are in evidence and effectively shielded from interfering factors. It is this feature of scientific activity that makes it reductive and piecemeal. Scientists construct models of the various capacities of the arrangement of the nomological machine under specific circumstances, but, under other circumstances, different combinations of capacities may be evident and a different law may be applicable. As a result, she argued that laws do not account for the behaviors of things in the world because specific laws hold only insofar as things with stable capacities are thrown together under appropriate circumstances. In her view, we need to examine the nature of capacities if we are to understand how things behave, and this problematizes empiricism and scientific realism, both which Cartwright considers to be kinds of fundamentalism.

Cartwright's criticisms were primarily directed towards the assumption that laws are universally applicable to all domains. In other words, she rejected the realist claim that independent existence is necessary to explain the success of scientific theories. Contrary to this claim, Cartwright argued that unless we have a model for a law, which provides correct predictions within experimental tests, then we do not have an empirical basis for the belief that the law applies. For example, on Cartwright's account, we have good reasons to believe that Newton's Second Law of Motion may well be applicable to the motion of a gyroscope in space because we have an accurate model for this, but we have no such model for accurately calculating net forces on a banknote blowing in the wind, and, consequently, the belief that Newton's Second Law applies to the banknote is merely the prejudice of a Newtonian fundamentalist. In the absence of an empirically adequate model for the phenomenon, Cartwright argued that we have no empirically based reason to think that Newton's Second Law applies to complicated phenomena such as the movements of a banknote in the wind. If the situation is too complex to model and experimentally reproduce as a nomological machine, and thereby successfully predict its performance, we cannot know whether this abstract law is relevant or there is a different law describing different capacities in that situation—or even whether it is governed by a law. If the law and model are both empirically limited to the case of the nomological machine in question, we do not have any *a priori* reason to claim that the law applies

---

[38] Cartwright (1999: 50-8)

beyond this machine (or set of machines). If we aim to be empirical then, according to Cartwright, we should reject the validity of the fundamentalists' extension of the remit of the law to particular complicated situations for which we lack any empirically adequate model, and we should accept the patchwork and discontinuous nature of the Universe open to scientific investigation in terms of nomological machines.

Cartwright's conclusion needs to be taken seriously by philosophers of science, for reasons that I shall explain below, and her interpretation of science shows us what an empiricist philosophy of science would look like, if empiricists paid attention to how science was done in practice. However, Cartwright is unable to explain the scientists' rationale for using experimental machines to explore complex phenomena in the first place. Even though Cartwright raised some interesting critical points about the ontology disclosed by experimental science, points which the realist should engage with and address, her argument neglected to discuss the concept of "mechanism" in natural science and the use of such a concept in extending and combining the models used in the laboratory to the wider world. Without such a concept, of course we would conclude that laws only apply to nomological machines. Such a concept is essential to connect the nomological machine with the complex phenomena of the natural world. Cartwright failed to explain the rationale of using machines to explore natural phenomena because she did not explain how experimenters connect their efforts in the artificial circumstances of the laboratory to the phenomena of the natural world. On this point, realist criticisms of empiricism seem applicable. Experimenters must claim the objects of theory pre-exist their efforts to explore them, and these objects and the laws that govern them must exist outside of the laboratory. Experimenters must be fundamentalists. This claim is itself unjustifiable and irrational if scientists were to accept Cartwright's philosophy of science. Hence, we must accept that, at some level, all scientists are fundamentalists because all experimental science presupposes mechanical realism.

Once we recognize the role of mechanical realism in establishing this reductive fundamentalism, thereby explaining the conceptual connection between the nomological machine and the natural world, we can see how it is essential for the existence of experimental science. We must acknowledge that, without such a concept, the classification of experimental physics as a natural science is completely unintelligible and arbitrary. We can also understand how the concept of mechanism is used to connect the closed (simple, isolated, repeatable) machine performance of the nomological machine—whereby the experimental apparatus can produce an empirical regularity of X changing with respect to Y—with the open (complex, interconnected, messy, and ever-changing) natural world. On such a conception, the only possible machines are understood to be only those machines that can be built out of natural mechanisms governed by natural laws, and nomological machines are the simplest machines that reveal those natural laws. It is this concept that connects the nomological machine with the Universe and the possibility of discovering the universal law(s) of Nature. Without a concept of mechanism, scientists would be unable to explain how the empirical regularities were produced during an experiment, either in terms of tendencies or capacities, nor would they be able to extend those explanations to the natural world, piecemeal or otherwise.

Cartwright does make the valid point that neither the intelligibility of any model nor the ability of physicists to reveal the capacities of nomological machines, in and of themselves, logically supports the truth-status of their causal accounts and theoretical explanations. But, unfortunately for her whole argument, it does not follow from the fact that predictable models are empirically limited to nomological machines (which are nothing more than basic sets of machine performances and their associated set of fundamental representations of how they work) that the Universe is shown to actually be a patchwork of such machines. What Cartwright has shown is that the scientific worldview is abstracted from a patchwork of such machines, and she is right to argue that when scientists treat natural laws as universal they are fundamentalists, but instead of rejecting fundamentalism on the grounds of empiricism, she should acknowledge the importance of fundamentalism to science and instead rejected empiricism as a philosophy of science. To understand the metaphysics of science, we need to explore the epistemological implications of basing natural science on a fundamentalist interpretation of nomological machines. Instead, Cartwright's analysis of experimental science only inadvertently reveals the extent that empiricism is an inadequate philosophy of science, and that the natural sciences, such as experimental physics, are not actually empirical sciences at all, in the empiricist's sense of the term.[39]

Does this mean that the realists have won the argument? No. At least, not quite yet. While realism does raise fatal problems for empiricism, which is unintelligible and arbitrary as a philosophy of experimental science, in turn, as Cartwright has shown, empiricism raises fatal problems for realism too. The empiricist is quite right: There is no logical reason to believe that it follows from the empirical adequacy and predictive success of any theory that any postulated theoretical mechanisms exist independently from the particular experiments in which they are applied. History has shown us, time and time again, that there is always the possibility that at the empirical level two different theories can agree but utilize different theoretical entities and mechanisms. Copernicus' heliocentric theory of the solar system presented a very different theoretical picture than the Ptolemaic geocentric theory, but they both had (more or less) the same level of empirical adequacy and predictive success.[40] As the contemporary empiricist Bas van Fraassen argued in his 1980 book *The Scientific Image*, many theoretical entities utilized in past experiments have become merely of historical interest, whereas the empirical knowledge those theories have produced has remained. Newton's theories of motion and gravitation in absolute space and time have been replaced with the Einstein's theories of motion in relative space-time. Newton's theories of motion and gravitation are inadequate outside of its realm of applicability, and the predictive success of Einstein's theory is more general and requires fewer assumptions about the Universe. Yet much of the empirical adequacy of Newton's theories of motion and gravitation remains, within its limits (low velocity compared to the speed of light in the presence of low masses), all while Newton's idea of absolute space and time has been recognized to be an abstraction or a fiction. Van Fraassen argued that this historical fact justifies skepticism about all theoretical explanations. If all past theories have been shown to be only empirically adequate within some range of limits and contexts of application, while using abstractions and fictions, then this provides a good reason to suppose that current theories will be shown to be empirically inadequate in the future, and the theoretical ideas they utilize will be shown to abstractions and fictions too. Some new theory will take its place. As van Fraassen argued, the literal truth status of any theory is not decided by the current empirical adequacy of that theory. It is highly probable that Einstein's theory will go the way of Newton's and become replaced with a more empirically adequate theory that will use very different theoretical entities.

Van Fraassen brings us back to Kuhn's idea of scientific revolutions, and the reasonable expectation that a new scientific worldview and fundamental theory of the Universe will emerge when the old scientific worldview and fundamental theories have been shown in the future to be empirically inadequate, flawed, and comprised of fictions.

---

[39] Rogers (2005)
[40] Kuhn (1957)

The historical contingency and plurality of theory suggests that there is not any necessary connection between the empirical adequacy of any theory and the objective reality of its referents. In other words, the empirical success of theory does not demonstrate its objectivity. It is possible that through trial and error, using theories only as a heuristic tool or resource—a cognitive instrument—in the design of experiments, we could eventually build an experimental apparatus that demonstrates the empirical adequacy of the theory in question, without understanding why the apparatus works in terms of any general or universal theory. James Watt, for example, patented his rotary steam engine in 1791 after over 30 years of trial and error experiments, 33 years before Sadi Carnot published *Reflections on the Motive Power of Fire* (1824) in which he described the ideal heat engine, and 63 years before William Thomson (Lord Kelvin) published *On the Dynamical Theory of Heat* (1854) in which he coined the word "thermodynamics" and set down the fundamental principles of the First and Second Laws of Thermodynamics. Over 100 years after Watt's invention of the steam engine, in 1896 Ludwig Boltzmann published his *Lectures on Gas Theory*, in which he set down the principles of statistical mechanics and the atomic theory of gases, and, in 1912, Walther Nernst published his theorem setting down the definition of the Third Law of Thermodynamics, with its abstract concept of entropy and the irreversibility of thermodynamic systems. James Watt did not need any knowledge of these theories and laws of thermodynamics to build a working steam engine.

Even if we accept that scientific theories are progressing in terms of their empirical accuracy, the historical development of science gives us good reason to remain skeptical about causal explanations and theoretical entities. As Popper argued in *Conjectures and Refutations* (1963), the acceptance of the progress of scientific knowledge prohibits any such certainty about the truth of theory. This is widely accepted by scientists and philosophers of science alike, and entirely consistent with van Fraassen inductive historical argument: it logically follows from historical examples of theory change that there reasonable doubt about the objectivity and theory-independence of today's unobservable particles and forces. It follows from this argument that there is no prerequisite for any *isomorphism* (one-to-one correspondence) between the structures of a theory and the structures of natural law for the theory to be empirically adequate. The situation is one of pluralism and plasticity. This is a serious problem for scientific realism. If the epistemological standard by which the empirical adequacy of any theory is determined necessitates that theory must be falsifiable and its accuracy tested in comparison with observation and measurement, this determination can be made independently of any realist commitments the proponents of the theory may or may not have. If van Fraassen is right about the absence of any rational basis for a commitment to scientific realism, we must reject it as a rationale for scientific activity.

Van Fraassen argued that successful scientific theories arise through competitive social processes to decide on their *empirical adequacy*.[41] Only the successful theories emerge, so we should not be surprised at the predictive and descriptive successes of current scientific theories. These theories are latched onto observed empirical regularities in Nature (e.g. objects of the same size and shape fall with the same acceleration regardless of differences in weight). Whether we grasp this under the unifying concept of gravity, or which theory of gravitation we find most intelligible and explanatory, are historically and socially contingent decisions. The processes by which we make these decisions are different from the processes by which we determine the empirical adequacy of a theory. For this reason, van Fraassen insisted that causal explanations should not play an epistemological role in "the scientific enterprise."[42] He adopted the instrumentalist line

---

[41] van Fraassen (1980: 19-40)
[42] *Ibid* 22

that the best that scientists could legitimately claim to achieve are empirically adequate descriptions, predictive success, and manipulative control. Theories are nothing more than instruments to clarify experience and order cognition under more comprehensive and general descriptions. If a commitment to scientific realism is unnecessary for science to continue, and it plays no essential part in the scientific enterprise, then it is a baroque cultural fiction overlaid over science. Hence, van Fraassen rejected scientific realism in favor of *constructive empiricism*. Theories remain ways of representing experience in terms general concepts, descriptions, and categories. Indeed, as Popper pointed out, scientists are limited to being able to test their best guesses, and these guesses are contingent and replaceable. Hence, according to van Fraassen, scientists should rest content with "saving phenomenal appearances"—requiring empirical adequacy from theories—and remove all explanatory causal-accounts from science. "Saving phenomenal appearances" involves ordering, structuring, and clarifying our experiences in specified contexts in terms of general concepts, descriptions, and rules. The best that scientists can hope to achieve—if the evidence is in favor of theory—is to further refine and develop theories of how things behave *as if its theoretical entities existed as real referents*, even if our theoretical representations of them are only approximations—but this does not support any deeper claims about the fundamental composition of the phenomena under investigation. Hence, for the empiricist, "the facts of experience" must have an epistemological primacy over "the causes of experience," which are contingent and replaceable "useful fictions."

Is this an end to the discussion? Have the empiricists won the argument over the realists? No. At least, not yet. The empiricist has assumed that we have a clear and absolute distinction between theory and experiment, so we are able to prevent the mixing together of descriptions and explanations during the scientific activity of testing theory by means of experiment. This assumption seems reasonable only while we operate at a certain level of abstraction of theory. If the distinction between description and explanation is blurred in experimental science—wherein explanations and descriptions are mixed together, and used to produce each other—then this blurring problematizes empiricism as a philosophy of science. If—as Hacking and Gooding have shown, and I have argued above—it turns out that it is impossible to maintain a clear distinction between description and explanation in actual scientific activity, this seriously problematizes the empiricist claim that experience has epistemological priority over theory. Van Fraassen rejected realist claims concerning the existence of unobservable theoretical entities (e.g. a gravitational field) and the notion of an objective natural world that is knowable through scientific activity, and independent from whether we know it or not. Allegedly, objective and mind-independent reality is—*by definition*—beyond the capacities of human understanding, and, therefore, the task of producing a complete and correct theory of objective reality is an impossible one. I shall resist the temptation to delve into the logical paradox that results when we make such a final statement about something we are supposedly unable to make any final statements about, and, instead, question van Fraassen's claim that scientific realism is quixotic and unnecessary for the scientific enterprise. Should scientists be rest content with "saving phenomenal appearances"? Should scientists and natural philosophers consider scientific theory be strictly limited to producing abstract formalisms—sets of mathematical equations and fundamental representations—that successfully predict observational results and provide cognitive heuristics but do not provide us with any knowledge of the causes of experience? Do explanatory causal accounts amount to nothing more than 'useful fictions'? Why should scientists *behave* as if the theoretical entities they are using are real referents? Why is this *false belief* necessary if realism plays no role in the scientific enterprise?

If we were to describe theories as fictions, as does van Fraassen, we would still require theoretical training and interpretation in order to ascertain what the empirical facts of measurement were (in all but the most trivial of experiments), even if we refrained from explaining them. For example, physicists have made the observation of the empirical regularity that hot objects in a cooler environment cool down, and cold objects in a warmer environment warm up, and formally abstracted and universalized this observation in terms of the Zeroeth Law of Thermodynamics. Yet this constitutes an almost trivial starting point for the study of thermodynamics. The physicist wants to know why this empirical regularity occurs, and explains it in terms of the flow and overall conservation of heat, which is further abstracted as a form of energy. Yet there is a big chasm between this empirical regularity and an interpretation of it in terms of the flow and conservation of heat. It takes quite an imaginative leap to cross this chasm. An even greater imaginative leap is required to quantify heat (e.g. in terms of calories or joules) and construct an apparatus to measure the conservation of that heat flow. On van Fraassen's account, it is hard to see how thermodynamics could have progressed from the Zeroeth to the First Law; let alone how the Second Law, with its esoteric definitions of work and entropy, could be part of "the scientific enterprise." We could not even explain the invention and use of a thermometer.

How does the physicist use these theoretical explanations when designing, building, and operating the instruments and apparatus required to make observations and measurements? Empiricists have ignored this question. By ignoring this question, the empiricist has presupposed the theory-independence or neutrality of the established technical and interpretive practices of contemporary scientists. By presupposing the availability of a set of theory-neutral and technical methods, instruments, and apparatus, each (somehow) available for empirical measurements and observations, the empiricist has ignored how instrumentation, education, and interpretation emerge from a cultural and historical background that determines the form and content of "the empirical" by filtering what may or may not be included as such. The empiricist has ignored the psychological, historical, cultural struggles required to make particular instruments and their interpretation intelligible as part of scientific activity, even when the empiricists invoke these factors to show the contingency and instrumentality of theory. If we look at how science is done in practice, however, the invention and use of instruments and apparatus to measure phenomena and make observations involves the use of explanations of how they work. The studies of the majority of the phenomena investigated by the studies of mechanics, thermodynamics, electromagnetism, radioactivity, solid-state physics, and quantum physics, for example, involve the interpretation of instruments and apparatus before we can have any experience of the otherwise invisible phenomena under investigation. If we only have experience of these phenomena as a series and set of changes of the digital or analog readings on calibrated instruments (oscilloscopes, graph plotters, gauges, or computer displays) made in relation to probes or detectors, then our experience of the phenomena is conditional on the existence of these devices. If these devices and the experiences we have by using them are only possible and made meaningful as measurements and observations in terms of explanations of how any of these measuring instruments and their display work, then the distinction between explanations and descriptions is blurred when making measurements using even simple instruments. Recognizing this point is an essential step towards understanding how natural science can be based on experimentation. Once we recognize that explanations are required to make observations and measurements at all, we should also recognize that "the empirical," when limited to the observation and measurement of changes that can be quantified and manipulated, is conditioned by "the theoretical" when designing, building, and using the instruments and apparatus to make those observations and measurements.

Furthermore, scientists need to share observations and measurements if they are to be considered by a wider scientific community as repeatable and testable, and thereby become scientific facts, whereby theories obtain their "empirical adequacy." This means that scientific instruments and apparatus must use the same conventions and standards of quantification if observations and measurements are to be commensurable (literally meaning 'common measure'). This means that scientific instruments and apparatus must be calibrated by using share rules and methods of calibration. Each instrument is calibrated using one or more measurable phenomena or quantities that are commonly accessible by members of the scientific subculture. Within contemporary scientific subculture, these would include weight, temperature, potential difference, time-signals, inductance, capacitance, thermal capacity, electrical resistance, phase, frequency, mass, magnetic field strength, tensile strength, electric charge, spin, etc. These are quantified in terms of arbitrary units (such as kilograms, metres, candelas, seconds, amperes, moles, radians, kelvins, newtons, coulombs, tesla, watts, etc.), with relations between them determined in terms of law-like behavior between numerical variables and constants. The calibration of instruments and apparatus would be quite meaningless without some appeal to theoretical interpretations to explain the choice of quantities and units.

As a result, the use of any instrument or apparatus involves accepting and utilizing a complex mixture of theoretical and empirical representations that are incorporated into practice through education and instruction. By claiming that measurement and observation using scientific instruments and apparatus provides direct experience—as if it were an experience that anyone could have—the empiricist simply has ignored the training and interpretation that is required to have those experiences. The empiricist considers potential differences (quantified as voltage and calibrated as a number of volts) as more empirical than electromagnetic fields only because the empiricist has ignored how past theoretical interpretations were used in context to build instruments or apparatus to use one phenomenon to measure another. What does voltage mean? If we explain it in terms of electromotive force or potential energy, we have to explain that abstraction in terms of "useful fictions," and so on. At each step of our explanation of these abstracts, we invoke the performance of a nomological machine to give it concrete empirical meaning (defining a volt in terms of the electrical potential between two points of a conducting wire when a current of one ampere dissipates one watt of power) and give that machine theoretical interpretation and significance. If we were to consistently adopt van Fraassen's empiricism then experimental physics would be a technical process of instrumentally using one set of fictional entities to investigate another set of fictional entities. If the measurable variables and constants used to produce scientific descriptions of observations and measurements in physics were not made in reference to "objective properties" and techniques of their measurement, it would be pointless to map out the empirical variation of one theoretical fiction used to make measurements and observation meaningful, against another theoretical fiction used to predict possible measurements and observations of measurable variables. The empirical adequacy of any laws that would be produced through such a process would be the pointless interrelation of different kinds of fictions and nothing more.

Does this speak to the psychological motivations of scientists? Modern experimental science could not be psychologically motivated by constructive empiricism and scientists would not rest content with it, unless it provides a meaningful explanation of how scientists design, calibrate, and use instruments to make observations and measurements. As realists have pointed out, seeking causal explanations for phenomena and revealing causal powers have been essential motivations for the historical development of scientific theories.[43] Yet, once we take the theoretical background of any instrument or apparatus into account, how can we treat potential differences as less fictional than electromagnetic fields? If we have no unmediated experience of either, wouldn't physics be nothing more than a technical way of writing fiction? If so, constructive empiricism would intersect with social constructionism[44] and post-structuralism[45]. Would scientists rest content with this? It seems to me that treating scientific theory as fiction does an injustice to both theory and fiction. Empiricists misunderstand the nature of both theory and fiction. They admit that the truth-value of theory is determined in relation to its use, but they only discuss theory in terms of representational and interpretive practices, and say nothing about how it is used in material practices. Scientific theory does more than describe and explain empirical regularities; it also predicts both quantifiable change (particular solutions derived from general formulae) and new phenomena (counterfactuals derived from general laws and mechanisms). This aspect of theory allows it to partake in discovery, as well as modeling natural processes and simplifying complex natural phenomena.

The history of physics provides us with numerous examples of empirical laws, such as Newton's Universal Law of Gravitation, Boyle's Law for ideal gases, Ohm's Law of electrical resistance, Raoult's Law of vapor pressure in mixed liquids, and Ryberg's empirical law for spectral lines. These laws only suggested that there is a simple relationship at work in complex phenomena, but they do not explain why it has that form. Newton found his own law quite unsatisfactory in this respect. Physicists, such as Newton, Leibniz, Boltzmann, and Bohr, were driven to find a theory, or model, that not only matched the empirical laws but also explained that model in terms of fundamental entities and causal mechanisms. Furthermore, given that the content of these explanations are made in terms of mechanisms, these explanations can take a functional role in scientific reasoning during the construction of further experiments. This gives physics an exploratory and developmental trajectory. Scientists seek to explain the existence of phenomena in terms of the same causal accounts they develop and use to design, build, operate, interpret, and modify the machine performances that are intentionally produced to model natural phenomena. The explanatory power of scientific models, theories, and laws to provide intelligible explanations of phenomena in the complex real world is central to the whole process of establishing the intelligibility of natural science. The fact that working scientists are unable to predict the behavior of natural phenomena, such as lightning, does not mean that they are unable to explain lightning in terms of the laws of electromagnetism, thermodynamics, mechanics, and optics. Scientists remain content to consider the phenomenon of a lightning storm to be explained even though they are unable to predict accurately the behavior of such a complex natural phenomenon. In fact, the "complexity" of such phenomena is itself postulated to explain their inability to accurately predict weather in general.

The motivation to provide causal explanations of the phenomena of the experienced world is irreducible to the empiricist's interest in accurately describing appearances (and the instrumentalist interest in the deduction of predictions). This motivation is an attempt to satisfy a basic human desire to explain and unify otherwise disparate experiences in terms of the same fundamental principles and causes. The purpose of measurement goes beyond testing theoretical predictions (which would be a pointless activity if we adopted empiricism). It is a route by which hitherto unexpected

---

[43] Harré & Madden (1977)
[44] Barnes (1974); Bloor (1976); Latour & Woolgar (1979); Latour (1987, 1991); Knorr-Cetina (1981)
[45] Foucault (1994, 1980) and Derrida (1967)

novel mechanisms could be discovered and a more general knowledge can be obtained. Empiricists have failed to recognize that an important goal of science is to produce explanatory theories that describe the law-like behavior of the causal agents that lead to the discovery of new phenomena of experience. It was for this reason that Charles Sanders Peirce criticized the empiricist assumption that accurate measurement was the essence of science because, as he put it, in novel experimental work measurements "fall behind the accuracy of bank accounts" and the determination of physical constants was "about on a par with an upholsterer's measurements of carpets and curtains."[46] Far from merely and arbitrarily allowing scientists to quantify patterns and establishing fits between theory and phenomenal appearances, theory also has a dynamic function in allowing scientists to explore the pattern itself, thereby finding new phenomena and exploring the complexity of both phenomena and theory. This dynamic is revealed through the application of theory in the context of discovery. The importance of this dynamic use of theory to explore patterns and reveal new phenomena and complexities needs to be recognized if we are to understand theory at all.[47] This explorative aspect of theory brings together contexts of discovery and justification, as the successful prediction of new phenomena and complexities gives theory increased truth-value within ongoing scientific activity, which, through feed-back relations, dynamically corroborates the representational aspect of theory and provides increased confidence in the explanatory aspect of theory. Hence we need to understand how theory is used in scientific activity in its three aspects: representational, explanatory, and exploratory.

  We also need to understand how theory allows "otherwise invisible phenomena" to be incorporated into experiments as a mechanism that can be used to explore other phenomena. For example, electromagnets and their associated "electromagnetic field" are used to contain and focus "beams of matter and anti-matter" in particle accelerators, which are used to explore "the fundamental constituents of matter" (i.e. "quarks" and "leptons") and the kinds of "particles" (i.e. "bosons") that pass between them as "interactions," transforming "particles" into other "particles" ("radiation" and "decay"), transferring "energy" between "particles, and "detected" by massive machines (comprised arrays of electronics, liquids, and gasses surrounded by electromagnets, connected to a super-computer), and, in turn, is used to explore representations of the fundamental forces of the Universe. It was for this reason that Hacking declared himself to be a realist about "otherwise invisible phenomena" when they are used as part of a technique to reveal other "otherwise invisible phenomena;" hence, he famously declared himself to be an instrumental realist about electrons when they are sprayed on niobium spheres to search for free-quarks.[48] I shall return to this declaration, but, for now, I wish to highlight the importance of recognizing how the practical or instrumental incorporation of theory into experiment is an essential requirement for science to progress at all.

  If we aim to understand theory and the reality revealed by experimental science, we need to examine the incorporation of theory in ongoing scientific activity, as it is done in practice. How should we understand the incorporation of theory into these kinds of machines, instruments, apparatus, and detectors, and the experiments performed using them? Both experiments and theories are only meaningful in relation to a cultural and historical background of theoretical interpretations of instruments and apparatus, and therefore we need to understand this background and how it has shaped "the empirical" and "the theoretical" aspects of experimentation. But we also need to understand how

---

[46] Peirce (1955: 330)
[47] Pandit & Dosch (2014)
[48] Hacking (1983)

experiments and theories gain meaning and truth-value from their use in ongoing scientific activity. Causal accounts of how any instrument or apparatus works are simultaneously representations, explanations, and predictions of functionality, and they are used by scientists during the design, construction, operation, and interpretation of instruments and apparatus. It is on the basis of such a causal accounts that scientists can claim to know what an experiment does, and are able to use it as a probe or detector in future experiments. Modern experimental science could not proceed without such causal accounts and any scientist who considers him- or herself to be an empiricist (in van Fraassen's sense) has disingenuously presupposed realist interpretations at some level in the design, construction, and operation of their experiments.

Scientists cannot perform experiments on "radiation" (itself a fiction according to van Fraassen's argument), for example, by restricting their accounts to "when substance A is placed near the Geiger counter, the current output increases by B" when designing the experiment in the first place. Constraining "the empirical" to the confines of the laboratory as a means to verify or falsify theories by testing their predictions and deductions against the controlled empirical regularities of the experiment, as orthodox empiricist philosophies of science have done, does not provide us with an account of how and why those experiments are constructed. Scientists utilize a causal account such as "if radioactive material emits energy sufficient to ionize a gas then this will be manifested as a current across a sealed tube of gas with a voltage across it. Therefore we can build a device to measure levels of radiation." The ability of scientists to construct experiments is dependent upon such causal accounts, which are used to reveal and explore "the empirical" via techniques, and the instruments and apparatus that comprise the experiment. Hence, empiricism tends to only account for experimentation in terms of a ready-made procedure of intervention and measurement, without any justification of why that procedure was chosen and performed in practice, because it has neglected to attend to the role of technology in experimentation. Not only do observations actively occur in the light of theories, within theoretical frameworks or paradigms, using concepts, visualizations and representations, but also theoretical accounts are implicit in the design, construction, operation, and interpretation of the instruments and apparatus used to make those observations. The theories and practices of modern science presuppose the possibility of the discovery of causes in their descriptions of phenomena because they require such interpretations of the experimental apparatus and measuring instruments when designing them. As a philosophy of science, empiricism falls short because it is premised upon the assumption that scientists have complete knowledge of how the experiment works, without any explanation of how this was achieved, while simultaneously insisting that any knowledge of causes is impossible and must be left out of science. It is unable to cope with novel (or revolutionary) experiments at all and cannot explain how and why experiments were constructed in particular ways. It is this inherent contradiction that reveals the inadequacy and irrationality of empiricism as a philosophy of science.

Novel phenomena require new concepts in order to have new experiences, and the application of new concepts implies speculative metaphysical assumptions regarding the nature of the phenomena, yet there is no reason to believe that these concepts will not be replaced in the future. If we seek to develop an intelligible philosophy of science, we need to understand how "the empirical" and "the objective" were metaphysically determined and developed within the technological framework of scientific activity. By reducing science to the systematic logical ordering of our sensory experiences, empiricism restrict "facts" to be statements of our immediate experiences made in terms of already given language. Thus it is impossible for empiricists to add any fact that cannot be expressed and logically analyzed in terms of already given language. It is impossible for empiricists to explain how novel experiences could be constructed in terms of novel expression and meaning because the restriction that they place upon legitimacy and intelligibility would not allow the processes involved in the construction of novel expression to begin. Empiricists cannot account for how new explanations and descriptions are made using new forms of representation and intervention, and, as a consequence, novel experimental research programs could not proceed according to empiricism, which is only able to provide a conception of experimental science during its Kuhnian normal science phase by uncritically accepting the meaning of the outcomes of experimentation as given to them by scientists. While empiricists like van Fraassen have overly-emphasized the historical significance of the revolutionary phase in his critiques of theory, wherein, time and time again, theoretical entities used to explain experience are shown to be imaginary and the scientific worldview has been radically revised and transformed, yet he conforms to science's normal science phase in his discussion of saving phenomenal appearances, and uncritically accepts the scientist's interpretation of the meaning of techniques and instruments as given. Van Fraassen's a priori rejection of the role that causal accounts play in "the scientific enterprise" of modern experimental science prevented him from producing an intelligible philosophy of science.

In general, if we accept that the use of causal accounts is an essential part of designing and building apparatus in experimental science, then empiricism, by rejecting the possibility of the knowledge of causes, cannot provide an intelligible account of how and why novel experiments are constructed and performed in the first place, and empiricism fails as a philosophy of science because the technological environment of the laboratory is unintelligible to immediate human experience. The sensorium of scientific instrumentation is a meaningless maelstrom of flashing lights, moving pointers, digital displays, graphs, and readouts to the uninitiated and untrained. There is nothing immediate about voltages, time signals, amplitudes, temperatures, and pressures read from instrumentation. Constructive empiricism fails because it results in scientists mapping out the empirical regularities that arise when one set of fictional entities (theoretical variables) are related to another set of fictional entities (calibrated units), and this results in a meaningless pursuit of writing fiction by mapping how one fiction relates to another. It is unable to justify how "the empirical" is identified and explored through experimentation.

What is the phenomenological character and ontological truth-status of the phenomena explored by making machines to disclose them? My argument is that in order to question the epistemological and ontological meaning of machines, we need to examine the motives for making and using them, and how those machines have been interpreted within the contexts of experimentation. What are the relations between experience and causal accounts? How do instruments and apparatus shape experience? One we recognize that a clear distinction between explanation and description does not occur in actual experimental work, we must also acknowledge that the vast majority of scientific research—involving feedback processes between theories and causal accounts of how the instrument or apparatus works—would fall outside of the remit of empiricism. Modern experimental science is not empirical in the philosophical sense; an experimenter does not rely upon direct experience but investigates the disclosure of interpreted instrumentation to direct experience. An intelligible account of science must provide an account of how this disclosure is possible and how it is achieved. Empiricists cannot provide us with an account of how observations of using instruments are obtained because they cannot consider as scientific the technical reasoning (made in terms of theoretical entities and causal-accounts) used in the interpretation of measuring instruments and experimental apparatus. Thus, they either cannot provide a meaningful account of observation within experiments in terms of scientific facts, or they are only able to provide accounts of observation by arbitrarily, passively, and uncritically accepting particular technical interpretations of particular measuring devices as given. By asserting that sensory observation founds all genuine knowledge, by presuming that only those statements that are derived from experience are legitimate or scientific, the empiricist has neglected to attend to the ways that experiences and the means of their production are constructed in the contexts of experimentation. By ignoring the technological framework, empiricism fails to understand the nature of "the empirical."

Theory—providing causal explanations of change and repetition in the performance of machines—is required to construct a meaningful experience of experimental apparatus and instruments, while observations and measurements are mediated by the technological framework and metaphysically interpreted in accordance with the precepts of mechanical realism. Experimentation is based on planned action, selections, and decisions within the context of the innovation and ordering of material and theoretical practices by utilizing causal accounts, and this both constrains and shapes scientific theories and how they are understood to be related to Nature. As Gooding showed from his study of Faraday's experiments, empiricism simply fails to recognize that exploration in experimental work requires speculative and constructive processes of visualization and communication that cannot be reduced to logical or empirical statements.[49] It is a social and interactive process in which experiences are made through trying to represent novel material practices and their outcomes in novel contexts in ways that facilitate communicating experience to others, and finding its confirmation in the ability of others to replicate those material practices and experiences, rather than passively received by the senses as raw data, and it is given cultural and psychological meaning through its historical foundations in metaphysical conceptions of technology. Empiricism is unable to deal with the processes of constructing novel experiences of novel phenomena because it fails to explain the ways that novel material, communication, and visualization practices are constructed when dealing with novel phenomena; it cannot provide an account about how novel experiments can be constructed in the absence of an established theory and techniques of observation and measurement.

Empiricism also cannot explain how theoretical entities "transcend" the particular contexts of their production or use, thereby allowing them to be used in the design, construction, and operation of different kinds of machine. If we were to accept instrumentalism (a descendent of empiricism) and claim that the meaning of theoretical entities lies solely in its value within the context of its use, this raises the question of how theoretical entities gain any utility at all as part of techniques to produce phenomena or as a means to explore them. On the instrumentalist's account, when physicists experiment on "electrons" in electric circuits they are not operating with the same kind of electrons "annihilated" in the LEP ring at CERN, those "deflected" in Thompson's cathode ray tube experiment, those "suspended" in Millikan's oil drop experiment, or those "sprayed" on Morpurgo's niobium spheres to detect free quarks. The instrumentalist is unable to explain how physicists are able to use the same theories, say QED, Special Relativity, or Maxwell's equations, in these distinct contexts. Nor how physicists are able to relate their different experiments to different theories of the electron. By limiting the "the empirical" to the context of use, theoretical objects are unable to transcend and unify the contexts and particularities of experience unless we presume a realist interpretation of those theoretical objects, as something objective, even if we admit that we (probably) have failed to understand them. Moreover, empiricism cannot consistently provide an account of

---

[49] Gooding (1990: 29-30)

how scientists fail to perform experiments according to theoretical expectations, except by asserting that it is only a matter of time and 'trial and error' before all theories fail to demonstrate their empirical adequacy. What determines whether a theory is empirically adequate or not? Take Murpuro's famous experiment to find free-quarks. According to the empiricist, this was a project that utilized fictional entities (electrons, charges, and electromagnetic fields) to attempt to find other fictional entities (free-quarks), so we must ask, what was the source of resistance to its success? If the proper use and identity of theoretical entities are completely determined in context then there is nothing to prevent an experimenter, in that context, from achieving success. One need only reconstruct the meaning of the performance of the apparatus according to one's intentions within that context. Other experimenters, in other contexts, could not object to this because one could merely argue that they were using the terms "electron," "charge," "electromagnetic field," and "free-quark" differently within different contexts, and that anyone who failed to repeat the observation merely had failed to reconstruct the context of use. Yet Morpurgo did not observe any evidence of the existence of free-quarks during his twenty-five year search for them. Why not? Any intelligible account of experimental science must provide an account of how experiments fail to perform according to theoretical expectations.

Empiricism cannot explain how either predictive successes or failures are possible, except by appeals to 'trial and error.' Hence van Fraassen's constructive empiricism is unable to explain how theories, supposedly based upon fictional entities, can provide predictive success and facilitate manipulative control at all, and this is further problematized when we take into account the causal accounts utilized in the construction of instruments and apparatus that allow experiences of "otherwise invisible entities" (e.g. bacteria or viruses) to explain visible phenomena (e.g. the symptoms and spread of disease). The realist has long since asked: if electrons and photons are merely fictional entities then how can empiricism explain the incredible predictive accuracy of Quantum Electrodynamics? We need to answer this question. In addition: how do the causal accounts utilized to construct instruments and apparatus to determine the predictive accuracy of theories like QED mediate and shape "the empirical" during acts of measurement and observation? Empiricists presume that technology has a logical and empirical character that is divorced from metaphysics, and reject speculative metaphysics as either nonsense or unverifiable because, by positivistic definition, metaphysics is taken to lack any [50]empirical content. While realists ignore the operational metaphysics that makes realism conceptually possible and they only consider the speculative metaphysics used to interpret specific theories and observations in the absence of the means to test theories and predictions, history provides many examples, such as seventeenth century atomism, that were untestable using the technologies available at the time. It is for this reason that even Popper acknowledged that various speculative metaphysical beliefs, such as Kepler's harmonics of the geometry of the planetary orbits about the Sun, led to significant advances in theory before there was the means to test those theories. Popper wrote:

"[There] is a least one philosophical problem in which all thinking men are interested. It is the problem of cosmology: *the problem of understanding the world —including ourselves, and our knowledge, as part of the world.* All science is cosmology... it is a fact that metaphysical ideas—and therefore philosophical ideas—have been of the greatest importance for cosmology. From Thales to Einstein, from ancient atomism to Descartes' speculation about matter, from the speculations of Gilbert and Newton and Leibniz and Boscovich

---

[50] Popper (1975: 206n, 277-8)

about forces to those of Faraday and Einstein about fields of forces, metaphysical ideas have shown the way." (Popper, 1975: 15-19)

By rejecting speculative metaphysics, empiricists have amputated a significant source of scientific speculation and have rejected the validity of the narrative aspect of scientific theorizing. This narrative aspect, as a kind of story-telling, is a cultural phenomenon that attempts to situate human beings within a world-picture: the scientific worldview. In all cultures, human beings have told stories about how the world came to be. This active social and psychological construction of narrative about reality is a part of what it is to be human. Scientists *qua* human beings are no different in this respect. Empiricists have made a crucial error by rejecting this essential aspect of science as unscientific or illegitimate. Theories remain unscientific (i.e. speculative and untestable), however, until the innovation of technological means by which they can be tested. The demarcation between a metaphysical speculation and a scientific theory, in these cases, is one of historical accident and yet the role that they played in the development of science was often significant prior to the innovation of the means to test them. Hence, empiricists have ignored what constituted "the empirical" and "testing"—and hence "scientific" and "falsifiable," along with ignoring the fact that "the objective world" and "scientific worldview" have been technologically mediated and constrained from the outset. Empiricists have misunderstood the nature of "experience" within science because of their uncritical acceptance of scientific observations alongside the Humean appeal to immediate experience as the foundation of knowledge. Yet, as Popper pointed out, the empiricist—by considering the problems of traditional philosophy as "metaphysical pseudo-problems" that are meaningless for empirical sciences—has neglected the fact that the central problem for traditional philosophy has always been a *critical analysis* of appeals to the authority of "experience."[51]

Once we recognize that "the empirical" explored by experimental science is disclosed via instruments and apparatus that utilize causal accounts, involving theoretical entities and mechanisms, we must also recognize that experimental scientists, whether or not they realize it, need to presuppose scientific realism at some level. When they claim to be empiricists, working scientists tend to be naïve realists who operate under a practical (as opposed to philosophical) conception of empiricism as a methodology, and they would not be able to perform experimental science if they were to become aware of the metaphysical underpinnings of their practices and reflect upon them.[52] On this view, working scientists have inconsistently adopted and intermixed both realism and empiricism, depending on the rhetorical or pragmatic needs of the context, and perform experiments via a metaphysical interpretation of technology that they term as "empirical research." Of course, an awareness of the paradigmatic and pragmatic beliefs of working scientists only shows us the internal rationale of experimental science, as an internally organized and rationalized human pursuit, and tells us nothing of the truth-status or rationality of this rationale. Demonstrating the necessity of the beliefs of any practitioners does not grant those beliefs any uncontestable epistemological and ontological foundations in correspondence to truth or objective reality. The scientific paradigm represents itself as a cultural means to gain knowledge because its practitioners presuppose that mechanical realism is "a self-evident truth." Without examining the objectivity and rationality of its operational metaphysics, experimental science remains an irrational pursuit—culturally classified as a natural science—even if its practitioners rest content with it.

---

[51] Popper (1975: 51-2)
[52] Feibleman (1982); Igov (2005)

Scientific Realism as an Internal Rationale

If we agree that realism offers a more adequate philosophy *for* science than empiricism, we still need to understand and question the metaphysical conceptions of "the physical" and "the empirical" that are required for realism to be intelligible as a philosophy *of* science. Yet, the metaphysical foundations of scientific realism have largely remained unexplored by scientific realists; they have either ignored the precepts of mechanical realism, or uncritically accepted them as "self-evident" and simply hitched their arguments for realism to "the empirical successes" of the natural sciences. However, some scientific realists—naïve realists such as Nicholas Maxwell, methodological realists such as Rom Harré, and critical realists such as Roy Bhaskar—have recognized the importance of exploring the metaphysical foundations of science for the task of advancing an intelligible philosophy for science, whereby a better understanding of the assumptions underlying its scientific methods and theories promises the improvement of science. Yet they have neglected to explore the metaphysical foundations of scientific realism itself.

Nicholas Maxwell argued that science is unintelligible as a human pursuit, unless we examine the metaphysical assumptions that are required for evidence and theories to be comprehensible.[53] For Maxwell, science is only possible because of speculative metaphysical assumptions regarding the ultimate nature of the Universe—assumptions that are required to make the Universe comprehensible as something that we are within and of which we can have knowledge—and he considered the role of philosophy to be one of revealing these assumptions. Scientists need to make judgments about the nature of objective reality from a position of complete ignorance in order to begin the inquiry into the nature of objective reality. Hence Maxwell argued that

"Metaphysics determines methodology. This makes it of paramount importance that a good basic metaphysical conjecture is adopted, one that corresponds to how the Universe actually is. A bad metaphysical conjecture, hopelessly at odds with the actual nature of the Universe, will lead to the adoption of an entirely inappropriate set of *methods*, and the result will be failure, possibly, of a peculiarly persistent kind." (Maxwell, 1998:5)

If theory choice requires persistent assumptions about the fundamental nature of the Universe and the scientific method, and there are not any empirical tests for these kinds of assumptions, then empiricism misrepresents the basic intellectual aim and methodology of scientists. If scientists do not seek the accumulation of empirical facts or descriptions of empirical regularities (or patterns), for their own sake, but, instead, seek to explain all phenomena in terms of the same fundamental principles or laws, then any adequate philosophy of science must recognize that theories are explanatory and offer causal accounts to explain phenomena, which entails metaphysical assumptions about the Universe, rather than aiming at merely providing accurate descriptions of natural laws, which can be tested by comparing predictions with experience. Maxwell's argument explains why complex natural phenomena such as weather patterns or the path of lightning bolts are considered by scientists to be theoretically understood in terms of basic physical laws even when they cannot be accurately modelled or predicted by theories. Maxwell's argument also explains why scientists would try to apply Newton's Laws of

---

[53] Maxwell (1998)

Motion to banknotes blowing in the wind.

The main target of Maxwell's criticism is *standard empiricism* (SE), which he considers to be the dominant philosophy of the natural and social sciences since the eighteenth century.[54] This epistemological doctrine holds that science achieves progress by testing theories impartially in the light of evidence *without making any permanent metaphysical assumptions about the nature of the Universe.* Maxwell argues that SE is untenable in practice because for any scientific theory there are (potentially) an infinite number of rival theories, which all agree about the same set of observed phenomena while predicting different unobserved phenomena. It would be an impossible task to explore all the predictions and the untested consequences of every possible theory, therefore, any choice in favor of any empirically successful theory would be arbitrary. If the doctrine of SE was adhered to in practice, science would be either an *ad hoc* process of arbitrary theory testing or it would come to a complete standstill. In practice, scientists utilize conceptions of simplicity, unity, comprehensibility, generality, and explanatory power when choosing between rival theories, but these conceptions entail exactly the kind of metaphysical assumptions about the Universe that SE prohibits. Hence Maxwell argues that the persistent failure of the traditional philosophers of science to solve serious and perennial philosophical problems (i.e. solving the problems of induction and deduction—how to relate theory to experience and *vice versa*—and thereby explaining how scientific progress is possible) is a result of the naïve or disingenuous acceptance of SE as orthodoxy by many scientists and philosophers of science alike. Without solutions to these problems, philosophers cannot demonstrate the rationality of science as a human pursuit, and scientists cannot rationally verify any scientific theory on the basis of evidence, given that it remains logically possible that (a) any theory will be falsified in the future; and, (b) there is a different theory that is at least as empirically successful.

Following Popper, Maxwell claimed that it is probably true that any given theory will be shown to be false (inaccurate) and replaced by a better (more accurate) theory in the future, in the light of further evidence and testing, even if the proposed theory is currently corroborated by evidence and testing. Hence Maxwell argued that if empirical success was the only standard for theory-choice, there would not be any rational basis for choosing one theory, given that it probably will be shown to be false, if sufficiently tested. On this Popperian line of reasoning, the empirical adequacy of any theory is itself indicative that the theory has yet to be fully or sufficiently tested. There would not be any rational basis for applying any scientific theory to any practical or technical problem, given that it is probable that the theory is false and any solution would be partial at best and doomed to failure when tested through its application in the physical world. It would also be impossible to accept any experimental result as "evidence," given that without a rational foundation, scientists would be unable to justify the generalization of any particular result to corroborate or refute any theory. Van Fraassen would be right: any theoretical entities described by any empirically successful theory would only be "useful fictions." Beyond the pedestrian, *ad hoc* collection of unrelated empirical facts, or describing empirical regularities of measurable variables and phenomena, it would be impossible, in terms of SE, to identify and justify rationally any scientific method or progress at all! The predictive successes of scientific theories could not be explained, their practical value would seem quite miraculous, and the explanatory value of theories would be considered irrelevant for the scientific enterprise.

---

[54] Maxwell (1984)

So far, so good.

Maxwell also quite correctly argues that, in practice, scientists choose between empirically successful theories in favor of theories that promise to simplify and unify descriptions or explanations of otherwise disparate phenomena. Scientists attempt to explain all possible experience and phenomena in terms of the same underlying pattern, principle, or law, and they assume that the Universe and a complete scientific theory of it both have a simple and unified nature. It is this metaphysical assumption that gives theorizing an exploratory dynamic—situating theory choice and testing simultaneously in contexts of justification and discovery—thereby, providing science with a progress-achieving methodology. The adoption of SE within the philosophy of science has been an obstacle to the development of an adequate philosophy of science, which would show how metaphysical assumptions about knowledge and the Universe are built into the aims of science. Hence, according to Maxwell, SE fails to provide a rational foundation for science and prevents philosophers and scientists alike from developing conceptual clarity about the role of conceptions of "simplicity" and "unification" within the scientific method and the aims of science. Maxwell argued that natural sciences like physics, biology, and chemistry are only intelligible *qua* sciences if we reject SE and critically examine the metaphysical assumptions that are required for theories to be comprehensible and lead to a growth in knowledge. It is essential that the philosophy of science reveals these assumptions because if metaphysics determines methodology then bad metaphysical conjectures will lead to inappropriate methods. Different metaphysical conjectures about the underlying nature of the Universe lead to alternative cosmologies, which lead to rival methodologies and theories. These can be compared by examining which methodology and theory better leads to a growth of knowledge and more clearly articulated assumptions about the comprehensibility of the Universe.

To satisfy his demand that the assumptions implicit in the goals of the scientific methodology are made explicit and increasingly simplified and comprehensible, Maxwell proposes a ten-level hierarchy of increasingly generalized metaphysical assumptions concerning the nature of the Universe. Supposedly, these assumptions become decreasingly problematical and less dubitable as one moves up the hierarchy. This ten-level hierarchy provides us with a system of conceptual filters, each based on an increasingly refined metaphysical assumption about the ultimate nature of the Universe that are used to select between theories on the basis of their simplicity and comprehensiveness, as well as their generality (range or scope of predictions and explanatory success) and empirical adequacy.

Level 1: *Evidence*. Empirical data taken from observations and experiments;

Level 2: *Fundamental theories*. All accepted empirically successful theories that explain phenomena in general.

Level 3: *Best blueprint*. The best available metaphysical interpretation of how the Universe is physically comprehensible, such as Newton's corpuscle theory of light and matter or Everett's Many-World view, wherein all phenomena and change are explicable in terms of some fundamental principles and unobservable entities;

Level 4: *Physical comprehensibility*. The thesis that all phenomenal change and diversity can be explained in terms of one kind of physical entity;

Level 5: *Comprehensibility:* The thesis that all phenomenal change and diversity can be explained in terms of one kind of entity, which is not necessarily a physical entity;

Level 6: *Near comprehensibility*. This thesis is that even if the Universe is not comprehensible it is nearly comprehensible to the extent that the assumption of complete (or perfect) comprehensibility leads to a better growth in knowledge than defeatism or making *ad hoc* experiments and observations;

Level 7: *Rough comprehensibility*. The thesis is that the assumption of approximate comprehensibility leads to a better growth of knowledge than defeatism or making *ad hoc* experiments and observations;

Level 8: *Meta-knowability*. The assumption that the Universe is meta-knowable in the sense that it is rationally possible to discover assumptions about the nature of the Universe that lead to improved methods for the improvement of knowledge;

Level 9: *Epistemological Non-maliciousness*. The assumption that it is possible to discover the general nature of the Universe by developing knowledge acquired in our immediate environment, even if the Universe does not exhibit comprehensibility or meta-knowability in our immediate environment;

Level 10: *Partial knowability*. This thesis is that the Universe is such that it is possible to acquire and possess knowledge of our immediate environment as a basis for action.

As we ascend this hierarchy from level 3 to 10, each assumption about the Universe is increasingly general and devoid of specific content, which, according to Maxwell, makes it less likely of being false. Corresponding to each level r (from levels 3 to 9) there is a methodological rule that requires that we accept the level r-1 assumption that best exemplifies the level r assumption and leads to (or promises) a growth in knowledge. This allows us to understand the assumptions implicit in current scientific methodology, each made explicit and increasingly simplified, and how these assumptions and their associated methods can be refined or changed as scientific knowledge is refined or changed. Maxwell argues that this system allows us to identify and explicate how scientists comprehend the Universe and how this comprehension evolves as science makes new discoveries, acquires knowledge, refines its aims and methods, and improves how scientific discoveries, knowledge, aims, and methods are understood. He called this *aim-oriented empiricism* (AOE) and considered it to be the only philosophically adequate basis for correctly identifying the progress-achieving methods of science and for understanding science as a rational means of inquiry. According to AOE, any accepted physical theory that does not explain the fundamental principles of all phenomena, change, and limits, must be false, even if it is empirically successful. For example, while General Relativity plus the Standard Model of Elementary Particle Interactions may well have huge explanatory success and predictive power, they are not unified and according to AOE must be false. A better theory must be possible. This explains the motivation for the search for a "deeper" unified theory from which General Relativity and the Standard Model could be derived as special cases. This motivation could not be considered to be rational in terms of SE, given the empirical success of General Relativity and the Standard Model within their different fields of applicability. Maxwell claims that AOE has been used by scientists, including Galileo and Einstein, to explicate and evaluate the speculative metaphysical assumptions used by theories to interpret "evidence," whether or not they know they are using it. Hence, Maxwell argued that the history of science shows that science has only been able to progress rationally when scientists have ignored SE, even when they paid lip service to it, and have put AOE in practice, even if unconsciously.

At first glance, it might seem that Maxwell has simply rediscovered the principle of parsimony (or Occam's razor), but his metaphysical system implies much more than this. When scientists choose simpler and more comprehensible theories over more complex and specific rivals, they are not merely choosing theories on the pragmatic basis of which one is easier to work with or more readily testable, but, much more importantly, scientists are assuming that, *ceteris paribus*, a simple and comprehensible theory is more likely to be true than a messy or incomprehensible theory, or a theory with a limited or special range of applicability or that requires additional *ad hoc* modifications and corrections when applied to different phenomena. This assumption entails an assumption about the way the Universe *is* and how we can come to know it. A theory is considered to be simple and comprehensible as a general theory when it has invariant content, form, and terminology across the whole range of phenomena to which it is applied as an explanatory and empirically successful theory. Hence Maxwell argued that the traditional philosophy of science laboring under SE has presumed an untenable relation between theory and evidence. By dissociating philosophical inquiry from the aims and methods of science from science, SE undermines the rationality of science, while AOE, according to Maxwell, provides a rigorous and fruitful conception of science that improves our understanding and practice of science, which has "fruitful implications not just for science, but for all inquiry and, in a sense, for all of life."[55]

Again, so far, so good.

Despite the correctness of his criticisms of SE and the traditional philosophy of science, however, Maxwell's argument is not without its own problems and underlying assumptions. Underlying his argument for replacing the orthodoxy with his ten-level hierarchy as a new philosophy of science is the presumption that once we have arrived at indubitable assumptions (that cannot be doubted without impeding the growth of knowledge) then we have a good reason to believe that we are nearing the truth. Obviously this is circular—begging the question about whether there is any growth of knowledge at all—and presupposes an implicit faith in science as a pursuit. It is an expression of commitment to scientific realism and reveals Maxwell's assumption that only a philosophy *for* science could be an intelligible philosophy *of* science. From the outset, Maxwell asserted that science is self-evidently successful.[56] He made this assertion without providing us with any conceptual foundation for the epistemological standards required to make this judgment in favor of science as a means to increase knowledge about Nature. This assertion allowed Maxwell to start with "evidence" as the first level of his ten-level system, upon which he built his hierarchy of metaphysical principles of "simplicity" and "comprehensibility." Yet, if we question the self-evident success of science, we should notice that Maxwell's system does not provide us with any account of how "evidence" is selected and produced. "Evidence" is the ground level—"Level 1." Questions about the means of the selection and production of evidence, and the epistemological justification of them as means, remain outside of Maxwell's system. This should be unsurprising; after all, he assumed the empirical success of science from the outset. However, if we are in pursuit of knowledge and truth, we should dig deeper and look beneath the surface of "evidence" and see how it is selected and produced.

Philosophers of science cannot simply accept a division of labor and allow scientists to produce and select "evidence" for them. If philosophers aim to base their

---

[55] Maxwell (1998: 2)
[56] *Ibid* 1

philosophy of science on anything apart from their faith in scientists, we need to know how the scientists produced and selected "evidence." Science does not exist in a cultural vacuum; evidence is produced and selected by human beings. Why did they make the choices and decisions they made? We need to look at how the selection and production of "evidence" is made possible and shaped by human relations and prior decisions (including modes of communication, cooperation, conflict, controversy, and hierarchies), via institutions (universities, research centers, textbook and journal publishers, governments, societies, media, and corporations) and "non-scientific" factors (cultural beliefs, language, political and economic power, funding, and access to resources), in a culture that provides shared intellectual resources and means of communication. We need to situate science within a culture (with its historical conditions and contingencies) that mediates and shapes how science is learned, communicated, and practiced, and thereby mediates and shapes how "evidence" is produced and selected, and interpreted in relation to theory. We need to understand the history, sociology, and anthropology of science.

Philosophers of science specifically need to examine the assumptions made about "the physical" and "the empirical" by scientists and natural philosophers since the sixteenth century, thereby determining (and reducing) what is to be included or excluded as "evidence" about the object of inquiry. These assumptions situate science within a subculture ('paradigm' or disciplinary matrix, to use Kuhn's terms) and the traditional philosophers have simply affirmed the value of the subculture and consider it a special human activity. Yet, any adequate philosophy of science must examine this subculture and look at how "the physical" and "the empirical" are connected through a metaphysical conception of technology in the context of selecting and producing evidence during experiments, using detectors and instruments, and making observations and measurements of objective properties and processes to determine the facts. Philosophers of science and scientists need to look at how technology has been interpreted and understood using these metaphysical precepts (implicit to the practitioners' beliefs as subcultural norms) within the development of scientific methods and practices. We need to take a step back and ask ourselves, metaphysically speaking, what must be assumed to be true for these interpretations of the meaning and significance of science to exist? What must be assumed to be true for scientific experience to be evidence of anything outside of the contexts of those experiences? What preconceptions about technology are necessary for it to be understood as a means to provide evidence about the natural or objective world and "otherwise invisible entities"?

Maxwell did not provide us with any account of the operational metaphysics that underlies experimental inquiry using instruments and apparatus as a means of producing evidence about the Universe. He simply assumed that technology does what scientists says it does. By presupposing that the "empirical success" of science is well-understood and self-evident, Maxwell's system presupposed mechanical realism as its underlying operational metaphysics, whether or not he knows it. This presupposition is necessary for the disclosure of "evidence" about the Universe on the basis of experiments using technology. It is an implicit "level zero" that allows experiments upon machines to be presented as disclosures of natural mechanisms, laws, and facts about the natural world. Otherwise, how could we conceive of these mechanisms and laws as existing independently of their contexts of production? Once we examine how the mechanical nature of the Universe is implicit in the thinking of traditional philosophers of science like Maxwell, we can make sense of their naïve realism in terms of their paradigmatic commitment to a mechanical realist subculture that underscores their faith in scientists as producers of evidence. The implicit commitment to mechanical realism and scientists is apparent in Maxwell's response to my essay "Metaphysics and Methodology: Aim

Oriented Empiricism," in which I pointed out that mechanical realism is an implicit level zero to his ten-level system.[57]

"Rogers is probably correct in holding that much of physical theory, such as nuclear physics, is most severely tested in the laboratory, but it is important to appreciate that even nuclear physics can be tested by, for example, astronomical observation. A recent example was the observation that the sun seemed to be emitting too few neutrinos, which led some physicists to question whether the relevant physics is correct, but which was eventually resolved by the discovery that neutrinos have mass. (Because they have mass, neutrinos, on their way to the earth from the sun oscillate from one type to another, unobservable type: hence the low observed flow of neutrinos from the sun.) But even if most physical theory is primarily tested and corroborated in the laboratory, this does not mean that special assumptions, such as mechanical realism, are required to apply physical theory to phenomena beyond the laboratory—in addition, that is, to the hierarchy of assumptions of aim-oriented empiricism involved in the rejection of empirically successful, *ad hoc* variants of the theories we accept." (Maxwell, 2009: 268)

Without any hint of irony, Maxwell appealed to theory (neutrino oscillations and mass) to explain how "the observation" of "the sun emitting too few neutrinos" tested nuclear physics outside the laboratory, as if this was some kind of immediate experience or observation that one could have by walking down the street and looking up at the sky. How did physicists make this observation? How was this experience possible? Maxwell does not ask these questions. He ignores technology and leaves it to the physicists to provide him with the facts of experience and their theoretical interpretation of them and their significance. Scientists have simply "observed" that "the sun seemed to be emitting too few neutrinos," and resolved this discrepancy with theory by the explanation that "neutrinos have mass," thereby justifying that theory and its assumptions. Yet, if we look at how physicists made this kind of observation, we discover that they made it using machines: a neutrino detector called the Super-Kamioka Neutrino Detector (Super-K) in Japan, or the IceCube Neutrino Telescope in Antarctica. The Super-K is a steel tank containing 50 thousand tons of water, located 1000m (3280.84ft) under Kamioka mountain; the tank surrounded by hundreds of photo-multiplier tubes and micro-processors, each arranged to detect Cherenkov radiation, which is theoretically understood as the light emitted by a charged particle moving faster than the speed of light in a medium (i.e. the water of the tank), the output of which is compiled through the computer analysis of time-signals, voltages, and detector position, allowing "the observation" and measurement of the predicted "recoil" of electrons from solar neutrinos. The IceCube is a cubic-kilometer sized cube of thousands of spherical sensors, buried in the ice 500m (1640.42ft) below the surface of Antarctica, with each sensor comprised of a photo-multiplier tube and micro-processor to detect Cherenkov radiation emitted from charged electrons moving faster than light in ice, again through computer analysis of time-signals, voltages, and sensor position. For both of these machines, the data is further analyzed using computers and theoretical models "to filter out" predicted solar neutrinos from those emitted from the decay of charged cosmic particles in the atmosphere. The computer programs written and developed to do this utilize theoretical assumptions to interpret the meaning of operational variables and the detector's performance. Clearly, the "observation" of "too few neutrinos" was mediated by complex technology, designed and interpreted using theory-laden models, and making that observation required conceptions and representations of how that machine works.

---

[57] Rogers (2009); Maxwell (2009)

The physicists working on a neutrino detector are trying to understand its performances in terms of the theoretical representations, such as "neutrinos," "neutrino oscillations" and "mass," taken from other physicists who developed the concepts and models of nuclear physics while working in laboratories on different machines, such as the Kamioka Liquid Scintillator Antineutrino Detector (KamLAND)—a concentric series of spherical chambers comprised of scintillating fluid in a massive balloon suspended at the core of a tank of oil, surrounded by inner layer of photomultiplier tubes and further contained in a steel tank of 3200 tons of water and an outer layer of photomultiplier tubes, and buried in a mine shaft in a mountain, surrounded by over 50 commercial nuclear reactors. Yet, according to Maxwell, no analysis of the role of technology is required to understand how scientists make these observations, so we do not need to think about why and how that technology is represented and interpreted as a means to make observations. According to the naïve realist, observing neutrinos is as straightforward as observing coconuts falling out of a tree, and we can ignore the fact that neutrinos cannot be seen by human eyes and must be inferred from the theoretical interpretation of machines. By ignoring technology and accepting the division of labor between scientists and philosophers, Maxwell could deny the need to examine mechanical realism or anything outside his ten-level system because (1) he assumes from the outset that nuclear physics is empirically successful and physicists have done what they claim to have done (or it is left to other physicists to show they are wrong); (2) if nuclear physics is empirically successful and physicists say that nuclear physics is part of natural science there is no reason to question this. Clearly, Maxwell's argument begs the question. It is based on a cultural faith in scientists. Hence his naïve realism—often hidden behind his "intuition" of the "self-evident" empirical success of science—is based on nothing more than a cultural prejudice in favor of science and vague appeals to the evolutionary (or practical) value of scientific knowledge.[58]

Naïve scientific realism is premised upon and justifies an uncritical acceptance of the neutrality of technology, as something objective and based on natural mechanisms and laws, which are assumed to be understood by the scientists who build and operate it. The naïve scientific realist believes that technology simply provides a technical means by which scientific theories can be tested through comparing observations and theory, but is unable to provide any justification for that belief apart from protesting that science would not achieve empirical success unless this belief was somehow justified. The naïve scientific realist also accepts that evidence produced through representational and material practices can relate the observable effects of unobservable causes with theoretical explanations of such changes in terms of natural mechanisms and laws, without any philosophical explanation of how either this kind of explanation or its "test" using machines is conceptually possible. The assertion that "empirical success" supposedly provides such theories with explanatory power is taken to be sufficient reason to assert that science would be irrational if we did not accept this.

Maxwell's arguments are much indebted to Popper's philosophy of science. Popper argued that if a theory based on unobservable entities produces accurate predictions then what that theory has to say about the unobservable world has a good chance of being more accurate and closer to the truth than a less predictively successful theory—or, at least, the assumptions it is based on are less likely to be shown to be inaccurate and false in the future.[59] This is what Gooding termed as *asymptotic realism*: where the objective truth is never reached, but the production of stable associations between representational and material practices that are capable of being communicable

---

[58] Maxwell (1984)
[59] Popper (1963; 1975)

and reproduced gives pragmatic reasons to believe in the proximity to the truth of those representations.[60] Predictive success is taken to be a reason for belief in theoretical explanations, which are taken to be truer (or more likely to be true) than other equally predictively successful theories if they explain things in a simpler and more comprehensible manner. Current scientific explanations may not be true but based on fewer assumptions about the Universe and have a more general range of application, and therefore are simpler and more comprehensive than past theories. Given that we have predicatively successful scientific explanatory theories—empirically successful theories—we supposedly have a reason to think of the assumptions about the Universe it has made as being justified as approximating the truth and thereby leading to a growth of knowledge. Scientific realists

"…not only assume that there is a real world but that also this world is by and large more similar to the way modern theories describe it than to the way superseded theories described it. On this basis, we can argue that it would be a highly improbable coincidence if a theory like Einstein's could correctly predict very precise measurements not predicted by its predecessors unless there is 'some truth' in it." (Popper, 1963: 192-3)

In other words, the predictive success of Einstein's Theory of General Relativity justifies (at least provisionally) confidence in its closeness to the truth—or, at least, providing some clue leading closer to the truth. When the precise measurements of time dilation (extension or contraction of an interval of time between events) between atomic clocks in different accelerated frames of reference (i.e. in different positions in the Earth's gravitational field) corroborate the predicted time dilation, as derived from its mathematical equations of time intervals with respect to position in an accelerated frame of reference with respect to another, we have good reason to believe that time, space, and gravity are more like the regions of space-time curvature as described by the four-dimensional invariant metrics of Einstein's theory than described by the Euclidean geometric of Newton's absolute space, time, and motion. When the measurements of the apparent position of the planet Mercury relative to the Sun are closer to those predicted by Einstein's theory than Newton's Universal Law of Gravitation, we have good reason to believe that the former provides better representations and explanations than the latter, and space-time curvature in the proximity to massive body provides a better (simpler, more comprehensive) explanation of acceleration due to gravity than does action-at-a-distance between masses. The fact that Einstein's theory is better than Newton's theory reveals that there must be something in Nature that makes it necessary to use (at least) four-dimensional invariant metrics to describe it, rather than three dimensional space and an independent dimension of time. Even when Einstein's theory is shown to be inaccurate or inadequate, and a higher-dimensional metric is required and space-time can be shown to be a special case or region, the greater proximity to the truth of Einstein's theory over Newton's remains. To deny this belief in *verisimilitude*, according to Popper would deny "the empirical success" of science and impede "the growth of scientific knowledge." Maxwell agrees with Popper.

While this may seem reasonable, it comes with philosophical problems. Due to their assumption of naïve realism from the outset, Popper and Maxwell ignore the cultural conditions and contingencies that shape what constitutes a simpler and more comprehensive explanation of any phenomena. At best, Maxwell's argument for "the growth of knowledge" takes the form of an "inference to the best explanation argument" and an application of the principle of parsimony, providing that we are already

---

[60] Gooding (1990)

considerably downstream from the cultural acceptance of mechanical realism. If a theory explains some "data" better than any other theory, by providing a simpler and more comprehensive explanation for the same "data," we supposedly have a good reason to think that it is true or based on simpler and more comprehensive assumptions about the Universe, and once we have already accepted mechanical realism, we can accept that technology is simply a means to test theories through observation and experience. By ignoring the metaphysical assumptions about technology that underlie experimental science, Maxwell's aim-oriented empiricism, like Popper's falsification and van Fraassen's constructive empiricism, uncritically accepts the meaning and significance of the outcomes of experiments, as given to them by scientists, without being able to explain how these outcomes obtained their meaning and significance.

However, once we recognize that science achieves its predictive success about the performance of technology—within artificial contexts of exclusion, selection, transformation, production, and reproduction—and all observations and tests of theory are technologically mediated, we can see that the objects of empirical success are the machines themselves, and any kind of naïve realism based on experimental science must presuppose mechanical realism from the outset. Once mechanical realism is presupposed, the successful implementation of a mechanism during an experiment is taken to be the actual interaction between otherwise "invisible" theoretical entities (such as particles or fields) in terms of their measurable effects (such as voltage difference or material phase changes between states of matter) on machine performances (such as changes read from dials and displays). It does not make any difference whether the experiment is a closed system (wherein the phenomenon under investigation is produced by a machine) or a detector in an open system (wherein a machine responds to something outside of itself); both situations involve technologically mediated experiences of the phenomena in question. These phenomena are not only "disclosed" by the technological activities involved in stabilizing machine performances—and we would not know of the phenomena's existence without those machines—but the "disclosure" is itself a production of theoretical interpretations and visualization in terms of "natural mechanisms" by which these "invisible entities" supposedly act upon the "matter" of a machine, and thereby gain empirical meaning via theory-laden interpretations and inferences of the performance of machines. The "reality" of physics remains something abstracted from the technologically mediated experiences of the apparatus, which has been given metaphysical significance as a means of disclosure of that objective reality, while "the physical" meaning of the experiment is inferred from the concrete material arrangement of the technology used. Henceforth, the Popperian work of experimental science is to engage in the abstraction of technological activity as a means to provide tests of the correspondence between theories of the invisible world of objective reality and the outcomes of representational and material practices that allow the "objects" and "mechanisms" of theories to become "observable" objects of experience. Hence mechanical realism provides an empirical constraint on what can be considered to be a scientific theory. Postulated theoretical entities are only acceptable within experimental science (whether described as particles, forces, waves, fields, or whatever) if and only if they can be used to interpret a machine performance via a postulated mechanism.

Natural science is directed towards the theoretical discovery of a complete account of the unchanging first principles of causal change of the Universe in terms of the mechanisms that govern the performances of the experimental apparatus. Metaphysics is directed to explain why the Universe exists at all, as a cosmic machine comprised of an intricate combination of mechanisms and fundamental components; the structure of the ongoing development of the technological framework of natural science gives it objective meaning and referents, even if any particular theory is a set of tentative and approximate inferences and hypotheses about natural phenomena, and these causal principles are metaphysically conceived as pre-existing and transcending the contexts of experiment and theory. If the technological framework pre-exists and transcends any particular experiment or theoretical interpretations of that experiment, then though technical education and training, the technological framework becomes embodied in practice and understood via the scientific worldview as means to access an objective natural world that pre-exists and transcends particular experiments and theories. This fundamental interpretation allows experimental science to be understood as participating in discovery of natural laws and mechanisms by discovering the laws and mechanisms of the technological framework. Through mechanical realism, knowledge of principles of change abstracted from the development of the technological framework have the characteristic form of a theoretical knowledge of how change happens in the world due to the actualization and exercise of natural mechanisms. The context in which these mechanisms are actualized and exercised may well be artificial inside the laboratory, but the possibility of their actualization and exercise is understood as determined in accordance with natural law. This mechanical realist interpretation of the world is only possible because of a metaphysical conception of "the physical" and "the empirical," and without this conception, the technological framework could not refer to anything outside itself. At most, without this conception, experimentation could only reveal the possibilities and limits of human labor and invention. Science would simply be a human activity comprised of the ongoing development and refinement of representational and material practices that exists within its own technological framework, as something objective without any referents outside of itself.[61]

Once we acknowledge the technological framework of experimental science, we can also recognize how the epistemological character of scientific knowledge has been reduced to causal accounts that explain natural phenomena in terms of technological "know-how." Once the ontological study of natural phenomena was reduced to the experimental sciences, the scientific method was reduced to the search for fundamental mechanisms and their technological implementation in future scientific research. Experimental science is premised upon the "how does it work?" question and it "tests" any answer to that question by attempting to produce and reproduce the postulated mechanisms in the ongoing development of the technological framework. Whether the postulation is considered to have been successfully corroborated or tested depends on whether it has proved to be useful for the task of disclosing its own instrumentality. Hacking termed this as *instrumental realism*.[62] The truth-status of any theory in science is deferred until judgments are made about the usefulness of any representations derived from theory in the extension of the ongoing technical work of experimentation. In this sense, the content of scientific theories remains situated within an ongoing research program and theories are only considered to be falsified if they lead to dead-ends within that program. This interpretation is consistent with Kuhn's. The operational metaphysics of mechanical realism entails that useful representations are given meaning within the

---

[61] Science can be compared with music. Music is the product of an objective, structured, and mediated human activity. The technological framework of music is embodied in practice as it innovates, produces, and reproduces its own structures and relations between representational and material practices. This framework transcends any particular practitioner or composition, and it is embodied in practice through education and training. Why do we not claim that music corresponds to some hidden objective reality or truth? Because we have not ascribed any metaphysical interpretation to music and conception of its possibility. Without such a metaphysical interpretation and conception, we rest content with ascribing the objectivity of music to the system itself, and, thereby, no matter what one might subjectively hear, it remains possible to hit bum-notes or be tone-deaf.
[62] Hacking (1983)

technological framework, even though, within the contexts of justification and discovery, scientific theories remain tentative, revisable, and pragmatic heuristics rather than absolute truths.

By taking technology into account, any adequate philosophy of science must move away from empiricism (classical, constructive, aim-oriented, or otherwise) and the naïve realism it entails when interpreting experimentation and the production of "evidence." Regardless of whether all scientists so happen to be naïve realists too, philosophers of science should not simply accept the outcomes of experimental science, and their interpretation by scientists, and thereby given significance and meaning as products of scientific activity. An adequate philosophy of science must move beyond the intuitionism of naïve realism, and the conformism it entails. Philosophy of science must move beyond limiting itself to the epistemological task of articulating the rules and methods of the justification of theory choice. Philosophers of science must look at how experimental science operates in both the contexts of justification and discovery, as historical and cultural technology, and how these contexts and the means to explore and make sense of them are produced and shaped within a subculture that grants metaphysical significance to its practitioners own representational and material practices. Philosophers of science cannot simply accept a division of labor and assume that scientists actually do what they say they do. Philosophers of science must look at the metaphysical interpretations of scientific activity, and, therefore, they must look at their own philosophy of science. What are the metaphysical assumptions of scientific realism? How do these operate in shaping or directing scientific practices and interpretations?

How can scientific realists answer these questions? They must show how realism has *methodological value* for interpreting scientific practices in the contexts in which those representational and material practices are justified when discovering new phenomena that have been predicted by theory. Scientific realists must move away from naïve realism and advance a methodological realism wherein it is shown how realism is assumed by the scientific methods used to produce and select evidence. Instead of basing arguments for scientific realism on it being "self-evident," as if this is the only intelligible interpretation of the empirical success of science, the scientific realist must show how the adoption of a realist interpretation of entities predicted by theory is instrumental in the production of the means to discover those entities. The scientific realist must show that scientific discoveries and experiences are methodologically dependent on adopting realism, and science would not exist at all, as a human activity, without a realist interpretation of its methodology and its results.

Methodological scientific realists[63] acknowledge that science is a human activity with a metaphysical interpretation of its significance, and they are also aware of the transformative role of historically situated and developed human agency in shaping how scientists understand and explore Nature. They acknowledge the cultural conditions and contingencies of scientific activity and experience in contexts of justification of theory choice, and they acknowledge the cultural conditions and contingencies of human psychology and relations in the contexts of discovery; as scientific realists, they share a cultural faith in science as a special means to discover new phenomena and powers, and gain knowledge about the world. They are aware that technology not only provides theory with its empirical confirmation in the context of justification, which leaves us with empiricism, but also shapes the trajectories and policies of scientific research and gives science existential meaning as an objective human pursuit that operates in a context of

---

[63] Findlay-Hendry (1995), Harré (1986), Boyd (1984), and Putnam (1985) are recent proponents of methodological realism.

discovery. On such a view, technology is the interface between human thought and the objective world. Accordingly, it is not only the simplicity or comprehensibility of any theories that determines theory choice, but it is the role of a realist interpretation in motivating and determining the strategies and content of the further development of technology in the search for theoretically predicted phenomena (e.g. black holes, as predicted by General Relativity). The discovery of new technological powers—new forces of production—bestows upon science its meaning and rationality as a human activity, and thereby technology limits, shapes, and empowers choices between theories, as well as determining what constitutes evidence of theoretically predicted phenomena.

For the methodological realist, scientific realism finds its epistemological confirmation as a foundation for the methodologies available to human beings to use shared technologies and methods to transcend subjective perspectives, and thereby achieve objective knowledge that will act as a corrective activity upon cultural and psychological influences and biases on our perceptions and interpretations of the world, with all their individual, social, and historical differences. Harré (1986) termed this as *policy realism*: science is a cultural activity, but for scientific realists it is a special kind of cultural activity. To this end, methodological scientific realism embraces the ongoing nature of science as a technological activity, and finds its confirmation within the technological power of science to enhance and expand human experience and agency while operating in the context of discovery. Through technology, science promises to take us much further than our limited bodies would otherwise allow. We can extend our powers of observation by using instrumentation such as telescopes and microscopes. We can send robots and probes to distant planets. It is this kind of technological power that, for the methodological realist, proves that science provides us with a wider and more inclusive worldview than our "inborn nature" would provide, and the methodological scientific realist hopes that it is technology itself that will provide us with a less subjective and more objective account of the world, as human beings learn to put aside subjective differences and instead embrace the shared possibilities of objective discovery through technological innovation of new instruments, detectors, and machines. It is this belief in technological power that underwrites the commitment to scientific realism, even if its adherents are not consciously aware of this commitment.

Methodological realism is a better philosophy of science than naïve scientific realism, but there are still philosophical questions that philosophers of science need to ask. How can we understand the methodological function of this metaphysics on how scientific activity is conducted? Do scientists need to be scientific realists? Does a scientist who adopts scientific realism behave differently from one who does not?

The methodological scientific realist argues that scientists seek a literal understanding of past and present theories, even if they remain skeptical about the truth-status of all theories, and the literal application of concepts underwrites their employment in the construction of new theories and experiments within the artificial contexts of the laboratory. The basis for this belief is that new theories predict and explain new phenomena, and without a commitment to scientific realism these powers would be inexplicable and unintelligible. This "empirical success" of science is the realization of Francis Bacon's dream of new knowledge offering new powers, and what underlies methodological scientific realism is not simply the identity (naïve convergence) between truth and explanatory power, but also the technological power to explore the Universe. Not only does this situates the methodological realist's interpretation of science in the context of discovery as well as the context of justification—scientific activity would be meaningless without discovery, and the context of discovery provides scientists with an internal rationale to pursue science and shapes how scientific activity is conducted and communicated—but also reveals how scientific realism entails a metaphysics of technology.

Accordingly, if the rationale for scientific activity is based on realism, and science achieves successes in both prediction and explanation, then the scientific realist considers it absurd to claim that science could achieve such successes without achieving some approximation to the truth of how the world *really* is, and, thereby science receives its own confirmation in its continuance. If we accept that science continues to make new discoveries we must be scientific realists and this acceptance would change our behavior *qua* scientists in how we conceive of the possibilities of discovery and exploring the natural world. Hence methodological realism not only provides an internal rationale for being a scientist in the first instance, but it also shapes how science is organized and presented as a cultural activity involving representational and material practices to discover natural mechanisms and laws, and communicate those discoveries. It gives content and direction to the aims of science. From the perspective of a person who adopts this internal rationale, the discoveries of science not only validate its underlying metaphysical precepts, but the intelligibility of science as an enterprise depends on accepting the truth of these precepts. Methodological realism is based upon instrumentality and intelligibility arguments for rationality and objectivity, even if any given scientific theory remains a contingent and falsifiable approximation. Once we accept that science is successful—anything else would be unthinkable, according to the scientific realist—then that success underwrites the realist motivation for being a scientist. This means that a scientist cannot consistently adopt anti-realist philosophical positions, such as empiricism (positivism or instrumentalism) or idealism, and make scientific activity intelligible and rational, because they wouldn't be able to practice science if they did. Consequently, it is at the technological level that the methodological realist adopts naïve realism and its faith in science as a special cultural activity.

Methodological realism requires and maintains a distinction between the appearance of the world (according to our sensory experience) and the reality of the world (open to experience through technology). Mechanical realism provides the cultural basis for this belief and is itself a modern manifestation of the ancient dream that everything in the world can be expressed and known, and derived from universal and eternal principles of change and persistence. However, while they do adopt naïve realism at the technological level and uncritically accept scientists own interpretations and inferences as givens, methodological realists do renounce the intuitionism of naïve realism. Rather than ignore technology, methodological realists affirm and celebrate the technological mediation of experience as something that imparts objectivity to scientific experience, and thereby it cannot rest its own conceptual foundation on appeals to inborn intuition or perception. Combined with the cultural acceptance of mechanical realism and the objectivity of technology, in lock-step with technological power's own "self-evidence," the relation between science and reality has become reduced to one of calculation and manipulation, prediction and control, and scientific realism finds its confirmation in the successful prediction of calculations and the discovery of new powers through using technology to manipulate matter and energy. The technological framework disappears from consciousness and into the cultural background—taken for granted as a given, something neutral as a means to an end. Henceforth the technological framework and its metaphysical interpretation are learned during technical training and education—and for the traditional philosopher there is no need to analyze it even this much—and the successful reproduction of that framework and its interpretation in theory and practice is taken to be the mark of successfully learning how to do science.

While the methodological realist is quite right to recognize the necessity of realism for the intelligibility of experimental and exploratory science, and to recognize their own cultural faith in science as an objective means of discovery through technology, they have failed to pay attention to the specific metaphysical foundations of scientific realism, and how they became implicit to scientific discourse, practice, and performance (all demonstrations and experiments are performances to others as a means of communication). This metaphysics underscored the Galilean categorization of the world into primary and secondary qualities or characteristics according to the sole criterion of quantification. The mathematical idealism of Galileo—which Alexandre Koyré misrepresented as originating from a Platonic metaphysics[64]—required the acceptance of the existence of a primary, absolute, complete objective world (being as it is in itself) to which any subjective modes of perception (the secondary qualities) could be related, at least in principle, to mathematical representations of natural mechanisms at work—a notion more Archimedean than Platonic, as I shall explain in the following chapter. The natural motion of each mechanism could henceforth be described in terms of geometry and quantifiable primary qualities that were open to modelling, measurement, and mechanical manipulation, and thereby confirmed or refuted knowledge through technological activity.

As a consequence of this metaphysics, the task of explaining our experience of the blueness of the sky is reduced to one of explaining the mechanical consequences of the interaction of mechanisms, such as the electromagnetic radiation of frequencies on molecules within cones and rods of cells in the retina, which convert the absorbed radiation into electrical signals transmitted along on nerves to the brain. Only that which can be measured and tested through technological activity can be included in explanation of phenomena, such as how we see the color of the sky, and thereby testing the correspondence of any such explanation of phenomena as revealed by using theoretical explanations to design, build, and operate machines or devices to demonstrate the mechanism at work. As such, scientific knowledge is produced through technological activity and understood in technological terms amenable to further technological activity, and reduced to those terms, even when theoretically represented as referring to natural processes, events, and phenomena outside the context of the experiment. Scientific knowledge remains provisional on contingent modes of human agency and labor, produced in technological contexts further given conditional and contingent cultural meaning and expression; yet science only maintains its objectivity and continuity through the innovation of technological activity, and the pragmatic and instrumental use of theory. Hence, without understanding its origin in mechanical realism, scientific realism would be seemingly paradoxical or founded on little more than a mystical appeal to a faith in scientists as witnesses of their own technical expertise. Naïve or methodical, the scientific realist's position is one of a faith in scientists as able to provide explanatory knowledge of the causal principles of change and persistence in the Universe, even though any "knowledge" so far provided by scientific activity has been short lived and replaced. Mechanical realism has allowed methodological realists to accept the paradigmatic shifts in scientific thinking—when the scientific worldview negates and transforms itself periodically—while also simultaneously holding contemporary scientific thinking to be closer to the truth because science continues to make new discoveries. The justification for this metaphysical faith in the objectivity of science begins with the commitment to the precepts of mechanical realism and finds its ultimate justification in the discovery of new experiences and technological powers. It is this faith that reveals the extent that scientific realism, as a philosophical position, is based on a psychological and cultural disposition that makes scientific activities and discourses meaningful in terms of

---

[64] Koyré (1992)

technological activity, even when scientific theories are incomplete, transitory, and falsifiable, and even when technological activity is taken as something "self-evident" and not worthy of philosophical examination, providing it remains innovative and provides new experiences and powers.

To a large extent, most scientists, whether they knew it or not, are methodological realists. However, does methodological realism advance far enough towards being an adequate philosophy of science? If we demand that any adequate philosophy of science must reflect on and question its own preconceptions and assumptions, and the contexts within which they make sense and have meaning, then the answer to this question must be no. Philosophers of science must remain critical of their own position, and, in so far as they affirm realism, they must be critical realists.

In *A Realist Theory of Science* (1975), Roy Bhaskar argues for a critical realist philosophy of science. Bhaskar begins his argument by showing that neither empiricism nor idealism provide intelligible and rational philosophies of science. He argues that material practice is central to the endeavor of experimental science. He stands alongside many within contemporary science studies, such as Hacking, Gooding, and Pickering, in this respect. Like many of his contemporaries, Bhaskar also argues that science is a social product and activity. Knowledge is not created in a vacuum out of nothing. It is produced by human beings for human beings. It can only be produced by means of production, which involves material practices, and revised understandings of those means are achieved via the transformation of existing insights, hypotheses, guesses, and anomalies, etc., during material practices. Experimentation changes how we think about Nature. Bhaskar also acknowledged that experiments are artificial and we need metaphysical assumptions to relate the closed system of the experiment to the open system of the natural world. He argued that the traditional philosophy of science had neglected to examine the role that a concept of mechanism plays in experimental science. Scientists seek to account for some phenomenon of interest in the open system by using a closed system of the experiment to produce an isolated regular pattern of events with the aim of realizing, exercising, and identifying a (set of) mechanism(s) most directly responsible for those events. Producing this explanation will involve drawing upon existing cognitive material (operating under the control of analogical reasoning and metaphor) to construct a theory of mechanism that, if it were to work in the postulated way, could account for the phenomenon in question. Scientists need to be realists about these mechanisms if the activities involved in identifying and explaining them are to be intelligible.

Bhaskar accepted that material practices are social activities, emergent from and situated within a culture; he also accepted that their existence and interpretation are historically transient and dependent on the powers of human beings as causal agents. How should we understand those powers? Bhaskar jumped the gun and assumed that the only possible answer to this question is that human powers are the result of realizing and exercising natural mechanisms through material practices. True to the traditional philosophers' division of labor, he asserted that it is not the task of philosophy to determine what the mechanisms are—that is the task of science—but, rather, "the function of philosophy is to analyze concepts that are 'already given' but 'as confused'."[65] The work of philosophy is to examine critically the questions "put to reality" and the manner in which this is done. However, for Bhaskar, philosophy is not simply to be the handmaiden of science. The task of philosophy is to answer the deeper metaphysical questions, what makes science possible? What must the world be like for those scientific

---

[65] Bhaskar (1975: 24)

practices and interpretations to be possible and exist?

Bhaskar raises these fundamental questions for any philosophy of science, and this lays the ground for a critical realist philosophy of science. However, Bhaskar jumped the gun (largely as a result of the division of labor he adopted) and assumed that scientists do what they say they have done, and, therefore, he granted them the final authority on the meaning of scientific practices and interpretations. He assumes scientific realism from the outset. Hence, according to Bhaskar, the referents of scientific discourse must be natural phenomena and events that pre-exist scientific activity, and only critical realism, which Bhaskar initially termed *transcendental realism*, provides intelligible and consistent answers to the above philosophical questions. Transcendental realism is a philosophy *for* science, rather than "merely" a philosophy *of* science, claimed Bhaskar, and he declared his intention "to provide a comprehensive alternative to the positivism that has usurped the title of science." He proposed his critical realist theory of experimental science as a third position to stand against both empiricism and idealism (1975: 8). A transcendental realist interpretation of science is necessary for experimental activity to be intelligible as a human pursuit, and it "is the only position that can do justice to science" because "without such an interpretation it is impossible to sustain the rationality of any scientific growth or change."[66]

However, Bhaskar did not simply assert naïve realism. Nor did he adopt methodological realism. He went further and took steps to examine the metaphysical entailments of his own scientific realist commitments. He distinguished transcendental realism from empiricism by arguing that a pattern of events produced in experiments signifies the existence of an invariant generative or causal mechanism rather than merely signifying regularity. He distinguished his position from idealism by allowing the possibility that generative or causal mechanisms referred to in explanations as something that may be real rather than always imaginary. Dismissive of idealism, such as Kantian and neo-Kantian positions, as simply being unable to explain why experimental science exists at all, the main target of Bhaskar's argument was positivism, as the descendant of Hume's empiricism, for which experience limits human knowledge to the knowledge of events. He argued that empiricism is unable to sustain either the necessity or the universality of natural laws because empiricism, while affirming the existence of things, events, and/or states of affairs, denies the possibility of any knowledge of underlying causes, powers, or structures. Empiricism restricts knowledge to statements such as "when A happens then B happens" and refutes claims such as "X causes B." He argued that positivism is guilty of "an epistemic fallacy" by reducing the question of "what is" to that of "what can be known," and, therefore, it cannot provide an intelligible philosophy of experimental science at all. Hence, according to Bhaskar, any intelligible philosophy of experimental science must acknowledge the existence and knowability of natural mechanisms as natural causes governed by natural laws.

Bhaskar maintained that there is an ontological distinction between natural laws and patterns of natural events—e.g. thunder and lightning—because the core of theory has a conception or a picture of a natural mechanism or structure at work—e.g. ionization—and, therefore, he asserted, scientific theory presupposes that natural laws must operate independently of the conditions for their identification in terms of patterns of events. The world outside the laboratory walls is a complex and messy place in which regularly repeated constant conjunctions (or regular patterns) of events A and B, which are such that when A happens then B always happens, are uncommon. Yet this kind of

---

[66] *Ibid* 15-26

uncommon event is required for the empiricist representation of experimental science to get off the ground, and, therefore, argued Bhaskar, an empirical science would largely only operate within the circumscribed confines of the closed system. It would not be able to justify itself as a natural science because the production of any constant conjunction of events that it was able to produce as empirical regularities would not be transferable from the context of production in the laboratory to the outside world in which the laboratory is situated. These empirical regularities could not be presented as the consequences of natural laws and there would be no purpose in producing them or attempting to gain knowledge about them. The experimental sciences would teach us nothing about Nature. Thus, for Bhaskar, once again jumping the gun and assuming realism from the outset, empiricism cannot provide any rationale of scientific activity because it cannot make the practices of experimental sciences intelligible as a realist pursuit. Hence, if scientists seek to isolate mechanisms and identify the governing law, and do not stop at noting the existence of constant conjunctions, then the presupposition of the existence of a mechanism and law must be prior to any attempt to identify or isolate them. For example, regardless of how nineteenth century physicists like Michael Faraday and James Clarke Maxwell represented the mechanisms and laws governing the behavior of electromagnets and electrical circuits, their efforts to represent the law-like behavior of electromagnetic phenomena could only be intelligible if they believed that the mechanisms and laws of electromagnetism exist prior to and independently of experiments and mathematical formulations. If the purpose of experimentation is to actualize the mechanisms that are governed by those laws, any intelligible account of experimentation must presuppose that natural mechanisms and laws are prior to and transcend the experimental activity that actualizes them.

Thus, for Bhaskar, the assumption of the efficacy of a law must precede human attempts to actualize and stabilize a pattern of events in material practices, if those attempts are to be intelligible. If the laws of electromagnetism, for example, are real independently of whether they are known to exist then we can presume that they exist prior to their actualization in the movements of the magnetic needle when a magnetic compass is moved next to an electrical wire. Furthermore, we can also presume that the actualization of those laws occurs, under those conditions, whether or not we perform the act of measurement at all. Thus empirical observations of the movement of compass needle when placed next to an electrical wire presuppose that the possibility of the causes of the needle moving pre-exist and transcend those observations and make them possible. Hence, Bhaskar's argument for the realist structure of experimental activity was based on the following five premises:

(i) Causal laws are ontologically distinct from patterns of events;
(ii) Causal laws are those aspects of reality which underpin, generate, or facilitate the actual phenomenon that we may (or may not) experience;
(iii) The intelligibility of experimentation presupposes that reality is constituted not only by experiences and the course of actual events, but also by structures, powers, mechanisms, and tendencies that makes those experiences and the course of events possible;
(iv) Knowledge cannot be equated with direct experience;
(v) An adequate account of science requires the presumption that reality revealed by science exists independently of the human efforts to reveal that reality.

His conclusion was that only a realist analysis could sustain the intelligibility of using artificially closed systems in experimental science to learn about the natural processes in the open systems of the world. It is for this reason that, according to Bhaskar, experimental scientists must be realists.

How did Bhaskar understand the relation between experiments and mechanisms? When an experiment has been set up so that only one mechanism, or a single set of mechanisms, operates then we have a closed system. Even though no system is ever perfectly closed, experiments can approximate a closed system sufficiently enough to satisfy the purposes of experimental science. The reality of the mechanism(s) disclosed by experimental activity is subsequently subjected to empirical scrutiny and the empirical adequacy of the hypothesis is compared to that of competing explanations. The concept of natural mechanism provides a metaphysical conceptual link between explaining changes in the real world outside the laboratory and explaining the constant conjunctions of events inside the laboratory. A mechanism is therefore an explanation for the repeatable processes and machine performances that provides intelligibility to causal accounts used to explain those processes and performances in terms of natural law. On Bhaskar's account, mechanisms can only be identified in closed systems and are restricted to the artificial contexts of experimental sciences, but can be used to explain the events in the open systems of the real world. Clearly mechanical realism was presupposed by Bhaskar's metaphysical argument for transcendental realism in *A Realist Theory of Science*, but, he gives it more explicit form in his definition of experiment in his later book *Scientific Realism and Human Emancipation* as "an attempt to trigger or unleash a single kind of mechanism or process in relative isolation, free from the interfering flux of the open world, so as to observe its detailed workings or record its characteristic mode of effect and/or test some hypothesis about them."[67]

Bhaskar argued an adequate theory of experimentation must allow for three kinds of "ontological depth": *intransitivity*, *transfactuality*, and *stratification*.

Intransitivity allows the possibility that the mechanisms identified by experiment operate prior to and independently of their discovery, and also that they are not changed by the processes involved in their discovery. While the transitive dimensions of experimental activity are the contingent practices of the experimenters and the current theories that they are working with, the intransitive dimensions are the mechanisms and laws that the experimenters are trying to discover.

Transfactuality allows the possibility that the laws of Nature exist and operate independently from the closure of the systems in which they also can be existent and operational in the artificial production of empirical regularities. The constant conjunctions of events in experiments are produced in accordance with natural laws that operate independently of the experiments that disclose them. Theoretical explanations explain laws in terms of the structures which account for constant conjunctions in closed systems, while they are applied *transfactually* in the practical explanation of the phenomena observed in open systems. Bhaskar proposes that his analysis provided a condition of the intelligibility of experimentation because the laws that science identifies under experimental conditions continue to hold as *transfactuals* (not empirical regularities) extra-experimentally, and this provides the internal rationale for practical, explanatory, diagnostic, and exploratory scientific work. The purpose of experimentation is to identify a universal law (within its range), which in virtue of the need to perform a disclosing experiment is not actually or empirically present, and allows experimental science to operate within a context of discovery. Once the ubiquity of open systems and the

---

[67] Bhaskar (1986: 35)

necessity for experimentation are appreciated, laws must be analyzed as transfactual—as universal (within their range)—but neither actual nor empirical.

Stratification allows the possibility that Nature imposes a certain dynamic logic to scientific discovery in which progressively deeper knowledge of natural mechanisms is achieved as "the strata of reality" are uncovered *a posteriori*. For example, the thermodynamic mechanisms invoked to explain the heating of water can be explained in terms of the bonding and elastic properties of molecules of water, which, in turn, can be explained in terms of the electronic structure of water molecules, and, in turn, the electrodynamic principles of quantum mechanics, and so on. The historical development of the scientific understanding of why water boils when heated (rather than freezing) in terms of "deeper strata" of explanation follows "the strata of reality" because it is necessary to explore higher-order strata before one can investigate deeper ones. Bhaskar claimed that this conception of stratification allowed him to isolate a general dynamic of scientific discovery and the development involving the identification of different strata of explanations given in terms of natural necessity, tendencies, and mechanisms. Hence Bhaskar argued that it follows from the stratification of reality that any adequate science must provide stratified explanations in terms of a multiplicity of levels of causal mechanisms. Even though one kind of mechanism may be explained or grounded in terms of another, it cannot be necessarily reduced to or explained away in terms of it. Such grounding is consistent with its emergence, so that the course of both Nature and scientific exploration are different than it would have been if the more basic stratum alone operated. The higher-order structure is real and worthy of scientific investigation in its own right. Any explanation that is (tentatively) accepted must be also explained and this further explanation must in turn be explained on the basis of a stratification of reality. The real multiplicity and stratification of natural mechanisms grounds a real plurality of sciences that study them. Physics explores deeper strata than chemistry, which in turn explores deeper strata than biology, and the laws that govern these higher strata cannot be reduced to those of the lower strata, even when they can be understood and explained as the consequence of them.

Bhaskar went even further. At the base of these three kinds of "ontological depth" was a causal criterion for attributing reality: the theoretical entities and processes proposed as plausible explanations of observed phenomena were to be established as real through the construction either of sense-extending equipment or of instruments capable of detecting the effects of phenomena. He presumed that accurate models in terms of functionality necessarily imply that they are accurate models of the phenomenon under investigation and, consequently, these models disclose the underlying mechanisms of the phenomena under investigation. However, given that Bhaskar has assumed realism from the outset—and therefore these underlying mechanisms must be interpreted as natural mechanisms—he was able to argue that these three kinds of "ontological depth" also provided two criteria for an adequate realist philosophy of science. Bhaskar termed these two criteria as *vertical and horizontal realism*. Vertical realism assumes that science is a progressive, continuous, and reiterative process of movement from manifest phenomena, through experimentation and creative modeling, to the identification of generative causes, which then become the new phenomena to be explained. Horizontal realism assumes the universality of the workings of generative laws or mechanisms (within their range) and is a statement of the independence of the mechanism discovered during experiments. The horizontal aspect supposedly explained why mechanisms may be possessed unobserved and the vertical aspect supposedly explained why they could also be discovered in an ongoing irreducibly open-ended process of scientific development.

This two-fold realism is committed to the belief that the powers or tendencies of underlying generative mechanisms disclosed through experimentation provides us with insights into the nature of things outside the context of the experiment. Without this belief, for Bhaskar, experimental science is unintelligible. It is therefore an essential and defining feature of experimental science is that it uses the closed system to find the otherwise hidden mechanisms that operate in open systems, as a means of producing causal explanations about the phenomena in open systems. Therefore it cannot be accounted for in terms of experience because, as he put it, "scientifically significant generality does not lie on the face of the world, but in the hidden essences of things."[68] Accordingly, the goal of science is to move from experience to the identification of "protolaws" and then to the identification of "laws of Nature," where a "protolaw" is a potentially non-random pattern and result of Nature (including those produced only in the laboratory) that is epistemologically significant. In other words, it is the postulation of a mechanism to explain a pattern of events. The crucial scientific transition is the identification of a generative mechanism or law-like structure that explains a "protolaw" and would ground a law of Nature i.e. a transfactual and efficacious tendency, understood as universal (within its range) but non-empirical, necessary, and discovered *a posteriori*.

Bhaskar argued that a scientific realist interpretation of experimental science is necessary *for* experimental science because only a scientific realist interpretation can make experimental science intelligible as an activity by showing how the metaphysical conception of mechanism and law connect the closed system of the laboratory with the open systems of the natural world.

Yet we need to go further and explain how Bhaskar's metaphysical conceptions were themselves intellectually possible. How is transcendental realism conceptually possible? Bhaskar described his interpretation of experimental science in terms of a metaphysical argument because this kind of argument is based on the question "what must be true in order for 'x' to be possible?" where 'x' is taken to be some self-evident fact about existence. Such metaphysical arguments are premised upon explanations of the facts regarding the evident (or actual) and conclude that there is a "more fundamental something" that is a condition for the possibility of these evidences (or actualities). Bhaskar started from the premise that experimentation occurs in science and asked: what must the world be like in order for this practice to be intelligible as a human pursuit? What makes scientific experiments possible? Bhaskar's answers to these questions were based on the following premises:

(i) Scientific experimentation exists;

(ii) Experiments are physical and not just mental;

(iii) Experiments involve causal interactions with the material world;

(iv) Causal interaction is only possible because we are embodied beings;

(v) As embodied beings, we are subject to the same laws that govern the material world.

Clearly Bhaskar presupposed that there are laws that govern the material world and these laws are natural laws. Hence, his conclusion was that the same natural laws that govern the material world also govern experiments via the realization and exercise of natural

---

[68] Bhaskar (1975: 227)

mechanisms in material practices. That conclusion presupposed mechanical realist conceptions of what human beings are, how human beings interact within the world, and of what the world is comprised.

However, as Bhaskar pointed out, there may well be alternative metaphysical arguments that explain the same thing differently and possibly better. Bhaskar offered us an alternative as an example:

(1) Science exists;

(2) Science discovers underlying mechanisms;

(3) If there were no underlying mechanisms then science would not be possible.

Unsurprisingly, he concluded that there are underlying mechanisms discovered by science. Of course, even if we accept premises (1) and (2) it does not follow that the "underlying mechanisms" discovered by science are *in fact* "natural mechanisms." Again, Bhaskar's argument presupposed mechanical realism from the outset, yet it is a metaphysics that he does not explicitly address. Bhaskar's argument is also a statement of faith in the truth of scientific discourse that presupposes naïve realism, albeit at a deeper level than methodological and naïve realism. His assumption of the necessity of a transcendental realist interpretation of experimentation for the intelligibility of science is based upon little more than an appeal to the internal rationale of experimental scientists and their own testimonies regarding the meaning and significance of their practices.

By taking this internal rationale of experimental science as demonstrative that it is justified by "the empirical success" of scientific activity, Bhaskar has conflated intentionality with actuality. This is a fatal move for a realist argument. It must be a criterion for any realist position that, in any practice, the intentions of the practitioners could be at odds with the actuality of those practices. It must be possible for someone to think that they are doing one thing when, in reality, they are doing something else. Otherwise no one could ever be fallible; no one could ever be deluded, tricked, or hallucinate. No one could act in bad faith, or be mistaken or insane. However, if we seek to avoid making this philosophical mistake, we must accept that it is conceptually possible for scientists to be able to continue exploring and developing their representational and material practices, while being wrong about their metaphysical interpretations of the meaning of those practices. Furthermore, given that there exists a cultural plurality of available interpretations of any set of representational and material practices, all of which give different meanings to those practices, there always remains a sufficient degree of ambiguity for those practices to be taken as successful by the practitioners while simultaneously being seen as otherwise by others. This ambiguity is necessary for a cultural anthropology of science to be possible. It is possible that experimental scientists could intend to reveal natural mechanisms through experimentation, and find the meaning of their practices in terms of that intention, yet only produce artificial mechanisms and only interpret these as natural mechanisms because of a culturally embodied metaphysical interpretation. It is only necessary for the continuance of the cultural practice of interpreting experimental science as a natural science that it remains internally intelligible in terms of mechanical realism and the practitioners interpret the artificial mechanisms that they produce to be natural mechanisms.

How can we understand this cultural interpretation without 'going native' and believing it ourselves? We only need to show the process by which this interpretation is made and we can explain experimental science as an enduring cultural phenomenon. We do not need to accept that any internal rationale is justified by the continuance of any cultural practices. We can understand the claims that practitioners make about their own practices, without sharing their commitment to their truths and values. For example, many religions have been practiced for thousands of years. Is it the only intelligible explanation of the existence and continuance of any particularly long-lived religion that it must be based upon truth? If we were to adopt Bhaskar's style of argumentation then we would have to accept that it was. After all, the devotee, no doubt, would claim that their representational and material practices were based upon the truth of their beliefs regarding the metaphysical significance of those practices. Rituals of purification, healing, exorcism, sacrifice, or burying the dead would be a good example of representational and material practices combined with metaphysical beliefs about the worldly (as well as otherworldly) significance of these practices. Many religious practices are culturally embodied metaphysical theories that appeal to some kind of intransitive cosmic order—a cosmology; hence, it is believed that acting in accordance with that order will bring worldly powers and material benefits. Otherwise what would be the point of petitioning God or gods via prayer? What would be the point of ritual and sacrifice? They would be unintelligible as human pursuits.

Devotees claim that their religious practices are successful (and necessary) and have done so for thousands of years. However, non-believers would readily claim that those practices were based upon cultural conventions, authorities, social power structures, and traditions, etc., rather than corresponding to any objective reality or truth. We could argue that religion has maintained its existence as a cultural phenomenon through the maintenance of certain social structures, powers, and beliefs, all of which can be explained in terms of human relations and psychological motivations, and find it intelligible despite the "false consciousness" we have ascribed to its practitioners. We could equally argue this way about the cultural and historical conditions for the existence of the natural sciences and still consider them to be intelligible as cultural phenomena, with their anthropological, psychological, and historical dimensions. On this view, the continued existence of the natural sciences depends upon the cultural maintenance of certain social structures, powers, and beliefs that underwrite the continuance of a specific mode of interpretation of the meaning of representational and material practices. We do not need to accept the authority or truth of these sets of social structures, powers, and beliefs, to make either religion or science intelligible as cultural phenomena.

As I have shown above, Bhaskar's transcendental arguments presume that experimental science is only intelligible as an enduring human activity if real causal mechanisms exist independently of science. It supposedly follows from the fact that science exists, endures, and is intelligible, that real causal mechanisms must exist independently from experimental science. One can imagine an analogous transcendental argument for a realist theory of shamanism:

(i) Shamanism exists;
(ii) It aims to achieve knowledge of, and access to, a spirit world for purposes of healing the sick, exorcising evil spirits, etc.;
(iii) It would not be intelligible as an existent set of practices, if it persists and does not actually achieve what its practitioners claimed that it did.

Presumably we should conclude that shamanism must necessarily achieve knowledge of and access to a spirit world that exists independently of shamanic practices. However, this argument—like Bhaskar's transcendental argument for a realist interpretation of experimental science—is circular and begs the question. It is a *petitio principii* because it presumes its conclusion: that the practitioners actually achieve what they intend to achieve. A similar argument could be made against Hacking's famous instrumental realist confession to be a realist about electrons because physicists claimed to spray them on niobium spheres in the search for free quarks. It seems to me that Hacking should also be a realist about spirits because witch doctors claim to use them to heal the sick. After all, forms of shamanism have existed for much longer than experimental physics, and, many people claim to have been healed by spirits. By both Bhaskar and Hacking's standards, we must accept the reality of the spirit world, if the endurance of a practice and the internal rationale of its practitioners are criteria for its truth and its continuance would be unintelligible if it was not successful.

Kant was aware of the difficulties inherent in applying transcendental arguments to the experimental sciences. Kant too appreciated the difficulties in applying transcendental principles to the particularity of experiments in physics (and chemistry). "How is physics possible?" and "How is the transition to physics possible?" were both central questions in his later work. These efforts were unpublished during his lifetime but have recently become available.[69] Accepting the products of the science of his day, Kant argued that the intelligibility and existence of the sciences of electromagnetism, thermodynamics, and chemistry presuppose the existence of the aether. Due to the obsolescence of the idea of the aether within contemporary science, the significance of Kant's posthumous argument for the transcendence of natural laws of electromagnetism and chemistry has been lost on contemporary philosophers of science. If we can put aside the contingency of Kant's conceptualization of this transcendental structure in terms of the aether of nineteenth century physics, we can see that Kant had realized the necessity of a transcendental conceptual structure of some kind in order to make the particularities of experimentation intelligible as revealing universal laws and part of natural science. This is an important realization. Even if we accept that it follows from Bhaskar's argument that real causal mechanisms exist and are discovered by experimental science, we still need additional conceptual work to determine whether they are natural or not. We need to understand the transcendental structure that gives the particularities of the experimental sciences their meaning. It does not immediately follow from the fact that a generative mechanism exists that it is a natural mechanism. It requires a presumption of mechanical realism to make this leap.

How is mechanical realism possible? My argument is the mechanisms produced during experiments are representations—abstract indices or codes for material practices—that emerged and developed through dialectical interactions between human interventions and machine performances within the context of the ongoing development of the technological framework of scientific activity. These mechanisms constrain and transform each other through experimental activity, and these mechanisms are given cultural meaning in terms of mechanical realism as being natural mechanisms governed by natural laws. It is the technological framework itself that gives experimental science its objectivity, but it requires metaphysical precepts to represent these technological objects as something that is caused by natural mechanisms and laws. Thus it is via a shared technological framework given metaphysical interpretation via a shared subculture that the particular practices of the experimental sciences gain their meaning in terms of general or universal laws. The creative intellectual and physical labor process at the heart of scientific activity in general and all technological innovation is analyzable as a *dialectical historical process* that cannot be reduced to the categories of "objective" or "subjective," as understood by empiricist or realist philosophies. Hence, in my view, "the electron," for

---

[69] Cf. Notes 22:282 to 22:452 in Kant (1995:100-199)

example, is a code used to unify and explain a set of related machine performances and theoretical interpretations, with their associated techniques and representations, which can be transferred between contexts as technological objects, which allows machines to share some of the same components and instruments within the same technological framework. It is the shared codes between machines through their components that give the technological framework its transcendence. It is the technological framework that transcends the otherwise distinct contexts of experimentation and gives them objective meaning. Mechanisms can be understood as coded unifying representations of the ongoing juxtaposition, interconnection, and unification of otherwise heterogeneous technological objects (machines, tools, instruments, etc.) within a technological framework; the development of which is an ongoing, dialectical process of integrating the experiment within the wider culture as a means of disclosure, measurement, and observation of otherwise unrealized and unexercised mechanisms. I shall discuss this further in chapter four.

Once we recognize the need for metaphysical concepts to interpret these representations in terms of natural mechanisms at work, we can see how these concepts are tacitly embodied in historically and culturally situated representational practices learned through education and training. I shall discuss this further in chapter five. Experimentation is a creative labor process that is not possible without interpretations and technologies that pre-exist any particular theories or hypotheses to be tested. From the outset, the experimenter and all possible experiments are situated within the technological framework with its associated expectations and limits, and its interpretation in accordance with the scientific worldview. This framework is not completely under the control of any particular experimenters or experiments, which explains why it cannot be understood by applying the dualistic categories of "subjective" or "objective." Yet the failure of any experiment and the existence of material resistance to human intentions can be interpreted as indicative of objective structures and laws. How can we explain this? Production is not purely an individual human activity, nor is it a natural phenomenon. Individuals must engage in disciplined representational and material practices (i.e. theoretical and technical skills) with tools, machines, and materials, in order to have any productive capacities at all. We learn those practices from other people, as well as through trial and error, in relations to challenges, expectations, and demands imposed on us by other people, and, as a result, representational and material practices are shaped and given meanings that transcend their immediate contexts. Individuals have no productive capacities whatsoever until they situate themselves within the technological framework through education and training in relation to other people, and thereby embody cultural interpretations of representational and material practices, along with expectations of success and its limits. Individuals must relinquish any possibility of absolute control to become technologically empowered and productive human beings. Otherwise individuals could never learn anything or have new experiences. This involves relinquishing absolute control over individual agency during a participatory relation to learning the machine performances of the apparatus in response to human interventions. One must surrender one's appeal to the authority of absolute subjectivity to successfully learn and embody the objective relations of the technological framework in practice.[70]

It is this act of relinquishing control that is called *impartial or dispassionate observation*, and *discipline*; it is this act of relinquishing absolute control that gives technology its sense of autonomy—its objectivity—as something that transcends the individual human being when she or he has learned how to act as a technologically empowered, causal agent

---

[70] This is also true of music.

within a material world. Experimental scientists do not have absolute control of the processes by which they perform an experiment because they do not have absolute control over the development of the technological framework itself and how their work is situated within it. Long before they have learned how to become scientists, people are already socially and historically situated beings within the technological framework of scientific activity. The products and representations of science riddle modern culture. In order to be able to build, use, and interpret any experimental apparatus, an individual scientific practitioner needs to be responsive to learn how to do this in a way that can be reproduced by other scientific practitioners, and relate that apparatus to the already-existing technological framework and theoretical interpretations of Nature, which is current within the scientific subculture, while exploring the space created by the pluralistic, incomplete, an open-ended character of technological innovation. Hence the technological framework of scientific activity can be treated as a cultural phenomenon that has its own history of innovation and development, which exists independently of any individual scientific practitioner, and perpetually finds its confirmation in the discovery, production, and reproduction of new powers. Its possibilities and limitations are incomplete, unstable, and discovered through the ongoing innovation and development of that framework.

Once we take into account that scientists must learn techniques and skills to become disciplined to be able to perform the representational and material practices required to design, build, operate, and interpret experiments, we must also recognize that scientists learn (often uncritically) the cultural significance of the experiment in relation to its historical background, thereby making it a meaningful experiment in largely unquestioned and unexamined terms. While technological activities and how to interpret their results are established through practice and experience, they are also transformed as the technological framework is refined and modified to resolve contradictions that arise from its heterogeneous construction, and this gives new meaning to those activities. One need only reflect on one's own experience of "electrical charge" in various contexts of practice and experience to see how those practices and experiences are mediated by technology, and have been shaped and transformed by technological innovations. Material and representational practices are brought together to explain the meaning of "electrical charge" by explaining what one observes in the terms used to explain how the means to measure "electrical charge" work. Otherwise how can scientists explain what they mean by "electrical charge"?

Technologies are stable sets of associated representational and material practices brought together by human efforts to produce specified results by effecting specified changes in material arrangements. Material and representational practices developed simultaneously, thereby bringing together theory and practice in the contexts of justification and discovery by bringing together heterogeneous components and combining them into a stable means of achieving a predictable outcome of a specific and quantifiable intervention. This is achieved when a prototype is invented as a synthesis of previously disparate devices through an often arduous history of trial and error, labor and struggle, and many failures, in such a way that transforms how its components and the world is understood. The electromagnetic motor is an example of such a prototype. Not only did this invention of this device occur step by step with the innovation of new representations to explain how it worked, and how to use it, but these new representations radically transformed how its components—magnets, conducting wire, and chemical cells—were understood. This opened up new possibilities for discovery and innovation within the technological framework, and the corresponding shift in the technological framework caused revolutions, quite literally. This act of *dialectical synthesis* simultaneously occurs within both contexts of justification and discovery, as scientists learn the contours between interventions and their experiences of their consequences, and use that know-how to further innovate new possibilities of research and discovery, thereby explaining them in technological terms that can be further implemented in ongoing research and innovation. When the ongoing experimental development of representational and material practices is transformed in novel contexts into the means of disclosure of their own operation in terms of underlying mechanisms, then new physics is said to have been discovered and observed. Subsequent explanations find their corroboration in terms of their usefulness for further innovations and discoveries. This is what is really meant by "the growth of knowledge."

This insight shows that the real task of science is to unify and complete the technological framework in which it is historically developed and given cultural meaning as an objective means of discovering natural mechanisms and laws, and thereby discovers its own object and its limits in the contexts of the ongoing development of material and representational practices, and their theoretical explanation in terms of natural mechanisms and laws. This task is the historical mission of science. At no point does the experimenter experience an unmediated objective reality outside of the technological framework. Experiences of objective reality are emergent and developed as the technological framework is developed, wherein the meaning of "the empirical" and "the objective" are both mediated and dialectically developed within this framework. In this way, the individual scientific practitioner is empowered to participate in the contexts of discovery and justification, but is also appropriated and transformed into a causal agent by this dialectical relation within technological framework itself; as a result, the outcomes of scientific activity remain contingent and tentative on the current structure of the technological framework and how it is connected via codes with the contemporary scientific worldview, and these outcomes form the points of departure for the current frontier of research and development. This is what Kuhn referred to when he talked about the paradigm as a disciplinary matrix.[71] The determination of the final possibilities of any experiment is itself emergent and developed within ongoing development of the technological framework. These outcomes are themselves fed back into the technological framework as technological objects used to further refine this framework. Furthermore, due to the plurality of possibilities inherent to all experiments—otherwise there would not be an experiment—the challenge for the experimenter is to explore the productive possibilities of the experiment itself. The experiment becomes its own object. The whole technological framework of scientific activity is itself an experiment into its own possibilities and limitations, and it explores these through the ongoing construction and refinement of itself.

As we have already seen, Bhaskar's transcendental realism presupposes mechanical realism because he presupposed that the necessary objects of experimentation are natural mechanisms that exist independently of any experiment in the open system (Nature) but remain capable of being produced using artificial means (technology) within the closed system of the experiment. However, the general ontological conclusion that he derived from his interpretation of scientific practices as being rational and successful begs the question. His transcendental realism is as naïve as Maxwell's because it presupposes that scientists do what they say they do. Yet, it is the concept of the rationality of any scientific growth or change that is at stake for any critical realist interpretation to get off the ground. Bhaskar (like Maxwell) has simply dodged this question. An interpretation of science is not inherently flawed if it questions, or even rejects, any notion of rational scientific inquiry based on an uncritical acceptance of the

---

[71] Kuhn (1977)

internal rationale of scientists. Anti-realist arguments cannot be criticized on the basis that they do not assume that science is successful as a realist pursuit. They are designed to question or challenge this assumption. The onus is upon scientific realists to provide a justification for this assumption because without it there is no rational basis for scientific realism. To criticize anti-realist interpretations of science because they do not provide the basis for a scientific realist interpretation is an unreasonable criticism.

This unreasonable criticism is apparent in Christopher Norris's realist argument against anti-realist interpretations of quantum mechanics on the premise that anti-realist interpretations of quantum theory must be flawed because they are not realist.[72] In my view, Norris argument not only begs the question, be he missed the anti-realists' point about the limitations to what physicists can legitimately say on the basis of quantum physics if they are to be good empiricists and only use the results of those experiments to support their claims about the quantum nature of the world. Quantum physics (alongside Relativity) shows us that *measurement* is a defining limit on how we understand the physical world and its properties. It would have been better if Norris had given an account of how measurements are made within quantum physics and shown that anti-realist interpretations of quantum theory, such as Bohr's interpretation, were premised upon a realist interpretation of the construction of the experimental apparatus that provided the experimental results that they used to support anti-realist interpretations. How could an empiricist explain the design of the Stern-Gerlach experiment? Or explain a black-body radiator? Or how to measure the photoelectric effect? As I have argued above, empiricist interpretations of experimental physics presuppose a realist interpretation of how and why the experimental apparatus and instruments were designed and built in the first place. The construction of experiments, such as the double-slit interference experiment, presupposes the existence of entities such as electrons or photons, and electromagnetic waves and their diffraction in the design and interpretation, and even if one adopts empiricism when regarding limits on the statements scientists can make about the ontological status of the measurable states of an electron or a photon, the fact that such an experiment is said to demonstrate wave-particle duality about electrons or photons entails realist commitments about how to measure those states and how to produce a beam of electrons or photons in the first instance. The Copenhagen Interpretation of this kind of experiment not only presupposed that the quantum properties of such systems exist, but that they can be produced and observed utilizing the properties of such devices as cathodes, phosphorescent screens, and photomultiplier tubes.

It is for this reason that I have been critical of Bhaskar's assertion that events are "categorically independent of experience."[73] I agree that the rejection of the empirical definition of events in terms of experience is necessary for a realist philosophy for science, but that it is insufficient in itself to show that such categorical independence actually occurs. The realist needs to do more work than this. S/he needs to show that *the means by which events and experiences are produced within the laboratory are also categorically independent of each other*. However, this categorical independence is not apparent in real experiments. It is blurred by the practical interplay between representations and interventions during the ongoing work of disclosing novel phenomena and attempting to make stable and communicable observational techniques and interpretations. Of course, it does not immediately follow from that absence of categorically independent experience within experimentation that all observation is subjective. Events and experiences are selected and brought together within the technological framework, and thereby constrained and

---

[72] Norris (2000)
[73] Rogers (2005)

reduced, and they can only be taken as categorically independent by making the technology involved transparent or simply ignoring it. This is achieved by forgetting its history and the labor processes involved in its construction. An accepted technique of measurement is required and events are determined from within the development of the technological framework of scientific activity to the extent that the experience of the phenomenon in question is constructed in terms of a set of measurements and their interpretation, and these are black-boxed as techniques, procedures, and operations (such as increasing the strength of a magnetic field by turning a dial or pressing a button). The phenomenon is defined by this set and is a *technophenomenon*. I shall discuss this further in chapters four and five.

Any reference to an unobserved event (e.g. the ionization of a gas as a sub-atomic particle passes through it) is made via codes: the connection between theoretical and technical causal accounts in reference to an observed event (e.g. the clicks of a Geiger counter). Bhaskar's realism treats these two kinds of reference as if they were causally connected because he assumes a tight and competent link between scientific observation and the technical causal accounts used to explain the instrument used to make the observation. He has uncritically accepted an isomorphic causal connection between the technical expertise and intentionality of scientists; as did Popper, Hacking, Harré, and Maxwell. They took scientific codes literally, as giving sense, but ignored their context of reference and how that sense was made. Scientists do whatever they say they do, according to realists. However, this assumption is inconsistent with their realist theories of science and leads to the circularity of their arguments. It also ignores science as it is actually practiced by scientists. If one accepts that our knowledge of the intransitive objects of experience is itself transitive, as Bhaskar does, then we cannot assume that there is a tight and competent link between technical causal accounts and interpretations of the event in the contexts of its production. The possibility of change prohibits any such certainty. If our interpretations change, which Bhaskar affirmed, and our skill at making interpretations improves, which Bhaskar asserted, then we cannot, at any stage, assume that our current interpretations are correct, and, therefore, we must address the practical interplay between experience and events, and their explanations and codes.

Bhaskar claimed that any adequate philosophy of science must grapple with "the central paradox of science" that science is a social product that is concerned with the "knowledge of things that are not produced by men at all."[74] Yet, my argument is that we can cut through the Gordian knot of this "paradox" by examining the technological framework as the object of science that is given metaphysical meaning in terms of natural mechanisms and laws.[75] The technological framework requires human participation to exist but it exists independently from any individual participant, and it is this independence that has allowed it to become understood as the objective referent of scientific activity through being a shared-but-reinterpreted objective referent. It is only when technology is ignored or taken for granted as "self-evident" that we are presented with "the paradox" that the objects of production are social products and yet not completely controlled by human beings. There is, in fact, no paradox at all if we do not assume that technology is simply "man-made" and a transparent means to an end. Obviously, unless human beings had thought that technological means could discover natural mechanisms then there would not be any experimental sciences. The experimental sciences are socially and historically contingent and conditioned. Yet, to bring something into the world does not imply that one controls it or knows it. It is in this sense that we can say that the objects produced through scientific activity depend upon scientific

---

[74] Bhaskar (1975: 21)
[75] Rogers (2005)

thought as a condition of their existence, but these objects are not purely mental objects (ideas or concepts) because their meaning is given and determined through the practical interplay between representational and material practices in contexts of production and reproduction. Hence, as objects produced in technological contexts and given their meaning within the technological framework, they are subject to a dialectical developmental dynamic, and their behaviors can shape and are shaped by scientific thought, even though are not controlled by the thought of any individual scientist. The behaviors of these objects are dependent upon and structured by the technological framework itself, which is itself transformed to accommodate prototypes and new powers, and this framework transcends and gives meaning to individual experiments or observations.

There is nothing more or less mysterious about this than claiming any creative act of making or material practice (e.g. composing a piece of music or inventing a new musical instrument) is learned by attending to things outside of the individual, and transforms how making or material practices are understood and conducted by individuals. Technological objects are the products of dialectical relations and their "uncontrollability" is the consequence of the fact that the processes by which these dialectical relations unfold during their invention are not under individual human control because "control" is an outcome of the unfolding of these relations in novel contexts using other technological objects for purposes that they were not intended. To invent something simultaneous involves taking things from the technological framework, bringing them together and synthesizing something new in novel contexts, and weaving it back into the technological framework, thereby transforming it and creating new possibilities and problems (spaces and contradictions), and this process may transform the prototype into a device or a means that its inventor did not imagine or intend. It requires work to establish a prototype, and success depends on what is expected as possible or not, at any given historical stage of the development of the technological framework, which itself dependent on prior productive and interpretative relations. Any interpretation of that prototype in terms of fundamental mechanisms and laws is itself developed in relation to the technologization of natural phenomena to transform them into technological objects amenable to measurement, manipulation, and prediction.

The technological framework must be understood as both a cultural and an historical phenomenon, which gives it meaning and significance. Once we understand that technology is itself the product of a dialectical labor process that involves the innovation, description, and explanation of our productive capability through the experimental and adaptive development of representational and material practices, we can see how the technological framework shapes future scientific thought and practices, while in turn being shaped by how science explores and understands the possibilities and limits of technology. Nature participates within this process and shapes it, placing limits and offering resistance, as well as offering resources and powers, but how it does so remains mediated and shaped by the process itself, and therefore remains a reaction or response to interventions. We have no unmediated experience of natural phenomena through scientific activity, and this activity itself depends on protracted historical struggles to resolve contradictions within the technological framework in which that activity is made meaningful as a means of observing natural phenomena. It is for this reason, within the context of the technological framework, that the development of scientific activity should be characterized as dialectical, rather than subjective or objective, as it unfolds as the result of efforts to resolve contradictions and resistances, and also provides new tools and codes for the further development of the scientific method and the reality it explores. The results of experimentation are discovered within the context of the dialectical innovation and development of novel technologies by bringing together otherwise heterogeneous technologies, representations, and materials, and incorporating them into the already existing structures and relations of the framework to discover, produce and reproduce new technological objects. The possibilities and limits of the framework are themselves products of the ongoing development of the framework, along with the goals of scientific research at any given stage of its development. The logic of scientific discovery becomes the stratification of the technological framework into a series of nomological machines, as described by Cartwright, and the laws abstracted from these nomological machines—each a fundamental representation of mechanism and law in motion—become the stratified laws of Nature and related to their theoretical interpretations within the scientific worldview.

The philosophy of science must grapple with the question of the nature of "the artificial" and its relation to "the natural." Metaphysical answers to this question are historically conditioned in relation to our productive capabilities and how we understand their origin. Both realists and empiricists have presumed that the answer to this question is that it is simply something that is "man-made." Such an answer is made without understanding what is meant by either "man" or what is meant by "made." Consequently, the realist is confronted by a paradox that he can only resolve by assuming mechanical realism or considering experimental science to be impossible. Yet we do not have to resort to subjectivism or social constructionism. Although we can account for scientific change by locating the structures of change within the structures of social relations, once we take technology into account then we can go further than that. It is possible to provide an anthropological interpretation of rationality in experimental physics that allows technical growth and change, within the productive contexts in which technical choices and selections are made, which are themselves situated within the technological framework as it is undergoing dialectical development. A pragmatic concept of "bounded technical rationality" sustains a notion of scientific growth and change, within contexts of technological activity, without requiring any commitment to the truth of scientific realism, providing that this concept is understood in the context of its mechanical realist heritage and we examine the dialectical development of the technological framework in the creation of a scientific culture and worldview. This provision does not require any commitment to the truth of those precepts because we only need to address their function within the establishment of the template for subsequent development of scientific thought, codes, and practices. It does not matter whether these precepts are true or not. All that matters is how they function within the discursive and technological interpretations and practices of experimental scientists. Natural laws or mechanisms revealed through scientific activity can be understood in terms of a pragmatic cultural anthropology as being abstracted rules and codes referring to the dynamic interactions between human interventions and machine performances, providing we understand the role that metaphysics plays in providing a framework to interpret those objects, and through which we explore the contours of the possibilities and limits of the technological framework itself.

Without realizing it, when realists like Bhaskar claim that realism provides a basis for rational principles of action, they actually offer us a pragmatic principle of action directed towards the achievement of technical knowledge of how the world works. The constant conjunction of events produced in the closed system of experiment is only necessarily governed by natural law if we claim that such pragmatic principles are natural principles. Otherwise we have merely used "the natural" as a metaphor for "the artificial," and vice versa, thereby substituting technical knowledge for a knowledge of "natural law" and renaming it as such. This is poetic and not rationally justified at all. We could alternatively argue that science achieves progress only by extending the variety of technological objects and representations at the disposal of scientists. Contemporary scientists have more techniques, materials, tools, machines, and instruments at their disposal than seventeenth century natural philosophers did. Furthermore, the contemporary scientists have the recorded efforts of the previous generations of scientists at their disposal. In a technological context, contemporary scientists are able to deal, on an everyday basis, with far more complex, sophisticated, and powerful machines, instruments, techniques, and tools than the seventeenth century natural philosopher would have been able to imagine. It does not immediately follow from this innovative productivity that the contemporary scientist has one more iota of knowledge about Nature than the seventeenth century natural philosopher (or an ancient Greek, for that matter). This notion of "progress" in terms of an extension of technological powers does not provide necessary and sufficient reasons to presuppose that it does, in fact, discover any natural principles that exist *a priori* to the practices of experimental science. It is this very notion that is at stake—or at least should be—for philosophers of science.

Alternatively, experimental science may well only achieve "progress" in the context of the dialectical development of the technological framework itself, allowing the stratification of technological innovation to be represented (metaphorically) as the discovery of the stratified ontology of Nature by its practitioners (and some philosophers of science) who remain under the sway of mechanical realism, but whether science has progressed in epistemic knowledge is the very question at stake. Bhaskar did not establish any realist argument for "a rational dynamic of change" or "progress." He merely asserted that there is one. Otherwise science is unintelligible, according to Bhaskar. However, this neglect to establish a rational foundation for realism is not the object of my criticism. Bhaskar cannot provide such an argument because science is unfinished and we do not have its conclusions at our disposal.[76] However, my argument is that it has not achieved its own *techne* (understood in the Aristotelian sense of the knowledge of the universal causal principles of change and persistence in contexts of making, which I shall discuss further in the following chapter), and, consequently, scientists do not know what it is that they have done and what they are doing, if we demand of them that they provide a complete causal account of all their interventions and their associated consequences. This is not a failure on the scientists' part. If they could provide an account, there wouldn't be any need for experiments. Science is itself an experiment. As long as science remains experimental, its "empirical success" remains still open to question. After all, we cannot say that we are nearing the truth, improving our approximations of the truth, until we know what the truth looks like. We can only claim nearness if we assume that there is a final form of truth, which seventeenth century natural philosophers metaphysically anticipated, and scientists are providing us with facts and theories about it. This is only possible if we anticipate the form of the truth by assuming that the theoretical interpretation of the successful innovation of any technique to make novel technological power is a step nearer the objective truth about an objective, underlying reality. The Popperian formula of "predictive success equals nearness to truth" in the context of explaining the results experimental science in terms of theoretical "otherwise invisible" structures and entities is premised upon a conflation of knowledge of natural causes with the explanation of technique, and it is the presupposed "self-evidence" of the rationality of this conflation that is the target of my criticism.

Traditional philosophers of science have presumed that technology enhances and extends the powers of the mind and senses without changing or directing either. The construction of theories is a purely intellectual affair for which the technology of

---

[76] Rogers (2005)

experimentation does not have any constitutive role. We are supposed to believe that the technological limits of measurement, control, and manipulation have no reductive effects on how scientists explore and theorize about Nature, despite measurement, control, and manipulation being fundamental criteria and components of theory testing. By ignoring how interpreting technologies in the metaphysical terms of mechanical realism made new interpretations of observations and experiences possible, traditional philosophers have difficulty explaining how science developed into a series of specialized disciplines and fields of study, each with its own sets of techniques and associated machines, yet is somehow connected to a unified objective understanding and representation of Nature or the Universe within a shared scientific worldview. Ignoring technology has also allowed the traditional philosophers to ignore how technological innovation also changed how natural processes were themselves represented and explained, as well as providing new opportunities to explore the natural world. This has left the traditional philosophers of science unable to explain how new conceptions of new techniques and methods of representation and explanation are possible—except by declaring them as "empirical successes"—as well as unable to explain how experimenters have qualitatively transformed the forms of representation and concepts of explanation so as to make them in terms available for technological disclosure, measurement, control, and manipulation during the technological activities of experimentation. Technological innovation has not only provided new means and ends, but it has changed how human beings think about the natural world as something open to technology, and confirmed and reinforced that way of thinking by testing it through innovating new kinds of technological activity.

How do scientists use and describe their instruments and apparatus to explore and explain the natural world? By examining this question, we can discover how scientists relate their experiences to the theoretical disclosure of an underlying reality that is considered to be causal to the scientific experience of natural phenomena within the context of ongoing technological activity. This examination requires that we look at science in practice—as the outcomes of a dialectic between material and representational practices fed back into its ongoing development and structure—and identify the presuppositions about reality that make these scientific experiences and their causal explanations possible. Mechanical realism constitutes a fundamental transformation in how human beings understood both themselves and the natural world by transforming the natural world into technological objects within a structured and stratified technological framework. This, in turn, has informed and constrained the human understanding—and imagination—of the possibilities and limitations of further technological innovation. Both representational and material practices inform and constrain each other, and by doing so discipline and shape ongoing technological activity. The objects explored through experimental science are the products of a dialectics of technology, which has been made invisible by viewing technology as a mere means to an end. While critical realists have quite correctly challenged the Is-Ought dichotomy of Humean thinking—given that truth is a value—as well as the inadequacies of empiricism as a philosophy of science, we should challenge the uncriticality of critical realists regarding the origins of the metaphysics of modern science and the naïve realism they take for granted when interpreting scientific experience and theories. This attitude of uncriticality is little more than conformity to the culture of the societal gamble on the goodness and rationality of the construction of the technological society. As such, without addressing mechanical realism and its implications, scientific realism remains irrational and conformist—and, as Cartwright claimed, fundamentalist.

Technological innovation not only transforms the contexts of discovery of new phenomena and the construction and justification of theoretical efforts to understand the causes of these new phenomena, but also allows for the production of "exotic" phenomena such as Cooper-pairs in Ultra-Low Temperature Physics or matter-antimatter collisions in High Energy Physics, for example, which could not exist outside of the experimental apparatus, supposedly due to the complexities and asymmetries of the natural world that either destroy such phenomena or prevent them from occurring at all. By doing so, scientists claim to explore "natural mechanisms" that cannot occur naturally due to the postulated interference from other "natural mechanisms" (such as heat exchange or symmetry breaking) or the absence in the natural world of a specific set of conditions under which such mechanisms would occur naturally, except through artificial means—such in the demagnetized copper-cooled experimental cell of a dilution refrigerator, or in the evacuated pipes surrounded by lasers and super-cooled electromagnets of particle accelerators such as the LEP-ring or LHC. The technological processes via which these exotic phenomena emerge and are detected are interpreted in terms that are postulated as being identical with *the counterfactual* natural processes (which would supposedly occur if natural conditions allowed them to), and the responses of machines (such as dilution refrigerators or particle accelerators) that the scientists use to interact with these phenomena and map out these responses in relation to their expectations. It is the dialectical processes of building, interacting, and mapping that are represented as empirical discovery through intervention and observation.

How is this fundamental representation of the machine possible? Once we take mechanical realism into account, scientific instruments and apparatus do not simply allow one to see what is "out there" or "invisible," but require a dialectical relation between how scientists already understand the phenomenon under investigation (or explore new phenomena via analogies) and how they interpret how the instrument or apparatus works in interaction with that phenomenon. These two interrelated understandings inform and constrain each other, and therefore mediate and change each other, while the scientists try through their labors to produce and reproduce representational and material practices in relation to each other, and interpreting the results of the experiment or detector in terms of an objective but otherwise hidden, underlying causal reality. How scientists understand the horizon of technological possibilities that is open to them, alongside their understanding of the limits and range of applicability of the technologies at their disposal, simultaneously informs and constrains how scientists theorize about the object of their search and the methods they use to find it. Scientists could not search for the Higgs Boson by consulting rune stones or chicken entrails tossed onto the ground, given that the theoretical form and meaning of the Higgs Boson can itself only be understood, interpreted, and used within the technological contexts of High Energy Physics, all of which have been historically developed and structured through the use of machines. How scientists theorize about these machines changes how experiments are designed, built, and operated, as well as represented and interpreted, and, in turn, designing, building, and operating machines modifies and changes how theories about natural mechanisms and laws are understood, interpreted, tested, and further developed. This is a transformative feedback loop—a dialectics between past and present representational and material practices—wherein new structures and objects emerge from the experiment as the stable result of efforts to disclose, understand, and demonstrate the causes of natural phenomena in terms of the performance of machines by explaining the performance of machines in terms of natural mechanisms and laws. This is as true for Galileo's exploration of time by using a pendulum, as it is for William Harvey's conception of the blood circulatory system in terms of a pump and tubing, or for CERN and the search for the Higgs Boson to explain the creation of matter from the performance of the Large Hadron Collider, or James Lovelock's conception of *Gaia* in terms of regulatory mechanisms, or SETI and the search for extra-terrestrial intelligence by using radio telescopes. Scientific practices and aims are shaped within the technological framework.

The Search for Extraterrestrial Intelligence is a case in point. Most of us probably are realists on the question of the existence of extraterrestrials or aliens. They either exist or not, independently of the human search for them. Whether we can detect radio signals from extraterrestrials is dependent on (1) whether they exist, and (2) whether they transmitted those signals. Those dependencies exist regardless of whether we choose to look for those signals transmitted presumably before human beings had the technological ability to build radios and radio telescopes. However, any "First Contact" via the detection of extraterrestrial radio signals is mediated by technology in the sense that an understanding of how the detector works informs and constrains our understanding of the signal itself, and any structure it may or may not have. Furthermore, natural extraterrestrial sources would need to be eliminated as potential candidates for the origin of a signal, and inform and constrain the explanation of the signal in relation to many other experiments, techniques, and models from astrophysics. This situates the signal within the technological framework; it pre-empts and limits our imagination about the form of an extraterrestrial that was capable of building radio transmitters, and excludes all those possible forms of extraterrestrial life that are not radio builders but instead communicate by some other non-technological means, such as telepathy. It also pre-empts and limits our understanding about how we would recognize an extraterrestrial sign when we heard one, and where we should look in the night sky. The means by which the search is conducted is both shaped by and in turn shapes our imagination and expectations regarding the nature of extraterrestrial life and how we might find it.

Technology has shaped the scientific understanding of all natural phenomena since the Scientific Revolution. Natural science and the natural world have been metaphysically understood since the sixteenth century in technological terms. Technology has informed and constrained the structure and content of scientific inquiry by informing and constraining the object of inquiry, within both the contexts of justification and discovery, in such a way as to think about the Universe as something technological and technology as something universal, and finding the confirmation of that way of thinking in the discovery of new frontiers of technological innovation. Mechanical realism has allowed the distinction between natural and artificial phenomena and processes to become blurred and dissolved, thereby available as metaphors for each other. Scientific methods and experiences have been developed and related to each other in terms of a mechanistic worldview—itself based on mechanical realist metaphysical precepts—that represents natural processes and technology in identical or symmetrical terms, thereby informing and constraining the "empirical" and "theoretical" aspects of scientific activity in terms of technological activity, and, by doing so, allowing the ongoing development of science as a process of constructing a technological framework of research superimposed over natural phenomena, and representing the natural processes discovered through this means of investigation in terms of the objective structures and relations of that framework. It is these technological and objective structures and relations that are metaphysically abstracted into natural mechanisms and laws, used to code technical activity, and thereby used to justify the objectivity of natural science through technological innovation, even if the theoretical content of any particular abstractions are tentative, revisable, and probably will be replaced in the future. Mechanical realism was necessary for the metaphorical relation between the natural and artificial to be taken literally as one of correspondence via the use of mechanistic models as analogies between the natural and the artificial. The ability of physicists to explain and represent a range of natural phenomena, such a lightning, sunlight, gravity, the properties of materials, or the behavior of lodestones in terms of natural forces, mechanisms, and laws, using representations of those phenomena in terms of atoms, waves, fields, physical constants, etc., did not just happen—as if by magic—from the direct observation of those phenomena. Physicists learnt to represent and interpret their experiences of the behavior of apparatus and instruments—abstractly termed as "experience" or "observation" by traditional philosophers of science—and new methods of representation and interpretation; these are themselves generated in relation to the dialectics of technology, but it is only by conceptually and interpretively transforming technology into a cultural means to make interventions and observations—to have new experiences—that the performance of machines could be culturally transformed into intelligible disclosures of natural causes, and thereafter presented as "self-evident" disclosures of an objective, underlying reality.

Once we historically situate the metaphysics of experimental science, we can see that, from the outset, the dialectics of technology was central to how modern natural science originated and developed, and this cultural enterprise was reductively interpreted and represented as the discovery of how the Universe works, thereby confirming science as a self-correcting system of knowledge that provides empirical success and new productive powers. The dialectics of technology, since the sixteenth century until the present day, has generated the development of many forms of technical codes and cognitive methods of visualization and explanation, such as mathematical diagrams and models, as well as methods of simulation, demonstration, and inference, all made in relation to the ongoing development of new instruments and apparatus. Mechanical realism has allowed these methods of visualization and explanation—themselves metaphors superimposed over the natural phenomena in question—to become literally interpreted as corresponding to natural laws and mechanisms, even if it is accepted that the products of these methods are tentative, revisable, and probably false. The abstraction of these products into theoretical representations forms and transforms the scientific worldview and connects it with the technological activity against which it is "tested." The dialectics of technology generates profound changes in the ability of scientists and natural scientists to explain and represent both natural phenomena and scientific theories, but, due to the implicit cultural acceptance of mechanical realism, this dialectic has been abstracted into one of *isomorphic* correspondence between experiment and theory, and it is this process of abstraction that has been taken by scientific realists to be the confirmation of science's "empirical success."

To put this in terms of ordinary experience, so to speak, we all have seen objects fall to the ground when dropped or pushed over. Those objects remain on the ground and do not move, if they are just left there. These are common, everyday experiences, familiar to people from all over the world and to modern and ancient peoples alike. However, there is nothing immediately self-evident to ordinary experience that compels us to explain them in terms of the consequences of inertia, gravitation, and mass. In fact, there is nothing self-evident to ordinary experience that compels us to explain these experiences in terms of natural forces, mechanisms, and laws at all. We have learnt how to do this. Furthermore, there is nothing at all self-evident in ordinary experience that allows us to connect those experiences of falling objects to the apparent movement of the Sun or the Moon across the sky. We have learnt how to do this as well. There is nothing self-evident that allows the experience an apple falling from a tree and the movement of the Moon to be unified in terms of Newton's Theory of Universal Gravitation, let alone in terms of Einstein's General Theory of Relativity. Yet, according to the traditional philosophers of science, by appealing to abstracts such as "simplicity," "comprehensibility," "reason," or "inference to best explanation," and by simplifying the history of science as nothing more than the ongoing comparison of theory and observation (if they bother to mention the history of science at all), the "explanatory success" of natural science is apparently self-evident; it needs no further explanation of how it is even possible for human beings to learn how to explain ordinary experiences in scientific terms and technical codes. If we look at how science is done on practice, however, this "self-evidence" can be brought into question.

How is it possible for human beings to explain lightning in terms of electromagnetism in relation to an abstract such as "potential difference"? What does "potential difference" mean? Voltage? What does "voltage" mean? It is not immediate to ordinary experience that lightning is akin to the electrical spark between two charged plates of metal. It is not immediate to ordinary experience that the shock one receives from a lightning bolt or an improperly handled or malfunctioning electrical machine is an "electromotive force," let alone *the same* "electromotive force." It took the eighteenth and nineteenth centuries for experimenters and natural philosophers to accumulate the efforts of building and using electrical and electromagnetic machines to explore how "magnetism" and "electricity" work in terms of interpretations and representations of how these machines worked, and interpret them as demonstrable aspects of a deeper and unified "electromagnetism," before this analogy became obvious and self-evident to nineteenth and twentieth century scientists and philosophers, and the stable performances of these machines could be abstracted and formulized as Maxwell's Laws of Electromagnetism by removing the technological framework of these stable performances from the scientific worldview, and subsequently refine these laws in terms of Quantum Electrodynamics via a new set of machine performances.

Both empiricists and realists have neglected the constructive role of technology in experimental science, along with its metaphysical interpretation. Both empiricism and realism are based upon a false ontological dualism between human activity and objective reality as the only possible poles of control at work in experimental activity. Empiricists ignore the naïve realism inherent in the design and construction of experimental apparatus and instruments, thereby arbitrarily claiming that experience of the performance of those machines and devices provides a limit to knowledge, and, therefore, objective knowledge is not possible. Realists, on the other hand, uncritically accept the empirical success and internal rationale of science, while also ignoring the metaphysical foundations of the technological framework in which scientific activity occurs, thereby declaring scientific activity to be a special activity capable of accessing objective reality, which is whatever scientists say it is. However, as I have argued above, arguments for scientific realism are circular and based on the mere assertion of historically and culturally situated faith in science. Realist interpretations of the rationality of experimental activity are based on the hidden presumption of mechanical realism and only provide us with an internal rationale. However, if the constructive role of technology is taken into account then we do not have to assume that either experiments are (subjective) human constructions or that they must reveal the (objective) laws of Nature that exist independently of human activity. We can understand the objective law-like structures to be the structures of the technological framework within which theory and experiment are situated and explored; consequently, this framework mediates scientific experience and observations, thereby filtering, constraining, and transforming how Nature is explored and understood. It is inherently reductive; thereby gaining its discipline and power.

If we take technology into account then we see how the truth disclosed by experimental science is neither purely a fiction nor self-evident. It is the dialectical synthesis (resolving contradictions and incoherence) of a trajectory of a metaphysically interpreted and incomplete technological framework of science that creates and transforms the reality it reveals by creating and transforming itself. The concealed object of science is the possibility of technological innovation itself. Science appropriates natural phenomena and transforms them into technological objects and operational variables, explores and predicts (re-)productive relations between these objects, represents those relations in terms of mathematical codes, models, and visualizations of mechanisms, and adapts and modifies itself to feedback as it either (1) explores the fullness of these models to disclose further confirmations of these quantified and functional representations of material practices, or (2) explores predicted new relations as counterfactuals in the context of discovery, and feeds these back into the technological framework as new technological objects available for future research or as transformations of Nature into forces of production. Science is a cultural experiment into the transformation of technology itself that explores its own possibilities and abstracts them into mechanical models and laws that it can metaphorically project over the natural world and test by feeding them back into itself. It is a metaphysical and poetical experiment that has been transformed through its embodiment in technology into a *cybernetic autopoiesis*, interpreted as an embodied poetical and productive, self-generating, self-referential, and self-correcting technical system of organized labor.[77]

This insight opens up the possibility of interpreting experimental science as a cultural phenomenon in terms of technology and free-creative labor communicated and made culturally meaningful through the scientific worldview. This reveals the naivety of the empiricist's demand that all metaphysics is purged from the philosophy of science. We simply cannot understand science if we ignore its metaphysical foundation and cultural meaning. The material significance of science indeed lies in its value for wider society as a means of technological innovation and the discovery of new forces of production, but it theoretical significance lies in how the possibilities and limits of innovation and production are understood in non-empirical terms. This shapes the future trajectories of science. It is within the technological framework of science that all scientists share realist commitments and a sense of objective connection with the natural world, and therefore all experimental sciences presupposes mechanical realism at some level. The empiricist's demand that the philosophy of science should be free from metaphysics is simply quixotic sloganeering and based on a misunderstanding of science. Not only do all truly novel theories require speculative metaphysical interpretations, but the whole enterprise of experimental science is founded upon the operational metaphysical interpretation of the technological framework itself. New experiences require new concepts and, if we are to understand those experiences and how they were possible, we need to understand the metaphysics that made those experiences possible. In the following chapter, I shall turn to the question of historical origins and explain how this metaphysics and the scientific worldview grew out of a fifteen century cultural interpretation of mechanics and its mathematical laws.

Objectivity and Technology:

Science is pursued by human beings from a variety of motivations to achieve a variety of ends. Science has intersected with technology through industry, commerce, education, politics, warfare, communications, medicine, mining, agriculture, and engineering, as well as reflecting and reinforcing cultural norms, ideals, and interests. Science is a cultural and technological activity in this sense. However, we cannot understand the essence of science

---

[77] Rogers (2005)

and its methodology if we take science as cultural and technological only this sense.[78] Science also involves producing and reproducing a cultural understanding of the nature of the world and the human place within it—the scientific worldview—and this has afforded human beings with explanatory power as well as new technological innovations. If we view the scientific worldview as a cultural product, we still need to raise the question whether it is only shaped by cultural factors and human relations (and therefore no different from religion or any other cultural means of producing explanatory narratives) or whether "something non-human" or "objective" shapes how human beings understand and engage with the world through scientific activity, and whether this makes science a special cultural activity. What could this "something non-human" or "objective" be? How does it shape how human beings understand and engage with the world?

The objects explored by science are encountered as products of the technological framework, interrelated with techniques and machines through a series of causes and effects, and the structures of those interrelations correspond with the structures of the technological framework itself. This is this correspondence that allows the object of inquiry to be something that is capable of being engaged with and followed in its sequences of responses to technological interventions. It is this capacity that provides experimentation with its objectivity and grants technology its instrumentality as a means of scientific observation and measurement. Experimentation involves representational captures that encounter and interpret the object of inquiry, but attendance to the objectivity of the technological framework within which the experiment is situated and from which it is comprised as it *disciplines* representational and material practices to cast them *in advance* as something that corresponds to the structures of the technological framework. This casting in advance involves metaphysical anticipations about the nature of the object and how that nature can be known, and thereby allows the technological framework to inform and constrain representational and material practices, and the possibilities for their theoretical interpretation. These metaphysical anticipations allow facts about the object to be discovered through technological activity, and furthermore allows theoretical predictions and expectations to be tested through calculation and measurement. The objectivity of any experimental result is itself secured against the possibility of calculable and measurable consequences.

It is this characterization of the properties of the object of inquiry in terms of the calculable and measureable has provided scientific methodology with a means of producing, selecting, and reducing the possibilities of experience in a way that constrains experience within the technological framework. The experience of the objectivity—the sense of the apparatus "doing its own thing"—of the object of experience is itself something that can only be understood in relation to theoretical explanations of techniques, which allow the "itness" of the object to be determined in relation to theoretical expectations and interpretations of calculable and measureable consequences.

---

[78] This point was made by Heidegger (1977c: 155-82); Heidegger proposed that science is the theory of the real. Heidegger elucidated "the theory of the real" by means of an etymological analysis. He analyzed the modern conception of "the real" in terms of "that which works" to show how performing and executing are central to the setting-forth and self-exhibition of reality. Heidegger used *arbeiten* and its compounds (*bearbeiten* "to work over or refine," *zuarbeiten* "to work toward," and *umarbeiten* "to work around or recast") juxtaposed with *wirken* ("to work"), in order to set in place the performative way in which modern science brings "the real" (as an object in a causal sequence) into presence. Modern experimental science involves working towards and striving after reality in order to capture and secure it. Theory as *Betrachtung* meant capturing, entrapping, and secure refining of "the real," in a sense reminiscent of Popper's use the metaphor of the net to describe the purpose of scientific theory to capture features of the world. Heidegger was aware of the capacity of nets to let things pass through as well as trap things. See Rogers (2005) for further discussion.

This not only gives scientific theory a concrete referent, but it is also allows the normative coherence of theory to be represented as (asymptotic) correspondence in a way that recognizes the role and importance of human creativity in science, in both contexts of discovery and justification, and how technology makes new experiences possible, while also seeing how the technological framework informs and constrains those experiences. By recognizing that this gives the object of inquiry both a transitive (social, cultural) dimension of theoretical explanation and an intransitive (technological, objective) dimension of calculable and measureable consequences of scientific activity, this allows us to explain how scientists "save the appearances" in a way that satisfies van Fraassen's empiricism, as well as incorporate the insights of Kuhn, Heidegger, Goodman, and Feyerabend, wherein the norms and assumptions of scientific inquiry change periodically thereby changing how the objects of science are understood and engaged with; it also addresses Bhaskar's realist insight into the two-fold dimensionality of experimental science. Changes in normative coherence corresponds to the "transitive dimension," while the technological framework itself gives science its "intransitive dimension." Thus the validity of classical physics was limited but not contradicted by modern quantum physics and relativity. Modern subatomic and space-time theory refined the technological framework of their respective researches to explore relativistic and quantum phenomena but did not invalidate the results of classical mechanics, thermodynamics, electromagnetism, and material science. It refined their limits. The specification of the limits of validity of theory was a confirmation of the objectivity of technological framework, and the refinements of the scientific worldview provide theory with new meanings, norms, and interpretations in terms of sets of fundamental representations, techniques, and kinds of machines.

It is the technological framework that gives machine performances in scientific experimentation their objectivity—as the calculable and measureable responses to technological interventions—and the incompleteness of the technological framework of the experiment provides the ambiguity and uncertainty necessary for there to be an experiment at all, and gives experience of "the changing of the changeable" in the apparatus its objectivity through the technological disclosure of the object of inquiry, and interpreting its deviation from theoretical expectations and prediction, as resistances or recalcitrance, and therefore as something "real" at work behind appearances. By anticipating the nature of the object of inquiry, thereby preconceiving it in technological terms from the outset, the possibilities and limits of representations are preselected in a way that shapes both theory and practice. In the experimental sciences, "the real" is what comes into being as "self-exhibiting" within the technological context of the experiment, and it is this constraint on the possibilities and actualities of how "the real" is encountered that gives theory a concrete referent and places restrictions on how theory is formulated and tested. Refining theory in relation to the "self-exhibiting" of the apparatus grants theory with its correspondence to reality completely in terms that are already-meaningful within the technological framework. This problematizes both realist and empiricist philosophies of science. Wrought by labor interpreted through metaphysical precepts, made possible and meaningful from within a culture, scientific theory is constrained by and in turn explains technology in order to disclose "the real." The objectivity of scientific inquiry is itself based on a metaphysical interpretation of the labor processes by which the technological framework is constructed, refined, and developed.

It is the phenomenological experience of the dialectics of technology that is at the core of the objectivity of ongoing research and scientific activity within the technological framework. The exploration of that phenomenon by mapping out the possible interactions between human interventions and machine performances is an experimental research project. Technique binds together an object with procedure and methodology; hence the technological framework of techniques and objects upon which it projects its template defines the methodology of any experiment and the ground-plan of action projected over the research project. Distinct experiments are defined in terms of related sets of technological objects, procedures, and theories in accordance with the posited purpose of the experiment, which is itself determined in relation to methodology and the technological framework. Each machine is an associated set of fundamental representations and technological objects within the technological framework, and it is the theoretical explanation of these machines and technical codes that connects the framework and the scientific worldview. This is as true for how rats are represented and understood in experimental biology as it is for how the motion of needles around wires are represented in experimental physics. Rats are disclosed by technique to be a set of machine performances (repeatable responses to interventions), and therefore produced and reproduced as a technological object (just as much as electromagnets or interferometers) available to produce constant conjunctions and disclose the operations of an underlying objective reality.

The methodology through which the properties of new phenomena are "discovered" through their technologization does not simply amass results, but uses those results to adapt itself to a new procedure through which those properties can be disclosed. Experimental sciences are verified by the successful development of its projected plan of action, by means of its methodology, into new procedures and experiments available as technological objects within the technological framework, to be transformed by implementation in future research by efforts to respond to the challenges and goals of other scientists. The specialization of science into specialized fields is not a "necessary evil" due to the increasing enormity of the results of research, but is a necessary consequence of the technological core of scientific methodology and how its refinements are incorporated and tested within the technological framework. It is the adaptation of research to its own results that, as an ongoing technological activity, provides modern sciences with objectivity and a cultural justification for the specialization and institutionalization of research. Any philosophical attempts to characterize experimental science solely as a means of "testing hypotheses" could not explain why experimenters choose their plans of action in the first place or why they consider their experiences to reveal an objective world that exists independently of whether any experiment is performed or not.

Ongoing technological activity provides any research project with the capacity for institutionalization in terms of its requirement to restrict each particular field of investigation to a specific set of technologies and associated objects, while sustaining its solidarity and unity within science in general in terms of the scientific worldview. The ongoing activity of technological activity builds the plan of action and its associated technological objects into all adjustments that facilitate any refinements of methodology that further the reciprocal checking and communication of results. In this respect, the scientific method is primarily concerned with the regulation of the exchange of talents and skills. Extending and consolidating the institutional character of the sciences in terms of ongoing technological activity secures the objectivity of methodology; this determines, at any given time, what is taken to be objective in research. The researcher is directed according to institutionalized and specialized projects appropriate to the technology in question. The negotiations at meetings, the information collected at conferences, the books and papers contracted by publishers, are all directed and organized through the institutionalization of specializations. The objects and measurements of empirical research are bound-up together with manipulative and representational techniques. The research scientist is not only forced to work as a technician to be capable of working effectively, but also how the scientists understand the phenomenon is itself defined in terms of the ongoing refinement of technological activity and its further specialization. Technological innovation and predictive success go hand in hand, and the establishment of "the objectivity" of new experiences and powers, and their theoretical interpretation and explanation, is itself a social problem of integrating new products of labor into the technological framework. Empirical investigation is itself a product of labor in which each progressive refinement of the technological framework is transformed into a technological object available for the future and further refinement of the possibilities of labor itself. Thus science cannot ultimately be complete because it has no end apart from itself as a means to discover its own possibilities and limits. Science is itself an experiment into its own possibilities and limits as it simultaneously establishes and differentiates itself in its projections of itself as the means to engage with and understand the world.

The development and specialization of science occurs by means of corresponding differentiation of the methodological projections of the technological framework that are made objective via claims for the rigorous application of specialized techniques, which are adapted and established through labor. Projection and rigor, methodology and ongoing technological activity, mutual requiring and reinforce one another, and these constitute the essence of modern science. The unity of this system is not contrived by relating experience and theory, but instead lies in the reduction of Nature to those aspects of Nature that lets itself be put at the disposal of technological appropriation and transformation, and consequently, the "Nature" revealed by science is determined in advance as a technological object instrumental in its own production and disclosure. The world is reduced to energy and topology, differentials and ratios, which are themselves reduced to forms and structures that are amenable to technological manipulation and measurement. Nature is reduced to those aspects of Nature that can be determined within a system of calculation, coding, and categorization that abstracts the world into a set of mechanisms and materials.

What is the Nature that presents itself in this way? Which parts of the world are taken to be instances of Nature presenting itself in this way? And, how is the set of mechanisms and materials related to the totality of all that exists without performing a metaphysical reduction? Once "the real" is identified in terms of the calculable and measurable aspects of phenomena, this reduces method to that which calculates and measures, and the whole of the philosophy of science becomes reduced to the justification of any particular method of calculation and measurement, which ultimately is justified by its usefulness in ongoing technological activity. This not only requires a cultural decision regarding what may pass as science by limiting certainty and knowledge to the calculability and measurability supplied by technology, but limits Nature itself to the sum total of possibilities inherent in the measuring and calculating powers of labor itself. What allowed Nature to be limited and determined by labor? Mechanical realism provided the precepts by which this reduction was conceivable and intelligible. The possibilities of calculation and measurement are inherent to labor and the technological framework. Technology establishes the relation between the Nature revealed by science and the natural world within which scientific activity exists, and by doing so, establishes methods of measurement and manipulation in relation to their consequences. From the outset, this understood "the objective" and "the real" in terms of instrumentality. Experimental science is perpetually directed towards transforming Nature into a technological object available for future research within the ongoing development of the technological framework. Within the technological framework are situated all fields of specialized endeavor, which are made necessary, structured, and differentiated through the dialectics of technology, which is realized in representational and material practices by the transformative power of labor itself. Our task is to understand this transformative power. To do that, we must look at its origins.

3

ORIGINS OF MECHANICAL REALISM

In the preceding chapter, I introduced mechanical realism and argued that the presumption of its precepts underwrites all the experimental sciences and the scientific worldview, and it is presupposed by both empiricist and realist interpretations of experiments. The technological activities involved in measurement and observation would be unintelligible and meaningless without the presupposition of mechanical realism, and it is necessary for science to have any empirical success. In this chapter, I shall explain how and why these metaphysical precepts came to underwrite the experimental sciences (physics, chemistry, and biology) which are often taken to be exemplars of the natural sciences—the so-called 'hard sciences'— and are widely believed to provide us with knowledge of the facts, and legitimate the test of theoretical efforts to explain those facts in terms natural mechanisms and laws. We need to examine how these sciences inform and constrain human experience and understanding of natural phenomena and their causes. We also need to look at the factors that allowed the knowledge obtained in the context of experimentation as being represented as the objective knowledge of transcendental laws and mechanisms that supposedly exist outside of the contexts in which they are produced and reproduced. This involves examining the historical conditions and metaphysical foundations of the natural sciences.

It is only via the metaphysical interpretation of the technological framework can it be related to the scientific worldview at all, and it is only via mechanical realism can theory refer to anything outside of the context of the experiment, while that context reduces the experience of natural phenomena to machine performances, thereby informing and constraining the possibilities of theory and how it is tested. The metaphysical connections between the scientific worldview and the technological activities involved in experiments are established through shared codes: subcultural representations and interpretations of how and why techniques work in the way that they do. The objectivity of science is established through these shared codes and visual and theoretical associations, with all their assumed metaphysical precepts and entailments, and it is in this sense that science can be said to operate under a paradigm. As an intellectual pursuit, science must be understood as a social paradigm that is learned through education and training, and developed through the agreements and consensus of people working within a shared community of practitioners. Once we recognize that the *sharedness* of these meanings, which gives credence to their objectivity, occurs within the context of the interpretation of techniques, instruments and apparatus, and the encoded performance of machines, we must also recognize that the objective referents and structures to which these representations and interpretations refer are those of the technological framework within which the experiment is situated and related to other experiments and their theoretical interpretation. From this perspective, the facts of experience obtained through experimentation are sets of associated reproducible machine performances and their technical interpretations, which then can be compared to theoretical expectations and predictions.

How did mechanical realism make experimental science thinkable as a natural science? To answer this question, we need to understand the origin of mechanical realism, which was at first, during the late fifteenth and early sixteenth century, a formal and explicit philosophical metaphysics proposed to solve conceptual problems in the mathematical science of mechanics; subsequently, in the sixteenth century, it became an informal and implicit cultural presupposition that underpinned the scientific worldview, with all its paradigmatic and reductive dimensions. By understanding how experimental science originated through the cultural acceptance of mechanical realism, we can understand how the scientific worldview has concealed the reduction of Nature to being that which can be disclosed and manipulated through technology, and therefore restricted how human beings can explore and know Nature as something mechanical and technological, while concealing these restrictions on human thought. To understand this metaphysics and how it came to be thinkable, and thereafter concealed under its own "self-evidence," while all other metaphysical interpretations were rejected as speculative or fictional products of the human imagination, it must be placed in its historical context to show how people thought about mechanics and natural phenomena *before* the Scientific Revolution began. Changes in the way of thinking about Nature, and how human beings could encounter it, not only paved the way for the rise of the new sciences and the Scientific Revolution as a cultural phenomenon (a new social movement) against the old way of thinking (Catholicized Aristotelianism), but also made alternative metaphysics seem quaint and superstitious, ridiculous and nonsensical, romantic or poetic, and imaginary. The metaphysics of experimental science became concealed behind naturalistic interpretations of how technology works.

Mechanical realism has its roots in the ancient Greek, Arabic, and medieval Aristotelian and Archimedean studies of alchemy, mechanics, anatomy, and astronomy, finding its first shoots in the philosophical ideas of the Aristotelian mechanists of the fifteenth century, and finally emerging in the Scientific Revolution of the sixteenth and seventeenth centuries as something "new." Galilean natural science became possible. The (albeit limited) successes of the two mathematical sciences of astronomy and mechanics inspired the mechanical philosophers to propose that the motions of the entire physical world could be completely described in terms of laws, mechanisms, and mathematics, long before there was any "evidence" that this was even possible. The faith in the new sciences preceded their "empirical successes." From the outset, the physical world was to be described as nothing more than matter in motion in geometrical space because this was exactly the same abstract components that comprised the conceptual basis of the rationalization of mechanical devices and their projection over natural phenomena. This was situated within cultural contexts and demands for advantage and power, measured in terms of the satisfaction of political, commercial, and military ambitions in Europe and the exploration and exploitation of Africa, India and Asia, and the New World.

Contrary to popular belief, the predictive successes of Copernicus' system over the Ptolemaic system were very much exaggerated, and the cultural acceptance of the new sciences did not occur because of the predictive success of Copernicus and Kepler's mathematical treatments of planetary motions, nor Galileo's demonstration of the orbits of moons around Jupiter.[79] The new sciences were primarily accepted because of their successful association with the innovation of mechanical devices and new powers within societies that valued the economic, political, and military advantages of technological innovations.[80] Appeals to the successes of mathematical astronomers were *rhetorically* connected to the instrumental successes of mechanists, as part of the movement towards

---

[79] Kuhn (1957); Westman (1975)
[80] Long (1997); White (1962); Kaufman (1993); Bennett (1986); Merton (1970); Rossi (1970)

a unitary conception of natural science, but these appeals always took the form of analogies and metaphors, even if couched in terms of literal truth claims, and the objectivity of science became reduced to its instrumentality. The precepts of the mechanical realist metaphysics were required (at least implicitly) to connect, via the mathematical science of mechanics, the astronomical phenomenon of planetary motion with terrestrial machines, and represent and interpret both as aspects of a unitary natural science via mathematical demonstrations of mechanical motions, and therefore the demonstration of success of one was the demonstration of the success of the other.

By understanding the mechanical realist foundations of the experimental sciences, we can see how it has informed and constrained how the meaning of the "physical" and "empirical" have since been interpreted and understood. The understanding of technology as "applied science" has concealed its metaphysical interpretation behind the scientific worldview, with its abstract theoretical formulism and causal explanations of machine performances in terms of natural mechanisms and laws, along with its fundamental representation of Nature as an objective material world (a sum totality of natural mechanisms and laws operating on matter, in space and time) explored through technological activity.[81] While the obsession with "efficiency"—the perpetual quest for "the best method"— was a construction of the nineteenth century, the idea of Nature as efficient finds its origin in the mechanical realism of the sixteenth century, when it was proposed that there must be one single "most efficient mechanism" in operation between any particular cause and its effect(s) and that mechanism is the one that operates in accordance with mathematical laws, and the "most efficient mechanism" was termed as "the natural mechanism." Henceforth, it was the allotted task of the experimental sciences to find it for any particular cause-effect sequence, and the distinction between "pure" and "applied" science is *merely* the distinction between finding the "most efficient mechanism" and implementing it in the organization of productive practices.

Once we address the extent that experimental science involves both the discovery and implementation of "the most efficient mechanism" in ongoing research and technological practices, we can characterize experimental science as simultaneously "pure" and "applied." It was only possible for empiricists to reject metaphysics as unscientific once science's own metaphysics had become concealed within the cultural norms of representational and material practices and their theoretical interpretation. This way of thinking has extended itself to all the sciences, from physics and chemistry to biology and ecology, and even economics, with its appeal to invisible mechanisms ("the invisible hand of the market" or "economic forces") and laws, and the social sciences, including sociology and psychology, especially those involving understanding human relations in mechanistic terms (often couched in the language of "tendencies" and "forces") and their organization into modern industrial society in terms of psychological or societal mechanisms, or in the case of cognitive psychology or neuroscience reduced to the biological mechanisms of the brain and sensory organs.

Furthermore, once we can see how the form, content, predictions, and tests of theory emerge from and are mediated by the technological framework, as a series of mediated disclosures and manipulations, we can see how representations of "how technology works" form and transform the structure and content of the scientific worldview, and, in turn, via techniques and their interpretation, are informed and constrained by that worldview, with its postulated horizon of possibilities and

---

[81] In his essay *A Question Concerning Technology*, Heidegger (1977a) argued that the definition of technology as "applied science" concealed its essence and was based on a positivistic misinterpretation of both science and technology. See also Rogers (2005).

expectations that are projected over the natural world. It is only by paying close attention to how experiments have transformed the scientific worldview, which in turn has inspired further experimentation and innovation, can we see that the object of scientific inquiry is actually the technological framework itself. This provides the means for science to provide its own methods of testing and self-correction. The totality of experimental science is an experiment into itself and its own possibilities and limits. Hence, the frontiers and limits of "pure research" are identical with the frontiers and limits of technological innovation.

How can we understand this technological framework and its relation with the scientific worldview? The traditional philosophers of science have largely agreed that it should be ignored. Theory precedes and anticipates experiment in the form of hypotheses, conjectures, or predictions, or some set of unifying general principles or rules, they tell us, and is a purely intellectual construction. Experiments are simply designed to test them. There is nothing else to see; please move on. It is almost as if scientific instruments and apparatus appeared by magic, or fell from the sky, complete with an instruction book for their assembly along with an operator's manual. The traditional philosophers of science, irrespective of whether they call themselves realists or empiricists, have assumed that experiments are nothing more than the means by which scientists obtain the facts about the natural world via observation, and also the means to test theories to predict the response of natural phenomena to human interventions. They share the same mechanical realist presuppositions about the nature of scientific instruments and apparatus, alongside the rather *ad hoc* and problematical assumption that techniques and technical knowledge are self-evidently theory-neutral. These philosophers have ignored the way that technological processes and interventions have shaped how the facts have been disclosed to experience in the technological contexts of actually making and performing experiments. They have presumed that the results of experimentation, reported to them by scientists as "empirical regularities," "observations," or "data," are simply "the facts" against which the empirical adequacy of any theory can be tested. When theories provide accurate description and predictions of observations and measurements, the technological activity involved in making observations and measurements is taken to demonstrate the empirical success of science; it is presumed that technology neither changes the nature of the phenomena under investigation, nor does it shape the form of theory and how natural phenomena are represented and explained.

This shared presumption has reinforced something of a dismissive or negligent attitude regarding the philosophical meaning and importance of how we understand technology, especially in relation to science. It is simply a lucky accident that science achieves its greatest empirical success in technological contexts and that theoretical research is driven and limited by technological innovation. The shared belief in this lucky accident has allowed the traditional philosophers of science to presume the division of labor between scientists and philosophers, with scientists working in the context of discovery (of the facts and new phenomena) and philosophers working in the context of justification (of theory and methods). Of course, this interpretation is only sustainable if one does not pay attention to how the technological activities of experimentation have actually been performed, understood, and represented in practice by working scientists. It is simply assumed by scientists and philosophers alike that empirical data is produced through the carefully isolated exercise of natural mechanisms, within the carefully isolated contexts of experimentation, and realized through human labors and the arrangement of materials in accordance with natural laws. The traditional philosophers tell us that whether this has been done successfully or not is a matter for scientists to decide among themselves.

The traditional philosophers leave it to the scientists to do all the work and provide them with "experiences" against which theories can be corroborated or falsified, and nothing further needs to be said about technology apart from that it is the logical consequence of the application of scientific knowledge and rational thought to human problems in the material world. This of course tells us much more about the neglect of technology by traditional philosophers of science than it does about the origins of modern science. When the traditional philosophers of science have even discussed technology at all, scientific instruments and apparatus are assumed to have simply been the products of *applied science*, without any real effort to explain what this term means. Even when philosophers of science have been aware that the character of "application" is an ambiguous or problematical one, they have maintained the traditional view that technology is applied science.[82] Traditional philosophers of science have presupposed that science provides us with a rational understanding of reality, without really explaining what "a rational understanding of reality" means; nor do they explain how one goes about obtaining it (except by accepting the interpretations of scientists).

Not only can we criticize the technological naivety (and blind faith) of the traditional philosophers of science, which is tantamount to an uncritical belief in the division of labor between scientists and philosophers, we can also criticize how they have ignored all the historical complexities and controversies involved in the invention, usage, and dissemination of such instruments and apparatus, all of which involved many people, all working within scientific institutions (universities, research collaborations, and laboratories) and other social organizations (governmental, commercial, and military). By ignoring the highly involved and complex cultural norms and modes of communication (including criteria of publication) and reproduction, the traditional philosophers have presupposed technological determinism and ignored the contingencies of science. Yet, if we do not follow suit, we can see that each technological object (tool or machine) has arisen from a culture within which such devices gained their epistemological meaning and technical value as means to measure and observe natural properties and quantities, and the means by which these values are gained are themselves contingent and conditioned by culture. The traditional philosophers have ignored all of the cultural factors and influences that shape representational and material practices, especially those involving novel contexts of language use and meaning that are necessary for scientific experiments and observations, alongside their theoretical interpretation, to exist at all. Without language, no scientific observation, measurement, or theory is possible; nor could it be shared by human beings. The cultural factors that shape language also shape scientific experience and theory, by shaping the conditions for cognition and its successful communication, also provide the means for the association and interpretation of representational and material practices.

The traditional philosophers have also ignored all the technical difficulties and controversies involved in making and communicating scientific observations and measurements in a way that allows them to be reproduced by similarly placed (and skilled) observers. Perhaps most perniciously, the traditional philosophers have perpetuated the illusion that such instruments and apparatus simply allow us to see what is objectively "out there" or "otherwise invisible," thereby making technology into a seamless extension of our inborn senses and abilities that should never be questioned, and, consequently,

---

[82] Nagel (1961)

Nature is what scientists tell us "it" is, even when scientists themselves say that all science is open to question and revision, and their knowledge is contingent and replaceable. By presupposing the same mechanical realist interpretation of technology, this presupposition has allowed the traditional philosophers of science to neglect the historical role and cultural meaning of technological innovation in shaping scientific experience and theory, and how Nature is represented, explored, and understood. Take solar neutrino physics for example. We are supposed to believe that the act of interpreting the fluctuation of instruments attached to photomultiplier tubes inside a deeply buried vat of water means that a few sub-nuclear impacts (neutrino events) had occurred within the water and this interpretation had instantaneously empowered the scientists' eyes to see waves of billions of solar neutrinos flux outwards from the center of the Sun and through the Earth. We are supposed to believe that observing neutrinos is a simple act of perception and accessible to trained human experience. It is not even necessary for observations to match theoretical expectations to maintain this belief in scientific experience. When the numbers output from computers don't tally with theoretical expectations, scientists and the traditional philosophers can explain away that discrepancy by invoking the idea of neutrino oscillations, as if that in itself was some truth vicariously corroborated by scientific experience, given to the philosophers by the scientists. Thereafter one can observe neutrino oscillations as simply as one could count and classify the different colors of walruses, as if observing neutrinos was no different from counting walruses on a beach.

If we aim to understand scientific experience and practices, we need to question how scientific experience is constructed and take a close look at how technology shapes the contexts within which scientific experiences and practices occur. We also need to look at the purpose of those activities and how they are conceived and represented. How is scientific experience possible? How do scientific practices gain their meaning? Rather than remain constrained by the traditional philosopher's epistemological demand that we limit ourselves to the context of justification of scientific knowledge claims—perhaps even reducing them to a series of logical operators (as if we are limited to a true or false duality when evaluating the meaning of scientific facts and theories)—what we need to do is look at what kind of knowledge arises from scientific experiences and practices. What is the object of scientific knowledge and how is it known? How are scientists able to connect their experiences and interpretation of the artificial context of the experiment and apparatus with the wider natural world outside the laboratory? In order to answer these questions, we need to examine the history of the origins of the experimental sciences, such as physics, biology, and chemistry, and how they went about inquiring into Nature by transforming it. This has been extended to the study of human nature through biotechnology and biochemistry, physiology and pharmacology, psychology and computer science, surgery and neuroscience.

Technology and its mechanical realist interpretation have allowed the scientific method and the notion of objectivity to become based on a presumed connection between the representational and material practices of experimentation and a material substratum governed by mathematical laws. Once mechanical realism had been assumed then the content, form, and testing of theory necessarily became reduced to the measurement and prediction of quantifiable variables and constants, while the structures of theory became reduced to those that can translate into mechanical ratios and operatives, thereby allowing them to describe the causes and predict the outputs of instruments and machines. Once mechanical realism was presupposed by scientists as *a naturalistic interpretation* then scientific interactions within the world became reduced to sets of technologically mediated disclosures and manipulations of an underlying causal reality.

The associated techniques and theoretical explanation of these disclosures and manipulations have informed and constrained human thinking about the natural world by reducing it to the production of sets of mathematical relations and their associated material responses and theoretical explanation. The association of these disclosures and manipulations with instruments, apparatus, techniques, and theories constitute the structures of the technological framework within which the experimental sciences are situated and the natural world is explored.

How did the presumption of mechanical realism become self-evident? Etymologically, the word *meta-physics* literally means *after physics* and was coined in the sixteenth century to catalog the book of Aristotle that followed his *Physics*. The subject of that book has since been taken to be the point of departure for philosophical inquiries into metaphysics, from which perspective the metaphysics of Aristotle's teacher Plato could be retrospectively identified and analyzed in Aristotelian terms, and thereafter Western philosophy could discover its own tradition of metaphysics 1900 years later. Metaphysical inquiry is premised on the question of what must be true about the world if we can know the first principles or causes immanent and internal to all beings (including human beings) and thereby understand why the world is the way that it is and what our place is within it. It is also a question of what must be true about human beings (and human thought) if we can have this kind of knowledge and think about the physical world in this way. The philosophical task of metaphysics is to identify this underlying (non-empirical or conceptual) truth about the causes and nature of the physical world and explain how and why human beings can know it and interact with it. The Galilean response to this philosophical tradition was to remove all further metaphysical speculation by reducing natural philosophy to mechanics via the metaphysical presupposition of the precepts of mechanical realism, and thereby explore the physical world via the mathematical representation of machine performances and their explanation in terms of natural mechanisms and laws.

This series of metaphysical precepts arose from the theoretical work of fifteenth and sixteenth century Aristotelian and Italian mechanists *before* the rise of Galilean physics, and from whom Galileo Galilei took the basic concepts and fundamental techniques for his new sciences of physics and astronomy. This set down the template for all subsequent scientific methods and research. It is also evident in the philosophical writings of Francis Bacon and Rene Descartes, as well as in the works of seventeenth century physicists such as Isaac Newton, Robert Hooke, and Robert Boyle, working and corresponding within an intellectual community of seventeenth century natural philosophers and experimenters. It reveals that the essence of the object of scientific activity must be understood as emergent from within a technological framework of representational and material practices, which are given cultural meaning as natural objects that reveal and confirm the fundamental truth of the scientific worldview, even though it may well need refinement or even radical reinvention from time to time, and this is tested through technological innovation and the further development of the technological framework. Scientific specialization and new fields of research follow the differentiations within the technological framework itself, not as some 'necessary evil' due to the complexities of science and the natural world, but as a basic corroboration of the framework itself.

To understand the origins of this technological framework, we need to examine how new and fundamental representations of Nature arose from the late fifteenth century Italian understanding of the Aristotelian science of mechanics, which allowed the mathematical laws describing the motion of the six simple machines (the wedge, lever, inclined plane, balance, wheel, and screw) to be represented as natural laws. These representations were later projected by Galileo over the phenomena of motion, as geometrical demonstrations, and provided the conceptual foundation of the new science of experimental physics by bringing together the categories of the natural and mechanical under the same concepts and explanations. This new way of thinking resonated with fifteenth and sixteenth century alchemists, mechanists, and the philosophical proponents of the cultural value of the practical arts. The sixteenth century understanding of methods of heating and distillation of substances, for example, were developed through alchemy and allowed new and fundamental representations of natural processes in terms of artificial process, as if there were no difference between the two kinds of process, except that the latter required human intervention and the former did not. Thus the source of heat or pressure, for example, became irrelevant—the heat or pressure from an artificial source being no different to that from a natural source—and it became possible to represent natural change in terms of artificial interventions, via constructed abstracts, variables, and constants, which in turn could be used in the design, construction, and operation of novel experiments to explore natural phenomena by replicating them as an ensemble of technological objects available for manipulation, and represented in terms of natural mechanisms and laws. These fundamental representations provided the cultural background for the conceptual foundation of the new sciences of chemistry, biology, and geology.

It was through the metaphysical projection of these new and fundamental representations in terms of variables and constants that allowed natural phenomena, change, movement, resistance, and processes to be represented and demonstrated in terms of the performance of machines (apparatus and instruments) within the closed and isolated confines of the laboratory experiment, or a public demonstration of the operation of a device. It allowed the operation of a machine to be conceived as the outcomes of natural laws and mechanisms that transcend and pre-exist the context of the experiment and the demonstration of these laws and mechanisms at work. Machines became understood as the realization and exercise of natural mechanisms in the juxtaposition and arrangement of materials, methods, and techniques within the technological framework of the experiment—itself represented as a demonstration of theoretical explanations and empirical success. In this way, the machine became a metaphor for the natural phenomenon, which was explored as an analogy, and technology became represented as a literal means of disclosing and manipulating underlying natural mechanisms and laws. This was a metaphysical and conceptual prerequisite for the scientific worldview, and, subsequently, empiricism and realism. It was only later through the innovation and development of new kinds of machines, new fields of physics, such as thermodynamics and electrodynamics, and later atomic and quantum physics, that it became possible to explore a stratified material world through technological activity, with each stratum represented in terms of deeper fundamental mechanisms and laws, thereby mechanics, thermodynamics, and electromagnetism connected to astrophysics and cosmology, and, chemistry; later, via chemistry, it was extended to biology and ecology, genetics and neurology, geology and planetary science. It is only once the work has been done that the results of that labor and technological activity can henceforth be abstracted and represented by scientists and traditional philosophers of science as "observations" and "the facts of experience" against which theory can be tested.

Experimentation grew out of alchemy and fifteenth century Aristotelian interpretations of mechanics, but the kind of knowledge of causes it promises is an asymptotic ideal that is never achieved in practice. Hence scientists claim to get closer to the truth without ever reaching it. It requires the metaphysical presupposition of mechanical realism to establish experimentation as a methodology within natural science, and it is this presupposition that was the origin of the new sciences and the point of epistemological rupture and incommensurability between the Aristotelian and Galilean worldviews. The proponents of these two worldviews did not mean the same thing by "Nature" and "*phusis*," nor "matter and "*hyle*," even though they are commonly translated to mean the same things. For the Aristotelian, experimentation can teach us nothing about *phusis*, as experimentation is artificial and occurs in contexts that differ from those of the natural world; for the Galilean, contemplation and dialectics can teach us nothing about Nature, as they are human in origin and subjective (based on mental responses to sensory input). For the Aristotelian, causes could not be understood without understanding how they were intelligible as causes; for the Galilean, causes were forces or mechanisms that occurred from without and could only be understood in terms of mathematical demonstrations and reproducing them in the performance of a mechanical device. For the Aristotelian, mathematical descriptions of experience were always incomplete and did not reveal the essence of phenomena; for the Galilean, only mathematical descriptions were objective, referring to the primary qualities (quantities, dimensions, ratios) and all other descriptions were cultural artifacts based on tradition, speculation, and secondary qualities. Aristotelians and Galileans had a different idea of how to experience the world and theorize about it; they had a different methodology to identify, categorize, describe, and explore natural phenomena, and explain them in terms of natural change and an overall cosmological unifying conception of the world. Aristotelians and Galileans simply did not mean the same things by the words "space," "motion," and "time."[83] Aristotelians and Galileans disagree on how to interrelate explanations and experience, or even what constitutes an explanation and experience, and they both seek a different overall comprehension of the world. In fact, they have different standards and rules by which they could judge what constitutes a comprehension of the world and how to investigate it.

This paradigm shift from Aristotelian *scientia* to Galilean *science* occurred within the context of the European desire for novel technological powers, especially when these provided increased military power and commercial advantage, and advanced a culture of warfare and navigation around the globe. It was the satisfaction of the desire for technological power that provided the condition for the cultural acceptance of the naturalness of the conceptual synthesis of the precepts of mechanical realism and the possibility of both mathematical natural science and modern scientific technology. It was only once this symmetry emerged that the conception of modern technology as a unitary phenomenon, manifest according to universal natural laws, became possible and intelligible. This was simultaneously a naturalization of mechanisms and a mechanization of Nature. The conception of technology as a process of unlocking and utilizing natural forces, causes, and powers became possible, thereby not only reducing Nature to something instrumental for human purposes but also reducing theory to the disclosure of Nature's instrumentality; thereafter technology became the only rational means to test and evaluate the objectivity of any representational practices or theory in terms of their instrumentality in the disclosure of power. Henceforth material practices were treated as revealing a unitary kind of relationship between "Man" and "Nature," in which "Nature" could provide means for its own domination by "Man," using his natural-born abilities of reason, perception, and material practice. Henceforth, technology was a mode of human agency that established a new conception of "universal rationality," defined in terms of practicality or instrumentality, determined in accordance with a concept of

---

[83] Heidegger (1939); Kuhn (1957); Feyerabend (1975); Goodman (1978).

"efficiency" and represented in terms of the realization and exercise of "mechanisms" through material practices.

With the increasing interest in the development of the mechanical sciences in seventeenth century Europe, for the purposes of enhancing technological powers, the mechanical worldview became dominant. Once this had become culturally "self-evident," technology could be ignored by natural philosophers as being nothing more than "man-made" and a "means to an end" achieved by the application of natural science to practical problems. The path was cleared for the notion of mechanism to become the dominant explanative trope—the basic fundamental representation—whether in physics, chemistry, or biology, or in social science, economics, or psychology. It became the basis for mid- to late twentieth century efforts to reduce all science to that of information and cybernetics. This monolithic explanatory strategy was symptomatic of the accelerated mechanization of European social organization towards the monolithic goal of achieving technological advantages for the competing European social elites, and was inextricable bound-up with the cultural equation between power and the good. This transformation was a profound shift from the contemplative scholarly logic and poetics of Aristotelian natural philosophy towards the construction of mathematically rationalized machines and novel technological powers. Contemplative and scholastic Aristotelianism was not "falsified"; it had become obsolete and irrelevant.[84]

Once the status of mechanics had been transformed from a banausic art to a natural science—through the mathematical projection of the six simple machines as geometrical demonstrations—then those first principles could be presented as "eternal and necessary truths." In combination with the mechanical realist metaphysical premise that "natural causes" were efficient, this transformation allowed mechanics to be naturalized. The distinction between the artificial and the natural was dissolved for particular aspects of technology: the fundamental principles of mechanical motion. That was subsequently taken as self-evidently true and there was not need of any further metaphysical argument. Once this had been achieved then the ontology of experimental science, based on mechanical apparatus and a mechanical world view, could achieve an epistemological legitimacy as a means of disclosing truth. Mechanical realism had transformed from *meta-physical* to *techno-logical*: experimental science was a means of disclosing the truth about the nature of Nature by manipulating it and the truth of this extraordinary assertion could be taken as "self-evident." Here we can see how scientific empiricism and realism discussed in the preceding chapter were made conceptually possible by a cultural reification of technological power underwritten by mechanical realism.

Given the implicit *naturalness* of this cultural interpretation—itself based on the cultural acceptance of science in terms of mathematical projection of mechanisms over all natural phenomena and representing the innovation of new technological powers as the test of any hypothesis or theory—the technical acts of writing the book of Nature could be ignored (as mere means) and it could be read as if written by God and read by Man. Power became the measure of objectivity. Taking mechanical realism into account shows us how the Scientific Revolution of the sixteenth century was a cultural revolution that was founded upon *the naturalness* of the establishment of mechanics as a mathematical science and the epistemological legitimacy of mechanics as a natural science, and the *goodness* of increased technological power and transforming the material world to better suit human purposes. This revolution involved a transformation of the conceptions of

---

[84] Cf. Grant (1978) for detailed discussion.

matter, cause, natural necessity, and the dissemination of the mechanical worldview, in parallel with rhetorical appeals to the practical successes of mechanics in terms of new powers and experiences, and connecting that with the manifest destiny of Man to conquer Nature by using Nature against itself.

Mechanics and the Origins of Mechanical Realism

In his preface to the 1st Edition of *Philosophiæ Naturalis Principia Mathematica* (1686), Newton was able to propose that mechanics should not be limited to the manual arts, but instead used to investigate "the forces of Nature" and "to deduce the motions of the planets, the comets, the moon, and the sea."[85] Newton's statement of the "Rules of Reasoning in Philosophy" are a statement of the precepts of mechanical realism.[86] The first rule is an epistemological reduction of the possible form of natural causes that demands that we limit our investigation of the causes of natural things to the causes we identify as necessary and sufficient mechanisms to explain the appearance of natural phenomena. Newton presumed that *simplicity* is a natural principle; consequently, simple mechanical explanations were the most likely to be true (*lex parsimoniae* aka. Occam's Razor). The second rule is a statement the invariance and universality of cause-effect sequences; consequently, we should assign the same causes to the same effects, and, for Newton, all causes of motion were mechanical causes (and he thought his own law of universal gravitation to be unsatisfactory and incomplete because 'action-at-a-distance' did not offer a mechanical explanation). The third rule that "the qualities of bodies are only known to us by experiment," while simultaneously insisting that "the qualities of bodies" known through experimentation also informed us about bodies upon which no experiment has been performed. This is a statement of methodological reductionism that allowed the universalization of the properties of bodies from those properties identified through experimentation. This assumption was necessary for Newton to assert that Nature is both *isotropic and homogeneous*; this is an essential assumption for all experimental science because, without it, the experimenter could not extend the particular outcomes of any local experiment to demonstrate the operation of a universal and natural law. Newton considered natural laws to be geometrical and geometry to be founded upon mechanics, and told us that geometry

"is nothing but that part of universal mechanics which accurately proposes and demonstrates the art of measuring. But since the manual arts are chiefly employed in the moving of bodies, it happens that geometry is commonly referred to their magnitude, and mechanics their motion. In this sense rational mechanics will be the science of motions resulting from any force whatsoever, and of the forces required to produce any motions, accurately proposed and demonstrated."[87]

Fourteen years earlier, in the *Origin of Forms and Qualities According to the Corpuscular Philosophy* (1672), Robert Boyle wrote,

"That which I chiefly aim at, is to make probable to you by experiments, that almost all sorts of qualities... may be produced mechanically; I mean by such corporeal agents as do not appear either to work otherwise than by virtue of the motion, size, figure, and

---

[85] Newton's "Preface to the 1st Edition" of *Principia*, pp. xvii-xviii (Cajori's edition and English translation, Newton, 1962).
[86] *Ibid* 398-9
[87] *Ibid* xvii

contrivance of their own parts (which attributes I call the mechanical affections of matter)."[88]

He went on to confess that he hoped to explain cold, heat, magnetism, and *all other natural phenomena* in terms of mechanisms. Boyle's mechanical realism is most evident in his explanation of the air pump and representing it as a means to disclose the mechanical properties of matter in relation to an (already presumed) homogeneous and isotropic space void of matter ("a vacuum"). It became possible to observe the effects of "a vacuum" by using an air pump. Once the technological production and reproduction of "a vacuum" was established by Boyle as something real, with its own physical properties, by showing how it explained the performance of an air pump (and the suffocation of a bird trapped in a glass bulb), it could be henceforth transformed into technique and technological object available for other researches as a tested and repeatable means of disclosure.

Once this fundamental representation of the mechanics of the air pump had been established by Boyle as corresponding to something real, and otherwise inaccessible to experience, and hence the air pump was available as a means to disclose "the mechanical afflictions of matter," the cultural acceptance of mechanical realism allowed the air pump to disclose an objective relationship between all matter and all space that existed everywhere and for all time. Subsequently postulated representations and their refinements of the description of this objective relationship could be presented as ready for testing in comparison with the refinements in the construction of air pumps and their performance. The representation of "a vacuum" could then be woven into the scientific worldview and the wider culture as a description of a fundamental and objective characteristic of the structure of Nature,[89] while the cultural test of the empirical success of this fundamental representation was its usefulness in aiding the design and invention of new machines for practical purposes, for pumping air and water, and for creating watertight and airtight seals. This weaving together of cosmological truth and practical utility was rhetorically achieved by Boyle in terms of his natural philosophy via the demonstration of the air pump as a machine that simultaneously produces a vacuum and is explained by it. The epistemological criterion for the test of this assertion was that it could be instrumentally functional in the subsequent innovation of further machines and also explain how that instrumentality was possible. This epistemological criterion became central to the whole methodological enterprise of experimental science, and shows how science and technology were inextricably intertwined in Boyle's work. Once the criteria for the test of theory were made in technological terms and followed the technological innovation of new machines, the truth of "the vacuum" was deferred until its future implementation in representational and material practices, which led to new phenomena, and further refinements and tests of theory in terms of its instrumentality. The value given to knowledge obtained from constructing and performing experiments remains perpetually provisional on its use in the construction and operation of future experiments, and the contexts of discovery and justification occur within contexts of technological innovation.

However, this epistemological criterion was not invented by Boyle. It was explicitly proposed seven years earlier by Robert Hooke in the preface to his *Micrographia: Or Some Physiological Descriptions of Minute Bodies Made by Magnifying Glass, with Observations*

---

[88] Birch edition, Boyle (1965) p. 13
[89] See Shapin & Shaffer (1985, chap. 2) for details about Boyle's public demonstrations of the vacuum and the cultural contexts in which these demonstrations worked as demonstrations of anything at all.

and Inquiries Thereupon (1665).[90] Hooke described his "natural philosophy" as the real, the mechanical, and the experimental philosophy. Hooke advocated the transformation of "natural philosophy" from the observational, experiential, and categorical, into a process of making interventions into natural entities and processes using instruments and machines to produce explanations of the sensible phenomena of experience in terms of fundamental mechanical interactions. Hooke explained that natural philosophy was necessarily premised upon an intimate relationship between mathematics and machines. He frequently used machines to present illustrations of "the common rules of mechanical motions" that he proposed as the mechanical principles of Nature. The mechanisms disclosed through the mathematical representation of the operation of machines could be taken to be simultaneously describing universal mechanisms of Nature and as available for the development of new techniques to further explore Nature.

Newton, Boyle, and Hooke's mechanical realism represented machines as having explanatory power about Nature. They were able to assert the dream of deriving the rest of the phenomena of Nature from the same kind of reasoning from mechanical principles—even in the absence of any confirming evidence that this was even possible. By the end of the seventeenth century, the mechanical worldview was refined in order to explain the technological development of machines in terms of increasingly fundamental mechanisms, which were disclosed through the technological innovation of new machines and their explanation in terms of mathematical and natural laws. Technology was represented as the consequence and mediation of all relations between "Man" and "Nature," as being the rational utilization of natural mechanisms in the ongoing development of material practices, and, hence, it was understood as the rational and objective application of the causal powers discovered by the experimental sciences. The acquisition and control of new technological powers and their explanation—the further stratified development of the technological framework—was equated with an increased objective understanding of Nature, even though that understanding remained provisional and incomplete at each and every stage of development.

The mechanical realist conception of Nature and its philosophical explication was made in the early seventeenth century by René Descartes in his *Meditationes de prima philosophia* (1641) and his essays *Discourse on Method, Optics, Geometry,* and *Meteorology* (1637).[91] In *Meditationes*, Descartes conceived all natural phenomena in terms of machines and asserted that

"there are absolutely no judgments in Mechanics which do not also pertain to Physics, of which Mechanics is a part or type: and it is as natural for a clock, composed of wheels of a certain kind, to indicate the hours, as for a tree, grown from a certain kind of seed, to produce a certain kind of fruit. Accordingly, just as when those who are accustomed to considering automata know the use of some machine and see some of its parts, they easily conjecture from this how the other parts which they do not see are made: so, from the perceptible effects and parts of natural bodies, I have attempted to investigate the nature of their causes and of their imperceptible parts."[92]

The usual interpretation of Descartes' philosophical project is that it is a form of

---

[90] Hooke (2003); see also Bennett (1986) and Westfall (1983).
[91] For an English translation of *Discourse on Method and the Meditations* see Descartes (1968); see Descartes (1966-76) for his essays.
[92] Descartes (1968) p. 285

skepticism, subjectivism, or egoism based on the mind-body or subject-object dualism.[93] The usual interpretation of Descartes' *cogito ergo sum* ('I think therefore I am') is that it refers to the thinking-being, "I," as the human subject. This self-declared center of thought supposedly placed the doubting subject at the beginning of philosophy in order to provide reflection upon knowledge itself and its possibility, thereby placing epistemology before ontology, asserting the primacy of the mind or ego, and defining a human being as a subjective thinking being. This usual interpretation ignores the historical context in which Descartes philosophy gained its resources, meanings, and significance. It also ignores Descartes own preface (the *Discourse on Method*) to his essays in which he explicitly recognized the impossibility of taking a human being outside of his society, or culture, and its customs and morals, and offered his own meditations as a series of conjectures and reflections on the nature of knowledge and how human beings can possess it. Descartes' work must be place within the context of an historical period in which a new assault upon tradition had begun. This assault promised to free human beings from the bonds of this tradition by freeing the mind of Man to think for himself in terms that could be demonstrated to be true through reason and experience.[94] Descartes' *Meditationes* contained his demonstrations of the metaphysical foundations for the epistemological method of mathematical intuitionism that would replace Aristotelian natural philosophy without appealing to an alternative ancient philosophy.[95]

Descartes' enterprise reflected the passion for this new intellectual revolution and it was an effort to reveal and clarify the metaphysics of the new sciences. This enterprise not only emphasized that clear and insightful intuition, or certain deductions, are the routes to knowledge, but also held that method in general is necessary for us to know any truths at all. This method was to consist in the order and arrangement upon that which "the sharp vision of the mind" is to be directed if truth is to be discovered. Descartes did not start his discourse with doubt because he was a skeptic about experience or existence, but in order to clear the way for positing his mathematical intuitions as the absolute ground and foundation of natural philosophy. Mathematical intuitionism had to be based upon its own ground, as a fundamental principle, and be indubitable. Descartes' philosophical work was to postulate and explain the special axioms required for mathematical intuitionism (*mathesis universalis*) to ground and form the whole of knowledge. These axioms needed to be self-evident and establish in advance what constitutes knowledge and from whence, and how, the essence of that knowledge is to be determined. Hence Descartes' *cogito* should be understood in terms of mathematical intuitionism, rather than subjectivism, when thinking is itself understood in terms of a cognizance of itself as something one realizes for oneself as something one already does. This formed an intellectual standard of clarity of thinking against which any idea could be measured in terms of what one already knows if one thinks about it. Descartes' formula *cogito ergo sum* should not be understood as an inference from the self-evidence of thinking because the *sum* was the *fundamentum* rather than the *consequentis* of thinking. This formula was posited by Descartes because it allowed his mathematical intuitionism to be presented as something independent from the past (customs and traditions) or whatever is given beforehand (appearances and experience) and constitutes its own self-justification as that which already lies within us, to be brought forth by reflection and reason.

---

[93] Heidegger wrote (1999b, p. 298) that this usual interpretation is a story that reads like "a bad novel" because it neglected the questioning of substance and the establishment of mathematical projection that was central to Descartes' philosophical project.
[94] Heidegger argued (1977b, p. 127) that Descartes' philosophical project must be viewed in the light of the Renaissance idea that "man frees himself from the bonds of the Middle Ages in freeing himself to himself."
[95] See Osler (1994, chap. 5) and Shea (1991).

This can also be clearly seen in his posthumous work *Regulae ad directionem ingenii*.[96] Descartes articulated the aim of his life's work to be the project of grounding mathematical intuitionism in terms of its own inner requirements by explicating its rules as the standard of all thought. Descartes' self-appointed task was a work of reflection upon the fundamental meaning of his mathematical intuitionism as a method. This reflection was concerned with the totality of beings and the knowledge of that totality. It was the metaphysical "I" that was presented as the special subject before whom all things present themselves as what they are, and the self-evidence of this truth was premised on the fundamental necessity of the belief that clarity is given by God, from whom all things receive their objectivity. In relation to "the subject," things could then stand as something else, as "objects," but only in so far as they were revealed through the method of mathematical intuitionism and the cognition of that which God had made clear to reason.

Descartes was committed to mechanical physics and natural philosophy, and, consequently, his aim was to establish a philosophy of Nature as a sum of mechanisms governed by mathematical laws, thereby he presumed that all natural phenomena could be explained in terms of matter and motion in geometrical space. As we can see from a letter that Descartes wrote to Florimond de Beaune (dated April, 1639), he described the new physics as nothing more than "mere mechanics."[97] Physics was the application of the mathematical science of mechanics to the task of describing and explaining all natural motion and change, thereby revealing the fundamental mechanisms and laws of Nature, and the task of natural philosophy was to explain how geometrical space and natural law related to matter and motion. Hence, Descartes argued that the Universe is a plenum and that the matter filling it is infinitely divisible, identical with geometrical space, and has only the property of extension. He argued that the property of extension could be understood in terms of *a priori* knowledge—known in terms of the act of mathematical intuitionism—and there is no need for any appeal to experience or observation. Consequently, the first principles of natural philosophy could be known *a priori* through reasoning, which would lead to the discovery of essences, as revealed through mathematical intuitions. Experiences and observations were only required to determine the contingent actuality of phenomena.

Not only did Descartes use skeptical arguments rhetorically in order to clear the way for his assertion of mathematical intuitionism as a foundation of indubitable and demonstrative knowledge about Nature, but he also used theological arguments for God choosing to be bound by the necessity that God had freely created in the physical world. Descartes' theological points of departure were continuous with the presuppositions and arguments of medieval theology.[98] His metaphysical arguments about God's creation of eternal, necessary, and universal truths were situated within traditional discussions about the absolute and ordained powers of God, and these played a formative role in the development and interpretation of the possibilities of the new science. Descartes appealed to the perfection of God in order to justify the possibility *a priori* knowledge of the mathematical laws of Nature: Given that mathematical truths are clear and distinct, and God is good and perfect, then mathematical proofs must provide truths of the physical world. Descartes used this reasoning to establish his characterization of matter in terms of geometrical extension, infinite divisibility, and primary and secondary qualities. These characterizations constituted the fundamental elements of the physical world within Descartes' natural philosophy. His derivation of the existence and content

---

[96] For an English translation see Descartes (1961).
[97] Descartes (1966-76, II) pp. 541-44; for an English translation see Descartes (1983), p. 52, n. 14
[98] Osler (1994), ch. 1 and 5; Garber (1992) pp. 148-55, Rubidge (1990) pp. 27-29; Funkenstein (1986) pp. 179-192.

of these laws from God's attributes required knowledge of God's attributes of perfection and goodness. God was the first cause of motion and always conserved an equal quantity of motion in the Universe according to laws of inertia and impact. Descartes (along with Galileo and Kepler) identified mathematical truths as eternal truths that were central to the natural order of the physical world, and, therefore the form of the mathematical laws of Nature were expressions of God's will and creation.[99] Consequently, Descartes' *a priori* arguments for the eternality, universality, and necessity of God's choice of mathematical first principles and laws and how we could encounter them were premised on mechanical realism, while the goodness of God became the foundation for the possibility of human knowledge of Nature as revealed through mathematical intuition.[100]

Descartes' arguments for God's creation of mathematical truths provided the theological foundation of his metaphysics because if certain fundamental mathematical truths are necessarily true—and God is good and not a deceiver—then we can have *a priori* knowledge of them through our God-given reason. These *a priori* 'indubitable truths' provided Descartes with a foundation for his deductive methodology and his epistemology, upon which mathematical intuition was the means to know God's choices and will. He was able to start from his conception of the *cogito* based on the components of doubting, thinking, and being, to argue the *cogito* was indubitable as a transferable standard of clarity by which the reliability of any subsequent knowledge claim could be made. Descartes argued that if this general rule was true, and it must be, then there is a necessary connection between that which is clear and distinct in our minds and the natural order of the physical world created by God. If the rule were false then God would be a deceiver; this would be in contradiction with the conception of God as a good and perfect God. Unless the Creator was some kind of demonic evil genius (like the Gnostic demiurge), we can trust that which is clear before our minds and our God-given reason and intuition. This standard of truth provided "a general rule that the things we conceive very clearly and very distinctly are all true, but that there is nevertheless some difficulty in being able to recognize for certain which are the things we see distinctly."[101] This cleared a space from within which his mathematical intuitions and reasoning were able to provide arguments to confirm his method of systematic doubting, the *cogito*, the existence of God, the existence of the soul, and the essence of matter.

However, for Descartes, knowledge of the laws of Nature was necessary but insufficient to explain any particular phenomena. As a consequence of the perfection of God, the same laws of Nature would govern any possible world that could be created by God, and therefore, in order to obtain knowledge of this particular world, we require more than the knowledge of the *a priori* mathematical laws of Nature. Even though the mathematical laws of Nature were eternal and necessary, the actuality of the phenomenal world was contingent because the particular implementation of the laws of Nature was contingent upon God's actual choices.[102] In Descartes' natural philosophy, the laws of Nature represent the totality of possibilities of God's choices of phenomena and mechanisms when making the actual Universe. Observation and experiment were required to explain the phenomena of Nature because we needed to know which of all the possible phenomena that God could have created—each consistent with each other as counterfactuals derived from the same mathematical laws of Nature—are actually existent in this world, and which of the several possible mechanisms compatible with the same general law, governed the production of the particular phenomena in question.

---

[99] Osler (1994) p. 127
[100] See Bréhier (1968); Curley (1984); Hatfield (1989); and Garber (1992).
[101] Descartes (1968), p. 54
[102] Descartes (1983) pp. xxvi-xxvii, 85. See also Garber (1978) and Clarke (1982).

While one could know the *a priori* laws of Nature from one's God-given reason and intuition, one could only know through observation and experiment which mechanisms God had used to make the phenomenon in question, as well as which phenomena God had made.

In terms of the possibility of human knowledge, observation and experiment were constrained in terms of what could be made or manipulated mechanically either in practice or in thought. Descartes' scientific method was to produce mechanical analogies (or models) derived from first principles that would produce the same phenomena observed to exist in the world. Observations and experiments could then be used to eliminate deduced mechanical models from the potentially infinite set and provide criteria by which judgments regarding which mechanisms were the actual mechanisms involved in the production of the phenomenon in question. By using "empirical evidence" to eliminate deduced possibilities, except one, Descartes hoped that the demonstrative character of his natural philosophy would be secured. Hence, experiments and observations were not designed to induce or validate universal laws of Nature, but rather to deduce a set of possibilities and to show how natural laws had been applied to produce particular phenomena via natural mechanisms. The purpose of experiments and observation were not to provide data for the induction of theories; nor were they designed to test them. Descartes' scientific method was to deduce possible mechanisms from *a priori* laws of Nature to propose explanatory mechanisms for phenomena, which could be (potentially) eliminated via observation and experiment until only one remained. Through reason, human beings could deduce the set of possible explanatory mechanisms from the mathematical law, and the purpose of empirical investigation was to eliminate all the unexercised possibilities. God was a mechanic in Descartes' theology. Hence, Descartes mechanical realist understanding of the mathematical science of mechanics constituted the basis for his understanding of how we could come to know the actualities (choices) of the creative will of God.

Furthermore, Descartes proposed that, given any mechanism or set of mechanisms, the mechanism at work in producing the phenomena in question could be disclosed through the act of building and operating machines in accordance with mathematical laws; these could be used to determine the truth of any explanation by attempting to reproduce phenomena *artificially*. Once we understood these divine actualities and choices, then, by reproducing natural phenomena artificially, we could become more God-like in our power to bring counterfactuals into the physical world, thereby granting ourselves new God-like knowledge, experiences, and powers. Hence, Descartes wrote,

"we can have useful knowledge by which, cognizant of the force and actions of fire, water, air, the stars, the heavens and all the other bodies which surround us—knowing them as distinctly as we know the various crafts of the artisan—we may be able to apply them in the same fashion to every use to which they are suited, and thus make ourselves masters and possessors of Nature." Descartes (1968: 78)

Descartes' natural philosophy intimately bound together the human ability to make things with the discovery and subsequent implementation of natural mechanisms, thereby synthesizing a conceptual understanding of the artificial and natural as both having isomorphic structures (the action of natural mechanisms on objects) and a unitary origin (natural law). By securing epistemological validity to reproductive power, Descartes was able to connect the knowledge of natural causes with the possession of technical skills. Henceforth, he epistemologically secured the validity of any explanation of natural phenomena in terms of the innovation of technology associated with the application of natural mechanisms in accordance with natural law, and argued that any failure to implement any postulated natural mechanisms in material practices was possibly a failure on the part of the craftsman rather than a failure of the explanation (Descartes, 1968: 91).

Henceforth, natural philosophers and scientists were able to make good rhetorical advantage out of the problem of secondary hypotheses. If we cannot know whether the primary hypothesis (the conjecture under test) was in error, or whether the error lies in any secondary hypothesis (the assumptions used in the test) used in the deduction of observable possibilities (predictions) then there is not any logical "test" of theory at all. The failure to implement a theory in practice can always rhetorically explained away by postulating an interfering mechanism, criticizing the theory, or by criticizing the experiment. Newton used similar arguments regarding his prisms.[103] If a prism failed to resolve the seven-colored spectrum then Newton would argue that the craftsman lacked the skills to make it properly. Newton's definition of a good prism was that it showed a seven-colored spectrum.[104]

Descartes did not work in isolation. In the early seventeenth century, an influential group of self-professed mechanical philosophers emerged and communicated with each other. As well as Descartes, this group included Beekman, Cavendish, Charleton, Digby, Gassendi, Hobbes, and Mersenne. These people established a community of writers who were dedicated to the establishment of the metaphysical foundations of mechanical philosophy, the promotion of the growth of the new mechanical sciences, and the opposition to Aristotelians and the occult. This community had a formative influence on the next generation of experimental philosophers, such as Newton, Boyle, Leibniz, Pascal, Huygens, and Hooke.[105]

Descartes attended Jesuit school with Marin Mersenne at La Flèche from 1604 to 1609, and it was largely through Mersenne's academy that Descartes work became known in France.[106] Mersenne's *Quaestiones celebrimae in Genesim* (1623) *L'impiété des déistes* (1624) laid out his criticisms of Platonism (among other 'heresies'), he defended Galileo in *La vérité des sciences* (1625), proposed his mathematical laws of music and oscillations in *Traité de harmonie Universelle* (1627), and presented his own experimental work on the oscillations of pendula in *Cogitata Physico-Mathematica* (1644). Descartes also frequently corresponded with Isaac Beekman, his teacher and friend since 1618, when Descartes was a soldier garrisoned in Breda. Beekman did not publish any major works but introduced Descartes to Galileo's ideas, and was greatly influential on his ideas and arguments.[107] Descartes also frequently corresponded with Constantin Huygens (father of Christiaan) and his friend Henry More, a Cartesian rationalist who rejected Descartes' mind-body split but used the idea of extension to prove the existence of an immaterial substance or spirit.[108] However, it was during Descartes dispute with Pierre Gassendi that his work became widely known. Gassendi's *Syntagma philosophicum* was published in 1658 and his disputes with Descartes greatly influenced how mechanics and natural philosophy came to be understood in the

---

[103] Cf. Schaffer (1986)
[104] This ambiguity, in itself, is an irresolvable logical problem for Popper's philosophy of science, and it undermines the rationality of his method of falsification as the demarcation of the scientific method from non-scientific methods. Whether any experiment corroborates or falsifies a theory depends on social and cultural factors as well as technical and logical determinations.
[105] Sargent (1995); Shapin & Schaffer (1985); Westfall (1971, chap. 4); Elzinga (1972); Yoder (1988), Westfall (1962).
[106] Dear (1988) pp.12-13; Garber (1992) pp.5-9
[107] Hooykaas (1972)
[108] Osler (1994) pp.118-119

light of its theological assumptions about God and God's powers (—see next chapter).[109] Support for Gassendi's atomic theory of matter and idea of empty space rather than Descartes idea of extension and a continuum, had been published by Charleton in his book *Physiologia Epicuro-Gassedo-Charltoniana, or a Fabrick of Science Natural Upon the Hypothesis of Atoms* (1654).[110]

Descartes' work was introduced to Thomas Hobbes by Kenelm Digby, mechanist and member of the Royal Society, and author of *In the One of Which, The Nature of Bodies; in the Other, the Nature of Man's Soule; is Looked Into; In Way of Discovery, of the Immortality of Reasonable Soules* (1644).[111] Taking Descartes' mathematical intuitionism as the starting point for his rationalism, but rejecting his theological assumptions, Hobbes was a staunch advocate of materialism and considered human beings (including the human mind) to be entirely material and mechanical in structure and nature, and his *The Elements of Philosophy* (1655) laid down ideas about the mechanical nature of the mind and language that 300 years later were to become central to computer science, artificial intelligence, and the computational theory of the mind.[112] In *Leviathan* (1651), Hobbes considered politics to be the ultimate human science and that human beings were best situated to know society because it is of our making, while considering Nature to be a substratum of material forces and substances, against which Man struggles to subdue it for human purposes during the construction of society.

Critic of both Descartes and Hobbes, the poet and writer Margaret Cavendish wrote extensively on natural philosophy and the new sciences, and her *Observations on Experimental Philosophy* was published in 1668.[113] Her fiction *The Blazing World* (1666) is considered one of the earliest examples of science fiction. With the exception of Cavendish, all of the "natural" philosophies produced by these "natural" philosophers were premised upon the operational precepts of mechanical realism. And despite Cavendish's criticisms against the dangers of mechanical realism's reductionism, this operational metaphysics underwrote the subsequent speculative interpretations of Nature in terms of contingency and necessity, the nature of matter, cause, and mechanism, and accounts of the ontology of the world. Mechanical realism had allowed the seventeenth century experimental and mechanical philosophies to be possible. It was then the task of the natural philosophers to build their speculative metaphysics upon those precepts. Their disagreements focused upon concerns about which speculative metaphysics provided the most intelligible account of mechanical Nature and squared with their theological commitments about the nature of God and the possibility of Man making his own paradise on Earth by using scientific knowledge and the practical arts. But, how was this way of thinking possible? Where did this idea come from?

Nowhere can this idea be seen more clearly stated than in *Novum Organum* (1620) by Francis Bacon. In this book, Bacon presented his interpretation of natural philosophy as a new experimental philosophy. He was critical of sixteenth century arts, intellectual sciences, and philosophy and reserved his praise for mechanics and the practical arts. He considered Greek science to be childish, due to their basis on "bland and specious generalities" that lead only to "disputes and scrappy controversies" and "almost stopped

---

[109] Lennon (1993) and Osler (1994)
[110] Sharp (1973)
[111] Foster (1988)
[112] Mintz (1969); Spragen (1973); Haugeland (1985); Fodor (1994); Dupuy (2000)
[113] Mintz (1969, pp. 3-5) and Kargon (1966, chap. 7)

in their tracks," and he praised the mechanical arts for progressing.[114] Instead, he proposed

"the production of a Natural History by making a history not only of Nature free and unconstrained (when Nature goes its own way and does its own work), such as a history of the bodies of heaven and the sky, of land and sea, of minerals, plants and animals; but much more of Nature constrained and harassed when it is forced from its own condition by art and human agency, and pressured and molded. And therefore we give a full description of all the experiments of the applied part of the liberal arts, and all the experiments of several practical arts which have not yet formed a specific art of their own."[115]

Knowledge was obtained by constraining and molding Nature—intervening and seeing how the phenomenon responds. Bacon argued that the mechanical arts were founded on "natural axioms induced from experience," discovered by observation and subtle, patient, ordered movement of hands and tools, and he praised them for providing a "variety of objects and splendid equipment," having "contributed to human civilization." If directed according to their utility, the mechanical arts were capable of growth and flourishing. He cited the clock as an example of "a subtle and precise thing that seems to imitate the celestial bodies in its wheels, and the heartbeat of animals in its constant, ordered motion; and yet it depends on just one or two axioms of Nature."[116] He considered the mechanical arts in general to be praiseworthy as the source of civilization and political advantage, and singled out the discovery of the art of printing, gunpowder, and the nautical compass. The mechanical arts were the noblest human pursuit and "right reason and sound religion would govern its use."[117] He put it quite dramatically (and reductively):

"There remains one hope of salvation, one way to good health: that the entire work of the mind be started over again; and from the very start the mind should not be left to itself, but be constantly controlled; and the business done (if I may put it this way) by machines."[118]

For Bacon, the practical value of the new sciences was central to the whole enterprise of experimental philosophy from its very beginnings. Bacon proposed that humanity could better its conditions and be liberated from the dictates of "its organic state" by using the mechanical arts—"the happy match between the human mind and the nature of things"—to provide inventions (such as gunpowder, the compass, and the printing press) that would establish humanity as the master of Nature. Natural philosophy was underwritten by the science of mechanics and a societal faith in the possibility of constructing a better world within which scientific knowledge would liberate humanity from natural forces and the pernicious influence of superstition. For Bacon, the new sciences promised such things as the prolongation of life, the mitigation of pain, the retardation of age, and the restitution of youth, as well as providing us with certain and objective knowledge. The new sciences were to create a utopia, within which the scientists, as the denizens of Solomon's House in *The New Atlantis* (1627), were to direct

---

[114] Bacon (2000: 6-7)
[115] *Ibid* 20-1
[116] *Ibid* 69-70
[117] *Ibid* 100
[118] *Ibid* 28

their efforts towards providing human beings with new powers, making new experiences possible, and bringing everything within our reach.[119]

From the beginning of the seventeenth century onwards, the new experimental philosophy premised upon mechanical realism promised that the technological manipulation of Nature would empower and benefit human beings. Once mechanical realism was presupposed, the task of acquiring and testing knowledge became one of refining representational and material practices, identifying and associating interventions and consequences, and explaining them as the realization and exercise of mechanisms in material practices, and applying this knowledge to practical problems and further experimentation. From the outset, the new sciences were represented as attempts to understand the nature of matter and natural forces by representing material practices in technical forms (geometrical proofs, diagrams, and tables) that identified a mechanism or set of mechanisms, and these technical forms were tested by implementing as representations in the ongoing refinement of the new sciences.

Once human beings reduced knowledge to forms that we can cognate, manipulate, and control, thereby treating natural phenomena as machine performances, the phenomena is reduced to that of the performance of the apparatus, and only those aspects of the phenomena amenable to this reduction are considered to be natural properties; this reduces epistemology to the justification of techniques and methods, and through which the natural phenomenon is appropriated and mediated within a technological framework of techniques and representations. This filters (informs and constrains) the possible representations and explanations available for "testing." Only those that can be manipulated, measured, and reproduced by the apparatus can be investigated. Theoretical explanations are "tested" via the reproduction of machine performances and the interpretation of them as something objectively before us, graspable as "a mechanical system" through representational and material practices. Theories are developed and given "concrete meaning" in accordance with the practical problems of ongoing scientific research, as well as the problems of wider society. This promises us the certainty and power of knowing and manipulating the world in terms of representations that we have created for ourselves on the basis that they have been instrumentally tested in material practices, thereby finding their confirmation in the new experiences and powers they unleash.

In so far as Bacon's experimental and natural philosophy advanced its methodology of testing hypotheses by building and manipulating machines then it appropriated values from the crafts, reduced knowledge to know-how, and relied on the skill and knowledge of the craftsman.[120] Via the cultural representation of mathematics as a universal and eternal science, alongside the patronage of monarchs and merchants, the new sciences emerged as the means to produce certain knowledge and the satisfaction of civic, commercial, and military ambitions.[121] The know-how of craftsmen and practitioners was represented in terms of mathematical and rational principles of mechanics, thereby transformed into "true knowledge." When coupled with the patronage of political, economic, and military powers, this cultural interpretation elevated the status of the mechanical arts out of the banausic and prepared the way for the Scientific Revolution.

---

[119] Bacon (1989); see also Price (2003) and Rogers (2006)
[120] Bennett (1986)
[121] Eisenstein (1979); Long (1997)

The awakening (enlightenment) of "Man" became represented within the Scientific Revolution and scientific worldview as being the recognition of the truth that mechanical realism provides the foundation for a rational understanding of the methods utilized to discover all natural structures, relations, and changes. The Scientific Revolution was much more than a struggle against the authority of the Church, the divine right of kings, and superstition. Of course, within the seventeenth century, it became a process of establishing a means of rhetorically using the mechanical worldview as a culturally shared standard by which human authority could be shown to be arbitrary, and a matter of custom and contingency, but these arguments were based on a fundamental faith in the progressive nature of scientific knowledge as being something that would liberate human beings from ignorance, squalor, fear, and vulnerability through science and technology.[122] Nature became represented as the objective source of all technological power, and something that must be overcome through the discovery of how it works; technology became the neutral arbiter of its own objective truth, and increased technological power became the standard against which all other natural philosophies could be criticized and rejected, while this standard was cloaked with and concealed under appeals to experience, reason, and knowledge. Through this metaphysical interpretation of how the practical activity of human beings is possible, Nature is represented as being implicated in its own modification, and this, in turn, led to the eighteenth century empiricism and the nineteenth century developments of technological determinism and positivism (once mechanical realism and the neutrality of technology had become "self-evident"). The rationality of the scientific worldview and the whole cultural project of developing the experimental sciences as empirical and natural sciences entails the presupposition of those precepts. Without the metaphysical precepts of mechanical realism, this epistemological interpretation of technological activity (along with the categorization of the experimental sciences as natural sciences) would be arbitrary—absurd and irrational.

From the beginning of the Scientific Revolution, the mechanical worldview was developed and refined alongside the innovation of methods, techniques, and instruments, and the relations of quantities and constants associated with interventions and consequences provide the mechanical realist interpretation of representational and material practices with its objectivity. Within the mechanical worldview, human beings and the modifications of the natural world that human beings impose upon the preexisting order of things are taken to be a part of Nature because human beings and our capacities to change our environment through interventions are taken to be natural. This naturalizes technological activity alongside the mechanization of Nature, and the limits and actualities of labor become the frontier for new scientific research and abstracted in terms of natural laws, whereby labor can be removed from natural philosophy and thereby reveal how Nature works.

However, mechanical realism did not begin with Bacon and Descartes. To understand how mechanical realism arose, we must attend to the influence of Galileo on the seventeenth century natural philosophy. How and why did this happen? To answer this question, we need to look at how and why the work of Galileo presupposed mechanical realism, and from whence he gained these precepts. Galileo's most well-known statement of his natural philosophy comes from *The Assayer* (first published in 1623) in which he asserts the necessity of mathematics to read "the Book of Nature":

"Philosophy is written in this grand book, the Universe, which stands continually open to our gaze. But the book cannot be understood unless one first learns to comprehend

---

[122] Rogers (2006)

the language and read the letters in which it is composed. It is written in the language of mathematics, and its characters are triangles, circles, and other geometrical figures without which it is humanly impossible to understand a single world of it; without these, one wanders about in a dark labyrinth."[123]

How should we interpret this mathematical intuitionism? From whence did Galileo draw his inspiration and intellectual resources?

Sources of Galileo's Mechanical Realism

Some traditional philosophers of science would have us believe that Galileo's commitment to geometry revealed an inherent Platonism, yet this does not explain how Galileo's reduction of the natural to the mechanical via mathematical representations was conceptually possible and intelligible, given that it is quite antithetical to Plato's philosophy with its concern for human wisdom and virtue. Yet, understanding the roots of mathematical intuition helps us to see how mathematical reasoning was fundamental in Galileo's natural philosophy, while also giving us some insight into why some philosophers of science have considered modern science to be founded on a Platonic or Neo-Platonic metaphysics.

This idea that Galileo's metaphysics was Platonic is largely based on the somewhat superficial insight that the dialogic form or style used Galileo was popular in the Renaissance, with its rediscovery of Plato, alongside how the mathematical intuitionism of Galileo's work compared with Plato's arguments for the truth of geometrical demonstrations as a standard by which the clarity of higher thinking (of the Forms of justice, truth, and beauty) could be judged. The shared belief in a heliocentric Universe between Plato and Galileo (following Copernicus) is also seen to be evidence of a shared metaphysical worldview.[124] These philosophers have ignored Plato's theory of recollection (and its appeals to reincarnation and experience of the divine realm of forms when the soul is between lives) and the mysticism it presupposed. The case of the Church's trial and execution of Brother Giordano Bruno for the 'heresy' of teaching the Copernican system is also cited as the sixteenth century Platonist predecessor of Galileo and a Renaissance herald of the new natural philosophy, but what is ignored is that Bruno was executed for the 'heresies' of his beliefs in reincarnation, denial of the Holy Trinity, and his claim that Jesus Christ was an imposter and Moses was a magician. It also ignores the fact that the Copernican system was taught in Catholic schools and universities at that time.[125]

Once we pay attention to the actual dialogues themselves, we can see that there is only a superficial similarity between Plato's dialogues and those of Galileo. Indeed, the legend "let no one who does not know geometry enter here" may well have adorned the lintel of the entry portal to Plato's Academy, but in Plato's *Republic* mathematics is considered to be a rudimentary part of a child's education designed to train reasoning skills and elucidate the nature of clarity, while being only a first step of the philosophical task of reasoning and the recollection of the forms of justice, truth, beauty, and goodness. Mathematical reasoning is always logical and true, and deduced from a few simple axioms by the use of reason—as Socrates demonstrated in *Meno* by showing that even an

---

[123] Galileo (1975)
[124] Koyré (1992); Whitehead (1997)
[125] Bruno (1998), first published in 1584; see also Kuhn (1957)

uneducated slave boy could deduce the axioms, postulates, and the whole of geometry if asked the right questions and encouraged to exercise his reason—but nowhere in any of Plato's dialogues is knowledge reduced to the mathematical, even though (as Socrates argued in *Theaetetus*) mathematical knowledge is true, universal and eternal.

The idea that Platonism is the metaphysical foundation of Galileo's physics is little more than a fiction based on appeals to Plato's intellectual prioritization of reason over experience, the heliocentric Universe, the importance of mathematics for the education of children. The claim that Galileo's physics was Platonic in origin is largely based on the similarity between the rational cosmos of the *Timaeus* dialogue and that of the seventeenth century mechanical worldview. The focus of this dialogue is on the idea of the divine creation of a rational universe and the significance for this on how we should understand the nature of justice (harmony, rationality), and this dialogue should be seen as one dialogue in an ongoing debate and demonstration of rhetoric (Timaeus' speech follows Socrates' speech in *The Republic*, and was followed by *Critias*) but that is all swept aside and ignored by the traditional philosophers of science. The idea of experimentation as a means to gain knowledge of the axioms of Nature would have been nonsensical to Plato, and for Plato there was an important mystical connection between reason and the divine—Socrates' *daemon* or inner voice—that is quite frankly absent from any of Galileo's arguments and proofs, and has no part to play in scientific methods.

Certainly Plato considered it absurd and naïve to believe that attending to experience and the practical arts could be the route to know the Forms. For Plato, only reason itself can reveal the forms to us by recalling knowledge that already resides in our souls. One cannot know the form of a circle from experience because there are no perfect circles in the physical world; the form of a circle, which is shared by all things circular, is brought forth by reasoning and recollected rather than abstracted from experience. If the form of a circle transcends the totality of circular things and it does not exist in the physical world, from whence is the form brought forth from by reason? Socrates argument in the *Republic* was that one encounters the Forms in a transcendental and divine realm which one passes through during the moment between death and rebirth. Reincarnation follows the soul's journey through heaven or hell, and the lottery of the Fates that connects the transcendental realm and the physical world through rebirth. The shock of rebirth leads one to forget that moment of transcendence, but one can recall encountering the forms through reasoning and it is only on the basis of this act of recollection that one can know them and see how the physical world only contains imperfect copies.

The character Timaeus in the dialogue *Timaeus* was likely to have been a fictional creation of Plato and it is certainly a bit of a stretch to claim that Plato was a proponent of the mechanical worldview of a rational cosmos simply because he put this worldview into the mouth of one of his characters. Why should we decide that Timaeus spoke for Plato in a way that Socrates or Critias did not? It is significant that the character Timaeus was a stranger, an astronomer from Locri (a city in southern Italy) and, in the text of *Timaeus,* we see that the Athenians (including Socrates) considered Timaeus' speech to present a strange and implausible worldview of the Universe as "a shaking machine," but was treated by all as a "likely story" and judged according to its rhetorical merits as a speech. There is no reason to believe that Timaeus' "shaking machine" was Plato's worldview, just as there is no reason to believe that Plato believed that the city of Atlantis sank into the ocean or that the Athenians were born of Earth and Hephaestus, even though these are also proposed as historical truths in the dialogue.

Largely the influence of Plato's *Timaeus* on the Renaissance was due to the fact that it was one of the first of Plato's dialogues translated into Latin and it presented the idea of *the demiurge*. This idea of God as a maker of the world (a divine craftsman, perhaps modeled on Hephaestus)—who created other gods who imperfectly copied the transcendental forms into formless matter—was an idea that certainly influenced Renaissance theologians (just as it had influenced second century Gnostic thought), and may well have influenced Descartes' seventeenth century conception of God, whereby the world can be seen as a machine made by God and governed in accordance with mathematical laws (or ratios) that God does not change. While this idea of matter as some formless, plastic stuff into which form can be imposed has clearly influenced the modern conception of the material world, especially within architecture, engineering, and artificial intelligence, Platonic thought had more influence on the philosophy of science than it has on science itself. Traditional philosophies of science share the presupposition of the Platonic view that rational thought, logic, and mathematics transcend the material world—even though the nature and possibility of that transcendence and its relation to experience remains vague and mysterious—but seventeenth century "natural" philosophers extended this presumption to assert that the primary relationships between the mind and the world are those of cognition, manipulation, and control. In other words, they technologized Creation to make the mathematical science of mechanics into the interface between Man and Nature, via a conception of God as demiurge, while simultaneously founding the objectivity of that relation on a Cartesian theology (after removing all references to God and making), and thereafter, in the absence of any other philosophical tradition to appeal to, declaring the mathematical intuitionism of Galilean physics to be based on Platonism.

Historians of science and technology, however, show that the most influential ancient sources for the science of mechanics were the works of Heron, Archimedes, and Pseudo-Aristotle, rather than Plato or Pythagoras, and if there is any Platonic influence, it is largely limited to the philosophy of science and may be found in the traditional philosopher's dismissal of technology to the realm of the banausic. This (false) conjecture about the Platonic origin of science has further concealed how technology has shaped the development from Galileo's mechanical physics to contemporary studies into quantum physics, genetics, molecular chemistry, biochemistry, astrophysics, artificial intelligence, cybernetics, robotics, and neuroscience. It would be better to describe the origin of Galileo's physics as Euclidean rather than Platonic (or Pythagorean). Euclid was greatly inspired by Theaetetus (the Pythagorean mathematician and fallen warrior discussed in Plato's dialogue of the same name) and almost a quarter of Euclid's *The Elements* (all of Book X and XIII) is concerned with Theaetetus' ideas about incommensurable lines and areas, and the construction of five regular solids (pyramid, cube, octahedron, dodecahedron, and icosahedron) within a given sphere. It was these ideas that inspired Johannes Kepler to develop his harmonics model of planetary orbits and law of the elliptical paths of the orbit of planets around the Sun. While the attribution of these ideas to Plato is a reconstruction of Renaissance Platonism and Neo-Platonism, this ignores the fact that Euclid's own sources pre-existed Plato and largely appeals to the belief that Plato was a mathematical rationalist who seriously proposed the mechanical cosmos of the *Timaeus* dialogue as his theory of the Cosmos.

How did Galileo encounter Euclid? One of the most valuable ancient sources for Galileo's physics was the works of Archimedes, rather than the works of Plato. Archimedes had influenced many thinkers throughout Europe. In the thirteenth century, Roger Bacon had invoked Archimedes against those who did not "dare to know" and, in the sixteenth century, Commandino wrote, "with respect to geometry no one of sound mind could deny that Archimedes was some god."¹²⁶ Archimedes' fame as an inventor of fantastic machines was widespread in the fifteenth century, largely through the account of Plutarch's *Life of Marcellus*, first published in the late first century A.D.¹²⁷ It was first published in Italian in 1571 (in English in 1688 and in German in 1852). Although famous for his mechanical inventions, there are no surviving texts directly attributed to Archimedes that contain his reputed devices; only second and third-hand accounts remain. Plutarch claimed that Archimedes destroyed all his designs for machines because of the ignobility and danger of such records.¹²⁸ His only remaining description of a mechanical device is his orrey to model the mechanical motion of the heavens. His surviving works are all mathematical and include the geometrical solutions to the sphere, cylinder, conoids, spheroids, the equilibrium of planes, spirals, buoyancy, quadrature of the parabola, the diameter of the Earth, numbers, square roots, irrational numbers, arithmetic, a method of integral calculus, the diameter of the Universe, probability, solids, center of gravity, and measurements.¹²⁹ There are references to lost works on polyhedra, numbers, balances and levers, gravity, optics, the mechanical motions of heavenly bodies, parallel lines, circles, triangles, and machines. In the *Method* (addressed to Eratosthenes—discovered by Heiberg in 1906, copied onto tenth century parchment with the final leaves written on sixteenth century paper), Archimedes described "a certain method, by which it is possible for you to get a start to enable you to investigate some of the problems in mathematics by means of mechanics" and wrote "for certain things first became clear to me by a mechanical method, although they had to be demonstrated by geometry afterwards because their investigation by the said method did not furnish an actual demonstration."¹³⁰ This mechanical method had wide appeal to the medieval and Renaissance mechanists and mathematicians, and Archimedes' principles were put into practice throughout northern Italy during the fifteenth and sixteenth century.¹³¹

It is beyond doubt that the works of Archimedes had a profound influence on Galileo. This profound influence can be read from Galileo's own words in *Du Moto* (*On Motion*), written in 1590 (but unpublished until 1883): "I cover myself with the protecting wings of the superhuman Archimedes, whose name I never mention without a feeling of awe."¹³² In this work, he presented his physical analogy of the balance to describe the motion of naturally moving bodies and for the motion of a body on an inclined plane and a pendulum. He wrote *On Mechanics* in 1593-4 (published in French in 1634 and in Italian in 1649) presenting Archimedean geometrical treatments of the simple machines. He started from his premise that all simple machines could be reduced to a problem of an Archimedean balance, which became fundamental to Galileo's physics.¹³³ This argument was based on the Archimedean principle that all machines operate on the same physical principles so a complete understanding of any one of them is adequate for the deduction of the mechanical properties of all the others. Having chosen the balance as fundamental and used it to derive the laws for an inclined plane, the lever, the windlass, the capstan, the pulley, and the screw, Galileo constructed a "dynamic equilibrium" method as the basis of all his physical explanations.

In 1586 Galileo had constructed a hydrostatic balance following Archimedes' geometrical arguments to determine accurately the relative amounts of two metals in an

---

[126] Clagett (1978) p. 1225
[127] Plutarch (1961); see also Authier (1995)
[128] Plutarch, (1961: 4-5)
[129] Heath (n.d.)
[130] *Ibid* 13
[131] Keller (1971)
[132] Galileo (1960) p. 67
[133] Machammer (1998)

alloy mixture, which he described (in Italian) in a paper published in 1644.[134] Galileo also studied the Archimedean concept of "the center of gravity" and wrote a paper (in Latin) on "Theorems about the Centre of Gravity in Solids." This paper was unpublished until printed in 1638 as an appendix to the *Two New Sciences*. In this paper, Galileo argued that by treating a falling body as a body rising, falling, or floating in a medium, then, in Archimedean terms, such a body was "reduced to weights of a balance."[135] He used this method in his treatment of hydrostatic phenomena in *Discourse on Floating Bodies* (1612) and in his *Dialogue on the Two Chief World Systems* (1632). He used this method to describe motion as separated into two independent horizontal and vertical axes to describe the fall of a body from a moving point as that of a parabola. He argued that the Earth could revolve around the Sun "without the breath being snatched from our mouths, nor birds being flung from out of the sky."[136] In his last work, *Two New Sciences* (1638), he concentrated on explaining natural motion using his experiments on motion of weights down an inclined plane. While it has been questioned whether Galileo actually performed any of these experiments, he used the pendulum experiment and the balance *reductively* as exemplars for the description of all natural motion, and therefore the fundamental representation of motion.[137]

Galileo probably first became acquainted with the works of Archimedes in 1583 through the Tuscan court tutor Ostilio Ricci, who was a pupil of Nicolò Tartaglia (1500-57), who, in 1543, had translated the works of Archimedes into Latin.[138] Tartaglia taught perspective, architecture, and in 1537 published his mathematical science of ballistics. Tartaglia's famous books are *Nova scientia* (1537) and *Questi et inventioni diverse* (1546).[139] He also taught mathematics, surveyed land, designed fortifications, made maps, and invented mathematical instruments. His studies included arithmetic, geometry, music, astronomy, perspective, and architecture. He had translated Euclid's works into Italian in 1543 and, in 1551, he published his Italian translation of Archimedes' *On Floating Bodies* and, using Archimedes' hydrostatics, Tartaglia derived and proposed a method of re-floating wrecks.[140] He had also studied and translated Pseudo-Aristotle's works into Italian and declared in *Questi* Bk. VII that mechanics based on the principles of weight was the cause of every ingenious mechanical invention.[141] He argued that arguments about Nature could only be based on experience, whereas abstract arguments about mechanics should be based on mathematics. This led him to assert that arguments based on mechanics were superior to those based on mere observation because reasoning based on mathematics was more rigorous than reasoning based on experience.

When observation and mechanics did not agree then the notions of "error" or "material hindrances" could be used to explain the discrepancy.[142] Tartaglia had invented the method of *transdiction*.[143] This form of inference constructs an explanation of a phenomenon in terms of an unobserved mechanism in order to explain the deviation of an observation from theoretical expectations. This method was central to the use of mechanics as an explanatory tactic, and, by adopting this tactic, Tartaglia had preempted the mechanical philosophers of the seventeenth century. Galileo frequently used this

---

[134] A translated version of this paper, "The Little Balance," can be found in Fermi & Benadini (1961)
[135] Galileo (1960: 38)
[136] Settle (1967)
[137] Naylor (1989); Cantor (1989)
[138] Seegler (1966: 4)
[139] Translations of parts of *Questi* Bk. VII and all of Bk. VIII in Drake & Drabkin (1969: 104-43)
[140] Laird (1986: 52-3); Clagett (1978: 508-607)
[141] Laird (1986: 52-3); Clagett (1978: 508-607); Wallace (1984: 203-5)
[142] *Questi*, fols. 78r-v, Drake & Drabkin (1969: 105-7)
[143] Osler (1994: 117); Mandelbaum (1964)

tactic; the discrepancies between the path of a cannonball and a parabola, for example, could be transduced as the mechanical consequence of the invisible force of friction. A subsequent mechanical experiment could be constructed to demonstrate friction, and, due to the presumed universality of such a demonstration—the presumption due to the conception of motion in terms of rectilinear and universal motion—it could be taken to have disclosed the reason for any failure of mathematical description to match experience in different experiments, while simultaneously affirming the truth of that mathematical description.

Tartaglia accepted the Aristotelian classification of mechanics as a "sub-alternated science" because its method was abstract mathematical demonstration but its subject was physical and consequently both mathematics and experience were required in the development of mechanics. He argued that mechanics provides knowledge to calculate the strength (*virtù*) and power (*potentia*) of any machine to augment the strength and power of men by any degree.[144] Tartaglia attempted to inscribe a formal mathematical treatment of mechanics by combining the statics of Archimedes with the dynamics and kinematics of Pseudo-Aristotle. He was unsuccessful because he could not consistently combine the Archimedean proofs based on equilibrium and the Aristotelian arguments based on velocities. However, Tartaglia had laid down the challenge to his sixteenth century Italian contemporaries.

Francesco Maurolico was the first of Tartaglia's contemporaries to take up this challenge.[145] He already had established his reputation in astronomy and optics and by translating and commenting on the works of Euclid, Archimedes, and Pseudo-Aristotle. In *Problemata mechanica cum appendice* he discussed the scope and classification of mechanics within the sciences. Maurolico died in 1575 and *Problemata mechanica* was posthumously published in 1613.[146] He listed mechanics (along with music, astronomy, perspective, geography, architecture, painting, sculpture, stereometry, and cosmography) as an intermediate science between the mathematical and the physical that was distinct from the secular arts, and considered mechanics to be a part of "contemplative philosophy" due to its mathematical aspect. He argued that the dynamics and kinematics of Pseudo-Aristotle's mechanics had to be based on "the doctrine of equal static moments" and consequently mechanics also had to be based on Archimedean principles.

It is this goal of combining Archimedean statics and Pseudo-Aristotle's dynamics that inspired many sixteenth century Italian mechanists. The sixteenth century Italian aristocrat and a military engineer Guidobaldo used Archimedean techniques to solve the problems set by Pseudo-Aristotle starting from the lever and then on to the rest of the six simple machines. Guidobaldo's *Le mechaniche* was published in 1581 and *Liber mechanicorum* was published in Latin 1577 and Italian in 1588.[147] He aimed to establish mechanics as a branch of rigorous axiomatic geometry and claimed that any machine based on such a mechanics would work in the real world. Another sixteenth century Italian military engineer, Giulio Savorgan was also inspired by Archimedes, innovated Italian town fortifications, developed mathematical explanations of mechanical advantage, and invented "Archimedean devices" to aid the lifting and transportation of heavy cannons.[148] He invented light, robust, and powerful lifting-gear based on spur gears,

---

[144] *Questi*, fol. 82v; Drake and Drabkin (1969: 111)
[145] Laird (1986: 53-5)
[146] The preface and several of the problems can be found in Clagett (1978: 784-7)
[147] Drake and Drabkin (1969: 239-328); Wallace (1984: 206)
[148] Keller (1975: 21-32)

worm gears, rack-and-pinion, block-and-tackle, winch and pulley, screw jacks and ratchet-jacks.

The Aristotelian mathematical science of mechanics was established in Italy through the influence of the university at Padua.[149] Since the fourteenth century, Padua had been a center for mathematical subjects (including astronomy, astrology, geometry, optics, and geography) and was the first Italian university in the sixteenth century to offer lectures in mechanics from the chair of mathematics. The introduction of mechanics into the curriculum at Padua in the 1560s began in the form of lectures on Pseudo-Aristotle's work. The elevation and establishment of mechanics from the banausic to the academic occurred at Padua through the influence of mathematically educated Aristotelian scholars such as Niccolò Lenico Tomeo and Alessandro Piccolomini. Tomeo was a professor of philosophy at Padua from 1497 to 1509.[150] He translated Pseudo-Aristotle into Latin and published a commentary in 1525. Piccolomini taught moral philosophy and mechanics at Padua in 1539.[151] Guidobaldo studied there in 1564 and Bernardino Baldi from 1573 to 1575. Baldi considered mechanics to be a "sub-alternated science" due to its physical subject matter described in terms of geometrical proofs.[152] In his view, mechanics was consequently of an equal status to optics, music, and astronomy. In his treatment of mechanics, he followed Maurolico and formulated his solutions to the problems raised by Pseudo-Aristotle in terms of Archimedean and Aristotelian concepts of statics, dynamics, equilibrium, motion, power, and impetus. Pietro Catena was the first lecturer in mechanics at Padua and gave lectures between 1564 and 1573. He was professor of mathematics from 1547 to 1576. His successor was Giuseppe Moletti.[153]

Moletti followed his Aristotelian predecessors and classified mechanics as "contemplative philosophy" of mathematical principles of statics, dynamics, and kinematics. According to Moletti, the task of mechanics was to demonstrate the most efficient means of performing the maximum amount of work with the minimum of effort. For Moletti, mechanics was a science and not an art because the geometrical first principles of mechanics were "necessary and eternal," whereas the arts were contingent upon human ends and values. The end of science was the knowledge of causes and truth, whereas the end of the practical arts was productive work. He argued that the first principles of mechanics were natural causes, mechanics was to be found in all the works of Nature, and these natural causes were governed by natural laws. Moletti transformed the traditional classification of mechanics from a "sub-alternate science" to a "natural science." While he still considered mechanics to be "intermediate" between the geometrical and the physical, he argued that it was based on both mathematical and natural principles and truths.[154] As professor of mathematics at Padua between 1577 and 1588, he paved the way for Galileo's mechanical realist physics. In 1592, Galileo succeeded Moletti as professor of mathematics at Padua.

---

[149] Schmitt (1975, 1983); Grant (1987)
[150] Laird (1986: 48-9); Rose & Drake (1971: 79-82)
[151] Laird (1986: 49); Rose & Drake (1971: 82); Suter (1969)
[152] Laird (1986: 56-7); Rose (1975: 248-51)
[153] Laird (1986: 60-2)
[154] Moletti's arguments can be found in *In librium mechanicorum Aristotelis expositio tumultaria et ex tempore*. Milan, Biblioteca Ambrosiana, MS. S 100.

Blurring the Aristotelian Distinction between Art and Nature:

With the exception of Archimedes and tales of the legendary Daedalus, there was a notable lack of texts on experimentation within ancient Greece.[155] The knowledge of Nature was distinct from the knowledge of the arts. For the ancient Greeks, art (crafts; the power of making) was distinct from Nature and belonged to human nature, as something god-given to human beings; or in the case of the myth of Prometheus, stolen from the gods and given to humans to make up for an oversight on the part of the gods. The origin of this power was something mysterious, divine, and magical, and largely taken for granted. It is no coincidence that Hephaestus, the Greek god of making, was born of the goddess of magic, Hecate, and of an unknown father. In ancient Greek, *techne* (plural: *technai*) loosely meant practical knowledge of the arts, crafts, or the organization of practical activity, such as farming or animal husbandry. Technique and methods applied in appropriate contexts would be included as examples of *technai*. It also meant "device" or "ploy" and had a similar meaning to "crafty" and "artful." In ordinary usage, *techne* referred to "cleverness" and "cunning" in getting, making, or doing all manner of things. It did not only apply to arts and crafts, but also as to trades, crafts, and skills of every kind. Athletics, warfare, commerce, rhetoric, and politics all involved *technai*. Techne involved a collection of tactics, stratagems, and tacit "know-how," as a loose collection of kinds of activity required to achieve specific ends or goals, and it was largely learned by mimicry, trial and error, and from experience.

This kind of usage is apparent in the 4th century BC Attic Greek writings of Xenophon—a student of Socrates and contemporary of Plato. In *Memorabilia* and *Oeconomicus*, Xenophon used *techne* interchangeably with *episteme* (certain knowledge of first principles or axioms) when discussing the arts of statesmanship and household management, and Xenophon's Socrates even treated both reason and virtue as if they were kinds of *techne*. Xenophon's Socrates considered *episteme* of the cosmos to be impossible (beyond human comprehension) and devoted himself to the questions of the limits of knowledge and how to live a good life. The emphasis on human wisdom and the limits of reason and knowledge are central themes for Plato's Socrates too, but in Plato there is a deeper and more mystical concern with divine harmony, justice, and the good. It was the focus on the good (especially in the light of the limits of knowledge and reason) and the blurring of the distinction between *techne*, *praxis*, and the intellectual virtues (especially *phronesis* or practical wisdom) that was later to become central to the ideas of the Stoics, such as Zeno, Diogenes Laertius, and Cato, and the development of pragmatic philosophies of human nature and how to live a good life.

Indeed, it was in the philosophical writings of Plato that *techne* was first treated as an abstract and formal kind of knowledge, which could be used as a theoretical and universal guide to govern making something or organizing practical activities with specific ends in mind. Plato used *techne* to refer to the general, abstract, and communicable first principles of making, measurement, and inscription for both cognitive and material practices of any kind. *Techne* and *episteme* (knowledge of eternal and necessary principles or axioms) were closely related by Plato when discussing art and knowledge in general, and were used interchangeably to characterize medicine[156] and mathematics[157] when quantities or ratios were involved. Plato used *episteme* to describe mathematical truth as eternal and necessary knowledge, while also using the word *techne* to describe mathematics (including logic, arithmetic, and geometry) as the highest form of art—something good

---

[155] Sambursky (1987); Waterlow (1982)
[156] *Charmides* 165c
[157] *Philebus* 55c-56d; *Gorgias* 450b-c; and *Ion* 532

for its own sake—while lower forms of art (carpentry, navigation, surveying) apply knowledge and mathematics to material practices to achieve specific ends—for the sake of something else.[158] *Episteme* referred to knowledge of fundamental causes, axioms, or principles, and was reserved to denote pure theory or any knowledge that did not relate to the material world in a practical or useful manner.[159]

Plato's Socrates often argued that all *technai* are involved with *logoi* (words, speech, reason, principles) bearing upon some specific subject matter of the art in question, even though some require a great deal of physical exertion and very little reason (i.e. horse riding, painting, or sculpture) and others require a great deal of reason and very little exertion (i.e. arithmetic, logic, or astronomy). *Techne* was the knowledge of all productive activities that could be reasoned about and taught, and any productive activity (*poiesis*) needed to be teachable through *logos* in order to qualify as *techne*. It was understood as the logical and communicable knowledge regarding the causal principles involved in making or doing something, thereby bound up with change within the physical world. *Techne* could either proceed by conjecture and intuition based on practice, training, and instruction (as in music, medicine, or agriculture), or it proceeded through the use of numbering, measuring, or weighing during material practices (as in tree felling, carpentry, and building a house). The mathematical activities of numbering, measuring, or weighing were taken to be the most truly *technai* because they were taken to involve the greatest precision and were more closely associated with mathematics and understood as *episteme*. Reasoned activities that operated by guiding acts of making through the use of mathematics constituted the *techne* of such activities; these provided the formal knowledge and rules by which material practices were performed, governed, and understood (*gnosis*). Only routine activities devoid of *logoi*, unreflectively based upon experience, mimicry, and habit (i.e. cooking or speaking), were considered by Socrates to be *atechnos* (devoid of art) and were learned through training (mimicry and repetition), experience (trial and error), and *praxis* (habit and routine). Of course activities such as cooking or speaking can be (and are) developed into arts, but for everyday purposes they usually are not. Such everyday practices were *alogos* (without words, reason, or principles). What such activities lack, according to Socrates, was knowledge of the *aitai* (intelligible causes) involved in what was made or done.

Aristotle, following his teacher Plato, also defined *techne* as a kind of knowledge of making or production that informed material practices by using general principles to guide those practices. He used the word *techne* to refer to any theoretical knowledge concerned with making that was explanatory, generalized, abstract, formal, and communicable.[160] *Techne* was an intellectual virtue (the other four being *episteme*, *phronesis*, *sophia*, and *nous*) that comprised "a true course of reasoning" (bringing together *techne* and *logos*) that guided stable dispositions to make particular things or bring about a state of affairs in a specific manner.[161] *Techne* was induced from otherwise unarticulated and particular experiences (*empeiria*) into communicable, formal, and general knowledge of the first principles (intelligible causes) involved in making or changing something; it was to be used to reason about how to make particular things in a specific manner in accordance with specific goals and appropriate tools and materials.

As an intellectual virtue, *techne* was inextricably bound up with an intellectual grasp (cognition) of first and intelligible causes that provided the kind of knowledge possessed

---

[158] *Gorgias* 503d-e; *Sophist* 253a-259e
[159] *Statesman* 258e; *Theaetetus* 186c
[160] *NE* 6.4; *Metaphysics* 1.1; *Rhetoric* 1.2
[161] *NE* 6.4.1140a11

by an expert (*technite*) in any one of the specialized crafts. As the general knowledge of the principles and causes of making, *techne* was the know-how and know-why of any specific art or craft. It was distinct from experience because *techne* was concerned with the explanation of experience in terms of general principles; *techne* is always concerned with the general and experience is always of the particular. Someone may have the experience that outcome B follows action A, but, without a complete account of why B follows A, that person would not be said to possess *techne*. *Praxis* (habitual practices collected together to do things) is learned from experience and mimicry, and be used to develop tacit (implicit, non-verbalized) skills and beliefs regarding the best way to proceed, but it was only when this acquisition of experiences and instruction had been inductively abstracted into "a general and true course of reasoning" that *techne* could be said to have been acquired. The craftsman needed to give a "reasoned explanation" of *praxis* before s/he could be said to be a *technite*. This "reasoned explanation" (or the why of know-how) was to facilitate the explanation (*logos*) of an artefact in terms of fundamental causal principles or rules.

Although *techne* was comprised of formal, communicable, general, and abstract principles of making, it was primarily learned through imitation of craftsmen, copying forms, and attending to the particularity of the appropriate materials. The organization of these materials, tools, and forms governs and guides *poiesis* (production) in accordance with its methods and techniques.[162] Aristotle argued that the materials and tools used in production were distinct from the *technite* as a causal agent, and *techne* was not contained in the produced thing or state of affairs that resulted from *poiesis*. The activities of *poiesis* bring-forth and terminate in a product, outcome, or end (*telos*) that was separate from them; a pot is something separate from pottery, the potter, and the clay. *Techne* generalized and abstracted the actualities of production and was outside of those activities. Within Aristotle's four fold causality of formal, material, final, and efficient causes, say in the production of a pot, it was the *technite* (the potter) who took on the role of efficient cause, the clay was the material cause, the intention to make a pot the final cause, and the form (idea, shape, design) of a pot the formal cause. As "a true course of reasoning," *techne* was taken to be contained within "the soul of the craftsman" (or "mind" or "intellect") as "a reasoned state or capacity to make," and, consequently, taken to be bound up with the maker and guided the hands to perform definite motions that moved the tools and shape the materials into the product. This motion embodied form (*eidos*) into matter (*hyle*) and produced substance or informed matter. Form could not be forced upon (or into) *hyle* because of the substance's active character (resistance, recalcitrance) in the reception of form, and the union between *eidos* and *poiesis* was directed by both *techne* and resisted by *hyle*.

Although, according to Aristotle, *techne* and *eidos* are in "the soul of the craftsman," the extent to which the *technite* could impose form upon the clay during *poiesis* was not entirely within the control of the craftsman (*technite*), and therefore *hyle* constituted a definite limit to the extent that *techne* could guide this union between form and matter. It was only to the extent that the intervention can be grasped by "the rational part of the soul" as form (*eidos*) that it could be known and a part of *techne*.[163] Perfect reproduction according to general principles and knowledge was quite impossible due to *hyle*, understood as resistance or the particularity of the particular, which, therefore, referred to limits on the applicability of knowledge to experience because general principle are incapable of being applicable to all particulars and experience is always of particulars.[164]

---

[162] *Metaphysics* 7.9.1034a10-11
[163] *Physics* 2.2.194a23
[164] *NE* 5.10.1137b13-15

The particularity of experience always evades complete capture by general definitions and categories. Theory is an incomplete guide to action and human beings must learn how to do something by doing it and attending to *technites* and experience.[165] Aristotle defined *techne* in terms of knowledge of the changeable and temporal; whereas *episteme* was reserved for knowledge of the eternal and unchanging. *Techne* was a general knowledge of the Being of Becoming—of *poiesis*, which, for Aristotle, was straddled on a continuum between particularity of experience and the generality of theory.[166]

The Aristotelian conception of matter as *hyle* should not be confused with the post-sixteenth century conception of matter as inanimate "stuff" (elements, atoms, particles) that is only changed by external forces or causes; *hyle* did not refer to a substratum of material structures (possessing properties and relations); nor any unifying conception of matter having specific phases such as solids, liquids, and gases (or atoms, plasma, and radiation). *Hyle* did not refer to an objective material substratum at all. Instead, *hyle* referred to an unknowable and incognate formless potential that is active in the reception of form by materials and is an ultimately mysterious response to human interventions.[167] How can we make sense of what Aristotle meant by this term? For Aristotle, no two lumps of clay were alike. *Hyle* referred to the particularity of any particular lump of clay, which could only be encountered through interacting with the clay, and did not refer to the properties of the substance called "clay," which would pre-exist encounters with humans. It referred to *the particularity of the particular*[168] or uniqueness of each and every experience of making a pot out of clay. It referred to the way that no two lumps of clay are identical and a potter is unable to make the same pot twice; the way that each and every pot, as well as the experience of making them, is substantively different even though they are all made out of the same substance, by the same potter, in accordance with the same design.

For Aristotle, it was this active and emergent particularity of the reception of form which captures the way that individual experiences of making things were all unique and could not be known completely in their particularity through the general *logos* of *techne*. Due to the generality of *techne*, this kind of knowledge remained abstract in relation to the particularity of individual experiences that could only be emergent *qua* particularities, and hence each experience of making a pot eluded complete and reproducible description. *Hyle* referred the phenomenological resistance experienced when trying to grasp production (*poiesis*) under a single set of principles (rules or instructions) that could be communicated from *technite* to apprentice in such a way that would bring something forth into being. *Hyle*—as the particularity of the particular— resists the imposition of the generality of form from having complete sway, and the *technite* had to be responsive (sensitive) to the way that *hyle* received form; *techne* only guides the craftsman's inscription of form into *hyle*, as the craftsman does not control the extent to which *hyle* receives form, but the extend that form is received into matter is outside of *techne* and the craftsman's control. The *technite* must attend to *hyle* and adapt in ways that will not be applicable to other experiences of making something, as each becoming of the being of whatever is produced is itself unique. The *technite* had to be responsive to the receptivity of *hyle* emergent in response to attempts to impose form that are phenomenologically manifest as the particularities of the particular of the experience of working with particular materials during attempts to impose general forms upon them. Just as much as the *technite* needs to know the appropriate forms, tools, materials, and how to combine and organize

---

[165] *NE* 2.1.1103a35
[166] *Metaphysics* 1.1.980b25ff
[167] *Metaphysics* 7.8.1033b20-1034a7; 7.9.1034a10-11
[168] *Metaphysics* 7.8.1033b20-1034a7

them, s/he needs to be responsive the particularities that emerge through *poiesis* that would not have occurred without the intervention of the *technite*. It was for this reason that Aristotle argued that perception (especially touch) was required alongside *techne* to guide the activities of making.[169]

Aristotle made a distinction between things that find their origin in the maker (*poieta*) and things that find their origin in themselves (*phusika*). A pot is brought-forth through the actions of the potter in shaping and drying the clay, whereas a tree is brought-forth in accordance with an internal principle of change (*phusis*) that operates from within a seed to generate its growth within its environment into a tree. When a goal (*telos*) was introduced through the activity of a *technite*, working towards the material realization of an idea or design, the source of change was outside and separate from the thing in which the change happens. Something could only be considered to be *phusika* when the source of change was immanent within the thing itself (such as the potential of a seed to grow into a tree). A skilled sculptor could carve a tree out of wood—with such skill as to reproduce a tree in every detail—but it would not grow, flower, bear fruit, or decay and die. A skilled sculptor could even make a simulacrum of a tree in every detail, including simulating the bearing of fruit and it falling from the tree, with each simulated fruit containing a tiny carved seed, but the seed would itself never germinate and grow into a tree. It is this idea of something coming out of itself *in accordance with its inner nature* that distinguished the natural from the artificial within Aristotle's thinking, but although Aristotle considered *poiesis* guided by *techne* to be distinct from *phusis*, he used his conception of *techne* as his primary analogy in his elucidation of his conception of *phusis*—as the absence of *techne*—while maintaining the autonomy of the latter.[170] *Phusis* is revealed by Aristotle in contradistinction to *techne*. He used *techne* to elucidate his conception of *phusis* as teleological (directed towards an end or purpose) when the *telos* (end or purpose) of something is the bringing-forth (*poiesis*) of its own inner nature in the giving of its final form to matter. The inner potential of the seed gives the form of the tree to matter as the seed sprouts and grows into a tree from within it environment or natural place. *Phusis* means that something is natural, in the sense that it comes out of itself within its place in an environment without needing interventions from human beings, wherein it reveals the cosmological embodiment of form into matter with all its imperfections.

For the Aristotelian thinker, the natural is understood as distinct from the artificial—the natural world is not man-made—and there remains much continuity with this way of thinking today in how Nature is identified as *first nature* in contradistinction to the *second nature* of a world transformed through human activities, such as those of migration, hunting and the consumption of food, reproduction, agriculture, deforestation, urbanization, industry, transportation, pollution, warfare, mining, engineering, and bioengineering. There remains a great deal of continuity in human thinking about Nature as something that pre-exists humans and remains outside of human control, yet is intimately and primordially connected with human nature as an inborn source of appetites and needs, our instincts and drives, and our ability to satisfy them through material practices. Much this way of thinking continues today in how human beings think of conservation and the preservation of wilderness, as well as how Nature *re-wilds* in areas deserted by human beings (e.g. within the exclusion zone around Chernobyl) or when the artificial environment begins to break down (plants growing through cracks in the pavement). Today it is quite commonplace to define wilderness and the natural world in oppositional terms to the artificial and the built environment, defining the nature of the natural as not being artificial (and vice versa), and therefore

---

[169] *NE* 2.9.1109b23
[170] *Physics* Bk.2, especially 2.2.194a22ff and 2.8.199a15ff

understanding Nature in terms of self-generation outside of human control, even if subject to human interventions and manipulation.

Aristotle further distinguished between *technai* that imitate Nature without changing it, such as a painting and sculpture of an animal or plant, and *technai* that change and complete Nature, such as medicine or agriculture, which realize the potential of human beings (to be in good health and well fed) in ways that would not just naturally occur if human beings did nothing. If left to Nature, human beings would get sick, starve, and die. *Techne* was the possession of the most helpless, unshod, unarmed, unclad, but highest animal who could, through *techne*, turn this weakness around, take advantage of *phusis*, and even complete that which *phusis* left incomplete.[171] Art imitates and completes Nature by attempting a union of form and matter that draws the *telos* (the end) to come forth from within. Through *technai*, human beings fulfil their nature, ensure their survival, and reach their potential. Thus, although *techne* could not provide a complete account and guarantee reproduction, and the *technite* must attend to the materials s/he works with, it was either through the imitation of Nature or directing *poiesis* towards the perfection and completion of Nature, that, for Aristotle, *techne* was rooted in an aspiration to complete *phusis* and thereby perfect the organization of human agency through arts, crafts, and politics.[172] It is this relation between *techne* and *phusis* that underwrote a great deal of European philosophical reflection on the nature of the artificial and the natural, and the distinction between them, beginning with ancient Greek and Roman philosophies, and continuing throughout the medieval and into the Renaissance.

The Aristotelian and Roman philosopher Pliny the Elder (Gaius Plinius Secundus) wrote in his *Natural History* (written in the first century A.D.) that perfumes and dyes made by human beings had conquered their natural counterparts in order to improve upon them.[173] The Aristotelian and Greek physician Galen wrote in his *On the Natural Faculties* (written in the second century A.D.) that medicine was an art in the service of Nature (*ars ministra naturae*) that perfected the body.[174] Medicine did something artificial because it forced things to act in ways that were contrary to their own inner tendencies, but, by doing so, it allowed human beings to overcome these natural tendencies and conquer disease, injury, infirmity, and even stall death. As well as the *Physics* of Aristotle, and his *Mechanics* (which was almost certainly not written by Aristotle), the geometrical mechanics of Pseudo-Aristotle, Heron, and Archimedes, several ancient texts dealing with the geometrical and general principles of craft practices were available, translated, and studied throughout the medieval period. These included the works of Vitruvius, Frontius, Plindy, and Pappus. Vitruvius (c. 1BC) wrote *De Architectura* on the theory and practice of architecture and the large-scale management of craftsmen and laborers. Heron also wrote detailed works on surveying instruments (*Metrica*, *Dispotra*, and *Catopica*) and practical engineering (*Pneumatics* and *Automatopoietike*), as well as his studies of mathematics, physics, and mechanics. Frontinus (c. 1AD) wrote *De Aquis* on the engineering and distribution of water supplies. Pappus (c. 4AD) wrote on mechanics to solve practical problems that sixteen hundred years later inspired Galileo and his solutions to the same problems. All of these works contained systematic collections of geometrical treatments, descriptions of inventions, designs, experiences, and accounts of established practices. If we consider technology to be the *logos* (rationale, accounts, principles) of *techne* then all of these works were *techne-logikos* (or technological) and provided explanations of how and why technology works. They do not merely constitute

---

[171] *Physics* 2.8; *Politics* 1337a1–2
[172] *Physics* 193b10 and 2.1.193a12–17; *Politics* 1.2.1253a2
[173] Bk. 21, XXII Healy trans. (1991)
[174] Brock, trans. (1947)

collections of accounts of trial and error tinkering, and, through these works, we can see how Aristotle's distinctions had set the framework for how technology was understood.

Historians have shown us that the experimental sciences of medicine and chemistry found their cultural origin in alchemy, magical ritual and practices, herbalism and the practical healing arts.[175] Ideas about experimentation into "natural processes" by using artificial means began long before the Scientific Revolution. These ideas occurred within the cultural rise of the practical arts and attempts to understand the nature of matter through human interventions, rather than dismiss it simply because it involved esoteric notions that we might find fanciful and nonsensical today. Alchemy was not merely something told about in tales and fables about mystical adepts and magicians, witches or sorcerers, but was something that was widely practiced by doctors, herbalists, and midwives, and there certainly was not any sharp distinction between alchemy and chemistry before the seventeenth century. It is through the Arabic and medieval developments in alchemy that we first begin to see the intellectual blurring of the Aristotelian distinction between the artificial and natural.[176] Arabic and medieval Aristotelian scholars were concerned about the status of alchemy; whether it was an art that imitated or perfected Nature, and whether its products were identical with natural substances or mere imitations in appearance. Their concern was about the idea of parity or identity between artificial and natural processes; for example, whether it was relevant or not if heating occurred naturally or artificially. Is the heat of a furnace any different in kind from that of a volcano? Can heat be understood in universal terms and any knowledge obtained about heat from a furnace is true of all heat regardless of its source?

By focusing on the consequences of processes rather than their origin, Arabic and medieval alchemists could represent the products of their interventions as being based on natural essences or principles, and, by doing so, they could also represent the artificial processes as having utilized and organized natural processes within artificial contexts. Alchemy was an art that not only sought to reproduce natural products in all their qualities (not merely their semblance), but also sought to complete and perfect the form and order of Nature by using that form and order to do so. While their thinking remained Aristotelian, the alchemists of the late thirteenth century and early fourteenth century blurred the traditional Aristotelian distinction between *techne* and *phusis* because they operated under the principle that things brought into being through art were artificial with regard to their mode of production but natural with regard to their essence.[177] This was grounded in an interpretation of the ambiguity in Aristotle's discussion of roasting and boiling in the *Meteorology* (IV 381b3-9), and whether cooking imitates natural heating. Late sixteenth century alchemists such as Bernard Palissy were increasingly working with a blurred distinction between the artificial and the natural.[178] In his *Discours admirables*, published in 1580, Palissy argued that the art of the potter and the natural generation of minerals were based on the same principles, and he proposed his theory of stalactite, mineral, and fossil formation using ceramic pottery as an explanatory analogy or trope of how they formed naturally. Was there is a meaningful difference between the two or were they essentially the same process? The decision that they were essentially the same process was a point of departure from the Aristotelian *scientia* and laid down the groundwork that preempted modern chemistry. It is only a small conceptual step from the idea that natural and artificial processes can be understood in identical terms to the idea that experiments

---

[175] Thorndike (1934); Yates (1964); Rossi (1968); Copenhauer (1990); Dalton & Park (1998); Newman (2004); Moran (2005)
[176] Newman (2004)
[177] *Ibid* 64-6
[178] *Ibid* 145-63; Dalton & Park (1998)

based on artificial processes and interventions can be used as a means to understand natural processes and laws. It is this way of thinking that became fundamental to the *modus operandi* of the natural and experimental philosophers of the seventeenth century.

Robert Boyle and Isaac Newton, for example, both considered alchemy to be a legitimate way of investigating natural processes and laws, and employed the same arguments framed by the art-nature debates of the thirteenth to sixteenth century alchemists in their formulation of the basic concepts of the experimental sciences.[179] Newton's alchemical writings show that he gave considerable thought to alchemy as based on natural and fundamental principles, as well as viewing it in connection with creation and the divine, and this deeply influenced how he understood physics as part of natural philosophy. Alongside his exploration of the divine and mathematical principles behind the architecture of Solomon's Temple in Jerusalem and using mathematics to decode hidden messages in the Bible, Newton's alchemical researches and experiments included the search for the Philosopher's Stone (a fabled substance capable of turning base metals into gold), the Elixir of Life (potion of immortality), and metallurgy, especially involving mercury and silver. These works remained unpublished throughout his life, largely due to the fact that alchemy was a capital crime (punishable by hanging) in England at that time. In writing *The Skeptical Chymist*, Robert Boyle ventured a critique of the Aristotelian concept of Nature and its influence on Paracelsus, and took the practices of alchemy as the starting point in his conception of the representational and material practices of the experimental sciences.[180] In Boyle's *notes on palingenesis*, he wrote down a method to regenerate dead toads or serpents within a rotting goose or duck carcass.[181] Boyle referred to this method in his essays "Essay on the Holy Scriptures" and "Some Physio-Theological Considerations about the Possibility of the Resurrection."[182] We may well consider this method to be fanciful or implausible, and likely to have been only a thought-experiment, but what we see in Boyle's alchemical writing is the thinking-through of ideas and concepts about life, natural processes, and change that underwrote the fundamental concepts and representations that he developed to establish epistemologically the norms of the representational and material practices of the new experimental sciences.

However, it was the two Aristotelian themes of imitating and perfecting Nature were brought together by the enterprise of alchemy with the goal of emulating Nature by transforming and purifying natural substances, as well as creating new substances. Alchemy was important for the development of debates among Arabic and medieval Aristotelians about the abilities and moral limits of human beings when using the arts to perfect Nature. While the ideas of transmuting base metals to gold had been an alchemical aspiration since ancient times, along with ancient ideas such as discovering an elixir for human immortality and or a method for creating the perfect human being, in the sixteenth century the advocates of alchemy (especially the followers of the sixteenth century alchemist and natural philosopher Paracelsus), synthesized a new and unified concept of process, and this raised questions about the moral limits of Man's new powers in relation to his essence. For some, alchemy promised the ultimate victory of human power over the natural world and the perfection and completion of human beings, while, for others, the assertions of alchemists infringed upon the power of God, aiming to turn human beings into a creator on the same level as the divine, and therefore was a satanic aspiration. What was the source of the artificial power realized through alchemy and invention? For

---

[179] Principe (2000); Cooke (2001); Dobbs (2002); Newman (2007); Calian (2010)
[180] More (1941); Moran (2005)
[181] MS Locke C44, Bodleian Library, Oxford University
[182] Hunter and Davis (2000)

the ancients and medieval, it was seen as something mysterious and magical—and for some diabolical—and as something unnatural, which promised to satisfy human needs and make marvelous inventions. The ancient Greeks gave us the story of Prometheus, stealing fire and the power of invention from the gods and giving it to men. In the Christian interpretations of the Fall, agriculture and animal herding (unnecessary while in Eden) had to be taught to men by angels, and were seen as something apart from wisdom, truth, and the knowledge of natural causes, which were part of the Kingdom of God. Ultimately, if the goal of alchemy was to replicate the act of the creation of life by God and perhaps even improve on it, wasn't it inherently blasphemous? During the medieval period, invention and innovation were seen as something demonic—the result of listening to the demon Belphegor who whispered ideas for new machines into the ears of men and distracted them from their fallen state—and alchemy was associated with witchcraft and devil worship.

Of course many of the exaggerated claims made by the advocates of alchemy were in inverse proportion to what could actually be accomplished, and few people today believe that it could create anything as fantastic and powerful as the magical golem or homunculus. But these fabled creatures were symbols for the limits of artifice and the folly of human beings in encroaching on God's power, rather than proposed as something that should be made, even if it were possible to do so. The Jewish fable of the golem was the story of the creation of an artificial man from the magical/divine power of words, with great strength and magical abilities, but lacking speech and intelligence.[183] The purpose of the golem story was to provide a parable of human folly and limitation when attempting to mimic the creative power of God, as the golem was always a failure in these stories. Mary Shelly's *Frankenstein* is a modern version of this story (wherein electricity takes the place of magic or divine power). The Arabic tales of the alchemical homunculus were of a very different creature.[184] The homunculus was created from an alchemical process involving human sperm being mixed with other substances inside an artificial matrix (a jar often was used as a substitute womb) to create the perfect human being. Typically, this story described an ideal conception of human being in terms of intellect and moral character, but often also granted the homunculus magical powers (flight, invisibility, telepathy, precognition, immortality, etc.). As a result of lacking "the imperfections" of human and natural beings, the homunculus struggled to live in the imperfect world, and ultimately sought its own death, and the story is a warning against using artifice "to perfect" creation. Goethe's *Faust*, Umberto Eco's *Foucault's Pendulum*, W. Somerset Maugham's *The Magician*, Peter Ackroyd's *The House of Doctor Dee*, Isaac Asimov's *Bicentennial Man*, and Philip K. Dick's *Do Androids Dream of Electric Sheep?* are modern retellings of the homunculus story.

Despite all its failings, alchemy was the prototype for the investigative and experimental sciences, which was practiced for the purpose of investigating natural processes by means of interventions performed using apparatus in a laboratory. The case of alchemy not only shows that medieval *scientia* involved experimental investigation into how to manipulate natural processes artificially, but, more importantly, discussions of the limits and powers of alchemy provided the background against which the debate about the possibilities and limitations of the nascent experimental philosophies of the seventeenth century emerged, and the new natural philosophers owed a considerable debt to alchemical literature and experiments. The new sciences of medicine and chemistry continued and refined the techniques of Arabic and medieval alchemists; the alchemical procedure of distillation became one of the fundamental methods of modern analytical

---

[183] Idel (1990)
[184] Pines (1996)

chemistry, and the alchemical goal of transmuting base metals into gold inspired experiments into compounds and elements, and the fundamental properties of matter, and has been achieved (albeit at high cost) during modern experiments in nuclear fusion. Alchemists were also able to color glass, invent new dyes and distillates, discover properties of metals and improve metallurgy, change the color of semi-precious stones and metals, clean gemstones and pearls, make artificial gemstones, invent basic electrolysis methods, and make imitation gold and silver. There were many intriguing precursors of the Periodic Table and modern ideas about the particulate nature of matter, the biochemical paradigm of life and disease, modern conceptions of medicine and physiology, and the scientific worldview. Even though alchemy may well have been based on fantastical cosmologies or esoteric concepts that we can see with the benefit of hindsight as based on fundamentally flawed or fanciful ideas, alchemy provided the template for investigation into natural processes by basing observation on experimentation and intervention, relating representational and material practices in contexts of justification and discovery, and the empirical idea of basing knowledge of natural principles on material practices. The alchemists' arguments for laboratory experiment and the replication of natural entities by using artificial processes led to the experimental philosophy of Francis Bacon, and alchemy was an important and fertile groundwork for the development of the new experimental sciences and the Scientific Revolution.[185]

The modern experimental sciences of physics and chemistry (and later biology) grew out of a long medieval tradition of debate and experimentation, in alchemy, mechanics, and the practical arts (especially medicine and surgery) which blurred the Aristotelian distinction between *techne* and *phusis*. However, the debates about the distinction between *techne* and *phusis* were not restricted to the contexts of esoteric Aristotelian discussions about the nature of alchemy. We also need to look the profound influence of Aristotelian interpretations of mechanics on the development of the experimental and natural philosophy of Galileo, Bacon, and Descartes (and later Newton and Boyle, *et al.*), especially in relation to the practical arts and the growing dominance of maker's knowledge as the knowledge of natural laws. The Aristotelian interpretations of mechanics and the mathematical treatments of mechanical devices had been developed in Europe from, at least, the thirteenth century onwards. In the thirteen century, Roger Bacon followed the Aristotelian distinction in his writings on the "marvelous motions" of mechanical devices, such as using a pulley to lift heavy weights, and the promise of using "the artificial" against "the natural" by building machines to overcome natural tendencies and resistance.[186] During the thirteenth century there emerged both generalized concepts of mechanical leverage and the view that Nature was a vast reservoir of forces and powers to be tapped and used according to practical know-how.[187] This suggests that the abstract conception of Nature as a resource that could be utilized by the practical arts began to emerge and become formulated some 300 years before the Scientific Revolution and over 500 years before the nineteenth century Industrial Revolution. The sixteenth and seventeenth century development of the new sciences was a continuation of the post-thirteenth century period of decisive development in the effort to use the forces of Nature mechanically for human purposes. This expansion required more resources and the continuing innovation of machines and techniques to enhance productive, explorative, military, and civic power.

---

[185] Rossi (1968); Cook (2001); Newman (2004: 256-71)
[186] Hackett (1997)
[187] White (1962)

Between the thirteenth and sixteenth centuries there were widespread innovations in civic, military, and economic technologies, as Europe began its expansion of political, economic, and military powers. The invention of the printing press accelerated the cultural dissemination of studies of mechanics and the practical arts throughout Europe, while they were rationalized and associated with the ancients through geometry in order to give them status of being "sciences." The mechanical and experimental sciences of the sixteenth and seventeenth centuries grew out of the contemporary mechanics, practical arts, and mathematics of the fourteenth and fifteenth centuries. The Aristotelian distinction between "the artificial" (*techne*) and "the natural" (*phusis*) had dominated thinking about mechanical devices and the practical arts throughout the medieval period and into the Renaissance.[188] Once the cultural status of mechanical invention had transformed from a craft to a science, the conditions were ripe for the formal and metaphysical construction of the science of mechanics as a natural science. It was this way of thinking about "the artificial" and "the natural" that was explored in the universities and centers of learning of medieval Europe, and greatly influenced the Renaissance. The intellectual developments and refinements of Aristotelian mechanics and mathematical representations of mechanisms were crucial for the emergence of the new sciences and the Scientific Revolution. The representations of natural mechanism used by Galileo, and more formally by Descartes (and later by Newton and Boyle, *et al.*) had significant continuity with those used in ancient Greek and medieval Aristotelian studies of mechanics, and they provided a template for the methodology of the new experimental philosophy.

Historians of science and technology have provided numerous examples of ancient and medieval measurements, experiments, and the use of mathematics to analyze mechanical devices and machines, to show that both ancient and medieval people had advanced the conceptualization of mechanisms in representational and material practices, and this discredits the "received wisdom" of the traditional philosophers of science that these activities are special to the scientific method that began during the Scientific Revolution.[189] Ancient Greece saw a great deal of ingenuity and innovation in mechanics and the practical arts, especially in architecture, music, irrigation, astronomy, metallurgy, engineering, and navigation, alongside the formal study of mechanisms and the geometrical demonstration of their fundamental principles.[190] Throughout the Roman Empire, not only were the practical arts studied in detail, particularly those involved with medicine, agriculture, and metallurgy, but architecture, engineering, and mechanisms were subjected to formal mathematical treatment.[191] Windmills and watermills were commonly used in thirteenth century Thessaloniki and throughout the Byzantine Empire, and these mechanical devices were the subject of detailed technical studies (involving new diagrammatic techniques and mathematical demonstrations) in the early fourteenth century, and often wind and water flow were understood as "natural forces."[192] The mathematical principles of engineering, architecture, irrigation and agriculture, alongside astronomy, medicine, music, and all manner of practical arts, were studied in detail in

---

[188] Laird (1986)
[189] Clagett (1959); Clagett & Moody (1952); Dugas (1955); Forbes (1955); de Santilliana (1961); White (1962); Wightman (1962); Usher (1962); Burstall (1963); Gille (1964); Kearney (1964); Rose & Drake (1971); Crosland (1975); Rose (1975); Clagett (1978); Lindberg (1978); Grant (1981); Schmitt (1983); Hill (1984); Wallace (1984); Bennett (1986); Laird (1986, 1991); Simms (1987, 1988); Sorabji (1988); DeVries, (1992); Long, (1997); Torrance, (1999).
[190] Rostovtzeveff (1941); Dugas (1955); Forbes (1955); de Solla (1959); White (1984); Usher (1962); Burstall (1963); Kearney (1964); Hodges (1970); Clagett (1978); Landels (1978); Hill (1984); Sorabji (1988); Torrance (1999)
[191] White (1984); Usher (1962); Burstall (1963); Hodges (1970); Landels (1978); Hill (1984); Torrance (1999)
[192] Rautman (2006); Brett (1939)

Ancient China.¹⁹³ The Sumerians and Babylonians are famous for inventing architecture, writing, astronomy, and mathematics, and possibly agriculture too, and the ancient Egyptians not only built the pyramids, but also advanced irrigation and other agricultural techniques, medicine, astronomy, and engineering, and connected these technological advances with their cosmology and idea of an ordered (and balanced) sun-centric Universe.¹⁹⁴ These cultures, their mathematics and cosmology greatly influenced the ancient Greek Pythagoreans (and later Euclid, Plato, and Archimedes), as well as throughout North Africa and the Middle East. It is interesting that the ancient Egyptian god Thoth, the Judge of the Dead, was also the god of mathematics, chemistry, magic, measurement, and writing, and later in Greek culture became associated with the god Hermes (the messenger, also connected with knowledge and the practical arts) who was later associated with alchemy, the Hermetic Order, and the occult rites and mystical arts of the Rosicrucians.¹⁹⁵

Throughout Europe, from as early as the thirteenth century, mechanical devices of all kinds were studied in detail and subjected to mathematical representation (geometrical proof or axiomatic demonstration) of first principles, and these formed the cultural and intellectual framework for studies of mechanics in the Renaissance. Their common argument is that the ancient and continuing medieval innovation of technologies and the fascination with mechanism contained the seeds of the sixteenth century development of mechanics and the new physics. The medieval modifications and criticisms of these works of became the points of departure for the mechanics of the sixteenth and seventeenth centuries.

The predominant mathematical problem for medieval mechanics was how to solve the problem of the six simple machines (the wheel and axle, the wedge, the balance, the lever, the inclined plane, and the screw). The methodological template to provide complete solutions for these machines followed the Aristotelian mechanics set down in Pseudo-Aristotle's *Mechanical Problems*. The only scholarly certainty regarding the authorship of the works of Pseudo-Aristotle is that it almost certainly was not Aristotle. They were possibly written by a student of Aristotle named Strato in the fourth century BC. Also, it is questionable whether it is even possible to construct a consistent mechanics from the few fragments on weight and motion scattered throughout the works of Aristotle.¹⁹⁶ It is questionable whether the Aristotelian mechanics had anything more than a superficial and nominal connection with the philosophical writings of Aristotle. In fact, Aristotle argued that geometry could not provide a complete description of movement because it lacked a continuity axiom.¹⁹⁷ The incompleteness of Euclidean geometry has been widely accepted by mathematicians and logicians since Frege published *Begriffsschift* in 1879.¹⁹⁸ Aristotle largely limited mathematics to having a use for *technai* such as astronomy, optics, and music,¹⁹⁹ and his physics has more in common with natural history than modern physics.

In the *Mechanical Problems*, Pseudo-Aristotle presented his formulations of the geometrical solution to the six simple machines by describing and deriving their motion in terms of the properties of circles. Since the mid-thirteenth century, many treatises

---

[193] Chatly (1942); Forbes (1955); Usher (1962); Hodges (1970); Landels (1978)
[194] Temple, (2000); Olsen (2009)
[195] Case (2009)
[196] Cartelon (1975)
[197] *Physics* VI, 1, 231a24ff
[198] Heath (1949)
[199] *Physics* II.2.194a7-11, *Post. Analytics*, I.7.75b14-20, and *Meteo.*, III.3-5

appeared that focused on the problems of kinematics and dynamics for mathematical and philosophical treatment also focused on the properties of circles. The thirteenth century mathematician Jordanus de Nemore used Pseudo-Aristotle's dynamics, as well as Arabic derivatives, in his book *De rationale ponderibus*, to tackle the problems of the balance, weights, and levers in terms of the circle, and his concluded with the general problem of geometrically describing all motion. By the fourteenth century, there were many books on the subject of the application of geometry to the problems of motion. John Buridan's fourteenth century book *Questiones super libris quattuor de caelo* included discussions of impetus theory, the possible rotation and motion of the Earth, the general law of leverage, the solution to the problem of the inclined plane, and the equilibrium of connected weights all in the context to providing a general geometry of motion. Both de Nemore and Buridan tacked problems that were later to become central to Galileo's work.[200] Medieval Aristotelian mechanics formed a tradition of attempts to construct a dynamical system based upon the geometrical projection of the circle over simple mechanisms. Of course, these early efforts utilized completely different conceptions of motion and matter from those utilized by modern mechanics. Buridan's Aristotelian notion of "impetus," for example, has no correlate in Galileo's mechanics and, in terms of modern mechanics, the early efforts made significant errors in their treatments of even simple mechanical devices. As well as struggling to find solutions to the general problem of motion, this system failed to provide a method to provide static solutions to the problem of equilibrium. But the failures of medieval mechanics are irrelevant for the question of whether Galileo's mechanics was a culmination of earlier efforts. After all, Galileo's physics also introduced conceptions of time and space that have no correlate in relativistic or quantum physics, and in terms of contemporary physics, Galileo's physics was also based on false assumptions and archaic representations. What we see in the pre-sixteenth century efforts was *the attempt* to describe the motions of simple mechanical devices in terms of Euclidean geometry. This set down the template for the methodology of subsequent efforts.

The fifteenth and sixteenth century Italian inventor Leonardo Da Vinci studied both Archimedes' mathematics and Pseudo-Aristotle's mechanics. Leonardo's work shows considerable efforts to bring together and integrate representational and material practices in terms of both Archimedean and Aristotelian geometrical proofs and demonstrations that the principles of mechanics and natural processes had an analogous explanatory connection. Leonardo regarded experience not only in terms of observation, but also in terms of an interventional exploration of "the processes of Nature" through chemistry, mechanics, and dissection, wherein natural phenomena were represented in terms of geometrical mechanisms and ratios.[201] His works included mathematical analyses of machines, in terms of mathematical mechanics, which were reduced, primarily, to the elements of force, impact, weight, and motion. Leonardo based his designs for machines and devices, including hydraulic devices, fortifications, weaponry, flying machines, submarines, the parachute, and the helicopter all on geometrical principles. Leonardo's design of the *Architronito* (a steam cannon) was based upon the drawings of cannon in *De Re Militari* (1483) by Valturius, who stated that the cannon had been invented by Archimedes. There is a lack of any supporting evidence for Valturius' claim, but a twentieth century physicist named Ioannis Sakas used Leonardo's sketches (as given by Valturius) to build this device in the mid-1980s, and it projected a missile (a 10 oz. tennis ball filled with hardened cement) to a distance of 150 to 200 ft. within seconds.[202] Leonardo's approach of exploring natural phenomena in terms of "natural mechanisms"

---

[200] Clagett & Moody (1952: 213-9)
[201] Dampier (1938); Clagett (1971: 490-1); Kemp (1981); Gulluzi (1987: 91-5)
[202] Simms (1988: 195-210)

experimentally, by applying geometry to solve problems in natural philosophy, was a continuation of medieval efforts in these directions rather than a radical or novel break from them.

This new *naturalistic* way of thinking about mechanisms can also be seen in the work of Giovanni di Guevara. Guevara was a Spanish noble from Naples, *praepositus generalis* of the Clerics Regular Minor, the Bishop of Teano in 1627, and a papal legate to Philip IV of Spain, who was also an Aristotelian scholar.[203] In his *In Aristotelis mechanicas commentarii*, published in 1627, he analyzed mechanics using both Archimedean principles and Pseudo-Aristotelian mechanics. He dealt with the principles of mechanics, the center of gravity, the simple machines, Pseudo-Aristotle's thirty-five mechanical problems, the scope of mechanics, and its relation with the other sciences. He defined mechanics as the art or science of applying geometrical principles to heavy and light things that must be moved or brought to rest artificially. Mechanics was based on the weight of the moved body and the strength of the mover (which could be an impetus or a machine) and it consisted in discovering the appropriate powers needed to move loads and to supplement Nature. Guevara's formal treatment of mechanics was in terms of "marvelous" (artificial) motion and rest, and each of these, in turn, was treated in the Aristotelian terms of "violent" (forced) and "natural" (according to internal principles) motion and rest.

In Aristotelian terms, violent motion arose from an external source whereas natural motion arose from the body in question. This distinction allowed Guevara to describe how mechanics and natural philosophy dealt with the same subject differently. He argued that natural philosophy was concerned with natural motion and rest whereas mechanics was concerned with marvelous motions and rest, which were comprised of both violent and natural motions and rest. He argued that natural motion was apparent in any motion that was produced by machines because even though all machines are put together and set in motion by violent motions, from an external source, how the machine responded was a natural motion. Human intervention was required to produce and activate any machine, but Nature played its part in how that machine operated or even whether it worked at all, and, consequently, the correct understanding of the operation of any mechanical device must be based on an analysis of the interaction between violent and natural motion and rest.

Galileo's Mechanical Realism

The Aristotelian interpretation of mechanics provides the background to Galileo's natural philosophy, but Galileo took the additional step of placing the methodology that "checked conclusions against experience" within the context of mathematical mechanics. Galileo frequently disregarded the primacy of experience and instead affirmed Euclidean geometry, inscribing motion solely in terms of the translation of a body from one geometrical point to another. Checking directly with Nature involved using a device understood in terms an intellectual *a priori* knowledge of mathematics, as being the only means of apprehending the truth about experience, and physics became *a priori* mechanical science in the hands of Galileo. Geometrical methods did not allow any investigation into quality and it is for this reason Galileo declared and dismissed the notion of quality as illusionary or secondary. His "insight" was that only by describing the experience of how a machine performs in the form of Euclidean geometry was one able

---

[203] Laird (1986: 65-7); Wallace (1984: 208-16)

to read "the book of Nature." It is this prioritization of geometrical reasoning over experience that has led to the claim of a Platonic origin of modern science, but Galileo did not base his theory on the knowledge of the forms of truth, beauty, justice, and goodness—nor did he inquire into the nature of the soul and the limits of human knowledge (leaving such questions to the Bible). Instead Galileo posited his science on the aspects of experience that can be described geometrically, and anticipated Nature in terms of a representation of Nature comprised of mathematical properties that can be isolated and treated as interacting components in a mechanical apparatus. His geometrically arguments were projected to disclose "natural mechanisms" by analogy with demonstrations using artificial devices. For Galileo, "checking with experience" required demonstrating the truth of his deductions of the mathematical fundamental representation of any natural mechanism involved practical experimentation because if one truly understood any natural phenomenon then one should be able to construct a machine to reproduce that phenomenon artificially.[204]

Galileo's metaphysical contribution should not be underestimated. Of course this contribution was in part technical. By innovating geometrical techniques to reduce all motion to simple mechanical motions—the motions of the balance and lever—Galileo was able to inscribe simple time-reversible mechanisms (such as the pendulum) in terms of Euclidean geometry and provide a mechanical determination of a variable (e.g. time), thereby allowing its quantification and universality of its measurement via using that mechanism. This was essential for the scientific method to get off the ground. However, the technical obviousness of this step conceals the extent that it was also metaphysical. By establishing mechanical realism as a basis for using mechanical devices as experimental apparatus "to discover" the mechanical principles of Nature, Galileo went further than had any of his Aristotelian predecessors at Padua. Not only were the motions of simple mechanical devices treated as natural and governed by natural laws, as Moletti had proposed, but they were also to be used to determine the mathematical "Laws of Nature" in operation outside the context of a mechanical apparatus. Instead of only understanding machines in terms of natural laws, as Moletti had done, Galileo was able to go further and postulate the use of machines to reveal and understand the natural laws governing all phenomena. All natural movements were to be treated as simple mechanisms at work, whereby the mechanization and quantification of experience in terms of the mathematical projection of relations between measuring devices. Galileo's *reduction* of change and causality to mechanisms shows that his mechanical realism was both the precursor to the seventeenth century mechanical worldview, the eighteenth century scientific worldview, nineteenth century scientism, and twentieth century cybernetics.

Galileo's method to "check" his theories against "experience" was a metaphysically interpreted technological activity. He used abstract mathematical designs to demonstrate the possibility of building mechanical devices to "imitate" natural phenomena. Success implied that he had "reproduced" the causal mechanisms supposedly "at work in Nature" when he designed a machine that reproduced the appearance of the natural phenomenon in question. Hence, he was able to claim that his mathematical deductions disclosed "natural mechanisms" when those deductions were successfully "embodied" in the design of a working machine. This is evident from Galileo's use of a pendulum to demonstrate his theory of motion[205]; his use of an astronomical sphere to demonstrate his theory about the Sun's rotation[206]; and steelyards

---

[204] Galileo (1960: 421)
[205] *Ibid* 152-3
[206] *Ibid* 348-9

and balances to demonstrate his theory of free fall.[207] Taking this into account, we can see that, from the outset, mechanical realism was presupposed by Galileo's methodology. This was an essential development for his new physics and provided the template for all subsequent experimental physics. Due to the metaphysics of mechanical realism, it became possible to move beyond conceiving of mechanisms and machines as a metaphor for the natural causes and processes in the natural world, while simultaneously understanding them *literally* as representing an objective, underlying material reality. Mechanics was an embodiment of mathematics in the world through ratios and quantities—and this is something quite antithetical to Plato's philosophy. In Galileo's arguments, mechanical realism had emerged into the seventeenth century as a substantive metaphysical position and constituted a set of precepts that could be applied to experience. Mathematical science required this set of precepts in order to appeal to *a generalized principle of operation* in Nature and correlate the motion of bodies, and their properties, with measurements and experiments.[208] Galileo's operational metaphysics of mechanical realism reduced the conception of "the physical" to identical with "the mechanical," and the mathematical motions projected upon the natural phenomena were the abstracted motions of the lever and the balance.

In *Two New Sciences*, Galileo defined his notion of force (*forza*) in terms of the lever and termed it as "mechanical advantage."[209] Galileo proposed the derivation of "most other mechanical devices" in terms of "the Law of the Lever" and Archimedes' geometry demonstration of the first principles of mechanical equilibrium.[210] The geometrical solution to the balance projected circular motion over the balance and provided the fundamental representation of the coupling of anti-parallel linear motions or forces. Orthogonal changes in motion or force—the transference between horizontal motion and vertical motion—could be described in terms of the geometrical solution to the lever and its fundamental representation (the moment of transference between vertical and horizontal motion described in terms of the inclined plane). The projection abstracted transference between circular motion and motions orthogonal to the plane of the circle could provide a fundamental representation of the screw. This set of fundamental representations provided him with a complete set of uniform mechanical motions to project over all natural movements in such a way as to allow the observer "to see" the geometrical essence of that motion, and paved the way for Newton's Laws of Motion and Universal Gravitation to be demonstrated in similar mathematical form as a geometrical projection over everything.

Galileo was able to develop the dynamics of the new physics that aimed to describe everything in terms of number, figure, motion, and causal mechanism. Hence, for Galileo, the precepts of his metaphysical mechanical realism encapsulated his conception of Nature and the methodology of natural science as the mathematical projection of the six simple machines. Galileo's arguments were based on a mathematical intuition that was needed "to see" the efficient causes (mechanisms) at work in the experiment, and these the necessary causes and fundamental mechanisms of Nature governed by mathematical laws. It was the balance that—as a metaphor and a model—became central to physical explanation and law, connected with mathematics through the geometrical properties of circles and triangles. The subsequent seventeenth century developments of the laws of conservation (mass, charge, energy), Newton's Laws of Motion, and the First Law of Thermodynamics, all were premised upon the metaphor of

---

[207] *Ibid.* 213-4
[208] Strong (1966)
[209] Galileo (1941: 124)
[210] *Ibid* 110-12; see also Heath (n.d.) pp. 189-220

the balance as a fundamental mechanical principle of Nature. We can also see this in eighteenth century understandings of hydrodynamics and pneumatics, and the study of solids, liquids and gasses, and in nineteenth century interpretations of thermodynamics and electromagnetism, necessitating the postulation of an aether as the medium for the motion of heat and electromagnetic waves, and even in nineteenth century interpretations of natural selection and economic forces, as well as twentieth century interpretations of ecology and climate change in terms of "natural balances."

Furthermore, by utilizing the method of transdiction whenever "the balance" failed to match experience, a projected invisible counter-mechanism could always be found and used to correct the discrepancy by projecting it as a posited counter-balance, which itself could be analyzed in terms of "the balance," and so on. This would suggest a new experiment to demonstrate the transduced counter-balance. This is common to all physics. Just as Galileo was able to explain discrepancies of the motion of a body from quadrature of a parabola in terms of "air resistance," and demonstrated the explanatory power of this transdiction by using the dropped weights and the pendulum thought experiments,[211] Wolfgang Pauli was able in 1930 to postulate the neutrino as a transdiction to explain the absence of conservation of mass, spin, and energy in observations of radioactive decay. Further experimental confirmation of the existence of neutrinos, could be retroactively projected over the initial discrepancies between observations and theoretical expectations as confirmations of the success of the mathematical projection of conservation and symmetry principles, thereby reinforcing the explanatory success of the trope of the balance. Once natural processes and change could be represented as the resultant interaction between balances, as balances within balances, and explaining imbalance in terms of an "external force" by acting as a lever, the seventeenth century mechanical worldview and the image of the Universe as the Grand Machine—a clockwork Universe, wherein each cog acted as a lever in balance with other cogs, each simultaneously acting as a lever and counterbalance.

The mechanical realist interpretation of natural motion in terms of the six simple machines not only made the mechanical worldview and the idea of the clockwork Universe conceptually possible, but it also provided natural philosophy with a methodology, wherein all phenomena could be treated as mechanical components within a greater machine, itself governed by transcendental mathematical laws, that could be investigated by reproducing those mechanical components in the laboratory. This provided the condition for specialization along the lines of solutions to the problem of transdiction, wherein each transdiction becomes a node in the branching structure of the sciences, and each demonstration of the existence of the transdiction through representational and material practices opens a new field of inquiry (e.g. the demonstration of the existence of friction simultaneously confirmed mechanical dynamics and opened the field of thermodynamics). All that is required to solve any discrepancy between observation and the mathematical representation is the postulation of a correcting mechanism and devise a further experiment to show the mechanical action of the correcting mechanism. Projecting this transdiction over the original phenomenon refined the model and provided an explanation—and opened the way for a new specialization, wherein each new sub-discipline could explore different components of the Grand Machine.

This not only formed the template for the scientific worldview, but allowed scientists to share the sense of participating in the same pursuit of gaining a theoretical

---

[211] *Two New Sciences*, pp. 252-6

and comprehensive understanding of *the same* Nature, while participating only within a narrow specialization and possessing knowledge of a limited and technical scope. It is this sense of sharing a scientific worldview that shows how science is a cultural phenomenon and a subculture (paradigm), thereby gaining its value as a means of gaining knowledge that transcends the experiences of any particular individual while finding its meanings within a community of practitioners. This worldview embodied the precepts of mechanical realism and made the experimental use of mechanical devices to ascertain the fundamental mechanisms of Nature possible, and gave the new science a dynamic as an ongoing technological activity of testing and refining mathematical representations. The subsequent speculative metaphysics of the seventeenth century (such as the Newtonian mechanical system of the world, Gassendi's atomism, or Descartes' mathematical continuum) each reveals a "paradigm shift" by introducing metaphysical interpretations that would have been nonsensical to previous natural philosophers, especially Aristotelians, but these new sciences and metaphysical interpretations all retained their unity via the scientific worldview through sharing the operational metaphysics of mechanical realism that had grown out of Aristotelian mechanics. Differing speculative metaphysical assumptions about Nature founded the subsequent mechanical philosophies, leading to different schools of thought (see next chapter), but they all presupposed the same operational metaphysical precepts of mechanical realism and conception of experimental science in how they understood the scientific method and worldview.

From the perspective of an Aristotelian natural philosopher, the mechanical realism inherent to Galileo's development of mechanics as both a mathematical and a natural science conflated *techne* and *episteme* in a way that makes a profound category mistake of confusing something artificial and based on material practices with something natural and based on eternal and universal logical axioms. Following Moletti and his Paduan predecessors, Galileo presented the *techne* of mathematical mechanics as the unchanging and eternal *episteme* of Nature. Thus, from its outset, Galileo's physics was *techne*-logical and *meta*-physical. It is true that this move was in part facilitated by the ambiguity between mathematical reasoning as *techne* and *episteme* within the texts of both Aristotle and Plato, which led to the abovementioned medieval developments in the Aristotelian distinction between Art and Nature in alchemy and mechanics, but it was also novel, given the reduction of Nature to the mechanical was something new. The cultural acceptance and "self-evidence" of this metaphysics grew out of the transformation of the status of the mechanical arts to mechanical science, via its usefulness, but it involved the reification of the abstract products of the mathematical science of mechanics as something objective, universal, and eternal, which was tested through technological reproduction and innovation of new powers. Henceforth, mathematics became the means of reading "the book of Nature" by building a simulacrum of natural phenomena out of mechanical components and describing the motion of that model in terms of mathematical laws. Securing this metaphysical foundation connected the *episteme* of Euclidean geometry with the *techne* of mechanics as a technological means to disclose the axioms of Nature that governed change in the world, and its test and validation was secured against the innovation of new mechanical devices as demonstrations of those axioms at work. The conceptual possibility of modern technology occurred simultaneously with the possibility of modern experimental sciences because of the emergence of this conceptual and metaphysical synthesis and unification of the origins of both natural phenomena and technological powers in terms of natural mechanisms and mathematical laws.

How should we understand scientific knowledge and its distinction from other kinds of knowledge? How did Galileo's new science differ from his predecessors? If we wish to understand the kind of knowledge obtained through experimental science, and how it differs from Aristotelian (and Platonic) conceptions of knowledge, we need to understand how *techne* and *hyle* give meaning to how Galileo understood the development of representational and material practices during experiments, and the kind of knowledge he hoped to obtain. What do *techne* and *hyle* signify in this context? How do these ancient conceptions of knowledge and matter relate to modern science? Given the fact that *techne* is the term for craft-knowledge, and the Ancient Greeks are not famous for their experimental practices, it may well seem odd to the reader to characterize the knowledge at work in highly technological modern experimental physics in terms of ancient craft knowledge.[212] My argument is that, in order to understand the role of *techne* in modern experimental physics, we need to examine the role of craft practices, mathematical practices, and *hyle* in experiments. The pre-Socratic usage of *techne* captures something of Hacking's use of the term *intervention* and the Aristotelian use of the word *hyle* captures something of Pickering's use of the term *material agency* to describe the emergence of resistances during ongoing experimentation (Hacking, 1983; Pickering, 1995). It also captures something of David Gooding's use of the terms "the participation of Nature," "recalcitrance," "phenomenal chaos," and "plasticity," in his description of the development of craft practices in the early nineteenth century experiments in electromagnetism (Gooding, 1990). Plato and Aristotle's definitions of *techne* were both premised upon knowledge providing the highest degree of communicability, precision, and repeatability, on the basis of "a true course of reasoning." This "true course of reasoning" is given in terms of the unchanging causal principles of change—as the knowledge of the Being of Becoming. This kind of knowledge is highly characteristic of the theoretical knowledge of the experimental sciences, which via concepts of natural mechanism and law, is represented in terms of the unchanging and universal principles of change. Since the mechanical realism Galileo (and his predecessor Moletti), the conflation of *techne* and *episteme* was based on the assertion of the universality and eternality of the mathematical science of mechanics. This has provided experimental science with an epistemological warrant for accessing the ontology of Nature by inscribing machines and mechanical ensembles in terms of causal mechanisms, and putting those inscriptions to the test by building and operating those machines.

This sense of the word *techne* is an important one for the characterization of scientific knowledge as an ideal aspired towards obtaining as a result of experimentation because "the true course of reasoning" in modern science involves the question "how does it work?" Answers to this question are given in terms of causal explanations and mathematical representations of natural mechanisms. These are "tested" by being used as guides to human interventions in productive and experimental labor processes. If we take *techne* to be characteristic of the ideal knowledge of causal accounts of the processes of building repeatable experiments, *hyle* provides us with a meaningful term to express part of the phenomenological experience of experimentalists of the way that experiments "do their own thing" and resist perfect reproduction. *Techne* is concerned with complete knowledge given in terms of "a true course of reasoning" and such knowledge would be the *end result* of experimentation. It is the *telos* of the experiment. It promises to provide the abstract, general, and communicable knowledge of how to repeat the experiment by identifying, isolating, and exercising the causes involved—themselves represented as

---

[212] According to Sambursky, *The Physical World of the Ancient Greeks* (Dugut (trans.), London: Routledge & Keegan-Paul, 1987) and R. Waterlow, *Nature, Change, and Agency, in Aristotle's Physics* (Oxford: Clarendon Press, 1982) there was a notable lack of experimentation in Ancient Greece for a period of over 800 years and that the works of Archimedes and the exploits of the legendary Daedalus were exceptional.

natural mechanisms and laws. It is for this reason that modern science cannot be understood simply in terms of "applied technology"—just as technology cannot be understood simply as "applied science"—because, through the use of mathematics, it represents itself as directed towards the realization and acquisition of *episteme*, and it thereby conceals the extent that it is an art that is metaphysically directed towards the acquisition of the *techne* of Nature.

Techne is the abstract knowledge of the relation between interventions and the outcomes, where *hyle* captures the way resistances emerge during experiments in response to interventions. In this respect, *hyle* captures the participation of Nature, recalcitrance, phenomenal chaos, and "plasticity," in the response of experiments to interventions. If an Aristotelian were to pay close attention to the phenomenology of how experiments are performed in practice, she could see that we can understand the object of experimentation in terms of an Aristotelean *hyle*, while the demand for the completeness of *techne* captures the asymptotic nature of the knowledge pursued by experimenters. *Techne* is concerned with complete knowledge given in terms of "a true course of reasoning," which in the context of experimentation would be discovered through experimentation and its end result. It promises to provide the abstract, general, and communicable knowledge of how to repeat and explain the experiment. Once we understand the techneic character of the causal knowledge sought by experimenters and how natural phenomena are reproduced and modeled in relation to building repeatable experiments, as a demonstration of their own truth and their repeatability is the test of objectivity, then we can also see that *hyle* provides us with a meaningful term to express part of the phenomenological experience of experimentalists of the way that experiments "do their own thing" and resist perfect reproduction, even under the most carefully controlled conditions.

On my account, *techne* is an ideal form of knowledge that is not achieved in practice, but promises the highest degree of communicability, precision, and repeatability through "a true course of reasoning" about the causal principles of how to perform and explain the experiment. It is this that gives science the sense of "approaching the truth" but never quite reaching it. In so far as experimenters can be said to aspire to gain this kind of knowledge, it must be gained at the end of the experiment (otherwise it would not be an experiment), and while experimenters claim to get closer to it, they never reach it. Complete and perfect knowledge is never attained, and scientific knowledge always remains provisional and tentative (and falsifiable). When "true course of reasoning" is given in terms of the realization and exercise of natural mechanisms, promising the knowledge of the eternal causes of change in the natural world, and the mathematical laws that govern them, this kind of knowledge remains practical in its usage but is *techneic* in its form even when given theoretical interpretation and meaning.

It is this sense of *techne*, as an asymptotic ideal, that is applicable to experimental sciences, such as physics.[213] When *techne* is the kind of scientific knowledge aspired towards through experimentation, "the true course of reasoning" involves the reduction of the inquiry into natural processes to the question of "how does it work?" When the mathematical operators of theory are given meaning in terms of natural mechanisms, this allows theory to be represented as *episteme* while in practice it is explored as *techne* and tested in terms of its efficacy as a guide and template for human interventions (representational and material techniques) in contexts of experimentation and technological activity. The Scientific Revolution presupposed a conflation of *techne* and

---

[213] Rogers (2005)

episteme that was based and justified on the implicit acceptance of mechanical realism, which underwrote the postulated universality and eternality of the mathematical science of mechanics and all subsequent physical theory. Mechanical realism provided Galileo's physics with an epistemological justification and methodology as a means to access the objective reality of Nature through the mathematical inscription and modeling of machines and mechanical ensembles in terms of causal mechanisms and their quantifiable effects on the motion of inanimate matter. Since Galileo, natural philosophers (and scientists) have disagreed with each other about the theoretical interpretation of those mechanisms, as well as disagreed about the philosophical meaning of scientific knowledge, but they have all agreed that these mechanisms are natural mechanisms and it is the task of experimental science to discover them and obtain knowledge about them.

What was the relation between mathematical intuition and experience in Galileo's experiments? By reducing phenomena such as motion and oscillations to include only those aspects of the phenomenon that are amenable to mathematical representation, say the fall of a cannon ball or the swing of a pendulum, Galileo projected Euclidean geometry over experience to filter out everything that was extraneous to his preconception of the essence of the objective reality of space, time, and motion. Galileo anticipated the nature of the object of inquiry and projected his mathematical intuitions over the phenomenon. The relation between Galileo's mathematical intuitions and experience is one of mathematical projection.[214] Understanding this act of anticipation is essential if we wish to understand science as a series of interventions and their interpretation. Measurement, observation, calculation, and relating representational and material practices are all fundamental aspects of experimental science, but these acts were also fundamental aspects of ancient and medieval crafts and mechanics. Experimentation as a means of acquiring information and testing cognition via a definite ordering of things and events has been a basic kind of experience and activity involved in all craft work, tool use, and material practices. This was familiar to ancient and medieval craftsmen alike. Ancient and medieval craftsmen worked with measurements, and often used mathematics and appealed to experience, and were no different to modern scientists in this respect. It was not the use of facts, experiments, measurements, and mathematics, and appeals to experience that was the fundamental novelty of the emergence of the modern sciences and these characteristics were not the defining characteristics of the Galileo's physics. It was the projection of mathematical representations over the phenomenon of motion and its reduction to only those aspects (properties) that were revealed through the projection were the defining characteristic of Galileo's science.

Galileo (Descartes, Newton, *et al.*) conceived the motion of each and every body as having one basic template according to which motion was nothing more than the determination of geometrical points, lines, and planes in uniform space and time. This basic template circumscribed its realm of application as both universal and uniform throughout all of existence, even though no such experience is possible for human beings. This allowed a level of abstraction over and above experience that was used to explain experience in term of unexperienced conceptualizations of phenomena that decided in advance what was to be learned from the experience of any phenomenon. For example, the conceptualization of a body moving under uniform and universal rectilinear (Euclidean) motion in space posited by Galileo (Descartes, Newton, *et al.*) as moving in a straight line at constant velocity forever until an external force acted upon it was one that

---

[214] In his 1929 inaugural lecture *Was ist Metaphysik?* and 1962 lecture *Die Frage nach dem Ding* Martin Heidegger considered mathematical projection to be the fundamental conceptual difference between modern science and medieval *scientia* or ancient Greek *episteme* (Heidegger, 1999b; 1999c). See Rogers (2005).

did not correspond to any experienced movement of a body on Earth or across the Heavens, and there is not any conceivable experiment that would or could bring such a body into direct perception as an object of experience. Mathematical projection within Galileo's physics (and subsequently Descartes, Newton, *et al.*) was premised on fundamental presuppositions and expectations that anticipated the phenomenon of motion and were placed over it in order to have particular experiences of motion that could be described by mathematics and corroborated through experimentation, and therefore it was a metaphysical act that determined in advance how space and time were to be understood and experienced.

How could a law that describes an imaginary abstract as a fundamental representation of things in contradiction to experience justify the new "empirical" sciences? Of course it is arguable that we could travel deep into the vacuum of space and there put Galileo's mathematic projection (aka. Newton's First Law of Motion) to direct experimental test, but there are two problems with this proposal. Firstly, Newton's First Law was widely accepted 250 years, or thereabouts, before our technological ability to perform this experiment was even possible, by sending a rocket out into deep space, and, therefore, we cannot claim that experience was a criterion for its acceptance. Secondly, even if we were to perform this experiment, how we could *prove* that the body was *in fact* moving in a straight line? Or that it would continue to do so *forever*? Or that it would do the same regardless of its direction or location in deep space? We need to acknowledge the irony of the empiricist's rejection of ancient Greek *episteme* and medieval *scientia* as mere metaphysical speculation, when physics—the exemplary 'hard' science—is premised on non-empirical precepts and projections.

Mathematical projection is a conceptual project of conceiving the essence of phenomena that skips over events to open an abstract (imaginary) domain wherein "the facts of experience" can show themselves in a way that confirms the correctness of this anticipatory act of mathematical projection or demonstrates its need for revision. Hence Euclidean geometry and Galilean physics was corrected by Minkowskian space-time and Einstein's physics, all while remaining mathematical projection. The template (or framework) of the event is superimposed over the event, informing and constraining how the event can be experienced, and thereby acting as a guide (or set of rules) for the selection, ordering, and unification of experiences; it reveals to experience how the phenomenon responds to this act of superimposition and it is that response that is taken to be objective. This has a two-fold significance. First of all, qualitative aspects of the phenomenon are swept aside and anything that does not fit within the projection is filtered out; secondly, there appearances (experiences) are passed over in order to reach the underlying objective reality—as it has been anticipated. The appearance of the phenomenon is an obstacle or a hindrance—not even a clue or token—to the task of obtaining the knowledge of its mathematical description and representation. Mathematical projection does not actually attend to the experiences of phenomena or events, given that it is brought to the phenomenon, and is only able to operate as a mode of disclosure of the essence of the phenomenon via its own metaphysical interpretation and anticipation.

Galileo (Bacon, Descartes, Newton, *et al.*) gave their commitments to attending to observation and experience of the phenomenon of movement of things in the world, denouncing their predecessor inattentiveness to experience, but in practice they skipped over experience in order to conceive of natural motion in terms of an abstract uniform and universal rectilinear motion (a grid of points and lines—what came to be known as a Cartesian frame) in Euclidean space and absolute time (operational outside of space). This metaphysical act demanded that motion must be properly conceived of in a specific way—within Euclidean space and absolute time—and what and how motion was represented and evaluated was something intellectual that was brought to the phenomenon and projected over it. It was axiomatic in its own practice and based on an intellectual framework through which experience was to be interpreted (and this was how the mathematical projection of space and time was understood by Kant in his *Critique of Pure Reason*, although he understood it as a structure of the mind itself). One need only go outside and watch the flight of a bird in comparison to a straight line to see how the straight line is something that is projected over experience and used to have the experience of the trajectory of a bird's flight, but it is not something that is internal or inherent to the flight of the bird, given that the axioms of Euclidean space do not exist anywhere in the world—they are projected over it.

The axiomatic abstraction of motion that was taken and posited through mathematical projection was of such a kind as to set things upon their foundation in advance by describing experience and explaining practice only in terms of fundamental propositions accessible to geometrical formulation and analysis. As an axiomatic project, mathematical projection anticipated the essence of things by sketching in advance the basic template of the permissible objective properties of natural phenomena and how they were to be described, alongside projecting the grid or frame of reference from within which those properties were mapped out and quantified. From the outset, the nature of scientific inquiry into motion was predetermined by the mathematical projection of motion in order to allow a line of questioning that posed conditions in advance to which Nature supposedly could only answer numerically, one way or another. Scientific experience is defined in terms of measurement and calculation, and experimental science is mathematical, because of this template projected over phenomena. Methodology and experience are reduced—premised on the reduction of how "the physical" was conceived from the outset—in order to establish measurement and calculation of the quantifiable aspects of natural phenomena as the defining characteristics of modern science since Galileo.

By conceiving of "the physical" in terms of a particular mode of intervention and disclosure appropriate to its axiomatic predetermination by the template, things were shown and understood in a way that was metaphysically prefigured in the project, and determined the mode of experiencing and studying phenomena in terms accessible to mathematical projection (i.e. measurement/observation and calculation/prediction). This template circumscribed the realm of the object's possibilities in terms only accessible to measurement and calculation and excluded everything else. It determined what could be considered as objective—as belonging to Nature—and what would be considered as subjective—as illusionary or irrelevant. Henceforth, natural phenomena could only be what they appeared to be within the template. This reductive act established uniformity and universality of space, time, and motion, and it also required and made possible the conception of the measurement of quantities as the essential determinant of things, and the calculation of outcomes as the test of theory. The mathematical projection of uniform and universal motion would *in the future* determine which bodies could be a part of Nature and explored through scientific activity. Experience was only meaningful in term of quantities that were already interpreted in terms of the mathematical projection of properties, such as inertia, mass, and force, and thereby available to measurement and calculation.

It is no coincidence that Galileo claimed that mathematics was the language of "the Book of Nature" (and Newton claimed his mathematical methods revealed *the system of the world*). Anything outside of the template was excluded as an *a priori*. Galileo's physics did not use mathematics because he conceived of ratio or geometry as revealing of transcendental forms accessible only to reason (as a Renaissance Platonist might have done), but, on the contrary, Galileo's use of mathematical representations and techniques was a *consequence* of the idea of mathematical projection as a means to grasp the underlying reality that explains experience in quantifiable and mechanistic terms. Galileo's application of Euclidean geometry (or analytical geometry by Descartes, infinitesimal calculus by Newton, or differential calculus by Leibniz) was only possible because of the metaphysical anticipation of the truth of mathematical projection implicit from the outset of Galileo's work. Henceforth, objects were determined only in terms of their positions in space and time, and this reduced and focused how natural law and change were understood in technological terms within Galileo's experiments.

This would explain why onlookers disagreed with Galileo's experience of the fall of bodies during his famous (and perhaps apocryphal) experiment of dropping weights from the tower of Pisa. They saw the weights hit the ground at slightly different times, while Galileo triumphantly upheld his view that they hit the ground at the same time. Mathematical projection made Galileo's triumph possible.[215] It was because Galileo conceived the motion of all bodies as rectilinear, universal, and uniform that he could conceive of acceleration as changing uniformly when an equal force affected it—thus cannonballs of different weights should fall at the same rate of acceleration and hit the ground at the same time—and any deviations could be explained (transduced) in terms of an external force such as air resistance or friction. This was also how Galileo could conceive of the motion of a body thrown onto a horizontal and smooth plane as being uniform and perpetual, if this smooth plane was extended infinitely in our imagination, when, of course, no such infinite and frictionless surface exists anywhere in the world. Galileo used this thought experiment to present his conception of the motion of a body in such a way as to allow the reader to cognate its truth for him- or herself. The mathematical projection of the template was tested, in the common, practical sense of the word, by using it as a methodology to calculate ballistic trajectories and tested against future measurements on the firing ranges of the naval shipyards and cannon makers.

Galileo's *a priori* conception anticipated the universal nature of all bodies and the mathematical was used to reflect what Galileo expected to learn from phenomena, and, consequently, reflected his own anticipations about Nature and the possibilities and limits of human knowledge. It was this anticipatory conception of the template that not only preempted the Newtonian characterization of the phenomenon of motion in terms of universal space and absolute time, laying the foundations for Newton's Three Laws of Motion and Universal Law of Gravitation, but also laid out the template of "the empirical" in terms of abstractions that were not directly accessible to experience. No bodies are special, every place is like every other, and no motion is special; force is defined only in terms of the change in motion it causes, and, ultimately, such changes are the only thing that is real. In this sense, Galileo's proofs set the groundwork of the new sciences and were the presentation of mathematical projection as a method by which the reader was to learn how to see geometrically for him- or herself in terms of intuitions about how things behave, truly and objectively. Ironically, Galileo's new science was as abstract, interpretive, and speculative as the scholastic natural philosophies that he (and Bacon and Descartes) criticized for their abstract, interpretational, and speculative nature, and is as metaphysical and non-empirical as the natural philosophies denounced by empiricists.

---

[215] It also explains the theory-ladeness of Galileo's experiments and observations, as pointed out by much of science studies post-Kuhn and Feyerabend.

Galileo's mathematical projection transformed the experiment into an object of explanatory representation of the veracity of the projection. This presupposed the mechanical realist interpretation of the machine as an interface between objective reality and human reason, and thereby establishing measurement and calculation as the epistemological rules by which the empirical success of theory could be demonstrated to anyone versed in Euclidean geometry. The "empirical successes" of modern scientific research were considered confirmed and corroborated only when successful measurement and calculation provided the test of "the truth" of any hypothesis or theory. Henceforth, scientists could only consider the calculable to be "the empirical," and only calculable aspects of natural phenomena could be categorized as its objective properties and relations. Galileo's "experience" was premised on both speculative and operational metaphysics. Its explicit assumptions about the constitution of the world—the reduction of the natural to the mechanical—allowed physical inquiry based on experimentation to be conceptually possible as a means to demonstrate the mathematical laws of Nature.

Galileo's assumptions did not occur in isolation. They were emergent during the developing European fascination with the possibilities and potentials of machines and materials, as well as a long-established Aristotelian tradition of mathematical mechanics. The mechanical science of Galileo was the culmination of Medieval and Renaissance developments of geometrical mechanics and technological innovations; it was not the radical break from his predecessors that it has been presupposed to be. The radical break his physics presumed and attempted to justify was the application of the precepts of mechanical realism to understanding the natural world, and it was this presumption and attempt that was novel, heralded the mechanical worldview of seventeenth century mechanical and experimental philosophies, and was a prerequisite for modern conceptions of science and technology. Henceforth, Nature was conceptualized in terms of universal mathematical laws, materials, mechanisms, necessity, and efficient causes, all of which could be discovered through technological activity. This premise allowed technological innovation (bringing novel technological powers into the world) to be taken as nothing more than taking advantage of those laws, and thereby revealing them. Technological innovation (the expansion of technological powers) could be treated as human participation in the natural order of things disclosed by the mathematical science of mechanics, and the use of mechanics in the world was the proof of its own truths. The metaphysics of mechanical philosophy was reduced, in accordance with the limits of the mechanization of natural processes, in such a way as to allow the invention of new machines to have the power of disclosing natural mechanisms at work. This was possible because the conceptions of Nature had themselves been reduced to that of mechanical processes. Once the fundamental principles of Nature were reduced to be the fundamental principles of mechanics, mechanics could be presented as the means by which the fundamental principles of Nature could be discovered, and technological innovation could be represented as the "empirical test" of the scientific worldview. Henceforth, science was the mathematical projection of the six simple machines over the natural world and the construction of machines to demonstrate the truth of that projection. The circle was completed. Mechanical realism had imposed upon future generations the limits of the reductive template of the scientific worldview, and its precepts were taken to be self-evident thereafter.

In the sixteenth and seventeenth centuries, the confidence in the human ability to construct and use machines to produce new technological powers grew to such proportions that it found expression in the reification of technology as the sign of the growth and objectivity of knowledge. The presupposition of mechanical realism allowed these reified machines to become transparent means of disclosure at the service of the intellect—of "Man." Machines could be used to disclose the mechanical principles of the Grand Machine—the Universe and everything contained therein—and conceived as a self-evidently rational exploration of Nature. Once embarked upon, this allowed two important desires to be offered a source of satisfaction: The first was the possibility of a comprehensible worldview of the natural world and how human beings are situated with it; the second was the promise of novel experiences and powers as the fruit of human labor and skill at making. Mechanics offered the second; the mechanical worldview promised the first. Appeals to technological innovation and new powers henceforth underwrote the validity of the new science and the metaphorical substitution between machine performances and natural phenomena.

The new sciences and the Scientific Revolution should be understood as a radical departure from the dogma of the Church and the Aristotelian scholastic tradition, and constituted a new cultural movement—a Zeitgeist, paradigm shift, or epistemological rupture, as historically-minded twentieth century philosophers such as Heidegger, Kuhn, and Bachelard have respectively termed it.[216] It fulfilled the Baconian promise of new powers and experiences, seemed to clear away the erroneous and superstitious thinking of the past, and promised objective knowledge and methods. It offered the possibility of "seeing the truth for oneself" in terms of principles and axioms that were "self-evidently and intuitively based on the facts of experience" and could be known to "any man of reason." It founded the spirit of intellectual liberation and discipline, and creativity and order that inspired the Enlightenment. But much more than this, it gave seventeenth and eighteenth century a method by which technology could act as the interface between "Man" and "Nature," understood as an objective and circumscribed realm which could be known through mathematical projection and tested through innovation. Via mathematical projection, not only did a new way of seeing the world become possible, but there also arose an obligation to the principles demanded by technology, which resulted in the development and differentiations into specializations and disciplines, each developed in terms of its own technical rules and methods, and henceforth "the Book of Nature" could be read through a division of labor, with each specialization reading its own division that corresponded with the objective divisions within Nature itself. In this way, technology simultaneously informed and constrained human experience.

The modern conceptions of science, mathematics, and metaphysics sprang from an already established metaphysical interpretation of Nature made in terms of mathematical projection. The representation of Nature as an independent and objective reality accessible via technology is only possible as a result of the metaphysical unification and interpretation of human experiences of the technical limits, possibilities, actualities, successes and failures of interventional material practices, with a conceptual understanding of how representational practices both guide and explain them. The seventeenth century focus on the mechanical worldview and the eighteenth century focus upon refining it and applying it to new realms were premised, from the outset, on the satisfaction of individual and national interests through technology, and this conditioned how the nineteenth century's Industrial Revolution was understood and developed. The direction of scientific research was immersed in and stimulated by the practical problems of the society for which the sciences were seen as both beneficial and progressive, and, as new machines were invented, the refinement of the scientific worldview was extended to encompass more of reality. It is this extension that is termed "a growth in knowledge."

---

[216] Heidegger (1939); Bachelard (1934/1985; 1938/2002); Kuhn (1962)

Once the nature of being human and rationality was defined and represented in terms of material conditions and technical agency then human interventions in the material world could become represented as the means through which human beings could discover human nature in terms of our technological powers and limitations. In so far as we are (at least in part) defined by our actions, technological innovation changes what we are by changing what we do and how we do it. It was for this reason that Hobbes (in *Leviathan*) claimed that the only thing we can truly know are the things we have made. Hobbes' individualism was itself premised on technology as something that is objective and universal that could mediate between "Man" and "Nature"—and other human beings—which provided the physical means for the possibility of human agency and knowledge, as disclosed by the sciences and practical arts. The identification of "Man" as an isolated and rational individual is also apparent in the writings of Bacon, Descartes, Bentham, and Rousseau. This individualism was itself a consequence of the beliefs in the primacy of the techniques of rational and reasoned discourse and practice, alongside a conception of technology as the interface between "Man" and "Nature" as disclosed by the science of mechanics and the practical arts. It established an image of "Man" as a being who was alone and perpetually struggling against Nature, for whom science, technology, and society were instruments in this struggle.[217] These conceptions of "Man" and "Nature" were a consequence of mechanical realism and the technical intention it inspired, and were the conditions for the Industrial Revolution and mass participation in fulfilling the technical intention to be represented as progress toward human destiny and the satisfaction of Man's needs. This aspiration and the possibility of its realization were premised on the same mechanical realist reduction found in Galileo's new sciences, Bacon's passion for the new sciences in terms of the practical arts being the means to discover the laws of Nature through material practices and new inventions, and in the philosophy of Descartes. It was this way of thinking that came to dominate the Enlightenment and how "rationality" was understood and put into practice.

The conflation between *techne* and *episteme* in experimental science promised "the growth of knowledge" and "empirical success," thereby satisfying Bacon's dream and allowing representations of human nature in terms of a rational material agent whose primary function is to improve the world through labor, and whose reward would be certainty and security, and new powers, experiences, and knowledge. Just as this way of thinking came to dominate the Scientific Revolution of the seventeenth century and the Enlightenment of the eighteenth century, it dominated nineteenth century thinking about Nature (such as John Stuart Mill's in his essay *Nature*, 1874), wherein everything that is artificial is also natural, and technological innovation demonstrated an increasing convergence between human perception and objective reality, and it is on the basis of such knowledge that human beings are best placed to understand themselves and society because it is of their own making. The greater degree of artificiality and technological advancement was represented as being a step closer to achieving a state of harmony and balance between "Man" and "Nature," and, consequently, between human beings and society. It was the foundation of the establishment of a universal humanity and rationality. Thus, even though the ongoing development of technology places human beings at a far remove from any natural and organic state of being (i.e. communal animality), we are born into a world that has already been changed by human beings in accordance to human purposes, and by presupposing that it is only the artifice which is implemented and developed in accordance with natural laws that has a chance of working; the human

---

[217] Georg Lukács (1967: 35), termed the individualism of the post Renaissance as "an individual consciousness à la Robinson Crusoe," yet it was this image of Man's struggle against Nature that inspired Marx's and Marxist thinking about science, technology, and society. See Rogers (2006).

understanding of what is involved in being rational and empirically successful was itself informed and constrained by technology. By the end of the nineteenth century and the turn of the twentieth century, the mechanical realist dream of explaining the totality of existence in terms of underlying mechanisms had become realized as a positivistic and technical task, bound together with technological innovation of the technical and scientific means to liberate and enlighten humanity by constructing a more efficient society. Its metaphysical origins had been forgotten.

4

THE TECHNOLOGICAL FRAMEWORK

Thomas Kuhn argued that the mathematical and empirical approaches to natural science emerged from the scientific worldview of the seventeenth century as two distinct "conceptual frameworks," which subsequently asserted themselves as the two distinct philosophical traditions of rationalism and empiricism (Kuhn, 1977). Developing Kuhn's argument, Margaret Osler argued that these two conceptual frameworks manifested themselves in terms of distinct "styles" of scientific narrative and practices, each governed by its own metaphysical and epistemological assumptions (Osler, 1994). Osler argued that these assumptions presupposed theological commitments regarding the nature and powers of God, and these two "styles" were exemplified in the correspondence between the natural philosophers Pierre Gassendi and René Descartes. The development of rationalism and empiricism into distinct conceptual frameworks followed their theological arguments about God's relation to natural laws and necessity. The rationalist Descartes argued that size, shape, and location were the primary qualities, and, since these were all geometrical properties, the essence of any material object could be known through mathematical reasoning. God was perfect and unchanging; therefore the mathematical laws of the Universe, created by God, were themselves perfect and unchanging. While Gassendi argued that God was absolutely free and could change the laws of the Universe whenever God chose to do so, Descartes argued that it was not a question of whether God could change these perfect laws because once God had created the perfect laws of Nature then God would not choose to change them. The empiricist Gassendi argued that all natural phenomena could be explained in terms of atoms of inanimate matter and their motion in geometrical space. This speculative metaphysics, based upon the ancient atomism of Epicurus and Lucretius, postulated that the Universe is composed of atoms and the void. Gassendi argued that atoms only possessed the qualities of size, shape, and heaviness, which can only be known empirically, and consequently the Universe cannot be described in terms of *a priori* knowledge achieved through human reason. His theory of natural philosophy was based upon the assumption that natural essences and laws are knowable only to God, whereas, for humans, objective knowledge are limited to experience and measurement.

How should we understand these theologically-based conceptual frameworks in the light of mechanical realism and technology? If these two conceptual frameworks resulted in two competing philosophies of science, rationalism and empiricism, and subsequent philosophical debates, say between Hume and Kant, and all the terms and distinctions they used, were the consequence of Gassendi v Descartes, then the distinction between the two is quite simple: the empiricists tended towards gathering the facts and practical problem solving, while the rationalists tended towards the development of grand theories of the Universe. These "styles" differed in the emphasis that they placed on whether empirical evidence or mathematical theory were primary for the interpretation of natural phenomena and the understanding of how the Creator created the world. However, we should heed Ian Hacking's word of caution when using the word "style" to characterize the difference between rationalism and empiricism.[218]

---

[218] Hacking (1982; 1992)

The term "style" should be reserved for the writing and presentation practices that have been developed to convey specific scientific narratives, rather than distinct philosophies of science.[219] In my view, due to the distinct set of visions and goals at work in these two dimensions, the term *teleological positing*[220] constitutes a clearer term than "style" to highlight the distinction between practical problem solving and grand cosmological theorizing. Teleological positing involves positing a goal or end of labor (intellectual or material) that is projected as the purpose or meaning of organized human activity for other human beings. Rationalism and empiricism were the formal outcomes of distinct teleological positings for human intellectual labor, but were developed in relation to each other and the scientific worldview, as the disagreements between their proponents were disagreements about prioritization of one teleological positing over the other. The teleological positings of theory and experiment are indicative of different human ambitions in response to the question of what could be achieved with the new sciences, as well as different speculative metaphysical ideas about the nature of the Universe and God, but they both remained grounded by the precepts of the same operational metaphysics and given the same objective meaning by the labors of scientists. I shall show in this chapter how theories and experiments, along with distinct narrative styles and forms of representation, occurred within *the same* technological framework of interconnected representational and material practices.

As I have already argued, it is this idea of usefulness or instrumental value of scientific representations and practices within the technological framework that underwrote the notion of scientific objectivity (and Hacking's instrumental realism). Osler (and Kuhn) missed the unitary operational metaphysics that underwrote these philosophies and situated them within the same technological framework, and, consequently, Osler (and Kuhn) were unable to explain how these two distinct conceptual frameworks were brought together within the technological framework as two aspects of scientific activity. Once we take technology and its operational metaphysics into account, we can see that mechanical realism was presupposed by both of these philosophies of science. These two traditions or conceptual frameworks were derived from the same mechanical realist metaphysical precepts, as two approaches within the same technological framework, and differed from each only in terms of their different speculative metaphysics (e.g. whether matter was continuous or discrete, or whether it could be known in its essence). Both approaches understood "the empirical" in the same way, given that "empirical evidence" was restricted to the measurement of variables and constants, which, at least in principle, could be measured using mathematically described and calibrated technological devices, apparatus, and instruments. This was the same regardless of whether these quantities were actually measured in practice, or only used in thought-experiments. Mathematical treatments were limited to forms that could be both abstracted from the observation of the performance of mechanical devices, thereby both informed and constrained by technology, and "tested" in accordance with their usefulness or instrumental value in the design, building, operation, calibration, and interpretation of such devices, and providing a representational connection between the manipulation of physical variables and operational procedures.

As the empirical and mathematical dimensions of the same technological framework are both brought together and unified by the framework, thereby giving both dimensions shared interpretations and meanings, the distinction between them pivots on the usefulness of mathematics to experiment or the usefulness of experiment to mathematics. These decision between them is reduced to the question of whether the

---

[219] See also Derrida (1967: 81-7)
[220] Lukács (1967); see also Rogers (2005)

emphasis should be on using mathematics to deduce testable predictions about machine performances (termed as empirical observations or measurements), or whether the emphasis should start with machine performances (again, termed as empirical observations or measurements) to induce and decide between competing mathematical theories or directions of technological innovation. Either way, the central constraint was that of the mechanization of the test of any quantifiable hypothesis or proposition in such a way that it could be part of the technological framework of the experiment. It is the interaction between both dimensions of this two-fold dimensionality within the same technological framework that is central to the development and refinement of the scientific worldview. Due to influence of the science of mechanics on the projected template of the new sciences, which defined how any assumptions could be tested, neither theoretical nor empirical teleological positings could be placed under experimental test in terms of the other, without invoking the very same assumptions that were supposedly being "tested." It would be impossible to set up any impartial experiment to facilitate making such a decision in favor of rationalism or empiricism. The question of whether we should prioritize theoretical or practical goals in order to promote the growth of knowledge is not something that can be answered by an experiment. Hence the teleological positing of labor has a fundamental function in making such a decision, and therefore we cannot remove human ambitions and ideals from the philosophy of science without simply making a partisan stance in favor of one conceptual framework over the other. This requires an anticipation of the essence of phenomenon, and whether it can be known or not, and the explanation of this anticipation requires metaphysical concepts. Rationalism and empiricism were both were based on the same mathematics and science of mechanics, assumed mechanical realism, and informed each other, refining their own positions in relation to the other; consequently there was not any mechanical means, either in deed or thought, by which a decision could be made as to the superiority of the one over the other. In this sense, both teleological positings can be seen as hypotheses about the potential of the whole of science, as a cultural enterprise, and reveal the totality of science to be an experiment into itself.

Mechanical realism had been used to justify the Galilean reduction of the phenomenal world into quantifiable properties that could be mathematically projected over human experience as the objective referent, and thereafter presented as the whole world—as a mechanized totality. The metaphysics of mechanical realism allowed machines to become both the concrete object of research into the natural world and the interface between "Man" and "Nature." This allowed machines to become a transparent (neutral, passive) means of "seeing" otherwise invisible natural mechanisms at work, as it was only when the natural mechanisms postulated in theory had been demonstrated to occur in stable machine performances, by their own instrumentality in organizing representational and material practices, that they were considered by scientists to be "tested" and "real." As a means of disclosure, the apparatus or instrument must be "grounded" on the interaction between mathematics and its machine performances, and therefore the form and content of both theory and experience have been transformed by the mechanical realist precepts and the techniques utilized to disclose those "natural mechanisms" at work in Nature. The forms and content of the "theoretical" and "empirical" aspects of experimentation are both informed and constrained—mediated—within the technological framework of experiment science and tested in terms of their instrumentality within the ongoing development of the technological framework. Once mechanical realism had become culturally established (which it certainly had become by the latter part of seventeenth century) then both mathematics and observation had become integrated through technology into experimental and mechanical natural philosophy, and we cannot consider rationalism and empiricism to have originated from distinct conceptual frameworks at all. The distinction is simply between the mathematical and practical dimensions of the same technological framework and its metaphysical interpretation. The mathematical dimension was more apparent in the rationalists (grandiose mechanical realists) such as Descartes, Galileo, and Newton, whose problems involved developing a mathematical description of the fundamental mechanisms of Nature; whereas the practical dimension was more apparent in the empiricists (modest mechanical realists) such as Boyle, Pascal, and Newcomen, whose efforts were directed towards developing particular machines in order to solve particular problems.

This is evident if we look at Newton's approach in *Principia*. Newton was an empiricist in so far as he argued that all facts should be induced from experiment and reevaluated in the light of further experiments. He was also a rationalist in so far as he argued that the demonstration of any truth should be deduced from mathematical first principles. *Principia* was already situated within the ongoing development of the technological framework of the new science and assumed mechanical realism as a given. This is also evident in Newton's *Opticks* (first published in 1704) where the lens is itself reduced to an optical lever that mechanically operates upon (otherwise) rectilinear rays of light, and the polarization of light is treated in terms of the wheel and the lever template projected over the phenomenon, just as Galileo had reduce the phenomenon of a swinging pendulum to a demonstration of the mechanical nature of time. This phenomenon could only be understood in terms of the projection itself (as Newton could not have any experience of the path of light, just as Galileo could not know time in isolation from events) and therefore revealed only in terms of the projection in relation to the technological framework.

The eighteenth and nineteenth century studies of mechanics, optics, thermodynamics, and electromagnetism were all further refinements of the seventeenth century scientific worldview that all emerged from within the same technological framework based on mechanical realism. How can these different mechanical, optical, thermal, and electromagnetic machines be characterized in such a way as to reveal a general principle by which practical experiments and mathematical theories can be linked and shown to be emergent from within the same technological framework? How was the methodological principle of mathematically projecting the abstracted motions of the six simple machines applied to different kinds of machine? I argue that this was achieved by epistemologically representing stable and repeatable machine performances as observations or measurements—each represented as the exemplary means to disclose a natural mechanism operating in accordance with a natural law—while maintaining their categorical ontological distinction in relation to the different kinds of materials from which those machines were built. Hence, thermal machines and electromagnetic machines can be said to realize distinct sets of natural mechanisms, exercised according to distinct natural laws, due to being built from distinct kinds of materials, but they were also said to reveal the same patterns of motion that could be treated using the same mathematical laws and thereby revealing an underlying unity and symmetry in Nature. The technological innovation of machine hybrids has provided the promise of a unified grand theory of everything, by bringing previously distinct forces together, once technology had been developed sufficiently to disclose it in terms of unified mechanisms and mathematical formulism. Of course this underlying source of unity and symmetry is the technological framework itself, but due to the precepts of mechanical realism it has become transparent (neutral, passive) as an objective means to disclose the natural unity and symmetry of the scientific worldview.

The distinct fields of eighteenth and nineteenth century physics all developed within the same technological framework that reduced natural phenomena to circles, anti-parallels, orthogonal reflections, levers, screws, and push-pulls, and thereby able to refine the same scientific worldview through the development of distinct mathematical laws of mechanical motion, each made in terms of the same form of mathematical projection, but projected over different ensembles of materials. The distinct fields of physics cannot be reduced to one another because they involve distinct sets of stabilized and repeatable performances of machines built out of distinct kinds of materials. Each set of representations is associated with a set of materials. I term these associations as *machine-kinds*, i.e. mechanical, optical, thermal, electrical, and magnetic machine-kinds, each with their own set of fundamental representations and mathematical theories.[221] Machine-kinds constitute what Cartwright termed as *nomological machines*, and, when interpreted in mechanical realist terms, they disclose what Bhaskar termed as *strata*.[222] The unification of any two machine-kinds, i.e. electrical and magnetic machines, is made in terms of synthesizing underlying representational and material practices in isomorphic terms to produce the same theory that predicts and explains "more fundamental" nomological machine performances, i.e. an electromagnetic machine-kind, built out the materials from both "parent" machine-kinds, and revealing of a "more fundamental" stratum of reality, and this is achieved by destabilizing the technological framework through innovation and subsequently re-stabilizing it in terms of a new and unified machine-kind and its fundamental representations. At the core of this ongoing process of synthesis and discovery is the dialectics of technology.

Via the cultural acceptance of mechanical realism, the historical trajectories of technological and mathematical innovation have been metaphysically brought together and appropriated into an ontological trajectory of discovery of deeper and more fundamental levels of matter and force through technological innovation and the further stabilization of the technological framework. This trajectory of synthesis of novel machine-kinds within the same technological framework and their projection as nomological machines over the natural world is termed by mechanical realists, such as Bhaskar, as the disclosure of ontological depth. This trajectory of the synthesis of strata follows technological invocation and technology becomes the objective referent and structure for a new theory, with a new set of fundamental representations, which can be used to explain and predict the performance of both newer and older machine-kinds because it shares the same technological framework and forms of mathematical projection. By representing his synthesis as a discovery of something that exists independently of the means by which it was discovered, and by making technology transparent as a mode of disclosure, the objective referent is represented as natural structures, capacities, or tendencies, which is then used to confirm the veracity of the act of mathematical projection through further technological innovation. Modern scientific research achieves progress by instrumentally implementing explanatory theories within the technological framework, itself developed and structured through the innovation of new machine-kinds as stable nomological machine performances, communicated via mathematical and visual representations, and metaphysically given meaning as an exploration of more fundamental strata of natural laws and mechanisms. These representations are "tested" by implementing them in the ongoing work of constructing and structuring the technological framework, and the success of implementing them in an ongoing research program is itself considered to confirm that these representations have some truth in them. Yet it is this process of implementation that has been ignored by the traditional philosophers of science. If we are to understand this process, we need

---

[221] Rogers (2005)
[222] Cartwright (1999); Bhaskar (1975)

to look at how the technological framework and the scientific worldview are related to each other.

The technological framework and the scientific worldview:

What does the term scientific worldview mean? Let's break this term apart. The part "world" refers to the totality of things or whatever exists, in its entirety. The part "view" refers to a cognitive grasp or "seeing" this totality. Hence, the observer—the one who grasps or sees—is imagined as outside of the world, at least as an observer, and is able to grasp the totality from a mental Archimedean point. This abstract point is entirely a product of the human imagination, but it is a fundamental intellectual act of reduction that is required for the possibility of having a worldview at all. Without having a worldview, without this Archimedean point, there could be no possibility of having a scientific understanding of the world.[223] Mechanical realism underpinned the scientific worldview and is the fulcrum, so to speak, upon which Archimedes would have stood to move the world. It is the foundation for the Scientific Revolution. Without it, we can innovate representational and material practices, and bring new technological powers and objects into the world, but we could not understand those practices in relation to an overall comprehension of a cosmological order and our place within it. It is this overall comprehension that is required to connect the representational and material practices developed and organized in the context of an experiment with a unified and communicable explanation of those practices in terms of natural laws and mechanisms, and understanding any knowledge obtained from particular experiments as providing a piece of knowledge about the Universe. Via the technological framework, the scientific worldview affords us the position of articulating, securing, and organizing knowledge of an objective world that we are simultaneously a part of and able to distance ourselves from in order to grasp it in its entirety. This allows us to measure and draw up the guidelines for interpreting experience in accordance with our powers to predict and manipulate things. "The world" is in place before us as a representation of all that belongs to it and stands together in it; as a total system that we are acquainted with as something that we can view and grasp, and are equipped and prepared to deal with through technological activity. When the objectivity of the scientific worldview is made credible through technological activity and the reproduction of mechanism, the Archimedean point is the technological framework itself. The scientific worldview presents to us as a totality that we can comprehend and engage with through representational and material practices, while representing the technological framework in terms of the realization and exercise of natural mechanisms, wherein these practices are associated and given meaning as engaging with the objective material substratum governed by natural law. The scientific worldview is itself a metaphysical act of projection of an abstraction as a template over human thought and experience used to guide specific sets of human practices (i.e. those involved in technological activity) and give them meaning.

Let's take a look at the concept of electrical charge, as an example. This idea was first proposed by William Gilbert in *De Magnete* (published in 1600) as an explanation for the behavior of a *versorium* (a simple mechanical device comprised of a rotating metal needle on a pedestal) when an amber rod rubbed with wool was placed next to it. Gilbert

---

[223] In his essay 'Science and the Age of the World Picture' (1977b), Heidegger considered having an intellectual grasp of the world, whatever is in its entirety, (*Weltbild* in German) to be a fundamental cultural characteristic of science and the modern age. See Rogers (2005).

coined the word *electricus* from *elektron*, the Greek word for amber.[224] *Electricus* was understood by Gilbert as a natural consequence of rubbing a piece of amber with wool or fur, and the repeatable behavior of the *versorium* suggested a natural mechanism at work. However, this term did not explain the phenomenon. Nor did it merely give us a new word for it. The term *electricus* and its explanation in terms of the rubbed amber causing the behavior of the *versorium* provided a repeatable means by which the phenomenon could be reproduced, explored, and explained by using a technological object to map out the contours of human interventions (moving the rubbed amber towards the *versorium*) and machine performances (the movements of the needle), thereby making a sensor. This technological context preconditioned how the phenomenon was going to be demonstrated, understood, represented, experienced, and shared with others. It took 60 years for Otto von Guericke to understand this phenomenon in terms of an electrostatic force, which he termed as *a cosmic virtue* in Book IV of *Experimenta Nova*.[225] The earlier chapters in von Guericke's *Experimenta Nova* were concerned with the construction and explanation of a series of modified and refined air-pumps as a means of demonstrating the existence of a vacuum and the corporeal nature of matter, and the rest of this book was concerned with magnetism. In Chapter 15, von Guericke described how he demonstrated and explored the cosmic virtue of *electricus* by building a machine which could both produce the phenomenon and act as a sensor. This device was a polished sulfur globe rubbed by hand and placed via a wooden stick in measurable proximity to small pieces of cloth, hair, feathers, etc. to map out the contours of attraction and repulsion in a way that could be reproduced by others by building the same machine and performing the same intervention. This demonstration of *electricus* was repeated by Robert Boyle in 1672 for the Royal Society. Boyle went even further and used his air-pump to show that repulsion and attraction crossed "a vacuum."

After von Guericke's first electrical friction machine had been further developed by Francis Hauksbee, at Newton's suggestion in *Opticks*, by rapidly rotating a glass sphere against a cloth belt, this new machine-kind provided the basic template for the further development and refinement of a series of increasingly sophisticated electrical friction machines throughout the eighteenth century by many experimentalists, such as Stephen Gray, William Watson, Andrew Gordon, Benjamin Wilson, John Canton, and Martin van Marum, and thereby provided the objective infrastructure for these machines to be represented as nomological machines. These electrical friction machines comprised the electrostatic machine family and were constructed out of mechanical and electrostatic machine kinds. They were developed in parallel to electric circuits and other electrical machines by Volta etc. after William Watson invented the first electric circuit in 1746. Since Stephen Gray had classified materials as "conductors" or "insulators" in 1729, the entire understanding of electricity was limited to how these machines were understood to work and the fundamental terms (and concepts and representations) used to formulate and articulate this understanding were developed through the invention and construction of these machines. The performance of these machines provided the empirical facts for George Ohm to abstract into what is known as Ohm's Law in 1827, thereby proposing a simple relation between current, voltage, and resistance. This parallel construction continued until 1839, when Michael Faraday built a series of experimental machine prototypes to show that the apparent division between static electricity and current electricity (and bioelectricity) was incorrect, and all were a consequence of the behavior

---

[224] The attraction of feathers and other small pieces of material by amber rubbed with fur was recorded as a phenomenon of interest by the ancient Greek Thales of Miletus circa. 300BC. Thales considered it to be a special property of amber.
[225] Healthcote (1950)

of a single kind of electricity appearing in "opposite polarities."

It was this development of fundamental representations and concepts through technological activity and innovation that provided the cultural resources for how electricity was understood as a natural phenomenon. To understand this, we need only to look at how Benjamin Franklin understood lightning as electricity after performing his famous kite experiment in June 1752. Franklin himself never wrote the story of this experiment. It took Franklin fifteen years, after the pioneering work in electrical circuits and electrostatic machines, to develop a vocabulary and set of representations to make sense of his experiment. The only record of it comes from Joseph Priestley's account, published fifteen years afterwards. Priestley's account who worked with Franklin on the manuscript (van Doren, 1938). Franklin considered his experiment to demonstrate "the sameness of electricity with the matter of lightning," drawing his terms and representations of electricity from the experimental work by Stephen Gray and William Watson *et al.* on electrical circuits and electrostatic machines and thereby identify the spark as evidence of "electrical conduction." Franklin used these shared terms when describing how his kite experiment demonstrated that the "the matter of lightning" is "electrical fire," which he followed the fluid model of electricity, which had itself been developed during the early experimental work on electrical circuits, and considered electricity to be comprised of two opposing forces.

"Make a small cross of two light strips of cedar, the arms so long as to reach to the four corners of a large thin silk handkerchief when extended; tie the corners of the handkerchief to the extremities of the cross, so you have the body of a kite; which being properly accommodated with a tail, loop, and string, will rise in the air, like those made of paper; but this being of silk is fitter to bear the wet and wind of a thunder gust without tearing. To the top of the upright stick of the cross is to be fixed a very sharp pointed wire, rising a foot or more above the wood. To the end of the twine, next the key may be fastened. This kite is to be raised when a thunder-gust appears to be coming on, and the person who holds the string must stand within a door or window, or under some cover, so that the silk ribbon may not be wet; and care must be taken that the twine does not touch the frame of the door or window. As soon as any of the thunder clouds come over the kite, the pointed wire will draw the electric fire from them, and the kite, with all the twine, will be electrified, and the loose filaments of the twine, will stand out every way, and be attracted by an approaching finger. And when the rain has wetted the kite and twine, so that it can conduct the electric fire freely, you will find it stream out plentifully from the key on the approach of your knuckle. At this key the phial may be charged: and from electric fire thus obtained, spirits may be kindled, and all the other electric experiments be performed, which are usually done by the help of a rubbed glass globe or tube, and thereby the sameness of the electric matter with that of lightning completely demonstrated."

Franklin erected a lightning rod on top of his house, which "drew" lightning down into his house. The rod was connected to a bell and a second bell was connected to a grounded wire. Every time there was an electrical storm, the bells would ring and sparks would illuminate his house. Franklin described the experiment as follows. The rod was

"fixed to the top of my chimney and extending about nine feet above it. From the foot of this rod a wire (the thickness of a goose-quill) came through a covered glass tube in the roof and down through the well of the staircase; the lower end connected with the iron spear of a pump. On the staircase opposite to my chamber door the wire was divided;

the ends separated about six inches, a little bell on each end; and between the bells a little brass ball, suspended by a silk thread, to play between and strike the bells when clouds passed with electricity in them. After having frequently drawn sparks and charged bottles from the bell of the upper wire, I was one night awakened by aloud cracks on the staircase. Starting up and opening the door, I perceived that the brass ball, instead of vibrating as usual between the bells was repelled and kept at a distance from both; while the fire passed, sometimes in very large, quick cracks from bell to bell, and sometimes in a continued, dense, white stream, seemingly as large as my finger, whereby the whole staircase was in-lightened with sunshine, so that one might see to pick up a pin."

Franklin used his experiments to show that electricity consisted of a "common element" which he named "electric fire," which is "fluid like a liquid" as it passes from one body to another, but it is never destroyed. Franklin proposed that "electrical fire only circulates. Hence have arisen some new items among us. We say B (and other bodies alike circumstanced) are electricised positively; A negatively; Or rather B is electricised plus and A minus ... These terms we may use till philosophers give us better." It was this use of "fluid" as a metaphor that allowed Franklin to make sense of his experiences and communicate them to others. This metaphorical substitution allowed him to create a visualization of the phenomenon and suggest methods by which it could be demonstrated and investigated in such a way that experiments on lightning and experiments on electrical circuits and electrostatic machines could be said to be experiments on the same thing. It was this unifying conception that allowed Franklin to propose and develop the idea of electrical charge and make sense of it in terms of the performance of machines such as batteries and the behavior of lightning.

Historians tend to portray "the winners" as if they were the inevitable consequence of some initial discovery (as if the electromagnetic motor was an inevitable consequence of Oersted's discovery that a compass needle was deflected when passed next to a live electrical wire). Electromagnetic machines were constructed and stabilized during the early nineteenth century as a new machine-kind and associated set of fundamental representations of electromagnetic motions through the efforts of many scientists across Europe. This involved considerable trial and error, personal conflicts and scientific controversy. The members of the electromagnetic machine-family were all the result of contingent human choices and efforts. On this point, the sociology of science and social construction theory provide us with many insights and examples that need to be taken very seriously. When a collection of new machines shares a common history of development and representations to explain their performances then they are members of the same generative machine-family. They are constructed out of all the possible permutations of members of the same machine-kind and they constitute a trajectory within the technological framework itself by creating new inventions along a distinct stratum.

Mechanical devices ranging from a simple pulley to the most elaborate clockwork device are all members of the same mechanical machine-family. Members of the family of mechanical machines are each permutations and combinations of the members of the mechanical machine-kind (the six simple machines) and the associated representations of the basic set of six fundamental representations of motion (given as geometrical abstractions or differentials). New machine-kinds start as hybrids between already existing but distinct machine-kinds. The steam engine is a hybrid between mechanical and thermal machine-kinds—synthesized together into the thermodynamic machine-kind—and can be represented as a machine to convert thermal energy into mechanical motion, or vice versa, and, by doing so, form the prototypes for the generation of a new machine-family of locomotive engines, pumps, cranes, etc. The steam engine allowed the development of the internal combustion engine and turbines. Electromagnets are hybrids of the electrical and magnetic machine-kinds, which can be represented as a machine to convert electrical current into a magnetic field or the motion of a magnetic within a coil to generate electrical current. Bringing together the steam engine and the electromagnetic motor allowed the invention of the steam turbine and electricity generation from coal, oil, gas, and nuclear reactors. Wind power and hydroelectric power became possible. These new machines and their associated representations did not occur in the flash of individual genius, as school textbooks often portray them. They were constructed as the result of protracted labor processes and struggles involving many people over a long period of time.

Electromagnetic machines share basic chemical devices (e.g. the chemical battery) and associated techniques, along with electrical and magnetic machine-kinds (electrical circuits and compasses) and sets of fundamental representations (such as "north," "voltage," and "current"), but the new representations (e.g. "the lines of electromagnetic flux") that were developed, as these new machines were built, involved human decisions and judgments about what could be successfully communicated as an intelligible model of how those machines work and what those representations mean. This reflected a cultural context within which what could be intelligible was shaped by the feedback processes involved in demonstrating the validity of the new representational and material practices in terms of the old practices and novel syntheses. It is only much later, after these machine performances and associated representations were further quantified and abstracted in terms of mathematical laws (e.g. Maxwell's Laws of Electromagnetism), after almost a century of struggles and choices, that they constituted what Cartwright terms as a nomological machine.

Prototypes such as Thompson's cathode-ray apparatus to measure the charge-to-mass ratio of cathode corpuscles and Millikan's oil drop apparatus to measure the quantum of electric charge were not as independent as Hacking and Bhaskar would like to believe. Via the concept of "charge," these experiments shared members of the same electromagnetic machine-kind in their construction, as both utilized the same sets of representational practices and material practices (techniques, calibration standards, and instruments). By examining the history of their development we can see that these two prototypes were members of the same electromagnetic machine-family, even if they were developed to determine distinct properties of matter. Their "independence" is itself a product of making the technological framework invisible and concealing the shared technical history of their innovation and development. "Independence" is itself a product of historical construction and revisionism, after all the work has been done, and is a consequence of mathematical projection and reductionism. Once we take this into account, we can see that the history of science is not only a history of discoveries and explanations of natural phenomena; it is also the history of the analogical generation and stabilization of machine-kinds, further developed into machine-families, and abstracted as nomological machines, which have been treated as transparent means of discovery of fundamental natural mechanisms and laws, and the machines themselves can be *retrospectively* represented as the products of the application of these fundamental laws.

I am not suggesting that the early experiments in electricity were conjuring tricks; nor am I suggesting that electricity is an illusion, or nothing more than a cognitive product of social consensus. My point here is that if we pay attention to the history of experimentation we can see that the traditional philosopher of science's view that experiment is nothing more than a means to test theory is a false view. Historical developments of experiments on electricity show that the basic terms and representation of physical theory were developed in technological contexts by building machines and making them work, and these contexts conditioned and shaped how theory was developed and its fundamental representations were given meaning. What does "charge" mean? We can talk of positive or negative poles, liquids, electrical fire, ions, electrons and protons, quarks and leptons, coupling constants and symmetry, but none of these terms explain what charge *is* qua a physical entity without an appeal to techniques and machines, wherein such terms are given meaning in terms of technical operations, calibrated instrumentation, and machine performances. The whole notion of electrical charge is itself a code—an abstract construct—developed within the technological framework, from which it gains its concrete meaning through ordering associations of techniques and machine performances, and its connection with natural phenomena is one of metaphorical substitution.

Technophenomena, the technological framework and the scientific worldview:

Modern experiments are technologically sophisticated projects involving a wide range of techniques, practices, machines, tools, skills, and knowledge. The objects experimented upon, such as electromagnetic fields, paraelectric materials, photons, nuclei, particles, electrons, quark-antiquark events, superfluids, neutrinos, etc., all require techniques and machines for their production, observation, and manipulation. Without those techniques and machines, we would not be aware of these objects at all. The relationship between scientific experience and these "otherwise invisible" objects occurs by technologically transforming the macroscopic objects of everyday experience such as wires and levers into a means of disclosure of an otherwise invisible underlying reality. Experimental scientists are concerned with macroscopic objects such as machines only because these technological objects disclose the underlying causal mechanisms that make machines work. The object of scientific inquiry is not the machines themselves; rather, the *technophenomena* that are produced by those machines, which, when represented as natural phenomena through mechanical realism and metaphorical substitution, are postulated as indicative of a deeper stratum of objective reality at work, and these phenomena are experienced by reading and interpreting the outputs of the machines involved. Without machines, technophenomenon could not be known and they are explained by making reference to the operational mechanisms involved in their own production and reproduction. The properties of technophenomena such as superfluids, the dynamics of phonons in crystals, the thermal capacities of metals, the properties of lasers, superconducting materials, neutrino events, the polarization of light, etc., are all disclosed through the mediation of machines, theories, and techniques. The establishment of scientific facts and theories about relations between properties requires putting techniques to work to explain machine performances. The machines, theories, and techniques put to work to make investigation of these technophenomena possible both inform and constrain experience and how we think about it. The observational aspect of experimental work involves the active technical application and modification of theories, methods, and techniques to design, build, and operate the experimental apparatus, and each observation is the result of efforts to stabilize the technophenomenon in question and interpret how it has been mediated, informed and constrained by the apparatus. Theories, techniques, and observations cannot be disentangled from each other, except in hindsight through philosophical and imaginary reconstruction, inference, and abstraction. Once we accept this fact about how science is done, and the complicated interactions between theory and practice, we can readily see that the traditional philosophy of science amounts to little more than revisionism and rationalization.

Take He-3 for example. Experimenters in Ultra-Low Temperature Quantum Physics do not directly experiment upon "natural" helium. They experiment on He-3. This is an isotope of the element helium. It does not exist in the Earth's atmosphere. It is the by-product of the nuclear industry and (to a lesser extent) the oil industry. Technological processes are required to identify, extract, isolate, store, and transport it. Physicists experiment on a purified sample of He-3, which is made available to them as an object, with its purity achieved and measured through technological processes that they did not design or perform. By what standard is "purity" defined here? "Purity" is defined in terms of an established and repeatable technique of purification, itself comprised of filtration, control, and measurement techniques, and, in order to know whether a sample is pure or not, the experimenters must do so in relation to that technique and their theoretical expectations of the performance of the object in response to their interventions. This anticipates what the essential properties of He-3 are—themselves only collected and understood in terms that are revealed by technological processes—and only that which passes through the projected template will be defined as "He-3" (to a percentage close to 100%) in accordance with how specific machines respond in interaction with it. Before "it" can be experimented on, helium is extracted, filtered, contained, and stabilized from a "natural substance" and transformed into a product; it is a technological object with its properties defined via techniques and machines, and the totality of that object is reduced in accordance with what can be measured, manipulated, and disclosed via machine performances. The reduction (extraction, selection, filtering, and containment) of the natural substance during its transformation into a technological object, and its further utility in the production of other technophenomenon (e.g. the scattering of Cooper Pairs in a superfluid) shows how the technological framework appropriates and mediates (informs and constrains) experience of natural phenomena, and the theoretical understanding of these technophenomena are built upon the stable and reproducible associations between representational and material practices in technological contexts. It is the set of machine performances of the experimental apparatus (a dilution refrigerator) in reaction to He-3, itself acting as a technological object, that comprise the technophenomena that are experienced and experimented on, and interpreted if it they were something given to experience as objective and natural.

These machine performances are defined in terms of "natural properties" because the purification technologies have been naturalized (made transparent, neutral, passive) as a means of disclosure. This took labor and interpretation, social agreement and consensus, and confidence in those interpretations and labor processes as they were developed in relation to other techniques and machines. These largely remain unknown to the current generation of scientists who experiment on He-3, which is black-boxed in its production as a technological object into an object with a defined set of properties, which are themselves understood as functions. In addition, the properties of He-3 disclosed by the experimental apparatus used by experimenters in Ultra-low Temperature Physics[226] are only those that can be disclosed via machine performances, such as those of dilution refrigeration and voltage resonance, for example, and these techniques of disclosure further mediate (inform and constrain) how the phenomenon in question is represented, understood, and further explored. Anything else will remain unobserved. The observed responses of He-3 to the interventions of the experiment are experienced as the set of technophenomena (which are represented as the sum total of "the empirical" observations of the properties of He-3) and compared with the theoretically interpreted

---

[226] For examples from Ultra Low Temperature Physics, see Fisher S.N., *et al.*, (1990b; 1995).

interactions between the apparatus and the expected responses of He-3. The responses of the apparatus are interpreted by the working physicists as the realizations of the transfactual quantum mechanisms that are independent from the experiment and are otherwise swamped (suppressed or destroyed) by impurities and higher energy interactions. When He-3 is studied by physicists for the sake of understanding the quantum properties of superfluidity at ultra-low temperatures, it is transformed into a technological object (a sensor) for use in the technological framework of the experiment as a means of disclosing these quantum thermodynamic properties, which are themselves understood as mechanisms. The use of He-3 as a technological object to disclose quantum mechanisms is taken to be objective on basis of its functionality within the technologies at the physicists' disposal, which are themselves emergent from the historically developed technological framework of experimental physics.

Once we recognize that the scientific experience of any technophenomenon is itself mediated (informed and constrained) by a set of associated representational and material practices, which, in turn, mediates how those scientific experiences are theoretically interpreted and explained, we can also see how the content of all such experiences are dependent upon and transformed by the technological framework within which observation and measurement occurs. Consequently, experience and its interpretation are dependent upon and conditioned by the history of interventions and interpretations made during the construction of the technological objects used in the construction of the apparatus, detector, or instrument. Paying close attention to the interpretive dimension of material practices shows us that machine performance, technical accounts, technophenomena, and technological objects cannot be isolated from one another; they emerge from and exist within the technological framework itself. Technophenomena such as the electromagnetic field, photons, and electrons are defined in terms of what they do, their performance, functions, and interactions in specified contexts of use, and these must be understood in relation to the expectations of the experimenters and the goals (teleological positings) of the experiment. Understood this way, the technophenomenon is neither purely abstract nor purely concrete. It is both. The performance and explanatory accounts are made "hand in hand" through their concrete implementation in the particularities of material practices.

Consider Hans Christian Oersted's famous experiment in 1820, using a magnetic needle to map out the contours of needle dips around the simplest electrical circuit: a wire connected to a chemical battery.

"...The opposite ends of the galvanic battery were joined by a metallic wire, which... we shall call the uniting conductor or the uniting wire... Let the straight part of this wire be placed horizontally above the magnetic needle, properly suspended, and parallel to it... Things being in this state, the needle will be moved, and the end of it next the negative side of the battery will go westward..."[227]

Oersted proposed on the basis of this discovery that an electrical current causes magnetic effects. However, it is not easy to reproduce this effect. The needle does not move in a clear and stable way. It is rather chaotic and messy, and it is difficult to witness the reported effect and keep the needle from touching the wire. Try it! You will discover that it takes quite a bit of practice before you can see the effect that Oersted proposed. There is clearly something happening, and the task facing the experimenter is to map out the contours between the movement of the needle in relation to its proximity and position

---

[227] H.C. Oested, "Experiments on the effect of a current of electricity on the magnetic needle" (in *Annals of Philosophy*, 16, 1820, pp. 259-276), p. 274.

next to the wire, but it is not a straightforward task in practice. The experiment tries to find a stable and repeatable technique to do this and produce a representation of what is happening with the needle. The experimenter needs to project a grid of possible paths of motion over the needle as it moves near the wire. After considerable difficulty finding a circular path around the wire, Oersted mapped out a concentric circular pattern of the lines of force and used this representation to show the magnetic field around the wire. As a technophenomenon, the magnetic field had become "observable."

Before a novel observation can be made it requires a new technique to reproduce the effect and communicate it to other people, along with an intelligible explanation of that technique, e.g. the magnetic needle will move if it is effected by the electrical current moved towards it. Where did these electromagnets come from? The electromagnet was not discovered by empirical appeals to ordinary experience or from a prediction deduced from theory. The electromagnet was invented in 1826 by William Sturgeon, thirty-five years before James Clerk Maxwell published his laws of electromagnetism in 1861. The invention of the electromagnet and theoretical interpretation of it in terms of the electromagnetic field took considerable effort on the part of eighteenth and nineteenth century physicists. Electromagnets are only available as technological objects for theorization because of considerable efforts by the experimentalists, such as Sturgeon, Oersted, Davy, Faraday, *et al*.[228] This involves using inventing new representations of how the effect occurs to explain both the novel technique and the effect itself. Those representations are projected over the phenomenon to map out and explain the interaction with it that leads to the production and reproduction of predictable changes and their stable responses. Each technique does not come out of thin air or the whispers of the demon Belphegor; it comes out of a culture, with a history of representational and material practices, and each and every technique is developed in relation to that history. Every technique needs to be related to already-established techniques, associated representations, and their expected consequences for new techniques to be intelligible as a technique. This relation can take the form of extension along the path of analogies and metaphors—the idea of an electric *current* or a magnetic *field* are metaphors, after all—as starting points, and, thereby allow techniques developed in one context to be used in another along these lines of similarity with other techniques and their associated representations. This situates the technique within the technological framework and allows intuition to take part in discovery. Otherwise experimenters would not know where to begin to experiment; nor would they be able to communicate what they did intelligibly to other people in a way that would guide others to repeat the experiment.

When we encounter difficulties repeating the experiment through what we might term as "material resistances," and instead encounter spontaneous and chaotic behavior, which deviates from the shape of the contours of responses that is projected over the phenomenon and through which we can experience it as a resistance, we encounter what the ancient Greeks named *hyle*. This is a response to our interventions and attempts to experience the responses of the materials in terms of the abstract forms that we project over those responses. These forms are our anticipations and expectations of possible paths of motion or resistance that materials will take in response to our interventions. Necessarily abstract and general, the set of possible paths are the totality of possible reproducible associations between interventions and responses, and it is through this

---

[228] See Gooding (1990) for a detailed study of the laboratory notebooks, sketches and diagrams, as well as the publications of eighteenth century physicist Michael Faraday. Gooding's study unpacks how Faraday invented methods to map out and represent electromotive force and electromagnetism in a way that other people could use these representations and associated techniques to reproduce the phenomenon in question.

projection that the experimenter is able to make sense and order out of phenomenal chaos or spontaneity of movement. The experimental process itself is a process of producing and reproducing the means of experiencing and knowing the motion of the phenomenon in the terms and within the framework of the projection, with its reductive, quantifiable, and repeatable power. It is inherently a process of learning how to interact with and experience the phenomenon as a machine performance, and thereby transferable between contexts. It is the process of ordering the phenomenon so that its power, what the medieval Aristotelians called its "marvelous motions," can be shown in a stable and comprehensible manner. Understanding the electromagnetic field as something natural that can be reproduced by constructing specific machines—particular arrangements of specific materials and interventional techniques—is only possible if one mentally follows the path of the sixteenth century mechanical realists and alchemists by considering the source of a phenomenon to be irrelevant for the purpose understanding its nature. Just as the heat produced by a forge was considered to be in essence the same kind of heat as that produced in a volcano, the electromagnetic field produced by a motor is the same kind of phenomenon as that produced by lightning.

It was not possible for anyone to have an experience of an electromagnetic field until people had put particular materials into specific arrangements and interpreted those arrangements as "generators" or "sensors," and they learned how to do this by experimenting with those materials, through intervening and seeing how other material arrangements respond. But, does this mean that the electromagnetic field did not exist before that time? Does an electromagnetic field exist independently of the machines that produce it and the detectors that register changes in calibrated instruments and the sensors to which they are connected? A caveat is necessary before answering these questions. Of course I am not denying that something has been brought into existence by these machines, and that something is very powerful indeed, but, as I have been endeavoring to explain so far in this book, the electromagnetic field that theories refer to and explain is understood and experienced by human beings through the very same representations and descriptions used by technology as a condition of the possibility of experiencing it in context. Once we recognize that the experience of any phenomenon occurs within a technological context and cannot be understood at all without recognizing that experience is structured and given meaning by technical representations and the methods by which we came to have those experiences, as experiences that we can make sense of and communicate to others, then we can see that an electromagnetic field is a technophenomenon, as an object of experience and manipulation, that is not only dependent upon the existence of technology for its existence, but that technology also reduces and structures experience to only those features that can be disclosed through it, and therefore also reduces and structures how we communicate, interpret, and explain those experiences to others. This does not challenge the reality of lightning or what we call an electrical shock. What I am saying here is that when we use terms such as the electromagnetic field to unify and explain phenomena, we are bringing together associated representations of machine performances that have been developed in technological contexts to explain techniques and make sense of the experience of the behavior of sensors and instruments, and giving them meaning under a single index or code (i.e. electromagnetic phenomenon). It is set of mechanistic representations that are projected over the natural phenomenon and the machine alike and thereby place phenomena and machines under the same categories.

The existence of electromagnetic field depends on a whole host of machines to produce and reproduce it, making it an object available for use and theoretical interpretation. It gains its meaning from technological contexts of discovery and usage and these contexts have their own history. Electromagnets are dependent on the existence of metallurgy and chemistry; the techniques to disclose and use an electromagnetic field are also dependent upon theoretical interpretations of the practices from which they obtain their meaning as interventional techniques to map out the contours of lines of electromagnetic force. This involves long and complex historical and social processes of innovating and utilizing specific technological objects in relation to other technological objects in order to invent such machines in the first place, and it also involves the development of representations and theories to explain how machines such as electromagnets work. These relations and associations comprise a stratum within the technological framework. The electromagnetic field was disclosed through the labor processes involved in bring together diverse technological objects into a single unified stratum of related and associated technological objects available for use to explain the performances of specific machine-kind (electromagnets) developed into nomological machines that were synthesized from other machine-kinds (electrical circuits, magnets, levers, etc.) for the purpose of disclosing a deeper level of fundamental natural mechanisms at work. It is only after all this work has been done that the technophenomenon of electromagnetism can be considered to be understood as a unified phenomenon available for theoretical interpretation and explanation in terms of a set of nomological machines abstracted into mathematical laws.

Every technological object is based on a set of more basic machine performances (voltages, time-signals, frequency resonance, etc.) unified by a fundamental representation (i.e. electron, charge, repulsion, energy gap, field, etc.) in such a way as to link theoretical interpretations with technical interpretations of those machine performances. In many respects, the teleological positing of the labor processes involved in coming to this understanding necessarily become black-boxed as a function: a component in another machine. As such, theoretical interpretations and expectations are inextricably bound-together with the concrete character of the performance of technophenomena as technological objects within a technological framework that defines what is a possible intervention or response, and what is not, within the context of human labor given cultural meaning as a natural science. It is the possibilities, actualities, and limits of the technological framework itself—its *alethic modalities*—that gives technological objects their objectivity. It is only after considerable work has been done can anyone say that they are realists about these technological objects. Hence when Hacking (1983) claimed to be a realist about electrons if physicists can spray them, he was a realist in an instrumentalist sense; his claim was premised on the objectivity of instrumentality. The objectivity of any technological object is itself dependent upon explanatory accounts (involving metaphysical interpretations) of how that performance was achieved and what it is taken to disclose. These accounts do not come from the phenomenon. Spraying electrons on molybdenum spheres is an act of interpretation of a manipulative technique made in relation to a machine built in order to disclose a particular technophenomenon; in this case the technophenomenon was that of "fractional charge," and the experience of this technophenomenon is only possible in relation to the meanings, interpretations, and techniques situated within the technological framework. The meaning of "fractional charge" is itself an abstraction and code (a quantified constant) related to particular representations and machine performances that would achieve their theoretical significance as indicators of the presence of "free-quarks" through the embodiment of that theoretical interpretation in the selection of techniques and technological objects collected together to construct the machine in the first place. The teleological positings and anticipations that preempted the experiment from the outset shape the interpretations of the performances of such machines during their construction and operation, even when the experimenters remain impartial or skeptical about the theory they are "testing." It does not follow from the stability of those machine performances that the interpretations of them are correct. Nor does it follow that it has any referents that exist outside of the context of the experiment. The existence of referents is what we are questioning. To preempt its answer is naïve and instrumental realism amounts to a whiggish interpretation of history.

Of course the scientific realist still might protest: How can you say that lightning and electricity are distinct? My response is that this is the very question that is at stake. How do we know that lightning and electricity are the same? "Physicists have measured it!" replies the scientific realist. Therein lies the rub. How have physicists measured it? What is the "it" here and how is "it" effected by the processes involved in measuring "it"? How did lightning become understood in terms of electromagnetism? Benjamin Franklin's discovery and the work of Michael Faraday *et al.* comprised of observational and experimental efforts to produce representations of the electromagnetic field, and did not respect a neat distinction between contemplative, theoretical, representational practice on one hand, and instrumental, technical, and material practice on the other (Gooding, 1990). When working out plans for experiments or particular interventions, using magnets, wires, and chemical batteries or electrical capacitors, Faraday would intermix representational and material practices and their interpretation, and would only consider it to have been successful when he was able to stabilize associations of representational and material practices in a way that allowed him to communicate a method to reproduce the experiment. Faraday's approach was entirely techneic in its practice, even though it was represented and interpreted in terms of the discovery of an underlying and otherwise invisible natural force. The lines of electromagnetic force were abstractions (a map) of the contours of machine responses to human intervention, e.g. how a magnetic needle hanging from a string moves when placed in proximity to an electrical circuit, and these abstractions are subsequently projected over natural phenomena on which no experiment has been performed, e.g. lightning.

The experimenter needs these guides to intervention in order to decide whether the phenomenon is behaving in accordance with theoretical expectations. A map of possibilities and their meaning is projected over the phenomenon in advance and the behavior of the phenomenon is decided in comparison with the map. The map is refined and modified through a process of mapping out the contours of human interventions and machine performances. This is a crucial preparatory step before theory can be tested against experience. It lays in advance out a plan of possible measurements or observations, and thereby connects techniques of how to perform particular interventions with representations of how they work, and it is this connection between techniques and representations that provides the starting point for theoretical interpretations. It is this step that situates every experiment in the technological framework as it draws its resources and analogies from that framework. When investigating otherwise invisible entities, the experimenter must connect a procedural plan of techniques and their explanatory representations with the responses of specific materials arranged in specific patterns. This sets down in advance the possible events that the experimenter expects to encounter during the experiments and use as a frame of reference to evaluate the performance of the apparatus in response to particular interventions (possible directions the pointer can move, distance of proximity to wire, strength of electrical current across wire, etc.). It is on this way that experimentation can be said to explore technophenomena, which are understood as machine responses to the imposition of fundamental concepts and representations gain their meaning within the projected plan of action.

Paying attention to the interplay between representational and material practices reveals how experimentation occurs through planned interventions upon objects in the world that are guided by the experimenter's conception of the object and the world. Through a series of planned actions, the object, the conception of the object and the world, and the experimenter's conceptions of how to intervene are all transformed in response to machine performances. Through experimentation, novel technological objects and the process of learning how to experiment on those objects are discovered simultaneously, while "already understood" phenomena are used as technological objects acting as devices and instruments to produce, interact with, and measure new phenomena. It is this insight into the ongoing, creative, and techneic process of experimental science that puts my position in opposition to scientific realism because it favors a more pragmatic, psychological, and sociological approach to understanding scientific research and the relation between theory and practice.

For the scientific realist, however, such a view is classified as "relativism," ant thereby dismissed and ignored. For the scientific realist, experimentation provides objective knowledge, by definition, and the object of that knowledge must be natural in origin, otherwise scientific activity does not make any sense. While I agree that experimentation does provide objective knowledge, the object of that knowledge is technology itself, and scientific knowledge is always relative to a stratum of technological development. A change in experimental practice may involve a change in understanding the nature of knowledge and its method of acquisition, but this always follows technological innovation and the development of new machines and techniques. Furthermore, interpretations of measurements can only be developed through experimentation in relation to interpretations of the techniques by which that measurement was made. It is for this reason that statements of the degree of precision of any measurement (and confidence in those measurements) are established in relation to evaluations of the sensitivity and "cognitive value" of the techniques used. Empirical inductive reasoning requires the applicability of concepts to objects, and, thus, if empiricism is to be successful, it requires the successful and complete refinement of those concepts in relation to the objective world via technique. However, such a process of refinement is never complete and the empiricist is dependent upon the work of others. These remain perpetually open to future refinement and transformation. Even at the level of making a simple measurement, such as measuring room temperature, there is always the possibility of future refinement and the development of new techniques and instruments. There is no such thing as "fixed data" in this respect. All "data" is acquired through the use of techniques, which inform and constrain the set of possibilities of measurement, and there is no such thing as a technique that cannot be refined.

This reveals the extent that mechanical realism (which underwrites the more abstract scientific realism) is a projection of the metaphysical anticipation of the achievement of a technical perfection that cannot ever be achieved in practice. It is in this context that the goal of experimental practices can be taken to be the achievement of its own *techne*, if we define *techne* as the complete knowledge of the causal principles at work in changing the world through technical activity. Objective knowledge is an imagined ideal that is associated with *techne* on the horizon of a projected anticipation of the complete causal explanation of the results of material practices. If the rationale for engaging in experimental science is to explore a reality independent of the activities to explore that reality, the rationality of that rationale is only possible from a removed and abstract level of anticipating the theoretical understanding of the completed processes of production and reproduction. It requires the prediction of the future to begin the process of transforming the present into that future. It does not matter whether we have traditionally interpreted those processes in terms of realism or empiricism. Any limitation to our philosophical imagination is nothing more than a psychological problem. However we attempt to rationalize it, the completion (perfection) of scientific research remains perpetually deferred in favor of ongoing work against which the *telos* of science remains an unreachable asymptote that is approached with each successive refinement to produce and test theory by developing a more refined and comprehensive understanding of the process of technological innovation, thereby realizing new cognitive and manipulative possibilities in future experiments by planning interventions in accordance with new teleological positings that emerged from within the technological framework as the result of past efforts in a way that can be communicated to other people and reproduced by them.

The asymptotic nature of this process reveals that it is a labor process directed by human imagination towards the achievement of the discovery of objective reality that presupposes a socially and psychologically conditioned and projected future, and its corroborations and falsifications stand entirely in relation to the technological framework. Imagined as the power that promises the realization of a hitherto unattainable technological limit, the asymptote of objective reality can be imagined to be known only when science is itself finished and compete, yet it perpetually shifts as new technological possibilities arise. In this sense, we can see how Popper's ideas of falsification and corroboration are attractive interpretations of "the logic" of scientific activity, while also seeing how the logic of scientific discovery is itself an entirely imaginary and cultural abstraction of technological activity. Experimental science, as both a historical and cultural phenomenon, is itself constantly undergoing change in its theories, objects, and techniques, and, therefore, change is an essential part of the process of scientific inquiry. This forms the basis for Popper's claim that science is "self-correcting" and "rational" because all scientific theories are falsifiable, yet his philosophy of falsification represents science in terms of an imaginary ideal against which no knowledge is ever achieved.

Progress in research is a technical, functional, pragmatic, revisable, and creative goal premised on technological activity directed towards a perpetually emergent and idealized anticipation of objectivity. Claims to increased accuracy in measurement are based upon theory-dependent techniques and therefore they cannot be rationally compared to any absolute standard. There is no fixed Archimedean Point outside of the context of ongoing technological activity. The pragmatic justifications of any claims to increased accuracy of measurements are always made in hindsight and are based upon the convergence between (or the mutual reinforcement of) theoretical and experimental practices in which the measured quantity is involved as a stable and reproducible machine performance. The basis for any pragmatic judgment that convergence has been achieved is based upon the achievement of stability and repetition, which makes the outcomes of experimentation into technological objects available to others. The notion that this somehow means that objective knowledge has been approached is based a conception of knowledge in terms of a context-bound technical knowledge that, from the outset, was directed towards the ideal achievement of complete causal account of the technological activities of experimental and theoretical practices. This explains how the technological framework develops in the context of novel experiments and achieves its correspondence to an objective substratum. A magnetic needle dangling from a string may be a simple (magnetic) machine, but it is still a machine. A coil of electrical wire attached to a chemical battery may be a simple (electrical) machine, but is still a machine.

Bringing together these members of distinct machine-kinds (electrical and magnetic) and overcoming difficulties and inconsistencies by synthesizing a hybrid prototype creates a new machine-kind (electromagnetic). By building all the possible permutations of synthesis and establishing their fundamental representations, with each machine acting as a nomological machine and means of disclosure, this new machine-kind constitutes a stratum within the ongoing development of the technological framework that is projected as corresponding to a deeper stratum of mechanisms and laws. The strata of machine-kinds (e.g. mechanical, optical, thermal, electrical, magnetic, electromagnetic, etc.) within the technological framework are interpreted as stratified objective structures of natural mechanisms and laws governing particular phenomena (e.g. mechanical motion, optical effects, thermal properties, the behavior of electrical systems and magnets, etc.). It is the resolution of problems and the construction of novel machine-kinds within the technological framework, along with their associated fundamental representations and theories, which gives the work of science a dialectical and progressive nature, and thereby creating a new set of problems to be solved by further research. The standard by which scientific practitioners judge their own objectivity is in reference to "the cognitive value" of their own judgments within an instrumental context of making and communicating intermixed representational and material practices. Cognitive value requires the achievement of social agreement between all (similarly placed) experimental practitioners and is made through the innovation of novel modes of reflection, discourse, representation, and material practice, and not through immediate intuition, deduction, or experience. This has nothing in common with the immediate experience required by philosophical empiricism for knowledge to be based on experience. It also has nothing in common with the immediate intuition upon which philosophical realism depends. It is only at the point of the asymptotic (unreachable) point of perfection that the object under investigation could be considered to be completely understood, absolutely stable, and functionally repeatable. Realism presupposes that this has already been achieved, simply because it can be imagined. Empiricism presupposes a level of training and skill, which it represents as if they are innate abilities.

Scientific experience is circumscribed as being that which is disclosed through socially accepted techniques of manipulation and representation from within the context of an already established technological framework interpreted in the explanatory terms and images of the scientific worldview. Objectivity is a technical pursuit and cultural aspiration that stands in opposition to the "self-evidence" of subjective experience because it must be demonstrated to another person by using mutually understood techniques of manipulation and representation, which is a condition for experience to be accepted as scientific experience. Subjective experience may well have a role in instigating new hypotheses but no hypothesis could not become part of science until it had been publicly justified via accepted techniques of manipulation and representation. Thus an observation made by an experimenter can only become part of science once it had been justified to the experimenter *and others* in terms of repeatable and communicable interventional techniques. Any proposed explanatory representations given in terms of "natural mechanisms and laws" can only be part of science if they are tested in relation to repeatable and communicable observational techniques. It is for this reason that experimenters write down accounts of which techniques they have used. Scientific journals would not accept a paper that merely recorded "observations" without reference to techniques. Evidence as only evidence of anything because of its association with an established technique. Experimental observation requires the development of "observational skills," manipulative and representational techniques, and if others are not able to acquire these skills then it is unlikely that the experimental observations will be widely accepted. Consider the case of "cold fusion." Fleischmann and Pons were unable to communicate a repeatable technique of how to observe "cold fusion," and, consequently, the validity of their work was brought into question. This is also the case for experimental efforts to observe gravitons and also emotional responses in plants.[229] If experimenters are unable to provide a reliable observational technique then the scientific confidence in their observations is undermined. Without technique, there is no starting point for scientific inquiry to even begin.

When scientists attempt to experiment upon novel phenomena, they develop novel techniques; they need to understand these phenomena and techniques in ways that can be communicated to one another and so agree about the object of their investigation. This involves coming to an agreement about what phenomenon is under investigation, what they want to learn about it, and how to proceed to learn it. These decisions are made as the investigation proceeds and are not completely fixed in advance. Experimentation is founded on a projected plan of action, which anticipates the phenomenon because novel phenomena require novel forms of communication and representation in order to reassure the experimenters that they are experimenting upon the same thing. This involves producing agreement about techniques and also about what was experienced when those techniques were implemented. If we are to grasp what an object is (i.e. an electromagnetic field) as an object of knowledge, we need to know how that object has been cognitively engaged with and how cognition was achieved. Experimenters learn how to articulate their experiences of novel phenomena along the way of experimenting upon them. Experimental investigation involves the progressive social organization of the research, the techniques, the resources, and the descriptions of experiences. This is a sufficient reason to reject the empiricist's appeal to perception, as if it is 'given' as something primary and immediate to individuals. This also gives us sufficient reason to reject the realists' appeal to intuition, as if it is 'given' as something primary and immediate to individuals. Once we understand that "perception" and "intuition" are complex human activities mediated by culture and technology then every act of perception and intuition in experiments require the justification to other people of the techniques used to make that perception possible and for that intuition to make sense. This situates the experimenter as a social being, and all scientific research is situated within and emergent from technical and historical contexts of the innovation and refinement of the technological framework within which scientific research occurs. Both empiricists and realists have misunderstood the practicality of theorizing, neglected the relevance of knowing-how to knowing-that, and that the interdependence of know-how and know-that is just as necessary to defending empirical claims as it is to explaining them.

Experimenters are not engaged in simply and passively registering "the facts of experience" when they are actively using techniques and making judgments about which technique to use when performing experiments and making observations. At each stage of experimental research, the experimenters publicly tie together techniques (both manipulative and representational) and machine performances into technophenomena that can brought into the public realm as reproducible technological objects, and the successful "test" of theories involves transforming those technophenomena into technological objects available within the technological framework for future use. This involves either the reinforcement or refinement of the scientific worldview as part of the reiterative process of drawing up a plan of action for how the research is to proceed, and it is this process of reinforcement or refinement of this plan of action that is implicated in refining the scientific worldview, via the further construction of the technological framework towards its (imaginary) teleological posited completion and perfection. Hence gravity waves remained an entirely imaginary consequence of Einstein's General Relativity until they were incorporated into the anticipated machine performances of the

---

[229] Cf. Collins (1985) chap. 3

Laser Interferometer Gravitational-Wave Observatory (LIGO) over a century later. The "observation" of "a gravity wave" transforms this technological object into a technophenomena that opens up a whole new field of experimental physics and astronomy along the trajectories of technological innovation.

Both the plan of action and the techniques implicated in the experimental set-up (the construction of the experiment, its operation, and establishment of its theoretical significance) can be known without a complete theoretical description of the phenomenon. Otherwise there would not be any point in performing the experiment. Nor would the experimenters be able to anticipate the phenomenon and devise a plan of action. They are more pragmatic and tend to consider an object to be understood when it becomes a stable and robust technological object available for future use. Once incorporated into technological activity, as a means to an end, it ceases to be an object for experimental investigation, except as an object used to investigate other objects, and becomes available as a technological object for future use. It is a stable part of the technological framework. It is at this point that Hacking would become a realist about it. Even when a novel experimental phenomenon is not theoretically understood, it still can be known via a technique. Does the double-slit experiment reveal matter to be comprised of particles or waves? That depends on the technique you use to observe it.

The scientific worldview is reinforced or refined by projecting the plan of action over the technophenomenon to provide an association between technique and a set of objects (the detector) and interpret that association as the outcome of underlying natural mechanisms in accordance with natural law. This is as true of the disclosure of "gravity waves" to explain the machine performance of large and complex machines like LIGO (two 2.5 mile long L-shaped machines separated by almost 2000 miles, each comprised of arrays of interferometers and lasers contained within an ultra-high vacuum), as it is true of "radio waves" as disclosed by Heinrich Hertz's experiments with electrical circuits; it is as true of the Higgs boson disclosed by the Large Hadron Collider as it is true of the constant acceleration of gravity as disclosed by Galileo's use of a pendulum. Techniques connect these objects with their theoretical interpretation to allow the investigation of the contours between human interventions and machine performances to provide information about underlying mechanisms and laws, and provide the data for the test of theory. Modern science anticipates and projects the plan of action required for the objective knowledge of Nature, which is defined in theoretical terms by calculation and prediction, and tested in technological terms by calibration and measurement. Theorists must not only calculate with precision in order to remain within the technological framework of science, but the theoretical terms themselves must be available as interpretations of machine performances in order to be tested through experimentation. Even the naïve realist should acknowledge this.

Representation and mathematics:

The electromagnetic machines of Oersted, Davy, and Faraday were not the mechanical devices familiar to Galileo, Descartes, and Newton. They were a novel machine-kind. Faraday was not trained in the use of mathematics and yet he is considered to be an exemplar of a modern experimental physicist. Was Faraday an exception? Or does Faraday show that physics is not actually mathematical in the common usage of the term? How did Faraday project his ground plans in such a way that his work was available for James Clerk Maxwell to use and explain with his mathematical laws of electromagnetism?

In order to answer these questions we need to take a closer look at how representational techniques and mathematical techniques are related in experimental work. How should we understand the relation between representation and mathematics? This has been a fundamental question for the traditional philosophers of science. Yet they have ignored technology. To understand the relation between representation and mathematics, we need to take a look at the technological framework within which science emerges and how mathematics is incorporated into representational and material practices. We also need to understand its connections with the scientific worldview. How should we understand the relation between mathematics and the technological framework? How should we understand what technology *is*? How does it relate to Nature?

As we have already seen, Newton anticipated the nature of space and time by the projecting geometry and provide a plan of calculation and measurement to confirm that projection, and thereby it became a self-contained system of calculation and measurement. Motion was defined as the homogeneous and isotropic change of position in a projected grid of space and time, while force was defined in terms of the magnitude of change of position in this grid. Every event in Nature is defined in advance as an event only in terms of how changes in position can be made or calculated within this projected grid, and this was guaranteed by restricting research in every one of its questioning steps, limiting it to revealing how natural phenomena behaved within this grid. All interactions had to be defined as magnitudes of motion and changes of motion within the projected grid of space and time that were quantifiable through measurement and calculation. Einstein's refinement of the grid by using different geometrical techniques was able to bring together electromagnetism and motion within the same grid of space-time events, modified through the Lorentz Transformation, just as the Newtonian world system was the refinement of the Galileo's abstraction and projection of the six simple machines within the Euclidean grid of space and time. How were the experiments in electromagnetism a continuation and refinement of the template of the Newtonian system?

Like Faraday, André-Marie Ampère's designs of early experiments in electromagnetism aimed at preventing the movement of magnetic needles and discs to reduce complex interactions to a fundamental motion with the aim of finding the particular configuration of materials and techniques that facilitated the production of that motion and the abstraction of the performance of the apparatus into a set of fundamental representations of that motion.[230] Why did Ampere choose this strategy? There are two reasons of interest here: The first was that he had adopted a common mathematical method of simplifying the problem in order to reduce it to forms available for mathematical formulism. This tactic was set-up to avoid the practically enormous task of finding solutions to the mathematical expressions for the complex technophenomena and associated set of representations produced by the experimental techniques of Oersted, Davy, and Faraday. This put Ampère in the "rationalist tradition," according to Kuhn and Osler. This strategy involved the mathematical projection of the Galilean reductive method and template that reduced all mechanical motion to that of the lever and the balance. This allowed Ampere to treat the problem as if it were one of ratios and simple differentials. The second reason was that Ampere had adopted a mathematical projection in an explanatory sense and projected the Galilean template over the phenomenon to invent a new machine-kind that could be represented as a set of fundamental ratios and simple differentials, thereby providing theory with a specific structure that allowed the

---

[230] Cf. Gooding (1990: 46-7)

new machine-kind to be projected as a nomological machine.

Michael Faraday, on the other hand, was destined to non-mathematical work. This would put him in the "empirical tradition," according to Kuhn and Osler. Yet, Faraday's work should be seen as an exemplar of mathematical projection. Experimenters make visual representations—which Gooding (1990) termed as *construals*—to show themselves and other people how to intervene in "the natural world," to create a technique by which others could understand the correspondence of these representations to experience, and reproduce those experiences through reproducing the representational and material practices of the experiment. How did Faraday know how to begin? Faraday's experimental work began in October 1820 with attempting to reproduce and map out the contours of Oersted's famous needle-wire-motion observation. The motion of a suspended magnetic needle around an electric wire was far from stable and well defined. It took Faraday considerable patience, practice, and skill to maneuver a needle around the wire (without it touching the wire) and map out its motions. Faraday construed the movement of magnetic needles around electrical wires in terms of circular motion; Faraday's notebooks showed the difficulties and efforts he made to invent construals of the performance of the novel needle-wire-magnet experiments.[231] After eleven months of experimenting with sideways motion, circular motion, and push-pulls, he managed to stabilize his configuration of the magnets and needles into the 1821 compact rotation apparatus. This produced revolutions! It was the stable predecessor of the electric motor. It was a hybrid electrical and magnetic machine prototype that produced stable rotations when it was connected to a chemical battery by metal wires. The 1821 machine was designed to demonstrate circular motion and an otherwise quite unpredictable movement was constrained to a circular path. It both enabled and constrained the spontaneous circular motion of a needle pendulum around a magnet. It is in this sense that Faraday mathematically projected the abstracted circular motion of the wheel and axle as the ground plan of the possibly stable motion to try, and his experiments on the motion of magnetic needles and electric wires were constructed to capture the circularity of that motion, within a template of representational possibilities.

By doing so, Faraday projected the ground plan of circular motion, but this did not adequately capture the movement of the "rotations" and needed further refinement. Faraday was aware that when he construed these "rotations" as skewed motion he was disclosing a non-Newtonian force, and Faraday acknowledged that the skew-aspect could not be subjected to the Galilean reduction. Faraday rejected the applicability of Galilean reduction to his novel machine, and developed a new fundamental representation of its skewed motion by using Ampère's work to develop a new construal of "the lines of force" in terms of the screw.[232] This construal of skewed motion is an abstraction of the screw that can be projected over the machine performance and thereby allow the phenomenon to be represented of an ensemble of differentials and ratios. Faraday construed that motion by using the screw as his fundamental representation and thereafter formal mathematical theory referred to this projection. The screw became a new non-reductive exemplar of "electromagnetic motion" and it was still one of the simple machines, thereby it allowed the refinement of the Galilean template, while also remaining consistent with it, and opened up the possibility of using this new machine as a nomological machine to disclose its own law-like behavior. Faraday's construal of his experiences in terms of screw motion of the "invisible" lines of force broke free from the Galilean reduction, while also remaining consistent with the mechanical realist methodology and worldview.

---

[231] Gooding (1990), chap. 5
[232] To see Faraday's construal of skew motion see Thomas Martin's transcription of the first part of Faraday's experimental record for September 3, 1821. Reproduced in Gooding (1990: 122-3).

By using this machine as a nomological machine to demonstrate the law-like behavior of its performances, Faraday was able to visualize the "invisible" force of electromagnetism as a screw-like mechanism.

For Faraday, the screw construal captured the provisional perceptual and interpretive possibilities of the responses of magnetic needles placed in the proximity of electrical wires and moved around them. The task facing Faraday was to show someone else visually how the tips of the needles moved and the paths they took. The visual record, such as drawings, sketches, and geometrical diagrams provided the means by which his personal experiences could be construed in a form available to other people. Faraday invented instructions—technical guidelines—on how to place the needle next to the wire and move it in such a way as to allow the responses to reveal themselves as a screw-like force, and show others how to construe his construals in such a way as to make stable, repeatable, and communicable methods to make the same observation.

As a nomological machine, it could make order of the real-time technical processes involved in making an observation, and the screw construal provided the template projected over the phenomena during the setting-up of further experiments. In the case of the early experiments in electromagnetism, the construed motion of the tips of magnetic needles, iron filings, and electrical wires were mapped out upon a space-time grid that was projected over the apparatus as "lines of force." The observation of these "lines of force" involved the space-time mapping of the interactions between moving a needle around a wire and the movement of that needle in response. It involved mapping-out these novel contours of human interventions and machine performances within the technological framework of the mathematical projection of the grid over a new juxtaposition of technological objects. The construals that were used to map out these contours were still circles, tangents, arrows, push-pulls, rectilinear motion, anti-parallel motion, and skew motion, as they were for the mechanical machine-kind (the six simple machines), but they were now projected over the movement of the magnetic needles near wires and inside coils, and required different methods to visualize how to make observations that can be repeated by other people.

Construals are selected on the basis of their heuristic, communicable, and instrumental value in making the movement and vibrations of the magnetic needle, dangled from a string next to and around an electrical wire. Construing involves a complex process of relating actions and imagination in order to simplify and grasp the whole process of making the observation. It links phenomenological experiences and abstract space by reducing the former to repeatable performances—what the traditional philosopher of science might call "a constant conjunction of events." This new fundamental representation of the machine in terms of the screw allowed mathematical techniques of ratios and differentials to be imported from the six simple machines to the novel kind of machine, the electromagnetic motor, and it is this non-Galilean construal of force that prevents the reduction of the stratum of the electromagnetic machine-kind to the mechanical machine-kind, and led to the inconsistency between the Galilean frame of reference and Maxwell's Laws of Electromagnetism, for which the speed of light in a vacuum is a constant, in inverse proportion to the product of electrical and magnetic permeability of a vacuum. The mechanical circular and rectilinear motions would not suffice for the novel machines. Faraday's screw-construal was a non-Galilean fundamental representation. At the theoretical level, this creates an abstract stratum of (non-reducible) mechanisms and laws. Maxwell's field theory utilized Faraday's construal of the tangential or skew motions of "electromagnetic lines of force" in terms of the screw. Once Maxwell had invented the *grad* and *curl* operators of differential calculus (and specifically invented for this task) then such motion could be described in terms of differential calculus, and provide calculations for measurements.

Faraday developed his methods to construe electromagnetism against a background of his previous efforts to construe how to visualize material practices and their outcomes. The acts of making novel experiences of novel interventions intelligible, such as the motion of a magnetic needle around an electric wire, need to be ordered in real space and time, by moving a real needle around a real disc. Construals couple together a technological object (probe) and a technophenomenon (response) into a disclosure of some mechanism at work within the projected grid in terms of "rotations," "tendencies," "pointing," "dipping," etc. These construals of motion—as provisional perceptual and interpretive possibilities—enable the earliest (pre-theoretical) stages in the interpretation of novel phenomena and have a heuristic function as a technique for exploring an emergent phenomenal process and revealing the underlying mechanism. Construals are pre-theoretical and practical, socially and historically situated, intelligible visualizations made to explain to others how to intervene, interpret novel experiences, and communicate trial interpretations. They can be compatible with several theories or with none. For example, when we picture and describe light as "rays," "waves," or "vectors," when showing someone how to observe the polarization of light through filters, we are using construals. They are a tentative and public means of visualizing and describing an otherwise "invisible" phenomenon (i.e. the motion and structure of light). Observers with different theoretical predilections can agree about salient aspects of the phenomena while disagreeing about their theoretical significance. How? Exchanging tentative visual and verbal construals about "the observed" and how to make "observations" requires negotiated agreement between similarly placed observers, who publicly construe and re-construe their experiences in relation to the construals made by other people during their attempts to have the same experiences of new and otherwise "invisible" phenomena. Attention to the use of construals in communicating techniques and experiences highlights the pictorial (rather than linguistic) aspect of scientific imagination and experimental work. The judgments regarding how phenomena should be represented are socio-technical judgments (made in relation to both other people and techniques) regarding the intelligibility of any technophenomenon and disclose how to make and use technological objects to disclose it to other people. Consensus between experimenters, dependent upon the successful exchange of observational and manipulative techniques, involves the dissemination of qualitative and pictorial representations of the phenomenon-as-a-process. Hence construals should be understood as being central technical codes to the processes of experimentation, and they do not permit either a monolithic fit with mathematical theories or a metaphysical commitment to determinism, realism, or materialism.

How do construals link with mathematics? If modern physics is inherently mathematical and experimentation is based on achieving stable representative and material practices then we need to understand how mathematics, cognition, and visualization are connected to technology. Derrida wrote that specific styles of writing, or *inscription*, were required for science to be possible (Derrida, 1967: 81-7). Bruno Latour and Steve Woolgar (1979) used Derrida's idea of inscription to characterize graph-plotting machines used in scientific work as *inscription devices*. Like Derrida, Latour and Woolgar treated science as a form of writing for which the aim was to produce text. However, they neglected to attend to the way that scientific inscriptions are fed-back into the processes of experimentation, as part of technique, in order to produce the inscription devices in the first place. We need to look at how this is done in practice. How are mathematical forms inscribed within the experimental sciences? Modifying Derrida's term *graphe*, I use the term *technographe* to denote any physical marks used in mathematical inscriptions, schematics, or diagrams used *technologically* in the design, modeling, and calibration of machine performances. Technographe are used for writing down mathematical techniques and inscriptions. A drawn circle, Arabic numerals, an equal sign, a differential operator, vector notation, matrix notion, Feynman diagrams, electric circuit diagrams, and grid coordinates are all examples of inscriptions constructed using technographe. They are the parts of mathematical writing used in constructing solutions, demonstrating proofs, calibrating mechanisms, modeling the performance of machines, and for designing machines. Geometrical proofs, algebra, analytical differential calculus, vectors, matrices, statistics, etc., are all written down, recorded, printed, and disseminated, through the use of technographe. Technographe are not the mathematical techniques themselves, in the same way that *graphe* do not tell us how to write or how to read. They obtain meaning as part of the inscription and interpretation practices used in those mathematical techniques, embodied in skilled mathematical practices, and they are situated within the technological framework. To understand the relation between representation and mathematics, we need to look at how this embodiment in practice occurs.

Let's take a look at Euclid's first proposition in *The Elements*[233] to construct an equilateral triangle by intersecting two circles: Draw a straight line AB. Construct an equilateral triangle on the straight line AB. Describe the circle BCD with center A and radius AB. Again describe the circle ACE with center B and radius BA. Join the straight lines CA and CB from the point C at which the circles cut one another to the points A and B. Now, since the point A is the center of the circle CDB, therefore AC equals AB. Again, since the point B is the center of the circle CAE, therefore BC equals BA. But AC was proved equal to AB, therefore each of the straight lines AC and BC equals AB. And things which equal the same thing also equal one another, therefore AC also equals BC. Therefore the three straight lines AC, AB, and BC equal one another. Therefore the triangle ABC is equilateral, and it has been constructed on the given finite straight line AB. All logical, right?

Actually, this is not a logical proof at all. Formally, in terms of modern logic, Euclidean geometry is incomplete because it lacks a continuity axiom in either the Postulates or the Common Notions. The first proposition remains unproven because it has not been demonstrated that the two circles actually intersect. From the perspective of modern logic, the Euclidean geometry available from antiquity to the nineteenth century was not a complete logical system. It was an art rather than a science. The basic postulates of Euclid's geometry, such as to describe a circle with any center and distance, draw a straight line from any point to any point, etc., are distinct technographe that can only acquire their meaning through repetitive practice. Euclid's geometry is a form of writing in which a set of primitive inscriptive practices constitutes the basis of the whole corpus. These practices are comprised of the inscription acts involved in the mathematical inscription of geometrical figures. In turn, each geometrical figure, once inscribed, becomes a distinct technographe that is used to inscribe further geometrical figures. The first proposition is inscribed by performing the technographic acts of drawing straight lines and circles. Proposition 1 provided the technographe to inscribe an equilateral triangle. This was used to construct further technographe. For example, Proposition 1 was used as a technique in the construction of Proposition 2 which, in turn, was used in the construction of Proposition 3, and so on.

---

[233] Heath, trans., 1952

A straight line and a circle are defined by Euclid in terms of acts of drawing and, consequently, we can only learn how to perform these practices by following instructions, performing the inscriptive act, and being informed by skilled practitioners of the art that the resultant is correct. Repetition is the route to learning. Each figure is a socially mediated (informed and constrained) artifact understood through education in how to perform the correct technographic inscription. It is a technological object available for further use, and only achieves its truth within the artifice of Euclidean geometry as a set of tacitly embodied practices and their products, and this is secured against a set of cultural norms and beliefs about the meaning of those practices. We are only able to intuit the self-evidence of these products once we have acquired the requisite artifice and have become skilled practitioners. Once this artifice is acquired, through education in this culture, then practices, reasoning, and intuitions are ordered within its framework as techniques and demonstrations of its own truth. The circularity of this, if you'll excuse the pun, is quite evident. The self-evident correctness of these exemplars is established by being able to use them and they are used because they are considered to be correct. Each proposition is proved by the act of inscribing it. Its self-evidence is a resultant of its practice and, consequently, Euclidean geometry is no more eternal and universal as is the social acceptance of the technographic inscription practices upon which it is based. Claims of eternal and universal truth are themselves premised on a totalitarian reduction of all that is into those aspects available for manipulation and control.

Each correct use of technographe is *an exemplar*, in Kuhn's (1970) sense of the word, because it constitutes a set of problem solving tactics that are learned, or constructed, by using them to solve problems, and, as a result, are taught through the imitation of teachers. These exemplars were technographically used in *The Elements* to construct geometrical treatments of angles, straight lines, ratios, circles, curves, areas, and solids, and each axiom is an abstraction and reification of a set of inscription practices. Its status as an *episteme* is achieved by its cultural acceptance amongst its practitioners (and anyone else that they can convince) on the basis of cultural claims for its completeness. This is what Kuhn termed as a paradigm, which can only be established in relation to a community of practitioners. These paradigmatic *epistemoi* are collected together and integrated by members of that community (passed on to the next generation through education) within a technical system as a fixed technological framework with a specified object-sphere (the geometrical figures, proofs, and theorems) and a clearly defined set of interpretations as to how to combine and relate them. Euclidean geometry is characteristic of both a *techne* and an *episteme* (Plato's and Aristotle's usage). It is the total set of related inscriptive practices that is presented as eternally, universally, and necessarily true scientific knowledge, with its own particular rules for organizing and categorizing knowledge, and inscriptional and representational practices. Foucault termed the totality of these rules as *episteme*; during distinct historical periods, which he termed archeological strata, different *epistemoi* dominate. In this sense, Euclidean geometry is an enduring *techne* that has been metaphysically presented as an eternal *episteme*. *The Elements* provided the technographic exemplars for the works of Archimedes, Apollonius of Perga, Nicomachus of Gerasa, and many others. This can be seen in the geometrical proofs of Archimedes and Apollonius, and Nicomachus' study of arithmetic based upon Euclidean ratios. These works, as well as *The Elements*, were preserved and disseminated from antiquity, through the medieval period, and into the present day, as collections of exemplars to be learned through imitation. The structures of these geometrical treatments were organized within the Euclidean template of axioms, postulates, propositions, corollaries, theorems, and proofs; they provided the exemplars for all subsequent geometry to emulate—the performance of geometrical proof is *mimetic*. Archimedes, Apollonius and Nicomachus all innovated new technographe and extended the Euclidean *techne* to include irrational numbers, projections, powers, series, and the geometry of ellipses, hyperbola, and parabolas, and thereby provided new technographe as exemplars.

The science of mechanics and the techneic use of technographe to inscribe the motion of the simple machines developed within the same technological framework, and it was this framework that was extended over new phenomena. The conflation of Euclidean geometry as a *techne* and *episteme* (in the context of the construction of the science of mechanics and the desire for novel technological powers) led to the emergence of mechanical realism within mechanics and the possibility of experimental science. The axioms of geometry have application in our world of experience because, through its practice, we have mathematically projected that application over parts of the world. It is through the embodied habitual practices— *praktognosia*[234]—of pre-conscious familiarity with the techniques of geometry that we are able to "intuit" the applicability of this projection. These axioms are abstractions within a technological framework that we impose upon parts of the world. This is not a structure of our untrained consciousness; it is the structure of the imposition of the mathematical inscriptive practices of the artifice of geometry. If *techne* resides "in the soul of the craftsman" it has been inscribed upon that "soul" through training, mimicry, and practice, and corresponds to what Kuhn termed as the disciplinary matrix. The "soul of the craftsman" is not something that we are born with; it is something that we embody through the social acquisition of a technological framework to the extent that we become so familiar with its practices that we are no longer aware of it. Our educated bodies have become situated within the technological framework of geometrical inscriptive practices. That framework, once embodied, becomes part of us and we become part of it. Art and the artisan reside in the same place: the body. Projecting the trained imagination performs the "outer sense" of mathematical projection. Our ability to have *a priori* knowledge of the truth of the axioms of geometry is dependent upon the embodiment of technique itself.

The mathematician's demand for formal rigor is always the demand for the formal explication and justification of non-formal practice. It is the demand for the logical abstraction of reified, skilled human labor. Yet formal axioms and the logically abstracted and encoded system in general will be both unintelligible and useless without the skilled practices from which it was reified. A logical proof for the constancy of the ratio of a circle's circumference to its diameter is meaningless if we are unable to draw and recognize a circle. It will not do to appeal to pure conceptualizations of a circle either because without the skilled practices from which those conceptualizations were abstracted there would be absolutely no possibility of applying them in the world. Application requires education of its practice and learning the meaning of practices require the acceptance of cultural assumptions and relations. Our capacity to ground geometrical imagination in self-reflective knowledge and present that grounding as the discovery of self-evident intuitions is itself a consequence of the pre-consciousness of technique (praktognosia) through *mimesis*. These praktognosic intuitions are only possible because the human subject is already well versed in the application of those techniques. The mathematician becomes sufficiently skilled in the art of geometry to the point where the art became invisible and its practice became innate and intuitive; once techniques are invisible (tacit or habitual) then all we see is the projection itself, as if it were the discovery of something "out there" in the world, or as a reflection of the "inner structure" of the

---

[234] Maurice Merleau-Ponty (1999) used the term *praktognosia* to denote the pre-conscious character of tacit and practical knowledge. It has parallels with Michael Polanyi's *tacit knowledge* (1958) except that tacit knowledge was situated in the educated intellect whereas, for Merleau-Ponty, praktognosia was embodied in the situated and habitual motility of the existential body-subject. I shall discuss this further in the next chapter.

conscious mind. It was for that reason that Kant, in his *Critique of Pure Reason* (first published in 1781), located the origin of that projection in the structures of the mind and its intuitions rather than in the art of geometry itself (Kant, 1964).

The technographic practices of geometry (drawing using a straight edge and a compass) were reified in response to the degree of unreflective application of those practices. This reification was itself the product of the challenge of imposing "rational order" upon practices, which were both informal (lacking in rigor) and disordered (inconsistent). The abstraction of those practices was a simplification of what was already taken for granted as being technically correct, in such a way as to induce a universal *techne* under which those practices would be integrated under a single theory of practice, which paid no attention to the particularities of practice. The move towards abstraction is an attempt to detach the technique from its application in order to generalize it from the particularities of use. This involves a synthesis of a diverse and divergent set of practices into a general axiomatic system that can be applied equally to all particularities. Under the *techne* of geometry the objects produced by the application of that *techne* would be abstract and idealized reconstructions of technographe that were conceptualized in terms of definitions and generalized as axiomatic principles. This reification allowed the axioms of geometry to be divorced from their practical origins and contexts of use.

Euclidean geometry is comprised of the total set of techniques of technographic inscription in accordance with general rules that were abstracted into a formal system of practice that could be represented as a complete *techne*. Yet, all mathematical abstractions are incomplete when applied to the particularities of experience in virtue of the generality of those abstractions. An additional set of guidelines alongside trial and error modifications are required to apply mathematical abstractions to experience, and abstractions must always be interpreted in their application. It is not the case that the formal system does not correspond to reality (as the realist would argue) and hence "the error"; if the system is itself incomplete and we are not in possession of a full and complete *techne* then we cannot know whether the system inaccurately represents experience or whether we followed the rules and applied the system improperly. In short, we do not know what it is that we have done when we have done it; nor do we fully know how to apply the general system to particular experiences in a consistent and coherent manner. We are forced to keep experimenting or give up. It has not converged to the asymptote of a complete and perfect system of representation, and, furthermore, until we possess a formal *techne*, we cannot consider the particular practices from which the system is abstracted to be homogeneous. They remain a heterogeneous collection of exemplars and tactics, each applied in context through imitation, and it is this heterogeneity that is the source of all incoherence, anomalies, inconsistencies, and contradictions. Heterogeneity is the source of all "error" and "resistance" between the application of formal mathematical systems, on one side, and practical experience, on the other. The problems that arise during that process are the results of the interference between heterogeneous objects synthetically brought together in the attempt to integrate them into a novel unified and coherent whole. The "failure" of any system is a consequence of its incompleteness, complexity, and disunity generating contradictions and anomalies. "Success" is a matter of achieving performative coherence rather than correspondence. The challenge of system building is to collect together and integrate heterogeneous objects into a homogeneous whole. It is this *synthesis* that creates new systems.

The problems that arose from attempts to project the Euclidean system over all phenomena resulted in what Kuhn termed as "a crisis." The inability to reconcile the Galilean frame of reference with the constancy of the speed of light as determined by

Maxwell's equations was one such crisis for nineteenth century physics. The resolution of this crises involved the invention of general non-Euclidean geometries (for which Euclidean geometry was a special case) brought with it a novel awareness. Not only were geometrical objects the products of our mathematical practices, but also they were arbitrary. Consistency was a product of the system, but there was nothing objective or necessary about the elements of that system. Dimensionality can be projected in any number of different ways, and there is nothing unique about any particular set of technographic practices used to draw and visualize dimensions. Any numbers of new technographic practices were possible (as Boltzmann, Lorentz, Minkowski, Gauss, and Einstein have shown), and we are free to construct any topology from any arbitrary set of axioms. There is not any "objective" space to which mathematical geometries must "correspond." Topology can only provide us with a coherent system of consistently mapping arbitrarily imposed axioms and definitions. The problem is how to relate these arbitrary spaces to the practices of experimentation and measurement. Any topology can be considered rigorous, in the disciplinarian sense of the word, providing that our axioms and definitions do not contradict one another when we attempt to combine them into a coherent system.

How is the "empirical" practical dimension of experimentation connected with the "rational" mathematical dimension? The case of Faraday's work provides an illuminating answer to this question. If we accept that the use of construals was central to Faraday's public reasoning process then we can readily see that this process was a technographic process even though Faraday was not a skilled mathematical practitioner. His socio-technical judgments about how to draw an electromagnetic field were premised upon technographic cognitions made within the context of a socio-technical learning process that allowed Faraday to realize for himself these cognitions as something brought by us to the learning experience. This is not a primitive or innate cognition but is a socio-technical cognition that was made "his" through the embodiment of artifice through innovative practices that explained the behavior of magnetic needles around electrified wires. It was premised upon a paradigmatic and praktognosic orientation towards objects from within the invisible technological framework of continually embodied artifice by someone who was already a skilled experimental practitioner. As Gooding pointed out, "construing creates 'giveness' of experience... [It is] a relatively stable but plastic interpretation of experience which guides further exploration and interpretation." (Gooding, 1990: 87) It is this "giveness" that is characteristic of mathematical projection. It is the "laying down of the ground-plan of Nature," in terms of the technological framework, while leaving it open for future refinement through further technological activity and modeling.

How are technographic inscriptive practices connected with models, interventions, and representations? So far, I have only discussed the pictorial and geometrical technographic inscription of machine performances. One of the crucial mathematical innovations of the seventeenth century was analytical geometry and differential calculus, and these allowed machines to be analytically inscribed. In order to understand how this was done, we need to look at *functives*.[235] What is a functive? A functive is an element of a function. Limits, constants, operators, and variables are examples of functives. A function defines the relationship between its functives and may, in turn, be used as a functive within another function. In physics, a function has to refer to a coordinate system in which the axes represent physical quantities and variables, which can be manipulated as technological objects, and it is this coordinate system that is

---

[235] Functive was a term used by Gilles Deleuze and Félix Guattari in their analysis of science and its distinction from art and philosophy (Deleuze & Guattari, 1994, chap. 5).

projected over the phenomenon under investigation. Physics proceeds in the face of the infinite chaos of existence by constructing a plane of reference from coordinate systems in order to slow down the disorder of chaos by external reference (or *exoreference*). Exoreference involves extrinsic determination of the meaning of the frame of reference. Physics is distinct from mathematics due to this extrinsic determination. Mathematics finds its meaning in the intrinsic consistency (*endoreference*) between functives, whereas physics has to extrinsically give those functives physical meaning in terms of something outside of the mathematical system. Exoreferences allows functives to participate in modeling and provides any coordinate system, composed of a least two independent functives whose relationship is the function, with meaning as a state of affairs or informed matter. Exoreference is necessary for the frame of reference to form a proposition that relates a physical state of affairs meaningfully to the mathematical system in question. This step allows functives to work as technical codes. For example, it is an act of exoreference that is required for the differential functive dy/dx to refer to the rate of change of pressure with respect to temperature. This gives it meaning as a physical state of affairs between pressure and temperature in a system that is extrinsically determined as a sphere of gas at constant volume. The function allows each dimension (axis, variable) to be fixed while the others are varied, and connects it to the variable operations (turning dials, pulling levers, or pressing buttons) of machine performances. It is the act of exoreferencing natural phenomena that allows mathematics to participate in modeling machine performances in terms accessible to technical manipulation, control, and measurement via calibrated instrumentation and technical codes.

Functives and technographe are meaningless without the technological framework of inscriptive practices, technical codes, and exoreferences in which they are situated. Mathematical methods (as procedural collection of mathematical techniques) are integrated within the technological framework via the exoreferences that connect functives to procedures via codes (i.e. increasing the strength of an electromagnetic field by turning a dial). This framework is given concrete meaning via its context of application. For example, Fourier analysis is a technique for analyzing complex "wave patterns" in terms of series of sine and cosine functions. It involves a collection of different technographe, functions, functives, techniques, and inscriptive practices. In order to be effective it must be embodied in inscriptive practices and exoreferences in the context of solving a range of particular problems. By applying Fourier analysis to the solution of an inscribed physical problem, say the solution to the Schrödinger Equation for electrons within a metal wire under a potential difference V, the solutions of this technique can be exoframed as expressions of physical states. The sine and cosine functions of the Fourier series can be taken to be the wavefunctions that are superimposed as the probabilities of the measurable behavior of the electrons. It is this act of projection that provides a connection between "the mathematical" and "the empirical," via technical; codes, while situating both within the technological framework of the experiment via acts of calibration, measurement, or manipulation.

By inscribing the contours of the interactions between human interventions and machine performances in terms of functives by using technographe, the contours can be mapped out in terms of operational and responsive variations to changing physical quantities. Exoreferences give this coded mapping its meaning and allows calculation to be central to observation. Hence, the physicist can slowly turn up the pressure acting on an experimental cell and read the variations in temperature from a calibrated thermometer, which can be compared to theoretical expectations. This can be recorded graphically and written down in the form of a differential equation. Differentials are particularly suited for inscribing machines and mechanisms, given that the differential (as a functive) relates variables in terms of a ratio of a rate of change of one variable with respect to another. These variables can be used as dimensions to construct an analytical framework in which the differential equations can provide the contours of the physical process under investigation with both qualitative and quantitative meaning, and it is the act of projection, underwritten by mechanical realist precepts, that allows this process to participate in exoreferencing.

The exoreferences that are projected over the machine performativity in terms of fundamental variations and dimensions are a part of a technological framework and give technological objects physical meaning as interventions and responses; hence changing specific variables (e.g. voltages or electric field strength) and be related to techniques and objective physical events (e.g. spraying electrons). These projected exoreferences are associated with the operational practices of the experimental procedures, and hence the projection of this framework is an anticipatory act of exoframing machine performances. It is this anticipatory act that gives experimentation its objectivity and correspondence to a transcendent Nature that exists outside of the context of the experiment that is said to govern its performance. The process of exoframing facilitates the process of writing out of the account (de-inscribing) both human interventions and machine performativity from the final narrative account of the results of experimental work. Exoframes can then be taken to be nomological representations of the physical processes involved in the experiment and abstract it in terms of mechanisms and laws. It is only after both human interventions and machine performances are de-inscribed from the final accounts (published as papers in scientific journals) that "the Book of Nature" can be read. But how does exoframing relate to Nature?

Exoframes and Nature:

Early experimental work in electromagnetism was an example of exoframing novel phenomena and modification of the mathematical projection to allow more freedom to construe the screw motion without the restriction of the Galilean reduction. Faraday *et al.* had invented a novel machine-kind and confirmed the methodology of mathematical projection. His work is an example of how the so-called rationalist and empiricist traditions were both aspects of the same technological framework. He projected this new machine-kind over the phenomenon in order to establish a series of interventions by which he could map out responses to those interventions by building machines and observing how they performed. This provided Faraday with a general methodology by which he could use technological objects (magnets, wires, needles, etc.), and an associated set of possible construals (rotations, screws, antiparallels, etc.), and procedure (build a machine and map out its motions) to discover general law-like behavior, which can be abstracted in terms of the technographic maps of the machine performances and abstracting these in terms of sets and combinations of simple motions, which are interpreted as "lines of electromagnetic force" or "contours of the electromagnetic field") and make these available as something to be explained in terms of physical theory. When these interpretations are fed back into the technical processes of technological innovation by suggesting new combinations of simple motions corresponding to possible novel machine performances, thereby giving the outcomes and sequences of machine performances meaning in terms of physical theory then these interpretations participate in exoframing experiments. This methodology sets in place the changeable aspects of the apparatus and allows them to change, as a sequence of calculable events, interpreted in terms of physical theory, in order to allow facts to become objective, fixed, and, hence, determinations of the constant and reproducible aspects of the changing of the changeable that can be made into rules and general principles interpreted in terms of the consequences of natural mechanisms and laws.

The construal of the movements of magnets and needles as the changeable motions of the six simple machines involves constructing the apparatus in such a way as to prevent it from making any movement that is not one of these simple machine motions. Hence I am not claiming that experimenters like Faraday were stage magicians who built a trick. I am claiming that he reduced the possible movements to that of one of the six simple machines and that his procedure, building a machine according to the methodology of the mechanical realist template, set in place and restricted the changeable movement to one of the six projected mechanical motions in such a way as to make it repeatable within the projected exoframe. It constrained and informed the phenomenon in terms that were made meaningful in terms of the exoframe, and thereby interpreted in terms of physical theory. This allows the technographic construals and representations of simple motions, used within technological contexts of mapping out the contours of machine performances in terms of constants and variables, to be abstracted into a mathematical law and interpreted in terms of the scientific worldview. After all the work of exoframing has been done, the empiricist will then claim that the law is a good fit to the phenomenon, and the rationalist will then claim that the phenomenon is a necessary consequence of that law.

In the context of all experimental sciences, regardless of whether experimenting on gyroscopes or genetically modified rats, experimenters' work at disclosing the changing of the changeable through the production of the stable and reproducible changes of those variable aspects of the functionality of the experimental apparatus. It is the process of mapping out the contours of the interaction between human interventions and machine performances by using those techniques and technological objects in relation to the development of ongoing representational and material practices, which provides technical codes, and interpreting this whole technological activity as instrumental in the disclosure of the underlying mechanisms and laws that make it possible. It is the task of the experimenter to determine in relation to these contours which aspects of machine performance change in response to particular human interventions, and how those relations inform and transform each other. The process of research is subsequently one of producing and mapping out the changes of the changeable in relation to exoframed human interventions using technological objects and the machine responses to those interventions. If experimenters want to map out the change in volume of a gas in relation to changes in pressure and temperature, for example, their work involves collecting together a piston, a Bunsen burner, a thermometer, and a pressure gauge, and varying each of the variables (temperature, pressure, volume) in relation to one of the others (while the third is fixed) to map out the contours of human interventions and machine performances. Those contours could then be presented as the manifestation (resultant, consequence) of the realization of the rules by which those variables are related "in Nature" via a "natural mechanism" operating in accordance with "natural law," otherwise known as "Boyle's Law."

Methodology sets down the set of techniques and technological objects that will be used by the research to disclose the "natural mechanisms" to be disclosed by that experimental research, and thereby organizes the experimental research program within the technological framework. By exoframing those contours, we could then write down the differential equation for those rules, and thereafter formulate a mathematical law, and interpret those laws in terms of physical theory. What the experimenter has done (or so the mechanical realist would claim) is create an artificial space (free of the chaos of competing mechanisms) by which those rules could be disclosed, and the theorist has found the underlying mathematical law that governs the relations between functions, variables, and constants evident during the experiment. By fixing these rules as the necessary consequences of natural laws, methodology is able to determine these laws in terms of the rules of calculation of the changing of the changeable that has been set in place by the technological framework of the experimental procedures. By projecting this framework over the natural world, whereby an act of metaphorical substitution is given physical meaning via a set of metaphysical precepts, this has made the experimental sciences intellectually and culturally possible as an intelligible part of natural science. Exoframed technological innovation circumscribes "tests" of physical theory, thereby reasserting and reconfirming experimentation as a part of natural science through experimental activity and its metaphysical interpretation, and it also circumscribes the observable realm and what can be known. Facts can only be made clear, as the facts, within the purview of concepts, rules and laws, and, therefore, research into the facts is intrinsically the application of concepts, rules and laws to technological activity because it is only through such activity can it be established that physical theory has been tested through verifying or falsifying the applicability of rules and laws to experience.

Experimental science is thus able to perform the "sleight of hand" that has been premised on "the self-evidence" of mechanical realism. Through methodology experimenters engage with Nature through a technological process of making a machine disclose Nature and represent what has been is disclosed the techneic realization of a set of technical codes of parameters and contours of the changing of the changeable and interpreting its possibilities and limits in terms of unchanging principles of change. The traditional philosophers of science have presumed, on the basis of the assumption of mechanical realism, that these are epistemic principles, while leaving the experimenters' de-inscription of human beings from the narration of the experiment itself, thereby removing the human participation from the account; this leaves machine performances as represented as the consequence of natural mechanisms and laws, rather than the responses to human interventions. The graphs and data tables published in scientific journals are abstracted sets of machine performances, given implicit cultural interpretation as a set of observations of a natural process, thereby revealing its patterns and limits as the results of mechanisms and laws. This shared and implicit interpretation is necessary to allow experimenters to remove technology from the final account of the experiment and present its *techne* as "natural law."

Research is made objective by encountering the phenomenon as it is disclosed by the technological framework, while procedure is freely directed to view the changeable within the structures of the technological framework that provides the horizon of the objectivity of change and limits required for facts to become present as concrete particularities of abstract and general laws. This is as true of parapsychology and the search for ghosts, or SETI and the search for extraterrestrial communications, as it is for the search for the Higgs Boson or mapping out human DNA.[236] Human beings do not have absolute control over the outcomes of technological activity. The disclosure of those uncontrolled aspects would not have been disclosed without human intervention, which anticipates, mediates, and shapes what is to be disclosed to human beings, given a shared set of cultural interpretations, even when what is ultimately disclosed is not controlled by human intentions. Whether ghosts or aliens are discovered, involves both the means by which those discoveries are to be made, and how those means respond in ways that are

---

[236] Heidegger (1977b: 120) considered this to be a fundamental aspect of all scientific research. See also Rogers (2005)

not under human control and design. What is *encountered* through the use of technology is not encountered as-it-is, but is encountered as the responses of technology to it and how we interpret those responses. Thus the Michelson-Morley interferometer could be built by human beings (namely Michelson and Morley) to measure "aether drift in the wake of the Earth's motion" and, to their surprise, discover that there was not any "aether drift" at all! They controlled the template of the experiment, and how it was performed, and it is this template which constrains but does not control what can be learned from the experiment. This provides methodology with an explanatory character in the context of discovery because it can present itself as a mode of disclosure of something all-pervasive but otherwise unencountered, such as a cosmological medium for thermal and electromagnetic radiation. Whether the null result discloses non-existence depends on the assumptions made about how to encounter it. Again, this is as true for the detection of an extraterrestrial signal or a voice from beyond the grave as it is for aether drift or gravity waves. Methodology accounts for an unknown (ET, ghost, aether, black hole) by means of a known (detector responses), and, at the same time, verifies that known (via technique) by means of that unknown (as an exoframe). Both the known and unknown are used to inform and constrain each other via the technological framework of the real-time activity of experimentation and the particularities of the performance of the machine, and, as a result, gives the experiment its sense and reference.

Consider the case of Newton's understanding of gravitation. Newton's Law of Universal Gravitation explained already established facts (objects fall to the ground when dropped, the Moon orbits the Earth, and the planets orbit the Sun in elliptical paths) in terms of an unknown (the universal force of gravitation) via a procedure (the calculation of paths using the inverse square law). Hence the unknown could be accounted for in terms of the known facts and these facts were verified as necessary consequences of the unknown. This can also be seen in Faraday's experiments. The known possible motions of electromagnetic machines would be explained in terms of "an unknown force" via a procedure of mapping out "the lines of force" using magnetic needles and electric wires. In the case of both gravitation and electromagnetism, "natural law" was established by mapping out the interactions between technique and technological objects because this informs and constraints by providing criteria for the anticipatory representation of the conditions under which the experiment can be performed. This framework is required to prevent the representation of the changing of the changeable experiments from being based upon "random imaginings," and, by being based upon nomological machines projected over the natural world, the representations sketched into that act of projection are interpreted as having objective, natural referents, even if these remain unknown or unencountered except by technological means. The planning and execution of experimentation, as a methodology given metaphysical interpretation, is supported and guided on the basis of assumptions about the phenomenon to form the basis for the means to encounter the phenomenon in such a way as to allow the facts, which either verify or falsify the law, to be induced into a general techneic form that is interpreted as a general epistemic correspondence to something that transcends the experiment (and supposedly exists independently from the experiment) and explains the response(s) of the apparatus or detector.

Due to the operational precepts of mechanical realism, this appeal to transcendence supposedly demonstrates the understanding of how the phenomenon is governed by "natural law" or something "out there." How the phenomenon is encountered shapes how the phenomenon is understood. Mechanical realism allowed the phenomenological experience of the particularities of machine performance to be presented as the means to disclose the phenomenon. In modern science, the ways in which experimentation is performed is dependent upon the particularities of what is being investigated and which type of explanation is required, and these are transformed through the dialectics of working out how to perform the experiment as a mode of disclosure of the phenomenon, and thereby explain it. However, it is only through the transformation of general conceptions of knowledge of the nature of Nature since the Scientific Revolution—which should be understood as a mechanical realist *coup d'état* over natural philosophy—that research through experiments became possible as a means of disclosure of natural principles and new natural phenomena, and exoframing could participate in encountering and discovering Nature.

Dialectics of Technology and the Technological Framework:

In his book *The Mangle of Practice*, physicist and sociologist of science Andrew Pickering presents an interpretation of scientific activity in which its results emerge from a dialectical relation between "human agency" and "material agency" that occurs on the interface of machine performativity (Pickering, 1995). His interpretation of modern experimental physics is that it is a performative and productive social labor process comprised of representational and material practices that are dialectically transformed during real-time accommodations to resistances. He argues that "material resistances" to "human intentions," and the accommodations made to plans and intentions in response to those resistances, demonstrate that materials are agents. Pickering argues that the strong program of the sociology of science does not provide an intelligible account of scientific practice because it neglects "material agency," which, in dialectical interaction with human agency, transforms scientific practice. Using Krieger (1991) as support, Pickering argues that physicists deal with the world as a field of agency with material dimensions and the scientific world is amply stocked with material agents. Human agency and material agency interact as "a dialectic of resistances and accommodations" in which machines are intermediaries that capture material agency as a particular combination of particular elements that acts in a particular way.

Using the example of Glaser's early twentieth century attempts to build a bubble-chamber, Pickering pointed out that Glaser had to find many different "solutions" to "the triggering problem," during the course of developing a working bubble-chamber that could detect "cosmic rays." However, each proposed solution failed, one after the other, despite Glaser's expectations of success with each new modification. In his critique of the strong program in the sociology of science, for which the results of science are social constructs, Pickering asked the following questions: if these "solutions" were socially constructed as "expected successes," and "the detection of cosmic rays in bubble-chambers" is also socially constructed, why should we see this sequence of failures? Where is the social causal factor here? Who is constructing the failures? How can the strong program in the sociology of science explain the unexpected failures in the history of physics? Why does failure occur when the prevalent social consensus and dominant authority expects success? If, as Pickering argues, along with sociologists of science, the interests and identities of scientific agents are at stake within scientific practices, why would human beings construct "failure" rather than "success"? The sociology of science can explain the extension of scientific culture and its successes, and how failures are constructed in contexts of controversy and tradition, but how can it explain the causes of failure when everyone expects success and all the participants' careers depend on it? Pickering wrote, "[i]t is clear that Glaser had no way of knowing in advance that most of his attempts to go beyond the cloud chamber would fail but that his prototype bubble-chamber would succeed, or that most of his attempts to turn the bubble-chamber into a practical experimental device would fail but that the quenched xenon chamber would succeed. In fact, nothing identifiably present when he embarked on these passages of practice determined the future evolution of the material configuration of the chamber and its powers. Glaser had to find out, in the real time of practice, what the contours of material agency might be." (Pickering, 1995: 52)

"Material agency" is defined as "simply the sense that Glaser's detectors *did* things—boiling explosively or along the lines of tracks or whatever—and that these doings were importantly separate from Glaser." (Pickering, 1995: 51) Pickering points to the state of affairs, which arose in the performance of the machine through Glaser's relation with the bubble chamber, was something that was not under human control. The machine "is the balance point, liminal between the human and the nonhuman worlds." (Pickering, 1995: 7) Science is a collection of powers, capacities, and performances that achieves expression in "captures of material agency" through "whatever [is] required to set machines in motion and to channel and exploit their power." (Pickering, 1995: 16) It is through this process that "material agency" is determined in terms of temporally emergent "resistances" and "captures," while "human agency" is temporally emergent as "intentionality" and "discipline." Machine and human performances occur simultaneously, and, consequently, in experimentation, the process of capturing "material agency" is one of "fine tuning" both human interventions and machine performativity in feedback relations with the other. The "constitutive intertwining" between "human agency" and "material agency" is continually and dynamically undergoing production through this dialectical process. Through practice and labor the contours of material agency are intertwined with modes of human agency in such a way as to inextricably mix them together as an ontologically and epistemologically productive "impure dynamic." Human agency, defined in terms of discipline and intentionality, is transformed and restructured throughout the process of trying to stabilize precise material configurations and "captures of material agency."

Physicists may, in the process of trying to succeed in achieving any original goal or project, end up succeeding in a different goal or project because plans and goal are revised; they are "subject to mangling in practice." Pickering termed these revisions as "accommodations" to "the emergent resistance" of material agency to human agency. Accommodations take the form of adjustments to intentions and practices, the adoption of alternative techniques, changes in the material configuration of the apparatus, or employing different expertise as a resource. These accommodations transform the original intentions and operate in a field of existing machines in such a way that the goals of scientific practice are emergent in relation to this field as they take advantage of prior "captures of material agency."

It is the relation between human agency and machine performativity, in which both are mutually modified through reciprocal tuning, which keeps human agency intertwined with material agency. The contours of material agency emerge as "resistances" and "captures" in response to human agency, but, these contours would not exist without human agency. Intentions and discipline are modified and transformed dialectically via accommodations to material agency, and are adaptations to it initiated by human agency, which was modified in response to its material consequences, thereby making possible new intentions and disciplines. This notion of a dialectical relation was central to Pickering's whole argument because "[t]he resistances that are central to the Mangle in tracing out the configurations of machines and their powers are always situated within a space of human purposes, goals, plans; the resistances that Glaser encountered in his practice only counted as such because he had some particular end in view. Resistances, in this sense, exist on the boundaries, at the point of intersection, of the realms of human and nonhuman agency. They are irrevocably impure, human/material hybrids, and this quality immediately entangles the emergence of material agency with human agency without, in any sense, reducing the former to the latter." (Pickering, 1995: 54)

Pickering proposes that scientific culture is a patchy, scrappy, disunity of diverse cultural elements in which scientific practice is nothing more than making associations between these elements. Material agency only arises as a result of scientific exploration finding new problems that arise when new machines are constructed to solve these particular problems. Pickering argues that the conceptual and material elements emergently arranged together and associated with representations through practice, which he calls "conceptual chains," link to representations of "captures of material agency"; these constitute the totality of articulable scientific knowledge (Pickering, 1995: 69). It is through this interweaving of conceptual and material elements that a representation of the experiment is constructed and linked within a pre-existent field of scientific knowledge, and changes that field. Concepts allow machine performances to be linked together with representations, spanning multiple levels of theoretical abstraction, and these are emergent through representational and material practices.

Pickering claims that this has an interesting implication for the epistemology of science. If the conceptual chains and representations associated with different machine performances constitute the totality of scientific knowledge then particular conceptual chains and representations associated with specific captures of material agency are incommensurable with one used on different machines. If post-1970s and pre-1970s particle physics used a different collection of machines and instruments then the conceptual chains and representations produced by these different phases of physics are incommensurable with one another and do not share common concepts and representations, and "the captures and contours of material agency" are also incommensurable, given that the associations between cultural elements emergent from the dialectic of accommodations and resistances are different, and therefore have distinct "temporally emergent ontologies."[237] As a consequence of this, Pickering claims that the content of scientific knowledge is nothing more than a series of "temporally emergent concepts and representations" associated with interactive stabilizations of material practices situated in a multiple and heterogeneous space of machines, instruments, conceptual structures, disciplined practices, and social actors. The two phases of particle physics are not doing the same thing and should not be compared to each other.

Pickering argues that technical knowledge, abstract laws, expertise, models, experiences, techniques, machines, concepts, etc., are elements, resources for mangling, that cannot transcend the dialectical process of experimentation and guide it in any way. He explicitly rejected the notion that there could be transferable skills, any general knowledge of machine building, experiences, or even guidelines, which could enable us to build machines. All of these are nothing more than heterogeneous elements for mangling; they do not shape the mangling process. In my view, it is this interpretation that is problematical, if we are to understand how scientific concepts and representations relate to the technological framework and provide us with a special kind of knowledge.

---

[237] (Pickering, 19951: 88-91). See also Pickering (1987), sections 6.4 and 6.5.

Pickering's interpretation left nothing that can help us to guide this dialectical process because there is nothing outside "the Mangle" and literally nothing that could enable us to know how to build machines.[238] It is all just happenstance. We feed our intentions into "the Mangle," which mixes up a load of heterogeneous cultural elements, transforms them, and spits out a new product as an element for future mangling. If that product fulfils our intentions then that is simply a matter of good fortune. In Pickering's analysis, the notion of "temporal emergence" meant that there is no substantive (non-sociological) explanation to be given for the extension of scientific culture. In other words, "things just happen." Pickering terms this as an "open-ended extension" and explains human participation in that extension as "a trial and error process." It is something that is inexplicable, in principle, because

"Nothing substantive in scientific culture or anywhere else… necessarily endures through and explains the process of cultural extension; everything in scientific culture is at stake in practice; there is nothing concrete to hang onto there." (Pickering, 1995: 111-12)

In my view, Pickering's claim that intentions, conceptualizations, and representations of machine performativity, and the heterogeneous elements available for experimentation, are all produced, interconnected, and transformed by the processes involved in actually building machines and performing experiments, is a fairly uncontentious claim. I also agree that the concepts and representations that link otherwise unrelated machine performances are produced through the real-time practices of experimentation. This is evident from how experimenters actually perform experiments. However, my complaint with Pickering is that he has neglected to examine how the products of the extension of scientific culture are actually fed-back into scientific activity in practice to inform and constrain the experimenters about how to proceed with new experiments and how to connect otherwise distinct representational and material practices within the technological framework of the experiment. Pickering does not allow the technological framework to inform and constrain the process of mangling in any way. It is all mere happenstance and opportunistic uses of "prior captures of material agency." By defining the ontological disclosures emergent through experimentation in terms of the interactive stabilization between "human agency" and "material agency," and limiting it to those stable relations as an *a priori* epistemological principle, Pickering's interpretation is somewhat positivistic. The claim that the pre-1970s and post-1970s phases of particle physics have different ontologies and are incommensurable is itself only possible due to a positivistic conception of epistemology. Such a conception makes particle physics unintelligible as a human pursuit because it obscures the conceptual need for metaphysical foundations to explain heuristically experience in terms of scientific models, via a concept of "natural mechanism," with those conjunctions of events captured as the set of associations between techniques and machine performances that comprise the technological framework of scientific activity. Pickering is unable to explain why and how experimental physics is culturally considered to be a natural science. It is a completely arbitrary designation. Although Pickering is correct to identify the object of particle physics as the emergent product of modeling machine performances, by neglecting the metaphysical precepts that underwrite the use of machines to explore Nature, Pickering misunderstood how models are used within human interventions to connect otherwise unrelated machine performances and interpret them.

Yet, as I have argued above, once we examine how technological objects achieve transfactuality via the technological framework then we can readily see that mechanical

---

[238] Pickering (1995: 102. n. 5)

realism unifies both phases of particle physics and allows both to participate in the disclosure of the same ontology, provided that this ontology is understood as stratified in parallel with the strata of the technological framework. Thus overlapping strata of distinct machine-kinds are methodologically unified by the same metaphysics that defines the modes of disclosure of the ontology of physics in terms of the ongoing innovation of machine-families, which are comprised of all of the possible machines that can be built via permutations of the same machine-kind. All mechanical devices no matter how complex are members of the mechanical machine-family (the simple machines) because they are all combinations of these basic machines, each of which acts as a component. Clockwork devices, locks, cranes and lifting apparatus, corkscrews, or pipe cutting tools, for example, are members of the same mechanical machine-family. Both phases of particle physics share some of the same components (electromagnets and ionizing fluids, for example) and the models (electromagnetic theory, solid state physics, etc.), utilized to interpret machine performativity through shared functives (i.e. charge, spin, and mass) in their exoframes of the machine performativity. In order for these two phases of particle physics to be incommensurable they would have absolutely distinct meanings—absolutely distinct exoreferences and technical codes—for these terms to have nothing in common whatsoever. Yet many of the components used in both phases of physics can be used interchangeably within the different experiments even when those experiments are given radically different theoretical interpretations. Technological objects remain transfactual and commensurable when they are used in different projects within the same technological framework to link different machine performances under the same theoretical description by using the same exoreferences. Terms such as "electron" can be used to bring together sets of exoreferenced functives, such as "charge," "mass," and "spin," to unify otherwise distinct machine performances via the same exoframing model. It is through the same technological framework of overlapping machine components, models, mathematical techniques, etc., that distinct experiments can be said to all participate in the same unitary methodology.

If projects share technological objects and technical codes then scientists can transfer their exoreferences and exoframes between projects and build up a stock of experiences, resources, and tactics, which can be analogously used as possible accommodations in analogous projects. Although this does not guarantee success, it does mean that experimenters can develop a bounded technical rationality, which is both shaped by and shapes the contexts in which it occurs, and which can constrain and inform their intentionality in making selections of possible accommodations and identifying potential resistances. Models are implicitly involved in the construction of the apparatus, the development of operational procedures, and the making of observations, and modeling is crucial for the labor processes of experimentation to be an intelligible means of the disclosure and implementation of natural mechanisms. Models provide the transfactual interpretations between otherwise unrelated machine performances, providing commensurability, and are used to guide technological innovation by analogically connecting machine-kinds within the development of a metaphysically conceived technological framework. This explains how scientists represent their ongoing technological activities as a stratified process of achieving ontological depth, and, through the analogical and metaphorical use of models, also shows how technological objects are transferable between experiments, achieving transfactuality and commensurability.

Intentions emerge against the paradigmatic background of clusters and constellations of available technological objects, expectations, and teleological positings for those objects. This background is situated within the technological framework from which the experiment emerged as a possible experiment that could be given cultural meaning as an experiment into something that exists outside of that framework and the contexts of human agency. If scientific practices share stable cultural elements, which have emerged from the same technological framework, these shared elements allow experimenters to acquire a cultural stock of technological objects, which enables them to choose particular accommodations as possible solutions to particular problems, and situates these solutions and problems within the shared technological framework as meaningful techniques and machine performances. The technological framework constrains and informs choices by associating and interconnecting machine-kinds with models and productivity in the design, construction, operation, and interpretation of new machine-kinds, or the innovation of a new machine-family member, and thereby opening up the possibility of new associations and innovations. Pickering did not allow this possibility in his analysis and, consequently, the choice of particular accommodations that particular experimenters made to deal with particular resistances is inexplicable to him except as *ad hoc* tinkering.

Take Morpurgo's experiment to discover "free quarks," for example. The technological framework provides a cultural background of technological objects and possible forms of this experiment, with its associated expectations and techniques, which connects interpretation of the meaning with the scientific worldview. This not only permitted Morpurgo to postulate a form of experimental apparatus as a means to search for free-quarks, but it also constrained how he could use the machine to proceed with this search. It is this notion of *constraint* that is absent from Pickering's analysis, and, consequently, Pickering could not explain Morpurgo's choices of "the accommodations to emergent resistances" during the process of building the experimental apparatus. For example, Pickering claimed that

"...Morpurgo found that the charges on iron cylinders seemed to drift overtime—from zero to e/10, for example. *Tinkering once more in material practice, Morpurgo found a new way to frame material agency*, discovering that he could achieve stable measurements, again of zero charge, if he spun the cylinders." (Pickering, 1995: 59, my emphasis)

By treating this new technical practice of "spinning the cylinders" as "a new way to frame material agency," Pickering ignored the extent that "measurements of charge" are techniques which are situated in a pre-existent technological framework that transcends and orders the particularities of experiences and practices via already established exoreferences and exoframes. That's why they are measurements of charge. What led Morpurgo to consider "that the charges of iron cylinders seemed to drift over time" to be a problem? And, what led Morpurgo to consider spinning the cylinders as a possible solution? Pickering answered the first question by appealing to "resistances of material agency," and could not answer the second question except by appealing to "ad hoc tinkering." Although he did account for why Morpurgo had to make an accommodation, Pickering cannot account for why Morpurgo chose the accommodation he did except by explaining it away as "tinkering." Why didn't Morpurgo try slaughtering a chicken and dripping its blood over the apparatus? It might have worked! I would suggest that the reason why "spinning cylinders" appeared to be a possible solution and "ritual sacrifice" did not was because the choices available to Morpurgo were constrained by the technological framework in which his experiment was situated. The drift, as a resistance, was a product of expectations and judgments of what good measurement techniques and experiences would have been in that context. Morpurgo only "tinkered" in this way because he was using an interpretive model of why there would be charge drift on the iron cylinders and what he could possibly do about it. This model was inherited along with techniques, component machine-kinds, exoreferences, exoframes, and other technological objects, through the extension of the analogical model within the technological framework. The spinning, as a possible solution—an accommodation—was more a product of the Morpurgo's expectations, technical experiences, and theoretical interpretations of "charge distributions" and "the properties of iron" than it was "mere tinkering."

According to Pickering, "material agency" emerges when human beings *actively* construct a new machine and *passively* observe the performance of the machine to see whatever "captures of material agency" occur.[239] This is apparent in Pickering's analysis of Morpurgo's observations when he wrote:

"…Morpurgo assembled his apparatus, switched it on, and then, *surrendering his active role*, stood back to watch what would happen—literally, through a microscope. Swapping roles, *the material world was in turn free to perform as it would*; the grains levitated and moved away from their equilibrium position when the electric field was applied." (Pickering, 1995: 79, my emphasis)

However, by claiming that there are two distinct and identifiable phases of scientific work, actively constructing the apparatus and then passively observing the results, Pickering has built the human-material distinction into his analysis and neglected the way that experimenters, like Morpurgo, *simultaneously* and *ambiguously* "actively" and "passively" perform experiments. The "active" choices that Morpurgo made in the construction of his apparatus were simultaneously "passive" responses to what was a possible choice, according to his current expectations of the alethic modalities within the technological framework, such as what was possible, impossible, probable, improbable, necessary, or contingent. The "passive" observations that Morpurgo made, after switching the apparatus on, were simultaneously "active" interpretations and construals of what he was seeing, made in terms of his current model of how the apparatus worked (i.e. that an electric field was being applied and how this should effect charged grains).

Pickering is aware that Morpurgo used an interpretive account of how the apparatus worked in order to "move from observations of the response of the grains to an applied electric field to statements about the electric charges carried by the grains."[240] However, Pickering neglected to attend to the extent that an interpretive account is also required to make observations of "the response of the grains to an applied electric field" in the first instance. Otherwise, why would we say that there was "an electric field" present and it had been "applied" by turning a dial, pressing a button, or whatever procedure we associate with the technique of "applying an electric field"? Pickering's neglect betrays his positivistic understanding of "the empirical" aspect of experimentation. Contrary to this, I argue that making observations involves simultaneously passively/actively interpreting what is happening during the experiment as interventions, observations, and expectations ambiguously interact. This is apparent when Morpurgo observed an anomaly. He was frequently confronted with anomalous grain motion in terms of his current expectations while attempting to make sense of it in terms of a model of how the apparatus worked. When he could not, he had to adopt an alternative tactic (which Pickering characterizes as "active") by modifying the formula for calculating charges on the grains by adding an additional term to the equation. However, contrary to Pickering, this tactic was also "passive" because the choice of techniques for the active modification of the formula part of the model is constrained in accordance with what Morpurgo perceived a possible choice to be in relation to what he could expect

---

[239] Pickering (1995: 21)
[240] *Ibid* 74

to measure using his apparatus, how he could expect to exoreference that modification as a physical mechanism, and simultaneously what he could expect to demonstrate by exoframing the apparatus in that way. The configuration and functionality of the apparatus, as well as Morpurgo's expectations of the functional and demonstrative potentials of his models, interpretations, and the apparatus, were informed and constrained within the technological framework.

Goals and expectations are an intrinsic part of any technology and are a pre-requisite for choices. Without goals and expectations we could not make judgments about which techniques, tools, or machines to use, and we would not expect them to work or break down in particular contexts and perform specific productive acts. Choices, decisions, and intentions are informed and constrained by the technological framework to the extent that the existence of particular technologies are ontological pre-requisites for the existence of particular choices, decisions, and intentions. The goals of labor bridge practices and experiences, revealing the extent that both are intrinsically defined in terms of each other, in accordance with models of efficiency and functionality, during productive processes emergent within the technological framework. Goals and expectations are emergent in the same way as any other technological object and cannot be localized in purely the human realm. Glaser could not have intended to build a bubble chamber if he were born into Galileo's culture. Nor could Morpurgo have intended to search for free-quarks by sacrificing a white ox at Stonehenge during the summer solstice. Neither would have been a possible choice within the historically emergent technological framework, given that the stratified structures of the technological framework required by those experiments would simply not exist.

The strata of the technological framework and their connection with the scientific worldview informs and constrains intentionality through bounded technical rationality because any experimenter has a technological constraint placed upon him/her to use only theories, interpretations, and intentions, which contain measurable elements and the choice of theory, interpretation, and intention is constrained by the limitations of the available measuring technologies. In the context of Morpurgo's experiment, he was constrained in his choice of technological objects according to the consensus of other physicists regarding what can or cannot be demonstrated by using particular techniques and machines. Morpurgo was constrained by the template of methodology and the paradigmatic consensus regarding "efficient techniques" and "achievable goals," and choices are always constrained to choose "efficient techniques" as the means to achieve "achievable goals." The technological imperative to innovate is central to experimental physics because it challenges human agents to innovate solutions to technologically produced problems and to explore the ontology of the world by increasing the available technological possibilities for future research. Not only does technological innovation inform and constrain—extend and challenge—intentionality by creating the possibilities for a whole range of potentially achievable new projects and goals, but also constraints and challenges intentionality when human agents maintain a perpetual deference to the future regarding the perpetual revisable estimations of appropriateness and efficiency of any technique, as well as the achievability of any goal. The "bounds" of technical rationality are themselves at stake and tested by experimentation and innovation. Thus Glaser and Morpurgo were challenged to invent their machines in accordance with the goals, expectations, and constraints of the technical framework in which they were situated and embodied, as a test of the alethic modalities and limitations of that framework. They were informed and constrained by the technological framework as soon as they took up these challenges.

Pickering required a concept of "material agency" because he left no space in his analysis for being informed and constrained within a technological framework that pre-exists and orders the particularities of experiences and practices of any particular experiment. Thus he maintained that the experimental apparatus—the machine—is an intermediary for the accommodations and resistances of the dialectic between human and material agencies. However, if we restrict our analysis of experimental physics to its performances, as Pickering implored us to do, then we do not have a phenomenological experience as machines as intermediaries except in terms of what is said about them. What we have are machines and interpretations of their performativity in terms of mechanisms, laws, and theoretical entities. It seems to me that Pickering described these machines as intermediaries between "human agency" and "material agency" only because he considered this step to be necessary to counter social constructionist accounts of science. However, my argument is that if we examine machines as technological objects within the technological framework rather than intermediaries then we can develop a rather different description of experimental science than Pickering does.

"Material agency" is an abstraction of the particularity of practices and experiences emergent through complex labor processes that are challenged to produce the knowledge of general principles of mechanisms and laws in terms of fundamental representations related (via exoreferences and exoframes) to the functionality and performativity of machines in order to make the natural world intelligible to human beings in terms of natural mechanisms and laws. The functionality and performativity of any machine is defined in terms of the larger organizational structures in which they are integrated in accordance with a teleologically posited goal. The experimental connection of teleologically posited machines with other machines within the ongoing construction of the technological framework generates "emergent brute resistances" in accordance with the degree of incoherence and contradiction within the technological framework. This "resistance" occurs when heterogeneous technological objects interfere with one another and is the consequence of the heterogeneity of the goals and possible choices to which they are posited as the means to an end. Coherence is achieved and resistance disappears when these heterogeneous technological objects are brought together *upon the anvil of practice*, so to speak, and forged into a single, complex technological object. It is the labor process itself that resolves the incoherencies and contradictions within the technological framework to transform it into a general and unified technological object capable of being transferred between contexts.

Technological objects are distinct but inseparable from one another in the technological contexts of experimentation and innovation. Machines are products and embodiments of intentions, expectations, beliefs, choices, and values, and as such—as technological objects—are transformative agents within wider scientific culture. The innovation of new machines, when used, generate new intentions, expectations, beliefs, choices, and values, through the powers, constraints, challenges, and demands, that they produce and disseminate through the scientific culture in which they are operational via the technological framework. Machines teleologically shape the productive directions and possibilities of labor, and it is in this context that machines can be said to have agency within their own right. For example, in Glaser's work, by attempting to connect heterogeneous components together with the aim of constructing a new unified machine, a new kind of particle detector, each of those components began as its own technological object (as a result of unification achieved through the labor of others) and the problem Glaser faced was bringing these heterogeneous technological objects together into a coherent and unified totality. The resistances arose through the problem of achieving stable cooperation of functionality and performativity as each technological object was juxtaposed and connected together. When heterogeneous technological objects are brought together, the particularities of each often disrupts the functionality of the others, and the outcomes are often inconsistent with the teleological posited goal such as building a particle detector.

Any inconsistency between technological objects cannot be identified with an "emergent material agency" precisely because it arises due to the fragmentation and disharmony of heterogeneity rather than from a unitary source, such as "material agency." Resistance is a consequence of components being brought together to perform functions for which they were not previously designed, developed, and stabilized to perform, and the task for Glaser was that of transforming their context of functionality and performativity into a unified, complex technological object—a particle detector capable of making new discoveries—within the technological framework. This task does not occur in isolation from the rest of the world. Hence, Glaser had the problem of building this machine, but he also had the problem of how to integrate coherently and consistently his bubble chamber into the wider context of particle physics. It is this two-fold problem that situates his work within the technological framework of particle physics, thereby informing and constraining his choices about how to proceed with building the machine. This involved not only integrating electrical components, glass tubes, and strange gases, but also involves integrating techniques, interpretations, conceptualizations, technical codes, political institutions, economic factors, social organizations, beliefs, values, and expectations, together within the technological framework as a stable, reproducible, and transferable means of production of "observations of cosmic particles." It is only by doing this, can new machines, like bubble chambers, be made to work as technological objects and become part of scientific culture.

This is especially apparent when we examine how "raw materials" are actually used in scientific work. Materials are integrated into the construction of machines in order to determine their properties on the basis of what that material does, as a component, to the other components and materials to which it is connected and acts on as a machine. From the outset, the intervention is bound-together with representational models of the material and the machine, thereby transforming a "raw material" into a technological object in accordance with human expectations and goals. "Raw materials" are organized within the technological framework, teleologically posited in terms of functionality and performativity, and connected with exoframes and codes to explain the structure of the machine in which it is a component. After all, what is a machine? It is not merely a particular configuration of "raw materials" (metals, plastics, glasses, gases, etc.) but it is a particular configuration of functions and performances that can only be emergent within contexts of implementation in accordance with goals (teleological positings). Machines are made to realize these teleological positings by reproducing functionality and performativity through integrating them into a coherent unity within particular contexts of use. Machines may be used in mechanical, mathematical, computational, social, political, military, biological, medical, scientific, analytical, sexual contexts, etc., and the "raw materials" are understood in terms of their functions and performances within these contexts. Functionality and performativity therefore are informed and constrained in context, emergent from the technological framework, and do not come out of some inner principle of Nature. The properties of "raw materials" are themselves the outcomes of organized labor set-upon otherwise heterogeneous technological objects, developed in different contexts, to gather them together, order them, work upon them, and integrate them into a stable, coherent, and unitary function and performance in accordance with the teleological positings emergent from the technological framework, and connected with scientific culture via associations between techniques and refinements to the scientific worldview. It takes historically situated effort, intelligence, and power to create, unify, and maintain new orders of functionality and performativity within a perpetually transforming technological framework, and interpret that effort, intelligence, and power in terms of natural mechanisms and laws.

Complex teleological labor processes are driven by the ongoing challenges of completing the technological framework by resolving contradictions and incoherence within its structures. Even the "raw materials" from which any machine is built are technological objects with their own histories of discovery, stabilization, and implementation. This is even true of substances that we take for granted as natural elements, but understand in terms of a long and protracted history of use and meaning. Take iron, for example. The properties of this element are identified in terms of its functions and performances in contexts of use, such as hardness, durability, tensile strength, availability, cheapness, etc., and all these functions have taken considerable work (over thousands of years) to organize within the ongoing dialectics of technology—thereby transforming iron from being "stuff" into a technological object available for use. Iron is itself understood in terms of its appropriateness and resistances within the context of a long history of heterogeneous teleological positings and labor struggles to achieve them. Resistances can be accounted for by examining the structures of intentionality and appropriateness in terms of means-ends relations, in the dynamic interaction between heterogeneous contexts of use, which may cohere with or disrupt each other in unpredictable ways, and, consequently, dynamically produce stable or unstable outcomes within different contexts. As such, resistances are the result of sociological, psychological, political, economic, and technical activities and goals, each with its own history of development and transformative change within the technological framework. To put it simply, iron ore is dug up out of the ground, but iron is made!

By treating "the source" of such resistances as the interaction of "human agency" and "material agency," Pickering has metaphorically substituted "material agency" for the wider context of the outcome of complex teleological efforts and struggles to integrate heterogeneous innovations into pre-existing technological and social orders. These efforts are inherently unstable because their interaction mutually transforms their functionality and performativity during the transference between contexts of use. This is the result of contradictions and incoherence between teleological positings, which are themselves emergent from the technological framework. Contra Pickering, the situation is not a dialectical relation between "human agency" and "material agency," but is the stratified outcome of dialectical transformations of the objects and structures within the technological framework itself. Whence the resistance? How can we make sense of the possibility and phenomenology of resistance without adopting scientific realism? What is the source of resistance? The problem facing us with this question is how do we locate and identify the source of particular resistances, in its unity, within transfers of heterogeneous technological objects between heterogeneous contexts of use. *Why do we think that there is a single source?* Resistance arises from the internal heterogeneity and incompleteness of the technological framework.

In the Aristotelian scheme this "resistance" is an example of *hyle* rather than *phusis*. The term *hyle* captures the sense of resistance to ongoing interventions and intentions in material practices. It is the phenomenological particularity of the particular and cannot be generalized between different contexts of use. It is a phenomenological response to human interventions that does not spontaneously come of itself. *Hyle* is emergent as a consequence of the attempt to impose idealized forms and intentions upon materials, and the particularity of that deviation from expectations is dependent on the content of use.

It is an emergent feature of material practice that occurs during the human attempts, guided by *techne*, to inscribe form into materials, which is neither controlled by human intervention, nor does it exist independently of human intervention, but it is a property of the context of *poiesis* guided by *techne*. *Techne* is an imagined (asymptotic) complete and invariant knowledge of what is considered as "a true course of reasoning" involved in bringing something into being through a communicable and universally repeatable act of *poiesis*. In experiments, the universalizability of any such act would depend upon its successful integration into the technological framework as a machine prototype, and its successful utilization in the extension of a machine-family or the innovation of a new machine-kind. Invariance, as the repeatability of a result through the repetition of intentions, interventions, material practices, and the "true course of reasoning," is itself *a techneic and poietic reproduction* of the transformative process from which it came about. A techneic theory (with its exoframes and exoreferences) is one that is used to imagine a specific object utilizing the same cluster of technographe, techniques, material practices, and machines. *Techne* is imagined to provide the contours, boundaries, limits, and prescriptions, to a set of alethic modalities that are given meaning within context and explained in terms of causal mechanisms and their effects. As the knowledge of the Being of Becoming, *techne* aims at unifying any particular practices, by generalizing the intransient causal principles of change within that work, and promises complete alethic knowledge as "the end result of experimentation," However, no such knowledge is actually possessed outside the imagination and the complete unification of practices and experiences remains incomplete. Complete and perfect reproduction is an imagined ideal that is never achieved in practice and experience. *Hyle* demonstrates that extent that the technological framework is incomplete and the particularities of the particular deviate from the law-like behavior required of *techne*.

It is this spontaneous and chaotic behavior that Pickering termed as "material agency." It is an abstraction of the phenomenological experience of machines "doing their own thing"—the way that their performances do not conform to expectations within the technological framework—and is not a pole or terminus of any dialectic. Pickering has misunderstood the dialectic of technology. It is only through particular conceptual structures, socio-technically constructed and inherited, on the basis of metaphysical presuppositions, that notions such as "material agency" can be intelligible as a conceptual device to explain the phenomenological experience of resistance when bringing together heterogeneous technological objects within a teleologically posited labor process. The interaction is between technological objects (interventions) and technophenomena (responses), and the dialectic of technology occurs in transforming the structures of the technological framework itself. In my view, we cannot make the process of building machines intelligible by analyzing it in terms of dialectic between human and material agencies. Building machines involves integrating a diverse and heterogeneous set of complex modes of agency, within the context in which any machine is a particular framework of interactions that interacts with other modes of agency, within the environment in which it is contained. As technological objects, machine are situated within a larger technological framework and scientific culture that order each other, and, in turn, are ordered within processes of gathering and ordering technological objects as standing-reserve to satisfy the never-ending challenges (teleological positings) of the construction and development of the technological framework, which transforms the world into materials and techniques available for future use, which are structured and developed by the dialectic of technology, thereby bringing new powers into the world and making new experiences possible. Resistance is emergent from the ongoing processes of labor and is only contextually conceived as a resistance until the coherent functionality of all agencies are produced and unified as an outcome of the innovative processes of labor in the creation of its own possibilities. At the core of the dialectic of technology is this creative labor process, which engages with an already-technologized environment, and it is this core that informs and constrains scientific experience, and, therefore provides the "evidence" or "data" for representation, conceptualization, modeling, and theoretical explanation and calculation.

The Framework of Experiment and Natural Law

Mathematical projection begins with laying down a template for the proposed law, as a basis for the experiment and planning out how procedure relates to the object-sphere of the apparatus. An experiment does not begin with a complete physical law which it then tests (as is often presumed by traditional philosophers of science), but it must presuppose the existence of a law which it aims to disclose and test. The methodology by which experimental science operates involves the presumption that there is a law, which can be disclosed by the proposed experiment (or series of experiments), and presupposes the forms of possible motions that would qualify as fundamental motions that can be disclosed by building and operating a particular machine. The presumption of the existence of a law is the basis for an experiment—its teleological positing and ideal—and the anticipations of how this law can be disclosed and tested inform and constrain theory. Setting up an experiment involves imposing a template of possible fundamental representations and conceptions of how the phenomenon moves and conditions under which these specific set of motions can be revealed, followed in their progression and combinations, and calculated to afford predications and tests.[241] This template has preempted what could be possibly learned about the phenomenon to such an extent that it has preempted the conceptual and representational conditions under which an empirical regularity would be produced, recognized, and accepted as such. It determines how "the empirical" is understood and identified from the outset. These representational and conceptual conditions also preempt how the experiment could be built in such a way as to be a controlled experiment upon something that is supposedly uncontrolled by the experiment. In other words, a fundamental decision has been made about what qualifies as a constant conjunction of events and the conditions under which an experimentalist could claim to have produced these constant conjunctions, and this fundamental decision is built into the template.

Via models, technological objects are transferable along the "genetic" lines of machine-families and their overlaps between kinds that give these technological objects transfactuality, while the performance of one component (e.g. chemical cells) can be related to changes in another component (e.g. electromagnets) via the postulated transfactual entity (e.g. the charged particle) and a model (e.g. the Standard Model of Elementary Particle Interactions). The challenge facing contemporary physicists is the creation of a new hybrid machine that brings together and unifies the fundamental laws, such as those projected as governing matter and radiation (electroweak interactions, strong nuclear forces, and gravitation), which gains its meaning as a means of disclosure via its historical construction within the technological framework. The ability to develop unified field theory into a Grand Theory of Everything is dependent on the successful incorporation of this new machine into the technological framework as a synthesis of all other machine-kinds and their associated representations, to create an ultimate machine-

---

[241] Heidegger (1977b: 121-3) considered this to also be a fundamental aspect of scientific research. See Rogers (2005).

kind and a final set of fundamental representations.

Following the above line of reasoning, we can see that novel experimental science creates and explores unions between previously distinct machine-kinds and their associated set of fundamental representations with the aim of reducing them to a single set of fundamental representations associated with the latest stratum of machines. The irreducibility of strata of machine-kinds and their associated representations is taken to indicate that there is some unknown underlying mechanisms and law from which all these strata could be derived, but the distinctness of the materials from which they are constructed prevents these machines being reduced to lower strata and, ultimately, the most basic stratum. Hence the stratum of electromagnetic machines required electrical devices and magnets, and while these more basic strata all required the development of mechanical devices, they cannot be reduced to this most basic stratum. Failures indicated that the structure of the technological framework is incomplete and a new stratum is required. For example, the failure to provide a complete quantum physics of DNA suggests that some new physics is required and the discovery of that physics and its mathematical laws requires the invention of new machines. Mechanical realism has allowed these historical lines of juxtaposition and unification between machine-families to be culturally represented as the genealogical lines of discovery and confirmation of experimental science as a fundamental natural science, even when the whole cultural project of experimental science is incomplete and led to many more failures than successes. The failures are mostly forgotten. The ontology of physics is abstracted from these interconnected lineages of successful prototypes and their associated clusters of techniques, materials, and models, as each is taken to be a refinement and confirmation of the scientific worldview. Highly complex machines (such as particle detectors)—as composites of the members of distinct machine-kinds (such as electromagnets and chemical cells)—are historically possible only because of the dialectical process between representational and material practices from within the technological framework, but it is the metaphysical interpretation of the dialectics of technology that leads to the creation and use of models as refinements of the scientific worldview, which in turn are used to guide paths of technological innovation and scientific research.

The "discovery" of law in the form of an empirical regularity explained by a mathematical function requires considerable effort on the part of experimenters and may take decades of work before it has been formulated. Discoveries are the outcome of labor. A constant conjunction is produced as a repeatable association between a technique and a machine performance, and each event is construed (conceptually and visually) in advance as one of the projected fundamental motions (mechanisms) associated with a machine-kind. Performing an experiment inherently and necessarily involves constructing machine performance as an ensemble of fundamental motions, mechanisms and machine-kinds. The permutations of all possible machines belonging to the same machine-kind constitutes a stratum within the technological framework, with technological objects, techniques, and mechanisms comprising the structure of that framework. The experiment is controlled to the extent that some kind of machine performance will be its consequent, as the ensemble is deliberately constructed in accordance with specific teleological positings of what the outcome of the experiment could and should be, but how the machine performs is not controlled. It is informed and constrained, but ultimately it does its own thing. This allows experiment to participate in discovery by refining, modifying, or even transforming the technological framework of the experiment in response to the particular machine performances of the experiment, and, given the novelty involved in bringing together diverse components to synthesize a new machine in the context of experimentation (otherwise there wouldn't be an experiment), this also leads to resistances, problems, contradictions, inconsistencies, anomalies, and unexpected consequences. These exist within the structures of the technological framework itself due to its incompleteness and heterogeneity, and resolving these requires new syntheses of stable representational and material practices, thereby qualitatively innovating how the machine in question is used and understood from within the technological framework, which is refined, modified, and even transformed to incorporate the new machine as a stable and reproducible technological object available for use by others.

It is the dialectic of technology that objectively grounds the process of moving from instability and resistance to stability and reproducibility by synthesizing new connections between representational and material practices, alongside new ensembles of associations between technological objects and technophenomena, in a context of innovation and discovery. The experimenter discovers the particularities of the performance of the machines, which have been built and modified according to the preconception of how they should perform, and how the experimenter can change and adjust practices and the arrangement of materials to learn how to make a stable and reproducible machine performance; thereby makes new syntheses and use these discoveries to demonstrate the intelligibility and usefulness of new representational and material practices through which the experiment can be understood and further developed as a technological object within the technological framework. It is the dialectical transformation of the technological framework itself that gives experimental science its objectivity and progressive character, even if the objects arranged and discovered through scientific activity are artificial products that remain uncontrolled by individual human agents, yet it informs and constrains scientific experience, and makes it possible, and informs and constrains any possible theoretical explanations of those experiences. I shall turn to scientific experiences and their relation with theory in the next chapter.

5

SCIENTIFIC EXPERIENCE AND OBSERVATION

Experimental science is a metaphysically interpreted labor process that occurs within a historically conditioned and developed technological framework. Experimenters resolve the contradictions within the structures of the technological framework by creating new syntheses of representational and material practices, making new technophenomena possible, first as the objects of experimentation and later transforming them into technological objects available for future use. When the stratified structures of the technological framework are abstracted into machine-kinds and machine-families, along with their associated set of fundamental representations and theoretical explanations in terms of natural mechanisms and laws, this requires new syntheses to resolve contradictions, thereby creating a new machine-kind and associated set of fundamental representations, and these new syntheses lead to the further refinement of the scientific worldview, which are tested by utilizing them in ongoing technological activity. As an ideal, on an imaginary asymptote, scientists aim at completing and unifying the technological framework—as a totality of all possible technological objects and their theoretical explanation—which would be the completion of experimental science itself and the end of all technological innovation. By aiming at completing and unifying the scientific worldview, as a comprehensive model of the Universe, the truth of science is deferred as an asymptotic ideal that could never be achieved in practice, thereby preserving the idealization of science as a self-correcting and practically useful activity tested through technological activity. It is this aim that reveals how experimental science is manifest as an ongoing technologization of all phenomena, perpetually providing technological objects for further experimentation and innovation, while it is represented and situated with culture as an epistemological mode of discovery in search of the (unreachable) *techne of Nature*.

From the sixteenth century onwards, Nature was thought of in technological terms, as a totality of laws, mechanisms, and objects; knowledge was represented as something *techneic*, while technology was naturalized, objectified, and given epistemological leverage as a means of discovering the hidden causes of natural change and resistance; scientific experience was reduced to the outcomes of making interventions and measurements. Regardless of whether the objects of experimentation are simple mechanical devices, electromagnets, steam engines, particle accelerators, or dilution refrigerators, natural mechanisms are abstracted from this ongoing labor process by *mapping out the boundaries of the alethic modalities of labor itself*, and these boundaries and limits are metaphysically projected, and therefore abstracted, as being those of Nature itself. While the relationship of technology to Nature remains metaphorical, through machine performances objectivity is made literal. Abstract mathematical models of machine performances are used to explain technological activity in terms of what is possible, probable, impossible, improbable, necessary, or contingent, and represent these alethic modalities as determined in accordance with natural laws. These alethic modalities of labor are analogically and metaphorically superimposed over natural phenomena outside the context of the experiment to predict machine performances in response to techniques, and representing these as tests of physical theory within the context of the experiment. These analogies and metaphors are required to connect the dialectics of technology with natural phenomena via clusters of theories, models, and hypotheses, and thereby test those theories, models, and hypotheses via further technological activity. In this way, representational and material practices physically remain within the technological framework of experimentation—as their object-sphere—and metaphysically verify science as the theory of the real. Thus scientific truth remains bound up with the calculation and prediction of machine performances—bound up with technology—while being metaphysically represented as connected with Nature and revealing (or approaching) a hidden (objective) truth, for its own sake. This two-fold understanding of the techneic character of scientific truth shows how experimentation simultaneously operates within the contexts of justification and discovery, and its results have practical application.

Making observations and models of how the experiment works are the products of the technological framework, and these transform how scientists understand science itself. These products are fed back into technological activity, thereby giving the historical development of the technological framework a stratified structure and its objectivity. This dialectic occurs on the technological interface of the interactions between human interventions and machine performances, moving from past to future efforts, and provides scientists with a guide for the further innovation of new prototypes, the development of new research programs, and a stratified sense of verisimilitude and ontological depth. The dialectical trajectory towards unification was built into the technological framework at the outset in the sixteenth century, but the metaphysical interpretation of its stages and structures allows scientists to "approximate the truth" via refining the scientific worldview. Hence physicists have been seeking to reduce all motion and change to one set of fundamental representations of natural mechanisms since Galileo attempted to explain all natural phenomena in terms of the six simple machines. The trajectory of scientific progress is one of technological innovation and the limits and structure of discovery constitute the strata of the technological framework itself. This trajectory towards a unified theory is itself an abstract mathematical projection over the whole cultural project of attempting to complete and unify the technological framework of science through its ongoing stratification and development.

Due to the assumption of mechanical realism, the technological framework is taken to be transparent as a means to discover underlying natural mechanisms and laws that supposedly exist independently of experimental science, even though the actual work of experimentation never occurs outside the context of cultural and contingent human choices, labor, and technological activities. Abstract mathematical functions are used to represent eternal laws of change and persistence within quantified machine performances, as technical codes and functives, which are in turn metaphorically presented as operational variables used in tests of theories of natural laws, even though scientists are often aware that any such theories are tentative constructs and are likely to be refined or replaced in the future. These theories are visualized by using models that correspond to quantified machine performances but are constructed in terms of cosmological principles. Maxwell's Laws of Electromagnetism and the equations of Quantum Electrodynamics are abstracted from the technological contexts of application in the construction of machines, while models based on those equations achieve their greatest predictive success within those contexts, and via mechanical realism are used to confirm the explanatory success of those models outside their contexts of predictive success. Hence Maxwell's Laws of Electromagnetism can be used to explain lightning storms even when they cannot be used to predict them.

Mechanical realism allows the closed-system of the machine to represent the open-system of the natural world. Technologically mediated representational and material practices are translated into socially mediated communication and visualization, reduced to an abstract mechanism, operational as a technical code in relation to an explanatory theory of the laws that govern natural phenomena, which allows techneic heuristics to be abstracted and represented as epistemic laws, even if their context of successful usage never departs from the context of technological activity. This is retrospectively justified through a process of successfully implementing these abstractions as physically modeled mathematical functions (exoframes and exoreferences) in the subsequent extension, innovation, and invention of their associated machine-families. By dialectically feeding them back into processes of guided technological innovation, which in turn transforms and refines the technical codes upon which the test or application of theory is based, experimental science is said to be validated as a self-correcting methodology to produce facts about the world and put theories to the test. The dialectical development of representational and material practices are justified in terms of mechanical realism as a process of the discovery of natural mechanisms and laws, alongside testing and refining those theories through implementing them within technological activity and extending the technological framework within which all experimental science occurs. Yet the dialectical process itself has become invisible and forgotten. By forgetting technological activity, labor, and human choices, "the Book of Nature" can be read. This is how science is said to progress and provide empirical success.

Experimentation and Labor:

How does the technological framework inform and constrain how science is practiced and developed? How has the technological framework shaped scientific experience? To answer these questions, we need to look at how the use of analogy is central to scientific thinking, how models and machine performances are connected, how technology acts as a metaphor for natural processes in laboratory contexts, and how transfactuality is achieved in scientific experience in real experiments.

Ralph Waldo Emerson wrote,

"Man is a shrewd inventor, and is ever taking the hint of a new machine from his own structure, adapting some secret of his own anatomy in iron, wood, and leather, to some required function of the work of the world." (Emerson, 1860: 169)

If we wish to understand "the work of the world," we need to understand labor itself, and, if we hope to understand modern science, we need to understand it as a mode of labor. All labor processes are social processes in which they are organized upon the positing of ends. The Marxist theorist Georg Lukács termed this positing as "teleological positing."[242] The labor processes of modern science, like all labor processes, have a purpose or goal. They have a teleological explanation in the sense that the work of science involves positing aims and ideals for the organization of other people to achieve specific outcomes of specialized labor processes. The labor process is directed from the outset to satisfy goals and overcome challenges, which are central to the setting-up of any experiment or any organized human activity, and these teleological positings are situated within the technological framework of the research program, and thereby are given

---

[242] Lukács (1978: 83-109)

meaning as intentions that are justified in terms of tests and refinements of the scientific worldview. Whether it is the discovery of a stable fusion reaction, gravity waves, or the Higgs Boson, these challenges can be posited to others as the horizon of possibilities upon which scientific progress can be achieved through labor and the further development of the technological framework. The practices and choices adopted in the execution of any experiment are made with the explicit aim of satisfying those purposes and challenges, while contributing to the research program in already established terms, and weaved into scientific culture by putting it forward as a technological innovation. Within the mechanical realist paradigm, an understanding of technique is knowledge and predicting machine performances is an understanding of reality.

Genuine novelty in experimental physics is the product of a technical complex of heterogeneous components and inventions combined into an ensemble, which is integrated with other ensembles of technological objects in order to produce and reproduce functionality within an incomplete technological framework. The labor process is directed towards achieving stable machine performances as the means to disclose "the phenomenon" through bringing together technological objects. All technological objects are complex and cannot be understood without addressing their connections and interactions with other technological objects and the teleological positings that they were constructed to satisfy within this framework, given that their meaning is understood in terms of their function within the framework. Converging and integrating their components into a stable and unitary center of functionality (what is often called 'black boxing') produces all novel phenomena in terms of stable set or sequence of machine performances. The labor processes of experimentation are the refining processes of stabilization directed towards producing and reproducing functionality as a technological object, and, via technical codes and operational variables, relating performance to theoretical expectations.

When labor is performed for its own sake then the experiment is an art directed towards its own self-perfection, for its own sake, which is represented as truth. It is this aspect of experimental work that is represented as "pure science." Science participates in discovery by discovering its own possibilities through metaphysical interpretations of artful labor. The purpose of the art of experimentation is to explore its own possibilities by making them happen. It is metaphysical in the sense that it aims to equate and unify these interventions and material practices under an operational principle derived from non-empirical conceptions of science, technology, and natural phenomena, and thereby explain how labor is possible at all. By showing how these metaphysical precepts are related to modern technology, through concepts of natural mechanism and law, we can reveal the paradigm from within which it is "self-evident" that modern science has been successful in the discovery of novel phenomena and achievement of predictive accuracy. By showing how the objectivity of science is itself dependent on a dialectical transformation of the technological framework, we can maintain our critical distance from the scientific realist's application of the concepts of "mechanism" and "law" to natural phenomena, and also maintain our distance from the empiricist's treatment of experience and observation as atomistic (fundamental and indivisible) and primary (natural and immediate). We can be constructivists about scientific experience and observation. The existence and progress of science does not support or refute the mechanical realist metaphysics that lies at its epistemological heart, which remains hidden from the view of the traditional philosophy of science, and, consequently, the technological successes of science cannot be used by scientific realists to support their claims about the empirical successes of science somehow implying that science is "getting closer to the truth." However, far from supporting a positivist or empiricist interpretation of science, we should deconstruct scientific experience and observation, showing how they are constructed through technological activity, and show that it is questionable whether a natural science based upon a metaphysical project of experimentation using machines and instruments is an empirical science at all!

Analogy and Scientific Thinking:

There are many examples of analogical reasoning and modeling in the history of science. Galileo used the orbits of Jupiter's moons as an analogical support for Copernicus' heliocentric solar system. Kepler used an analogy between musical harmonics and planetary orbital geometry. Newton used a terrestrial projectile as an analogy for the Moon. Bohr used the heliocentric solar system as an analogy for the hydrogen atom. The free electron theory of metals, the kinetic theory of ideal gases, the Ising theory of ferromagnetism, the Standard Model of Elementary Particle Interactions, the Big Bang theory, etc., are all examples of models constructed via the use of analogies to explain phenomena. Francis Bacon regarded the use of analogy as essential to scientific thinking.[243] Since Bacon, the investigation and observation of resemblance and analogies has been developed in scientific reasoning and explanation to provide us with a sense of the unity of Nature, interconnections between specializations, and a foundation for all scientific modeling. Analogy is a fundamental technique for enhancing intelligibility and facilitating communication by comparing the novel phenomenon to something familiar and drawing it in terms of visual representations, such as drawing a magnetic field as lines of force or arrows. It was by using analogies such as "curves of force"—which later became one of the fundamental representations in electromagnetic theory—that allowed nineteenth century physicists like Michael Faraday to visualize the phenomena under investigation, construct models, and develop novel techniques, experiments, and communicable experiences.[244] The analogical feature of model building is essential for the heuristic of scientific discovery to occur at all by providing explanations of novel phenomena in terms of familiar narrative and imagery. Through verbalization and visualization, the novel can be innovatively related to the familiar in such a way as to suggest familiar techniques to explore novel phenomena. In this way, a model can be used to make sense out of a phenomenon by relating it to things that are already considered to be pragmatically known and technically understood. This connects future experiments with past efforts and gives some sense of continuity, as well as a heuristic value for providing inquiry with templates and starting points.

Analogies allow a new phenomenon to be explored using the techniques and representations that have been established as means to explore invisible phenomenon via visible phenomena. An example of using an analogy in this way would be the wave theory of light, which was demonstrated using light waves analogously to ripples in water to explain phenomena such as the refraction and diffraction of light. This allowed the construction of a new model of light. The use of such a model provided physicists such as Young and Fresnel with transferable techniques by which they could build an apparatus to demonstrate the wave-like nature of light. Thus, analogy allows otherwise distinct technologies to be used in novel contexts. Once established through these demonstrations, the analogy was taken literally—as being real—and became available for further analogous exploration of new phenomena such as the diffraction patterns in the electron double-slit experiment. The analogous use of the wave theory of light to visualize

---

[243] Bacon (2000: 47; 144-7; 180-1)
[244] Gooding (1990), chap. 4

(and exoreference) the wavefunctions of Schrödinger's equation and QED integrals in terms of probability-waves are examples of this, and allowed for the development of probability-wave theory to connect the behavior of photomultiplier tubes with "electrons" via the use of a probability-wave diffraction model to explain observational events in the physical context of the double-slit experiment in terms of "wave-particle duality." This shows how the analogy with the Young and Fresnel experiment could be developed and permitted probability-wave theory to be postulated as an intelligible model of the "unobservable" quantum world, and provide a new conception of the role of "observation" in determining the behavior of the quantum world. The analogous relation of the technical definition of "probability-wave" in mathematical theory with water waves communicates and represents the perceptual phenomena via the machine performance of the apparatus in order to make technology and mathematics intelligible as means of disclosure of underlying reality. The heuristic use of the probability-wave theory to interconnect logical, mathematical, and computational analysis of "sub-atomic phenomena" provides a representational framework in which calculations and observations can be related (via exoframing) and also a logical structure through which thought-experiments can be "thought through." This allows a model to be used to make predictions by calculating numerical values, in relation to any variation of factors, and also to derive new hypotheses, theoretical conceptions, and possible observations, and relate these to material practices and thereby corroborate that it has "some truth in it."

The analogical connections between technological activity and natural phenomena via clusters of theories, models, and hypotheses are very important for experimental science to progress. They allow the explanations of the dialectical relations between techniques and machine performances to be understood as part of the methodology of scientific research, which takes the form of the development and application of techniques and technological objects used as instruments for the further development of science. Thus "the electron" (along with its associated techniques of manipulation and measurement) can be transformed from an object of research in its own right to becoming fed back into novel experiments as an instrument (a technological object) to investigate novel phenomena, such as "wave-particle duality" or "free quarks." This dialectical process of transforming the phenomenon into a technological object allows scientists to create new experiments—drawing analogous connections between technological objects—and also allows postulated "natural mechanisms" to be transferred along with the technological objects (sets of associated techniques, instruments, and machines) across the boundaries between otherwise distinct specializations. This allows distinct specializations to inform and confirm each other via the analogy. It also allows otherwise distinct natural sciences to inform and confirm one another. For example, in this way physics can inform and confirm geology by using techniques, instruments, and machines in the laboratory to model the formation of the Earth in terms of chemical elements, pressure, temperature, and time; or inform and confirm astronomy by comparing the performances of a nuclear fusion reactor to the phenomenal appearance of the Sun. This allows techniques, instruments, and machines to cross over the boundaries between these otherwise distinct sciences, such as physics and geology, or physics and astronomy, and it also creates new fields, such as geophysics and astrophysics, while confirming the unity between all sciences and specializations.

The metaphysical abstraction of the dialectic between representational and material practices to explain techniques and observations in terms of "natural mechanisms" also allows the creation of new machines through analogies by bringing together otherwise heterogeneous technological objects within the technological framework. This allows the innovation of prototypes to occur through analogical reasoning and provides the technological framework itself with its own internal logic of development. Each prototype is a hybrid constructed through trial and error processes of bringing together and integrating heterogeneous technological objects from other distinct kinds of machines, each forming its own stratum, and creating a new kind of machine—the prototype—by doing so. This creates strata of technological innovation (new machine-kinds) within the structures of the technological framework (e.g. electromagnetic motors being developed from electrical circuits, magnets, chemical batteries, mechanical devices, metal working, etc.), wherein new strata of prototypes emerge from the existing strata of the technological framework, and are further developed in all their varieties and permutations (i.e. into machine-families). This stratified evolution of increased complexity (what Stephen J. Gould termed as *punctuated equilibria* when describing the evolution of species) gives technological innovation an ontological sense of discovery and developmental continuity with past efforts, while generating new strata which cannot be reduce to the old strata nor explained in terms of their rules and structures.

For example, Millikan's apparatus to measure the charge on the electron was used as an analogy by Morpurgo to invent his apparatus to measure the charge on free quarks, with the transfactual connection secured via shared electromagnetic components in both machines being represented in terms of the same technical codes, such as "charge" and "potential difference." What could be demonstrated about integral quantum charge (e) could be demonstrated about fractional quantum charge ($1/3e$). Morpurgo's use of Millikan's oil-drop experiment as a model for a novel apparatus to search for free-quarks shows how the analogical use of a machine can suggest ways to associate "well-understood" experiments with new experiments.[245] The apparatus was postulated in terms of an analogical association between otherwise distinct theoretical entities (electrons and free-quarks) via an analogy between machines. This was a bounded technically rational strategy because what Millikan had already established about discrete electronic charge using his oil-drop apparatus, Morpurgo could potentially establish by analogy about discrete fractional electronic charge (a theoretical property of quarks), and consequently use Millikan's apparatus as a model for an apparatus to search for free-quarks. Even though his search for free quarks was unsuccessful, the means of conducting the search was established in relation to an already established experiment. At each and every stage of construction, his choices of components and their interconnection were analogously situated in the technological framework by using a model to refine and develop that framework, weaving together analogous theoretical representations of machines.

Techniques for the identification, measurement, and manipulation of "charge" were transferred by analogical use of models because both Millikan's and Morpurgo's apparatus shared technological objects that were members of the electromagnetic machine-kind, such as capacitors and electromagnets, and, through the use of shared quantifiable representations (exoreferences) and technical codes, such as "charge" and "potential difference." In the context of these shared machine-kinds, they could provide commensurable measurements and meanings to interpret the performances of both apparatus in commensurable terms that situate the new apparatus in the technological framework of ongoing research into the properties of matter. As a result, due being successfully situated within the technological framework as a technological means to detect free quarks, Morpurgo's apparatus demonstrated to the scientific community that free quarks do not exist when it failed to detect them after over two decades of failed attempts to do so. The analogy was sufficiently secured to give the scientific community

---

[245] Morpurgo (1972: 5-6)

confidence that such a machine should have detected free quarks if such entities actually existed. It is the analogy that connects machine performances to theoretical explanation.

Another example would be the comparison between the Glaser bubble chamber to detect "cosmic rays" and the LEP "antimatter-matter annihilating" ring at CERN. These are very different machines for particle detection. However, they do share important exoreferenced functives and technical codes such as voltage, current, magnetic field strength, momentum, energy, etc., without any significant variance in usage between the different machines, and they also share the same set of fundamental representations used to explain how these machines work. They share common representational practices, such as the use of differential calculus, statistics, algebra, arithmetic, SI units, etc., as well as shared representations of mechanisms, such as ionization, radioactive decay, electron and photon interactions, electron and positron production and annihilation, spin and energy levels coupling, virtual particles, etc. They also share functives such as charge, mass, spin, force, momentum, energy, half-life, etc. and their designers used these as technical codes to construct the machines in the first place. Physicists also utilized Special Relativity, Maxwell's equations, SI units, and the Periodic Table, etc. to provide a universal standard for description of how these machines work. They also used the same construals (e.g. particle, wave, track, spin, interaction, etc.) when providing visualizations of the underlying mechanisms that govern the performance of these machines. Furthermore, computers are connected to the LEP-Ring detectors and run reconstruction programs that technographically represent the "data" as "bubble chamber" pictures of "tracks" on the computer screen and allow a commensurable visualization of the machine performativity. The four detectors in the LEP-Ring are hybrids between liquid-based detectors (such as the bubble chambers) and the gas-based detectors (such as the Geiger counter). These two prototypes were related through construals of their performativity in terms of the mechanism of ionization, and the LEP particle detectors at CERN (DELPHI, OPAL, ALEPH, and L3) are, to put it crudely, not much more than 30 foot by 30 foot massive barrels of thousands of modified liquid- and gas-based detection cells, surrounded by a massive electromagnet and connected to the LEP-Ring. Each cell, when triggered, transmits a voltage peak, a time signal, and an ID number. That's all! Computers, reconstruction techniques, and models do the rest.

The use of the "ionization" representation brings together and connects chemical and electromagnetic machine-kinds, making the bubble chamber (cloud chamber, drift chamber, etc.,) and the Geiger counter (xenon tube, arc lamp, etc.), the detectors at CERN, and the LEP-Ring members of the same machine-family. They are linked historically and technologically as innovations of the same project (i.e. the physics of the motion of interactive charged particles in electromagnetic fields through liquids and gases), that was projected using the same ground plan (mapping out the connections between the geometry of kinds of events with the material conditions of those events), using permutations of the same machine-kinds (electromagnetic and chemical machines). The Large Hadron Collider (LHC) is an extension of this project via analogous connections between the physics of one kind of fundamental particle and another, which is itself possible because these analogous connections share the same history of technological disclosure and theoretical refinement. Through mathematical projection all these machines are related back to the abstracted motions of the six simple machines, but they are not reducible to them. It is this stratified transfactual membership of the technological framework that is metaphysically conceptualized by the mechanical realist as a process of disclosing stratified ontological depth when the development of any new machine-kind is made transparent as a mere means to discover a more fundamental stratum of reality.

Mechanical realism allowed the historical innovation and development of machine-kinds and their associated sets of fundamental representations to be visualized via analogy as the process of following the logic of the stratification of Nature. When metaphysically connected with the scientific worldview via analogous representational practices and models, the abstraction of the dialectics of technology in terms of "the physical" or "the real" provides science with an epistemological sense of developmental continuity—what Bhaskar would term as stratification and ontological depth—via the transfactual connections between otherwise distinct technological objects. These connections are dialectically synthesized during the construction of prototypes, by bringing together machine and other componential technological objects and synthesizing a new set of machine performances and an associated new set of fundamental representations, all exoframed and exoreferenced in terms of models of natural mechanisms. It is in this sense that we can understand the transfactuality of the technological framework itself as being the result of the use of overlapping machine-kinds and their associated set of fundamental representations during the invention of otherwise different machines by allowing representational and material practices to be shared between otherwise distinct contexts, and, by doing so, synthesize a new stratum of structures within the technological framework across analogical contexts of development. Models and theories can pass across these overlaps between machine-kinds, developed through the innovation of prototypes, allowing one machine to model the other, even when the machines are experimenting on different materials and used to explore distinct phenomena. This interpretation of the history of science shows that the unification of otherwise distinct fundamental laws involves the creation of a new hybrid machine-kind by bringing together other machine-kinds and stabilizing them as a set of predictable machine performances in response to modeled techniques, which is itself driven towards the completion of the technological framework as a unified framework via the resolution of contradictions and the creation of new syntheses. The dialectic of representational and material practices is inherently one of unification in the face of incoherence, inconsistency, trial and error, and contradiction between heterogeneous technological objects. Failures far outnumber successes. Analogy plays a synthetic role in bringing together heterogeneous domains of experience and allowing them to cross over boundaries between distinct specializations and bring them together as members of the same machine-family, but there are no guarantees of success. Machine performances do not always conform to theoretical expectations, but should they conform to their teleological positings—guided and limited by the technological framework—then they deemed a success.

Throughout the history of science, new models often emerge when two previously distinct disciplines or fields are brought together through analogous use of distinct technological objects, but there are many dead ends. For example, cosmologists trying to understand "dark matter" have sought a mechanism by which it can be detected. In contemporary quantum physics, "superfluidity" (itself an analogy) can be taken to be an analogy for "a vacuum" in which changes in "the AB-boundary position", between "the A-phase" and "the B-phase" of "superfluid He-3," can be taken to be analogous to "the symmetry breaking" that "occurs because of certain cosmological strings of dark matter" that are "predicted" by "the Inflationary Phase of the Big Bang" model (itself based on an analogy between space-time and an expanding gas).[246] It is because of analogical reasoning like this, that machines such as dilution refrigerators (previously used to map out the properties of superfluid He-3) can be used as possible "dark matter" detectors and the two previously distinct specializations of ultra-low temperature quantum physics and theoretical relativistic cosmology can be brought together. Whether

---

[246] (Bradley (1995a); Bäuerle, 1986a, 1998)

this will be ultimately successful depends on many factors—including contingent human choices—and it cannot be determined in advance. The trajectories within the technological framework are experimental and must be situated within that framework in relation to the culture of human efforts, choices, relations, and the whole project of refining the scientific worldview through technological activity and analogical reasoning.

Models, Metaphors, and Machine Performances:

We need to look at how models are constructed in relation to the labor processes and the research programs that circumscribe them. Modeling the performativity of machines requires positing causal relations and mechanisms that are relevant to the particular goals of that research program and testing them through further technological activity. The transference of models between members of the same machine-family is a process of cross-checking and modifying both the models and the performativity of the machines as analogies of each other. This process transforms the posited causal relations into a more abstract form and leads to their generalization as proto-laws to be further tested and refined through technological activity. The performance of one particular machine can be related to the performance of another particular machine through these generalized proto-laws when these machines are members of the same machine-family. For example, the physicists working on DESY in Hamburg, the DELPHI detector at CERN, and the SLAC machine in the United States related their experiences through the Standard Model of Elementary Particle Interactions and, by doing so, could say that they were investigating the same "fundamental particles" even when they were skeptical about the truth of the Standard Model. This occurred despite the fact that this general model needed to be interpreted, refined, and modified within context to be applied to those different machines in order to develop "independent" tests of that general model. This was possible because the same kinds of heuristic methodologies (statistical, interpretive, and diagrammatic methods) were learned by the physicists (via learning about exemplary experiments and theories alongside standard methods and conventions) and guided how they applied the general model in those different contexts. It is because of this shared culture that the physicists are able to generalize and unify their experiences as experiences of the same kind of events by relating their experiences in terms of the same set of fundamental representations and basic material practices, and, as a consequence, transcend the particularity of the specific machine performances of their specific particle detection apparatus. It is only in relation to these shared representational and material practices—situated within the technological framework—that these physicists are able to translate shared technical codes and their particular experiences of the products of their labors in into general observations of the same kinds of particle interactions and events, and either confirmations or refutations of the same general model. This endows the scientists' experiences with autonomy and allows their experiences to corroborate or refute those of other scientists working in different contexts on different machines.

In all but the most trivial experiments, different models are built into the design at various stages of research and development over a number of years, in a modular fashion, each allowing the experimenter to manipulate parameters of the apparatus that are already understood in theoretical terms (e.g. magnetic field strength) to explore via a technique (e.g. adjusting the magnetic field strength) to manipulate the phenomenon under investigation (e.g. the magnetization of copper at ultra-low temperatures) in terms of a model (e.g. quantum polarization of copper molecules), which later can be built into an apparatus (e.g. dilution refrigerator) to manipulate and explore some other properties of matter (e.g. helium-3 at ultra-low temperatures) in terms of modeled techniques (e.g. the measurement of ultra-low temperatures) and novel models of the behavior of matter under certain conditions (e.g. superfluidity). Parameters become manipulated by pressing buttons and turning dials because technical codes have been used to connect "raw" inputs or human interventions (e.g. changing voltage, resistance, and current by pressing a button or turning a dial) with "data" outputs or machine performances (e.g. the display on instrumentation or computer outputs). Models have been used to translate both interventions and observations into physical variables, thereby relating technical codes to theoretical explanations. These models are selected not only in relation to the teleological positings of the research project but also in accordance with historically conditions anticipations and expectations of the best way to integrate and situated the new apparatus and its results within the technological framework of contemporary science.

It is through the calibration of instruments and apparatus that models are built into any experiment at every level of its design, construction, and operation. Modifiable exoframes have a capacity to accommodate development and change in the configuration of the apparatus and the understanding of technical codes and how the phenomenon can be manipulated. Models also guide modifications of the apparatus to accommodate discrepancies and problems that arise in achieving stable machine performances, in a way that allows historical interpretations of the apparatus itself and all of its modular components to be built into it through calibration methods and conventions. This whole process is concealed through "black boxing" when prototypes become available to other research projects. In novel experimental work, there is considerable latitude to understand any discrepancies or errors in terms of modifications to the exoframes, technical codes, and operational variables used to quantify the performance of the apparatus. When given physical meaning as exoreferences, these are transdictions (such as Galileo's use of friction to explain the deviation of the path of real objects from the geometrical prediction). While purists may well term these transdictions as "fudge factors," they provide a mathematical model with enough plasticity to be applicable to the concrete contexts of actual experimental work, and suggest some new mechanism that can be explored as an object of research in its own right. This allows for the quantification of properties (an essential factor for measurement) and represents the phenomenon as a set of mathematical relations between functions, variables, and constants that are open to refinement and increased predictive accuracy, even when any predictions are qualified as "approximations" or given statistical weighting, and no ontological commitments are given to any theoretical entities investigated during the experiment. This also allows the experiment to be situated against proposed extensions of the technological framework via an apparatus to demonstrate and measure the proposed transdiction and the accuracy of modifications to the exoframe.

Epistemic and techneic uses of models are interconnected in ambiguous ways, via exoreferences, as metaphors used to model the complete physical experiment as a mode of disclosure by connecting the manipulative procedures of experimentation with quantifiable "physical variables." As Hacking, Gooding, and Pickering have all correctly observed, this blurs the distinction between representational and material practices during the design, construction, and operation of the apparatus. Both the configurations of the exoframed apparatus and the current model are refined and related to the collection of machine performances (disclosed as a technophenomenon) in order to establish correspondence between the model and experience on the premise that a theory "must have some relation to the truth," even while scientists remain skeptical about the reality of postulated theoretical entities or require transdictions to explain the failure to achieve predictive accuracy. This interpretation is itself a metaphysical interpretation of the ongoing process of the refinement of the technological framework. The machine performances, of course, are eventually predictable using the latest refinement of the exoframe, after a long, dialectical struggle to achieve stable and communicable associations between representational and material practices. Scientists then publish their results. Considerable "black boxing" results from this process. As science moves on, a former object of experimental research can become "black-boxed" and operate a function in a modular unit of a new apparatus as a form of intervention to explore the new object of research. Electrons can be sprayed in the search for free quarks, by turning dials and pressing buttons. The dialectics of technology transforms what was once a technophenomenon into a technological object available as an instrument or component to explore different technophenomena, thereby corroborating its truth-status in virtue of its instrumentality in ongoing technological activity.

How does this happen in practice? Scientists abstract, picture, and generalize real processes and deal with those abstractions, pictures, and generalizations rather than the real processes themselves. Only the most naïve empiricist would claim that a model should perfectly describe experience in every detail. Only the most naïve realist would claim that a model perfectly corresponds to reality. If a model were to reproduce experience or reality in every detail then the model would be useless. It would reproduce reality in all its details and would tell us nothing more than our untrained experiences would. It is essential that a model selects salient features from experience or reality. It must also explain reality in terms of causes and fundamental mechanisms, which we do not directly experience. It is inherently the outcome of human selections, which are made in relation to goals and expectations, made from with contexts of usage and only later abstracted as "an approximation to the truth." When constructing models, scientists have to choose between making the model simpler (easier to work with) or making it more realistic (complex), but they also have to make selections in relation to how the model is going to be used, given the technologies and materials at their disposal, and the aims of the research program. As well as making these selections, the intellectual labor processes involved in making any model involves explaining to other people how to abstract the apparatus as a set of mechanisms, operations, and input-output relations, which involves exoframing, technical coding, and the isolation of operational variables. This is necessary to explain the technical manipulation of the apparatus in terms of the model, alongside explaining the model in terms of the manipulation of the apparatus. The dialectic between representational and material practices results in making the model as simple as is pragmatically (or tentatively) acceptable for the purpose of continuing the research project, while relating physical variables and constants given by the model (e.g. charge and mass) to the manipulable parameters of the apparatus (e.g. voltage and current) via specific operations (pressing switches or turning dials).

The purpose of a model is to explain and to simplify experience in ways that can guide ongoing technological activities. The greater degree of overlap between experiments—the greater number of connections via shared machine-families—the greater degree of transferability of models between those particular contexts. Furthermore, once we understand that the purpose of a model is to explain a real process (by relating technological objects to natural phenomena via a concept of natural mechanism) by selecting the essential properties and variables of that process and showing how the relate to the inputs and outputs of the apparatus, the estimations of the truth of that model are themselves made within that context of usage and related to other models. Models allow the processes of experimentation to continue through relating representational and material practices to "physical variables" and "physical properties" at various levels of technological activity, which is the process by which machine performances can be collected together into a unified technophenomenon. The techniques and devices needed to isolate and manipulate these "physical properties" are themselves developed in relation to models. These models are represented as having "some truth in them" when they are technically successful in ongoing research in predicting and manipulating specific technophenomena.

Hence a model must be understood as a heuristic and pragmatic tool that is technically evaluated and re-evaluated according to its instrumentality in making representational and material practices useful and intelligible within the context of exoframing ongoing research and labor; it becomes metaphysically secured to its epistemic value for understanding Nature via the explanation of its instrumental value in relation to the techniques and codes used within the context of ongoing technological activity. Exoreferenced functives are used as technical codes within the exoframed apparatus to relate, through operational variables, the calibrated instruments and machine performativity of apparatus with the "underlying reality" that is being investigated by providing visual, descriptive, explanatory, and intelligible representations of the "changing of the changeable" within the apparatus and instrumentation to specific machine performances. This whole process is "black-boxed" into a series of associated and interpreted inputs and outputs, given theoretical meaning in terms of mechanisms and exoframed as a set of functions, and related to specific technical operations and procedures. In this way, particular machine performances (often no more than voltages and time signals) can be interpreted as disclosing specific physical events (such as the detection of a neutrino). Hence Nicholas Maxell can say (without any hint of irony) that neutrinos must exist independently of technology because scientists say they have successfully detected them. Ian Hacking can declare himself to be a realist about electrons when physicists say they spray them.

Yet, once we take the metaphysics of mechanical realism into account, we can explain the instrumentalist foundation of this kind of realism. It is the consequence of the culturally established metaphysical link between mathematical theories and the technological activities of experiments that justifies the use of exoreferences to connect the quantified machine performances of the apparatus and calibrated instrumentation with models of a "natural mechanism" at work, but this has become hidden in cultural interpretations of representational and material practices as being "self-evident" acts of observation. This implicit metaphysics appropriates a model of machine performativity as something with epistemic value—as a mode of disclosure—and gives machine performances meaning in terms of physical variables, with different models being used at different levels of this process of constructing technical codes and their procedural association with operational variables. Mechanical realism allows technophenomena to act as metaphorical substitutions for natural phenomena when they are used to design and construct the exoframes for instruments, experiments, and computer simulations, as well as to calibrate measuring devices, and interpret experiences in terms of models, codes, and variables that were themselves developed in novel experimental contexts within the technological framework. Shared abstractions between contexts are taken to be indicative of some unifying law or formulism. For example, Laplace's equations serve as a mathematical model for quantifiable change for diverse phenomena such as gravitation, electrostatics, electricity, elasticity, and liquid flow. It is the use of such models as analogies that unifies and informs otherwise distinct fields of scientific research, while their common exoframes and exoreferences are built into the apparatus and instruments from the outset as differential and integral ratios of manipulable and operational variables, and thereby relate the structures of the technological framework with mathematical methods. The transferability of the same mathematical formulism between distinct contexts is possible because of the transferability of the differential operator as a representation of a mechanism. This is taken to indicate some underlying common structure within Nature simply because the underlying common structure within the technological framework is ignored when technology is treated as a neutral (naturalized, objective) means of disclosure.

What is also largely ignored by the traditional philosophers of science is the (human) being for whom a model can provide truth at all. An explanatory model can only be said to approximate the truth and maintain its "ontological realism" through the commitment of its adherents and allies if it is an intelligible model for those beings. There are important existential conditions that provide intelligibility constraints on "the truth," and these constraints have their own historical and cultural conditions in relation to the kind of (human) being for which they are truths. A metaphor is only be taken literally true because it is an intelligible metaphor for a being that has learned from other people language and how to engage with the world. Experimenters skip over the phenomenon and interact with the metaphor as if it were true because they can understand the phenomenon (e.g. light) using the representations given by the metaphor (e.g. a water wave). If we postulate that light is *really* comprised of waves then the wave theory of light has been metaphorically projected over the phenomenon of light in order to make the behavior of light intelligible and the experimenter engages with the metaphor of light as a wave rather than the phenomenon itself. This possibility is culturally situated. Metaphors that work as substitution in one culture do not necessarily work in another. The intelligibility of any metaphor has cultural conditions and constraints, and we cannot hope to understand the growth and explanatory success of the experimental sciences without looking at the culture from which they emerged. The demonstrative and educational use of electrostatic machines in eighteenth century Europe and America[247] to suggest a connection between electricity and lightning by postulating a heuristic connection between such machines and the means to explore the natural phenomenon is an example of metaphorical use of models in relation to material practices, but without specific cultural developments, such as the acceptance of mechanical realism, such a heuristic would be unintelligible and would never be considered as having any truth or meaning in it.

The metaphorical use of models also allows scientists to explore the postulated truth while remaining tentatively skeptical about the final truth-status of the theories that they use to derive models. Such models are used instrumentally as a technique and heuristic without any formal commitment to their degree of correspondence to reality. Many physicists use quantum mechanics in this way. It facilitates calculations and predictions, but the physicists do not necessarily commit themselves to its truth. It remains a metaphor judged on its instrumental value. However, this pragmatic distancing of themselves from the truth-status of the model is itself only possible because of the insatiable cultural faith that there is a better model to be had. Even the most skeptical empiricists can be said to be critical realists in this respect. However, in an ongoing scientific experiment, the apparatus is itself constructed according to models of performativity, how those components will relate to one another, and how they will interact with the phenomenon under investigation, and hence throughout its construction metaphors have been built into it and taken literally at many different levels.

What is a metaphor? How do we understand the meaning of this word? Since Aristotle coined and defined the word *metaphora* (from the Greek *metapherein* 'to transfer')

---

[247] Hackmann (1986)

in his book *Poetics* (1457b6-9), a metaphor is usually defined in terms of a "deviation or displacement from literal meaning." This is how a modern dictionary such as the *Oxford English Dictionary* would define it too: "noun, a figure of speech in which a word or phrase is applied to something which it is not literally applicable; a thing regarded as symbolic of something else." However, a notion of "literal meaning" or "literally applicable" is extremely difficult to define and maintain, without entering into circularity, except by appealing to vague notions of conventional usage and tradition. How do we understand the symbolic relation between the phenomenon and the metaphor? This is made even more problematical when it is pointed out that the definition of metaphor in terms of deviation or displacement of the meaning of a word from its literal usage is itself a metaphor because the meaning of words cannot literally move.[248] This ambiguity in the meaning of the word metaphor is further heightened when comparing two things because we are not exactly making a literal correspondence between them in terms of a likeness (as we do when making an analogy between them) but rather superimposing one over the other and replacing one with the other. We are making a substitution.

This understanding of the use of a metaphor as being a substitution can help us understand how scientists use models. A metaphor preserves the sense of the model not being the phenomenon while simultaneous being taken to be a substitution for the phenomenon. It involves equating the previously unequal—making two things identical—while simultaneous preserving the sense of their inequality.[249] By using a model as a metaphor, scientists are able to treat the model as if it was the phenomenon, reducing the reception and representation of its being and presence by interacting with the phenomenon as if it were the model, which is already structured and interpreted in technological terms as a mechanized ensemble of manipulable variables and calculable outcomes. This act of substitution is exactly what Heidegger termed as mathematical projection.[250] The substitution of the model for the natural phenomenon is necessary for scientific thinking to be possible, and this act of substitution has been naturalized (culturally normalized) since the cultural acceptance of mechanical realism. To the extent that scientists (and some philosophers of science) are not aware that they make this act of substitution, their interpretations of their own representational and material practices are based in paradigms, in Kuhn's sense, and utilize tacit knowledge, in Polanyi's sense, and are habitually embodied in perception as praktognosia, in Merleu-Ponty's sense.

Ways of Seeing:

Clearly, natural science is a cultural activity that is open to anthropological research and sociological investigation.[251] It is in this respect that natural science and its results can be said to be social constructions. In this sense, modern science provides a series of tentative metaphors—as substitutions—to create a new way of seeing the world and giving it meaning.[252] Modern science does not differ from poetry in this respect. It generates new ways of seeing the world—a worldview—often via appeals to simplicity and intelligibility, and even beauty, in its search for fundamental truths and insights. Science is metaphysical poetry given technological expression in machine performances. However, as substitutions, models allow the development of the technological framework by allowing

---

[248] Ricoeur (1987: chap. 2)
[249] Richards (1965)
[250] Heidegger (1999b); see also Rogers (2005)
[251] Latour & Woolgar (1979); Knorr-Cetina, (1981)
[252] Black (1962); Goodman (1978)

transactions between distinct contexts of representational and material practice to transfer analogically across machine-kind boundaries by using models from distinct projects as templates. Within the technological framework, this connects science as a way of seeing the world with technique and established representational and material practices. These transactions between contexts within the technological framework extend the technological framework, creating new technological possibilities and problems, and extending the horizon of scientific imagination, bringing new ambitions and challenges. Starting without any certainty or knowledge, except the acceptance of mechanical realism and the anticipatory act of mathematical projection, substitution allowed scientists to utilize models metaphorically to promote them as a plausible (intelligible) means of making and communicating scientific experience, and testing those models through the invention of novel machine performances to give them physical reality and reveal their (closeness to) truth. It allows scientists to tentatively explore the alethic modalities of technological activity—metaphorically and metaphysically substituted as being a means of disclosure of underlying reality—through the ongoing dialectics of representational and material practices, thereby simultaneously exploring the reality that they create and creating the reality that they explore. If science can be described as metaphysical poetry—a way of seeing the world—experimental science could be described as a metaphysical techno-poetics (a metaphysical form of sculpture—a branch of robotics) in the sense that it realizes the truth of its metaphors by embodying them in machines, each acting as automatons that have been built and set in motion as metaphorical demonstrations of the underlying reality that makes the machine possible. As metaphysical techno-poets, like Pygmalion creating Galatea, experimental scientists create their mechanical simulacra of Nature, but instead of invoking the name of Aphrodite and creating for themselves a lover, as did Pygmalion in the Greek myth, scientists invoke the name of Truth (perhaps Aletheia herself) and set their machines to reveal to them Nature's secrets. Still bound by their alchemical roots, scientists seek new powers by learning how to reproduce artificially the natural conditions through which those new powers would come forth naturally if those conditions occurred in Nature. Bringing forth those new powers becomes the confirmation of the truth of the simulacrum from which those powers emanate.

Viewing science as a cultural way of seeing problematizes empiricism as a philosophy of science. Models are built into the apparatus and relate its components with realist interpretations of how models were used to design, build, and operate the apparatus as a means of providing observations. By using models metaphorically to accommodate natural phenomena within the technological framework, scientists are able to use models to provide intelligibility, cognition, and articulation, while relating variables within the apparatus literally in terms of functional sequences and relations, which are connectable to operations and procedures via models and exoframes built into the apparatus during its design and construction. It is this ambiguity between the metaphorical and literal use of models—which is concealed when the whole apparatus is black-boxed as a neutral technological object—that reveals how a working scientist could consider themselves to be an empiricist even though their apparatus requires realist commitments if it is to be intelligible as a means of disclosure of anything at all.

Transfactuality and Scientific Experience

Transfactuality should not be of any surprise given that it was built into the technological framework from the outset and has been continually maintained by the experimenters' use of the same fundamental sets of representations and technological objects to interpret their experiments. This is a consequence of using the same shared general models and machine-kinds to design, build, and operate seemingly independent machines within the same technological framework, all the while relating and abstracting that process in terms of the shared scientific worldview. Without shared general models and representations, those experiences would be incommensurable because they would lack any common frame of reference, and therefore the use of a general model is a precondition for any division of labor within ongoing experimental research programs to remain meaningful and useful to scientists in general, even if a general model is open to refinement or even replacement. The use of a general model is also a precondition for the subdivision of any research program into particular experiments and further sub-specialization. The use of general models also allows scientists to depersonalize their experiences and represent them as the experiences that anyone would have (providing that they were familiar with the general model) and therefore represent them as being impartial or objective. It is this depersonalization of the final accounts of the outcomes of representational and material practices that allows them to be translated into abstract "results" or "data," which in turn allows scientific work to be represented in terms of objectivity and value-neutrality. It also explains how scientists can consider their work to be empirical, and how the traditional philosopher of science can refer to their observations as "self-evident" validations of his realist or empiricist sentiments about the relation between experience and theoretical entities.

Depersonalization facilitates the removal of the particular labor processes that were necessary for scientists to have those particular experiences; the observations made by scientists can be presented as if they were passively read from the outputs of the apparatus. This is an example of reification, in Lukács' sense, and also what he termed as "phantom objectivity."[253] The experiments and theories of modern science—when they are presented as autonomous and transfactual in terms of a general model—can only be divorced from the challenges of technological activity and the teleological positings of the labor process itself by presuming the validity of the precepts of mechanical realist metaphysics and black boxing how models have been built into the apparatus. This is an expression of the conflation of *techne* and *episteme* that has been central to experimental science since its outset in the work of Galileo. The general causal accounts of the stable products of technological activity can be henceforth presented as candidates for the universal knowledge of the eternal and fixed efficient causes of Nature, given in terms of the general mathematical laws that describe the mechanisms responsible for phenomenal changes in machine performances in response to human interventions. It is at that point that both human agency and machine performativity have been completely written out of the scientific narrative. This ultimately is a reification of the culture from which mechanical realism and experimental science emerged.

The frontiers of new experimental sciences are the new challenges of technological innovation. The already stabilized technological objects of prior efforts achieve their objectivity to the extent that they are available as technological objects for further scientific research. The teleological positing of research is the teleological positing of the technological framework itself, which perpetually promises novel technological objects and associated sets of fundamental representations that will challenge the labor processes of scientists to disclose deeper ontological strata through the innovation of new material practices. The frontier of science attains an ontological "superiority" over the past efforts that its promise is entirely dependent upon. Novelty has cultural value in the context of discovery. This order of rank is entirely one of the "superiority" of challenges.

---

[253] Lukács (1971: 83)

There is a cultural preference for the future over the past, which is an expression of the Baconian desire for novel experiences and powers. The dialectic of technology has an irreversible temporal dimension between the past and future powers of technological objects, as they are explained in relation to refinements of the scientific world view, itself given in terms of sets of fundamental representations and mathematical theories via the transformation of the structures of the technological framework. New strata of machine-kinds, historically situated as hybrids within over-lapping machine-families, provides science with the "cutting-edge" or "state of the art" of ongoing technological activity in the developments of the technological framework that are manifest as new research programs, fields, and specializations through the subsequent innovation of this new machine-kind into a new machine-family of further differentiated machines by working through all the permutations of the new machine-kind and its associated set of fundamental representations. From within the context of this ongoing work, the history and culture that gives it meaning and makes it possible are treated as utterly irrelevant and forgotten, and empiricism and positivism became possible.

This ongoing work allows the creation of distinct techniques via the innovation of distinct machine-families and their associated set of fundamental representations, each understood as a specialization within a shared scientific culture, through the creation of the means to create and manipulate novel technophenomena. For example, the boundary of the object-sphere of High Energy Physics (quarks, leptons, bosons, etc.) is circumscribed by the permutations of particle detectors and accelerators innovated as a machine-family through the extension and synthesis of the same component machine-kinds and sets of fundamental representations, each invented and developed as the means to detect and measure fundamental particle interactions; even though, in time, the "particles" that the early generation of machines detected and measured (e.g. mesons, protons, and neutrons) were taken to be not fundamental after all, and these once "cutting edge" particle detectors became black-boxed components (or forgotten analogies) in the development of a new generation of particle detectors. Hence, the machine performances of the LEP ring accelerator at CERN and the linear accelerator at SLAC in the United States share the same history of particle detection and are both represented as means to disclose the same kinds of interaction between "fundamental particles." This sense of sharedness—their transfactuality—is necessary for them to be commensurable means to test and refine the same models and theories of how the new generation of "fundamental particles" interact and decay. This possibility of transfactuality is a consequence of the fact that the same models and technological objects were built into both machines through the processes of stabilizing the convergence and connection of different permutations of the same machine-kinds during the design and construction of those machines. It is a property of the technological framework itself. It does not follow from the fact that these machines are different that they are independent. There are merely different ensembles—different configurations—of the same machine-kinds, and, as such, are members of the same machine-family and related to each other through these associations. These machines are taken to be independent or autonomous (and confirming each other's results) because their shared component machine-kinds and their associated models have become transparent as technological objects within the structures of the technological objects, and their instrumentality for the further development of the technological framework is taken as further corroboration of the empirical success of experimental science and the available set of fundamental representations used to refine the scientific worldview. They constitute the structure of the technological framework itself and the dialectic between representational and material practices transforms that structure to create new machine-kinds and families.

Realism and empiricism both conceal the teleological positing that is inherent to all experimental science from within the technological framework and its dialectical development. By representing technological activity as the disclosure of natural processes—or as "applied science"—the cultural commitment to mechanical realism has anticipated and interpreted the purposes and challenges for which the experiments were set-up to satisfy from within the technological framework as being "self-evidently" meaning as techneic means to disclose the episteme of Nature. This claim of "self-evidence" hides the beings for which the machine performances are means to disclose truth and power, and, as a consequence, not only hides the cultural choices that have been made regarding the human inquiry into the natural world, but also hides the metaphysical relation towards Nature that provides the precondition for those choices. It is as a result of this concealment that technology became the natural mode for human inquiry into the natural world, and, simultaneously, became concealed from conscious reflection. This is how physics became the herald of a specific mode of engagement with Nature designed from the outset to reveal it in technological terms. In the case of physics, all physical processes are reduced to explanations given in terms of an abstraction such as "energy," a variable that only has meaning it terms of its measurability and future use, and the whole Universe becomes reduced to being a technological object, understood in terms of the quantified abstraction of the possibility of change (potential energy). Even where energy is not available for work that unavailability itself is transduced in terms of entropy and the Second Law of Thermodynamics. The stabilization of the scientific experience of any technophenomenon as a transfactual natural phenomenon produced under artificial conditions is itself the result of contingent activities and choices made during the efforts to stabilize the machine performances and their associated representations that (re-)produce the technophenomenon and explain it in terms of natural mechanisms. It does not logically follow from the ability to do this that any technophenomena experienced in this way were waiting to be discovered or actualized in accordance with natural law. That step requires mechanical realism. What is discovered is how to stabilize a reproducible and communicable association between representational and material practices, and use that to explain experience in scientific terms. On this view, without the presumption of mechanical realism, J.J. Thompson could not have claimed to discover "the electron" as something "out there" waiting to be discovered. Instead, he discovered how to make "the electron" an intelligible explanation of particular machine performances and secure the means of its disclosure as a technique within the technological framework. He made a particular scientific experience possible.

When *technai* (even when understood as an unobtainable ideal) are represented as the knowledge of "natural laws," the scientific experience of technophenomena such as "electrons" are limited to their instrumentality within interconnected strata of machine-kinds and machine-families for which "electrons" are utilized metaphorically as exoreferences in the inscription and interpretation of machine performativity, but taken literally when used to confirm realist sentiments and epistemological commitments to "the empirical success" of science. Yet, once we are aware of the technological framework within which experimental science emerged and is developed, we can consider "electrons" to operate independently of particular theories about "electrons," but not independently of the technological framework itself. Hence we need to make a distinction between technological activities and their epistemic interpretation. We can also consider "the electron" to exist independently from the specific technological contexts in which they are instrumental as the means to test particular theories or explain particular scientific experiences, given that the representational and material practices associated with "the electron" have been developed and are used outside those contexts, but "electrons" remain existentially dependent on the technological framework and its innovative extension through the dialectic between representational and material practices within the sum total of all of its contexts of usage. The dialectical emergence of "the electron" as a technophenomenon and its dialectical transformation into a technological object available for future use are fundamental aspects of how "the electron" is tested as a theoretical entity and also secured against objective referents within scientific experience on the basis of its instrumentality within ongoing technological activity. Given that the scientific experience of machine performances associated with "the electron" as a technophenomenon are exercised only in the contexts of use, it is technically irrelevant whether it is a "real and out there" fundamental corpuscle of matter because "the electron" is defined by how it is used as an exoreference to quantify and explain particular kinds of machine performances. Once transformed into a technological object, "the electron" has a concrete reality that should not be divorced from the socio-technical processes and contexts in which it is stabilized and utilized as something at the disposal of further research and scientific activity. On this account, "the electron" does not have any reality outside of the technological framework at all.

To use an analogy, this means that electrons are no more or less real than musical chords. Their existence depends on the existence of particular modes of technological activity, developed and practiced for generations, but they transcend particular contexts of use within the technological framework, wherein such technological activities were structured and developed. In other words, the electron is no more "objective" or "real" than a G-7 chord. It is "real" when we build a particular kind of instrument and "strum" it in particular kinds of ways in accordance to how building and using that instrument has been developed within the technological framework. Hacking should have also been a realist about G-7 chords because they can be played on a number of different stringed instruments. However, why aren't musical chords and electrons given an equivalent ontological status? The electron is not given an equivalent ontological status as a G-7 chord because the latter is not taken to have a correspondent referent in the natural world. The chord is taken to be artificial. This representation is simply a matter of cultural prejudice, habit, and tradition. Science and music are not only two distinct technological frameworks, with their own histories of development, but they have distinct metaphysical interpretations within post-sixteenth century Western culture. Science is given greater ontological significance than music because its technological and representational activities have been interpreted in accordance with mechanical realism. Hence we can understand the technological framework of experimental science as the historical product of culturally contingent labor processes—none of which conforms to empiricism—and we also can understand how and why scientific realists confirm science as the theory of the real, while musicians do not confirm music as being anything other than an artificially developed tradition of material and representational practices. Even though the phenomenology of usage and experience may well be similar when reproducing an "electron" or "G-7 chord," the presumption of mechanical realism allows the production of technophenomena such as "the electron" to be represented the realization of a fundamental division of matter that is a consequent of invoking natural laws by designing, constructing, inscribing, and interpreting machine performances. No such mechanical realist interpretation of "the G-7 chord" is made, yet we can understand its reality in terms of the technological framework of music within which it exists, without treating it as something fictional, ideal, or subjective. Once a particular machine performance is stabilized as reproducible, it requires the presumption of mechanical realism before "the electron" can presented as a defined and self-contained technological object utilizable for the exploration of a "deeper and more unified strata of reality." It is only on the platform of this metaphysical presumption that objective reality of "the electron" is confirmed through its instrumentality and Hacking can be a realist about it. In this way, instrumentality is taken by scientific realists such as Bhaskar to be a "deeper" exploration, disclosure, and discovery of natural mechanisms, causes, and laws. Henceforth, for naïve realists like Maxwell, its objective existence is "self-evident" and there is no need to discuss how we encounter it as an object of experience.

Scientific Experience, Observation, and Technological Mediation

Novel experimental science is not simply a method of hypothesis testing or theory falsification. It is a metaphysically interpreted technological activity in which observations, models, expectations, and techniques are developed simultaneously in the context of making the experiment work as a mode of disclosure. Modeling, achieved through the connection of visual representations and mathematical functions, allows a model to act both technically and metaphorically by simultaneously modeling how something works and what something is. Thus the process of modeling is an ambiguous two-fold epistemic and techneic process within the context of developing stable associations between representational and material practices, with the aim of constructing stable communicable techniques and observations within both the contexts of justification and discovery, thereby making scientific experience and observation possible. This process involves the dialectic between representational and material practices at every level of scientific research, wherein both kinds of practice interact with and shape each other as the working experimenters attempt to construct stable scientific solutions as perceptible and intelligible to both themselves and the wider scientific community.

Models allow for human imagination, reasoning, argumentation, negotiation, visualization, and intuition to be active in bringing "a natural mechanism" into the public realm by making "it" into an object for manipulation, cognition, and conception in terms of familiar macroscopic objects and relations. This use of modeling connects perception, imagination, and intuition, with "natural mechanisms" via technical operations and procedures, thus connecting the technological framework and the scientific worldview via technical codes and models in such a way as to facilitate visualization and ways of seeing "otherwise invisible" objects. One can press a button and spray electrons. It allows scientific experiences and observations to be translated into different contexts of ongoing and innovative representational and material practices, and the success of a model in scientific work is rhetorically supported by reference to its usefulness rather than its truth—on the pragmatic basis that "it is a model which works." It remains techneic, situated in the context of developing and tentatively evolving practices, apparatus configurations, and skills, but by virtue of its instrumental success it is treated as if it has some probabilistic epistemic correspondence. Probabilistic arguments require attendance to already established beliefs, assumptions, prejudices, and opinions if the arguer is to be successful in establishing the analogy as "probably true" because they are attempts to establish a similarity, a likeness, in terms of the familiar and already accepted. In order to secure probabilistic epistemic correspondence, a model must be rhetorically secured to the already established conventions of a community. This involves using imagery, metaphors, assumptions, predispositions, and values (which the community already has) in order to present the model as convergent and coherent with the community's conventions. By cohering with already established conventions, the model has the character of a "naturalistic argument": the community is more likely to accept what is "possibly true" as "probably true," and what is "probably true" as "true."

In *The Art of Rhetoric*, Aristotle termed arguments of this type as *enthymemes*. These are similar to syllogistic arguments from which *episteme* could be logically reasoned, but instead of deducing the "necessarily true" these arguments construct the "probably true."[254] This construction involves the rhetorical skill of knowing how to persuasively manipulate one's audience and discourse in terms of their established prejudices and conventions. For Aristotle, it was a *techne* and a counterpart of the syllogistic dialectic of philosophy. The success at achieving this depends on perceptions of credibility, appropriateness of the enthymeme, its inspirational quality and utility, and the difficulty (and risk) of refuting the speaker. In the scientific community, this requires widespread distribution of enthymemes through credible media (such as accredited or peer-reviewed journals) and also the refutation and avoidance of any criticism (by anticipating and pre-empting criticism by discrediting critics). This means that scientists are engaged in the public acceptability of their work when making their choices about the direction of that work. Scientists do this by trying to secure their work within the already existing conventions of the scientific community and the structures of the technological framework of experimental science itself. In Kuhn's terms, such an argument must be made within a shared paradigm and there would be a set of exemplar arguments that would have a high chance of working, providing its associated methods and techniques can be demonstrated to cohere within the disciplinary matrix of science. A scientific model is constructed out of visual, verbal, and mathematical analogies—mathematically projected over the phenomenon as a substitute for it—and, in order to qualify as a scientific model, it has to facilitate the derivation of measurable quantities that can be manipulated and calculated in the context of experimentation. This "empirical constraint" means that the possibilities available for any particular model are limited by the scientist's expectations of measurability and testability. This constraint is constructed according to expectations of cost, in terms of credibility risk and available resources, and perceptions of the limitations of available technology. The expectation of what is measurable or testable is decided in reference to the already established labor processes, the economic support available to the experimenters, and the conventions of measurement of the wider scientific community. This constraint situates the available choices of models to be consensually commensurable within already established practices of the technological framework and the available methods to construct exoframes and exoreferences. It is in relation to this axis of commensurability that perceptions of measurability and testability are constructed in relation to the perceived functionality of available machines. It is this *enthymemic functionality* that acts as a constraint upon theory and practice.

It is this constraint that explains why scientists will not consider any phenomenological description to be empirical unless that description has been produced within and validated by the technological context of experimentation. Experiences in experiments are metaphorically reduced (itself an act of substitution) to those of exoframed machine performances, which the experimenters interact with as if they were the phenomena, and therefore the object of experimental investigation is itself a metaphor—presented as a modeled technophenomenon—of the natural phenomenon under investigation. This show how there is a "gap" between the experiment and the natural world that can only be "bridged" within scientific experience by using modeled technophenomena literally, as metaphysically underwritten metaphors—as substitutes for natural phenomena. Scientific experience is defined within experimentation via a technological constraint upon the form of any models used and what the possibilities of

---

[254] A syllogistic argument takes the following structure: If All men are mortal and Socrates is a man then (necessarily) Socrates is mortal. An enthymemic argument takes the following structure: If Plato wrote about a philosopher named Socrates and Xenophon also wrote about a philosopher named Socrates then they (probably) wrote about the same philosopher named Socrates.

use could be, which reduces what the form of a scientifically demonstrable ontology could be and technologically mediates (informs and constrains) the experiences of the working scientists.

This technological mediation informs and constrains how scientists visualize, interpret, conceptualize, and abstract their experiences in terms of models and theories. Any model or theory is invalidated if it cannot be subjected to experiment; it must remain within the boundaries of manipulations and demonstrable context using publicly acceptable techniques and representations. Only those properties that can be quantified and reproduced can be a part of any scientific experience, model, or theory, and only those properties of phenomena that can be situated within the structures of the technological framework can be considered to be a part of science at all. Thus the "natural properties" associated with copper, for example, are those and only those properties that can be quantified and reproduced within a technological framework. Instead of exploring natural phenomena, the experiment replaces them with technophenomena modeled in terms of a fundamental set of representations, given as a set of possibilities for producing exoframes, exoreferences, and abstract visualizations, and understood in terms of how the properties of copper can be arranged and manipulated through technological activities such as compression, heating, electrification, magnetization, etc. This reduction is essential for scientific modeling and experimentation to be located within the technological framework and connected with representational and material practices. By doing this, scientists are simultaneously taking the model literally and metaphorically by treating the technologically mediated model as being an approximation of the phenomenon and also as describing technophenomenon as the objective properties of the natural phenomenon. The model has replaced the natural phenomenon through metaphorical substitution, while being treated as a mere instrument within the ongoing labor process of scientific research.

This use of metaphors can be seen in many cases of experimental quantum physics. The situation is even more ambiguous because it is a situation where scientists only experience of the phenomena under investigation occurs in the laboratory and nowhere else. There are no experiences of entities such as "superfluids," "neutrino oscillations," or "tau-leptons" that occur outside the highly technological environments within which the experiment is situated, and the act of observation itself is interpreted as changing the nature of the phenomenon. The phenomenon becomes understood as a set of possible observations and nothing else besides. The natural phenomena and technophenomena are recognized as identical; they are taken literally as one and the same phenomena. As such, these machines do not straightforwardly constitute metaphors because their performances are not substitutions of the phenomena but they are interpreted as constituting the context of the emergence of phenomenon and the sum total of our possible experience of the phenomenon. This highly advanced and technologically mediated science has the reverse situation where "the disclosure of a natural phenomenon" (e.g. "dark matter detection") becomes a metaphor for the machine (e.g. a dilution refrigerator). The machine becomes invisible as "the disclosure of the natural phenomenon," while remaining a technological object that can be used either literally (by being black-boxed as a modular component) or metaphorically (along the lines of overlapping machine-kinds) in the development of other machines and research programs. The "natural phenomenon" has substituted machine performativity and operates as a metaphor for the machine. This transforms the machine from being an automaton into being a means of disclosure. In this way, the outputs of the machine can be visualized as revealing a subtle quantum manifold, such as the phase-boundaries between the two phases of superfluid He-3, and discovering a world that would be evaporated in the sizzling heat of a snowflake.

Ways of World-making

All theoretical cognition in experimental science takes its departure from a background of material practice that precedes the theoretician. Theory is bound up with a mode of being-in-the-world that cannot be divorced from the cultural background and material practices that makes that mode possible. It was only in virtue of being anchored, located, and directed within this cultural background that the technological framework has become associated with the scientific worldview, but it is only in virtue of its instrumentality in developing the structures of the technological framework does the scientific worldview gain its objectivity. The metaphorical use of models and technological objects is an extension of this cultural background; as a consequence, it is a socially constructed reality, but technology mediates and shapes the social relations that construct that reality, given that society requires material practices for its continued existence and technology is required for scientific experiences to be even possible. Metaphors allow a disordering and re-ordering of social imagination in such a way as to say something about one thing in terms of another, but this gains objective literalness through its usefulness in ongoing technological activity. This is why we need to understand how science is emergent and located within a technological framework via the dialectic between representational and material practices. The dialectics of technology gives scientific experience its objective referent and conveys the sense that experimental science is a self-correcting activity. It makes scientific experience possible.

Moreover, as a techno-poetical way of seeing, through machine performances, scientific experience is an experience of the presence of something objective, unexperienced but real. The machine performance discloses a fundamental mechanism to experience. The techno-poetical use of models and machines corresponds what Kuhn termed as the revolutionary phase of science, when new concepts and representations are being created to deal with anomalies, inconsistencies, and contradictions between machine performances and theoretical expectations. The revolutionary phase is completed when a new machine-kind and its associated set of fundamental representations has been established within the technological framework and associated scientific worldview. During its normal phase, when scientists are using and developing models as literal representations in ongoing technological activity, the new machine-kind and its associated set of fundamental representations are used to construct a new stratum by working through all the permutations of the machine-family. Both of these phases occur within the technological framework through the dialectics of technology, due to the incompleteness of the framework with its heterogeneity and internal contradictions, metaphysically directed through labor towards the completion and unification of the technological framework and the scientific worldview. Anomalies, inconsistencies, and contradictions are the result of the incompleteness and heterogeneity of the structures of the technological framework and its associated scientific worldview. When scientists are committed to any particular historical stage—stratum—of the technological framework and its associated scientific worldview, they can be said to be working under a paradigm, in Kuhn's sense, as if the technological framework was complete and the scientific worldview corresponded to objective reality.

As techno-poets, scientists are directing their creativity towards making aspects of the world intelligible. They metaphorically situate these new ways of seeing as objectively corresponding to something "out there" on the basis of their claim that they are genuinely engaged in attempting to make parts of the world intelligible in novel and interesting ways. By allowing selection, emphasis, suppression, reduction, and re-organization of the components of novel subject matters to be made in terms of other subject matters, metaphors are essential to the development of new ideas and are not merely "dispensable graces and ornamentation."[255] Metaphors are essential for the development of novel technical languages from established technical languages and ordinary language.[256] However, what scientists have missed—perhaps as the result of the sense of familiarity that occurs when working under a paradigm—is that these parts of the world are artificial and have transformed the way that the scientists relate to the natural world. Scientists as techno-poets not only change how the world is seen, but, by experimenting with the technological framework, by making new machines and new discoveries possible, they change the world.

In this respect, scientists must be seen as engaged in a technological activity that is located, anchored, and directed from within a specific culture that is itself dialectically transformed when new technologies become available because of experimentation. Science is a way of world making that is shaped and directed by situated human beings for the benefit of human beings. However, when we begin to lose the sense that what has been made equal is not equal, when we lose the sense that the "model" and "phenomenon" are not the same, then our metaphors cease to be metaphors and become literal and assertive truth-claims. This positivistic strategy uses rhetoric to move from poetics to construct the enthymeme and establish something as plausibly true in terms of the current conventions, current perceptions, beliefs, standards, values, presumptions, dispositions, assumptions, prejudices, etc. The transformation of the metaphorical innovation of language into literal usage is itself the product of the dialectical transformation of the cultural background of conventional standards of language use and the representations available to communicate how one understands and has experiences of the material world. Culture, hence, shapes human behavior in pursuit of discovery and exploration, as well as how novel experiences are made intelligible and communicable as experiences. Hence the stratified development of the technological framework and its associated scientific worldview, achieved through the dialectics between representational and material practices, can be represented through realist commitments as discovering the stratified logic of reality and producing the means by which human beings can gain new knowledge and experiences of that underlying material reality. The strata of material reality explored through scientific activity are the stratified structures of the technological framework and its technical codes given metaphysical meaning in terms of the approximate knowledge of eternal and unchanging laws; their concrete meaning through machine performances and productivity. Via the construction of the analogy as an enthymeme, experimenters are to give the experiment literal meaning as means to have an experience of an underlying reality at work.

It is in this sense that science participates in world-making. As techno-poetics and a way of seeing, scientists create the scientific worldview—it is an *autopoiesis* of alethic truth. Yet, scientists also bring new machines into the world as a demonstration of this alethic truth. By "world-making" I mean the two-fold sense of making a worldview and providing the world with world-changing prototype machines (such as the electric motor, the atomic bomb, and the genetically modified organism). Scientists have brought a new

---

[255] Black (1962)
[256] Hutten (1958)

world into being. They have changed the worlding of the world, to speak poetically. The nature of *poiesis* has been transformed. The performative and productive art of experimental science explores itself as truth in terms of a "self-evident" series of disclosures of its productive and performative possibilities; the abstractions of which are metaphorically substituted as "Nature," which is tested through technological innovation and the discovery of new alethic horizons of scientific experience.

Constructing science as techno-poetics does not support either realist or positivistic interpretations of science. Of course, positivists and realists could dismiss this as "relativism," but by paying close attention to how technology mediates and shapes scientific knowledge, through the dialectic between representational and material practices, we can respond to this kind of dismissive nominalism. While the interpretation given above could be described as "relativism," given it acknowledges the cultural constraints on interpretation, enthymemes, and metaphors, it also highlights the material constraints placed on the scientific worldview, theory choice, and models through technological mediation and the structures of the technological framework. As Nelson Goodman put it,

"What I have said so far plainly points to a radical relativism; but severe constraints are imposed. Willingness to accept countless alternative true or right world-versions does not mean that everything goes, that tall stories are as good as short ones, that truths are no longer distinguished from falsehoods, but that truth must be otherwise conceived than as correspondence with a ready-made world. Though we make worlds by making versions, we no more make a world by putting symbols together at random than a carpenter makes a chair by putting pieces of wood together at random." (Goodman, 1978: p.94)

However, my emphasis is different than Goodman's focus on representational practice. When we take technology into account and understand science in terms of the dialectic between representational and material practices, we see how technological activity places severe material constraints on how scientists put the scientific worldview together, just as they place severe material constraints on how carpenters make chairs. These material constraints shape how we understand what it is that we are doing in the world. Truth cannot be divorced from the beings for which it is a truth without making it unintelligible. It arises as truth through a mode of disclosure of itself as truth. It is brought-forth as truth (*aletheia*) and belongs to production (*poiesis*) and technical causal knowledge (*techne*). The establishment of realism occurs when models are taken literally instead of metaphorically, and positivism occurs when scientists engage in the literal application of models during the normal phase of scientific activity. The realists simply imagine an Archimedean point for themselves; the positivists operate within one technological stratum as if it was the only possible stratum that somehow had been given ready-made to scientists and there had never been any other, nor would there be any other. These philosophies of science are both products of reification, in Lukács' sense of the term, and they involve the construction of *episteme*, in Foucault's archaeological sense. The *techne* induced and abstracted from the dialectic between representational and material practices are disclosed as *episteme* via the concealment of technological activity from our accounts of how this knowledge was possible, and this knowledge is abstracted from the structures of distinct strata of technological development. Once technological activity has been reified, as something objective and determined by natural laws, via the cultural acceptance of mechanical realism, its products can be represented as "abstract general principles of how Nature works" and represented as yet another confirmation of the "empirical success" of science. It is due to this process of reification that experiments can be seen to test hypotheses, falsify theories, and that the phenomena in question are represented as the product of a set of "natural mechanisms."

The metaphor of "natural mechanism" has been established in our culture for at least 500 years and is the conceptual cornerstone of experimental science. The cultural establishment of experimental science as natural science was simultaneously the rhetorical establishment of the literalness of the concept of "natural mechanism" and the neutral transparency of experimentation as a means of disclosure. The establishment of the method of experimentation as a road to truth was only possible because of the metaphysical precepts of mechanical realism and the reduction of the world to those properties that could be quantified and manipulated through technological mediation. Henceforth, techniques and machines could be treated as transparent means to the truth about natural mechanisms, and scientific narrative could be then presented as directly read from "the Book of Nature" rather than written by human beings. The dialectics of technology becomes reified as objectivity, secured against the productivity and mediation of the technological framework, which has become something concealed from view as being simply man-made and the result of the rational application of natural laws to practical problems.

Embodying the Dialectics between Representational and Material Practices

Nicholas Maxwell argued that scientific theories and models must be intelligible and comprehensible.[257] Yet, apart from making a rather hand-waving argument about how our biological evolution somehow gives us the ability to understand the world in scientific terms, Maxwell forgot that human beings are historically and socially situated within the world, and our thoughts and practices are conditioned by the society and culture into which we are born. Contrary to Maxwell's naïve realism, this world and how we are situated within it are not reducible to our biological condition and natural history, even if we accept that we do have biological conditions and limits. We have developed culture and society. How we experience the world and how we test our knowledge against experience are both culturally and socially conditioned. How we understand our biology and evolution are culturally and socially conditioned.[258] Our natural history has a social history. Once we accept that there are intelligibility and comprehensibility requirements for theory and model choice—and metaphors—then we should not ignore that our explanations are culturally and socially conditioned. What is considered to be intelligible and comprehensible in one society and culture is not necessarily so in all others. We need to examine science relativistically, as a subculture that has been historically developed and structured through human relations and choices. Scientifically persuaded human beings are embodied and situated beings. Scientific models and theories need to be understood as the outcome of a historically emergent and conditioned embodiment of this subculture of relations and choices. Scientific realists have simply asserted that this subculture is somehow special and their commitment to science is a reflection of their commitment to their society as an open society.[259] It is a commitment to the cultural superiority of Western Civilization.[260]

---

[257] Maxwell (1984; 1998)
[258] Foucault (1970); Haraway (1989)
[259] Popper (1974); Polanyi (1964)
[260] Horkheimer & Adorno (2002)

However, as I have argued above, modern science cannot be properly understood without examining how technology conditions and mediates the scientific experience of the natural world. Science must be understood as emergent from and developed within a technological framework that transcends all scientific experiments and theories, mediates them, and makes them possible as scientific experiments and theories. I argued above that we need to understand how this technological framework and its metaphysical and metaphorical association with the scientific worldview have been structured and developed through the dialectics of technology. Yet most working scientists tend to operate in a state of innocence regarding their history and the dialectical relations that make their research possible. How can working scientists participate in the development of the technological framework without being conscious of that participation? How is the dialectic between representational and material practices embodied? Obviously this dialectic does not occur in isolation. It occurs within a wider world that exists outside the laboratory. It has economic relations. Scientists have to earn a living. Scientists are engaged in economic exchanges from the outset. Scientific activity is bound up with participation in research programs, and the technological development of research programs requires money, materials, and personnel. How scientific research is pursued and developed, and how Nature is represented and understood, reflects the values of the political economy within which science is situated and available as a cultural activity. Experimentation is a highly developed form of technological activity that, for all but the most trivial experiments, requires cooperation, resources, skilled practitioners, and complex machinery. Scientists need to posit the value of the experiment for other people, who themselves have non-scientific criteria for evaluating whether they will support scientists in their work. The "internal" trajectories of research are conditioned by the teleological positings of the "external" world of commercial, political, and military ambitions that the research has promised to satisfy in exchange for materials, equipment, personnel, and space. Thus the teleological positings for any experiment is two-fold: the experiment must be presented as having an "external value" (i.e. value for economic, military, or political purposes) and as having an "internal value" for the ongoing activities of research (i.e. value for scientific purposes). It has an "external" value in which the cooperation of non-scientists is required to provide the resources and facilities for the construction of the conditions under which an experiment can occur. It also has an "internal" value in which the cooperation between scientists is required for the successful design, construction, operation, inscription, and interpretation of the experiment. Each experiment must be set-up as instrumental for the satisfaction of these two kinds of values, which are built into the set-up and bounds the estimation of what constitutes technically rational skilled judgments regarding the criteria for success and how to conduct the research. Scientists must explain the value of the experiment in relation to different alethic modalities of expectations and predictions of what is possible, probable, or even necessary for the solution of problems, as well as different understandings of what are the crucial problems that need to be solved.

Thus every experiment is set-up in relation to several paradigmatic backgrounds against which it is posited as an exemplar or solution within the "internal" trajectory of ongoing research and the "external" trajectories of the wider world's desires for further technological innovation and power. From within these trajectories, particular exemplary experiments and solutions are publicly presented as necessary (or crucial) for the further progression of those trajectories. In this sense, experiments are suggested by overlapping paradigmatic backgrounds of ongoing scientific research and technological innovation. Scientists are challenged by these suggestions to set-up and perform the experiments that are represented within these overlaps as being necessary and valuable. Once these challenges are taken up by scientists, this generates a further series of related experiments that emerge as necessary and valuable and scientists will be challenged to construct and perform them next. These experiments will be considered as tests and corroborations of the initial experiment they are successfully built and conducted, but the process of building them creates new problems and suggests new experiments. It is in this sense that historical understandings of the set-up and trajectory of experiments are *necessary* for us to have an understanding of science, but we can see how the test of scientific truth is itself perpetually deferred to its utility in solving future problems, providing future challenges, and satisfying two-fold "external" and "internal" values (which shows how science is value-laden and value-driven). The trajectories of the dialectics of technology are emergent as teleological positings from the technological framework. The highly complex machines, materials, and infrastructure needed by scientists require a division of labor in order to be successfully implemented within the research program, but the research program itself must be successfully positioned within existing relations of production in accordance with its promise of providing new forces of production. Internally, scientists are defined in terms of their roles (i.e. physicists, engineers, technicians, mathematicians, students, theoreticians, and computer scientists) within the research program, but externally they are ordered according to their value as technological objects available to satisfy the demands and challenges of scientific research. Each participant is defined according to the postulation of their value within the complex social organization of labor structured in accordance with the purpose of the whole. The object of their labor is not just "research" but is also the satisfaction of the economic relations of production within which the whole research program is situated as a means to an end. Technology operates on machines and humans alike.

Experiments emerge from the operations of technology upon labor and materials, through enframing them in accordance with the challenges to which they have been set to achieve and to quantify how Nature is revealed when it is set-upon in this manner, and are not simply tests of theory—in Popper's sense. Complex experiments constitute technological objects in themselves—used for a variety of different purposes—and are not simply rational orientations towards testing theoretical predictions. Where "theory testing" plays a role in an experiment, it is as a challenge within the complex of interconnected challenges set-up by technological innovation, which "tests" the usefulness of theory for the purpose of satisfying those challenges. If complex experiments can be said to be "a test" of something then they are tests of the productivity and creativity of the ensemble of social actors and materials that are gathered and challenged to construct and perform the experiment. It does not immediately follow from the success of an experiment that any theory utilized during it experiment was correct (or even approximately correct). All that immediately follows is that yet another challenge has been undertaken and completed, and that the experiment itself was useful for satisfying the goals or problems it had been postulated as the means to satisfy or solve. The technique of "spraying electrons," for example, was available as a technological object for challenges such as the attempt to discover "free quarks" and the effectiveness of "spraying electrons" is determined in relation to its availability as a technique. It is in this sense that the embodiment of the dialectics between representational and material practices is itself structured by *challenges imposed by the technological framework itself*, and these challenges must be understood in relation to a society.

However, we also need to understand how the dialectic between representational and material practices is embodied as a developing cognitive and intellectual understanding of natural phenomena and how scientists understand how to explore the natural world. We need to understand how human curiosity and ingenuity are satisfied through embodying this dialectic, consciously or unconsciously. Modern science is situated between two aspects of productive activity: the "internal" aspect of the production of intelligible explanations of machine performativity in terms of natural laws and mechanisms, which can be understood in terms of cognitive and intellectual relations, and the "external" aspect of the production of innovative transformative powers and machine prototypes, which can be understood in terms of economic relations and bringing new powers into the world. Indeed, these aspects inform and condition each other, given that economic relations condition how science is developed and intellectual relations condition the motivation to pursue science at all, which provides technological innovations that shape and transform the productive forces available to society, and thereby shapes and transforms the economic relations within society, and the new possibilities for future scientific activity.

In order to understand how science is developed in terms of cognitive and intellectual relations, we need to examine how models and theories are made intelligible and communicated as explanations of scientific activity and the natural world. We need to examine the complicated processes of embodiment involved in learning how to have experiences of otherwise "invisible" processes and entities. This involves understanding how someone becomes a skilled experimenter. Any new research student entering a modern scientific laboratory for the first time enters a highly complicated technological environment populated by already established practitioners. The student is a novice despite his or her cultural familiarity with everyday technology use and textbook science. The student's previous experiences and acquired knowledge of basic techniques and scientific theories are insufficient for the purpose of making this new technological environment intelligible. S/he requires further training and education. S/he is "initiated" into the research, as s/he familiarizes her- or himself with the experimental apparatus, techniques, procedures, theoretical models, computer simulations, and "the way things are done" within the laboratory. S/he learns how to navigate around this new technological environment. This also involves learning how to relate to the people who not only already inhabit that environment, but they also mediate how the student encounters and learns about that environment and the people within it. The student also has to learn how the work within the laboratory is related to the wider scientific community. S/he has to familiarize her- or himself with "the current literature" and "who's who in the field," as the definition of "the field" and "the state of the art." In this way, s/he intellectually embodies the paradigm or disciplinary matrix, to put it into Kuhn's terms.

The student learns what the laboratory's projects and aims are—its challenges and teleological positings. S/he learns how the labor of the laboratory is social organized and divided between all the scientists and technicians involved, and s/he gains her or his sense of identity within the laboratory in terms of these labor relations. The student also learns, as an equally important part of learning "how things are done," the social dynamics within the organization of the laboratory, and how to orientate her- or himself within everyday working practices and labor relations. This involves learning "who is best at what," "how to approach so-and-so" in order to obtain their help, how tasks are organized, how meetings are organized and resolved, and even how to join in with the laboratory members' sense of humor. As s/he participates within the labor of the laboratory, the student learns who are their allies and competitors, its communications and relations within the scientific community, and its economic exchanges with the wider world in which it is situated. In short, the novice student is socialized as a competent laboratory member by being orientated within this specialized technological environment, and it is through embodying the disciplinary matrix—to use Kuhn's term—through this process of socialization that places the student on the path to becoming a skilled experimenter.

Although students often attend general theoretical courses, learning both general mathematical and theoretical techniques, which provide the general models and methods required to connect the work in the laboratory with the work in other laboratories and science in general, most of the training will be in the context of the specific laboratory work. The student needs to rapidly make the highly complicated technological environment intelligible through the embodiment of the already established laboratory practices, techniques, and interpretations of machine performances. S/he learns the methodology. The first stage of this process of technological orientation requires a great deal of "black-box" abstraction of the laboratory's activities and the experimental apparatus into sequences and associations of settings and procedures. These range from the operations of devices and instruments in specified circumstances to the procedures for recording operations and readings in laboratory notebooks. The novice is orientated as a competent practitioner by being taught specific acts as technical operations with physical significance. It is through familiarization and repetition, in terms of "when this gauge reads A then turn that dial to B because this performs function C within the experiment" instructions, that the technological environment is embodied as a set of technical codes with their associated techniques, procedures, and interpretations of their physical significance. This socializes the student into "how things are done" in terms of technological activities that are represented in terms of cause-effect sequences and "natural mechanisms."

In this way, the novice learns how to operate the experimental apparatus, interpret its results, and become a skill experimenter. This form of learning is entirely one of attendance on the part of the student and instruction (both formal and informal) on the part of the already established practitioners. The term "attendance" denotes certain attitudes on the part of the student and means more than simply being present. It denotes an intellectual attitude of *listening, concentration*, and *care*. It is used in the same sense as when someone attends to their duties or attends to someone else's needs. This term implies that the student is engaged in more than a passive learning relationship within the laboratory. This is why we need to take into account the intellectual motivations of scientists when trying to understand how productive relations of science are embodied and further developed, and how this embodiment informs and constrains the pursuit and satisfaction of intellectual motivations. The student is actively orientating her- or himself through making the technological environment and its teleological positings, in which s/he is a participant through labor relations, both intelligible and her/his own. In this way, the student participates in seeing the world through labor.

How does this inform and constrain the student's scientific experience? The empirical constant conjunctions between machine performances and the interventional techniques are exoframed from the outset with theoretical descriptions and representations of "natural mechanisms." Models are used to exoreference techniques in terms of "natural mechanisms" as an essential part of making the process of experimentation intelligible. It is also essential to translate technical codes and machine performances in terms of operational variables and quantize experience. It is in this manner that scientists gain experiences of "natural mechanisms at work" against which the predictions of theories can be "tested." There is nothing primitive or intuitive about this process. The student needs to enter into a highly complex process of socialization in order to learn how to make the performance of the experiment intelligible in terms of interpretive models, mathematical laws, theoretical conceptualizations, and visual representations of "invisible physical processes" that are claimed to be occurring within the closed system of the experiment, and at every level this learning process is technologically mediated. A student needs to learn from other people the technical codes—that "by adjusting the yellow valve marked B, the pressure of the helium-3 flow changes," "by switching on circuit A, the magnetic field is turned on", or "when digital display C reads D, the phonon absorption rate has reached a significant level." Once "physical properties" can be "actualized" by pressing switches or turning knobs, which tacitly relate specific techniques and procedures to "adjusting energies," or "changing quantum states," or "lowering temperature," or "initiating magnetic polarization," etc., the student becomes a realist about the theoretical entities and operational variables exoreferenced in the models that exoframe the apparatus and the technophenomena it produces. This is what Hacking termed as instrumental realism, and, if technology is treated as a neutral means and technological mediation is ignored, it is manifest as the kind of naïve realism we see in a great deal of the traditional philosophy of science.

The operations and interpretations of the apparatus become embodied in a habitual and unconscious manner through the repetition of everyday procedural routines, and, through this habitual acquaintance with the apparatus, the competent experimenter has embodied knowledge of the apparatus—*praktognosia*, as Maurice Merleau-Ponty termed it—and, during the process of becoming competent, the technological environment has extended the perceptual capabilities of the experimenter "to infer" and "to see" otherwise invisible natural mechanisms and entities at work.[261] It is in this way that the student is able to develop personal knowledge—in Polanyi's sense of the term—as an implicit, intellectual knowledge of how the apparatus works in relation to models and theories as a technological means of disclosure. In becoming a competent experimenter, the student learns how to understand the apparatus for themselves—*mathesis*, in Heidegger's sense—through the development of personal knowledge.[262] Inspired by Merleau-Ponty, Polanyi, and Heidegger, Don Ihde discussed how instrumental realism becomes embodied through the habitual sedimentation of technical relations within the laboratory setting.[263] Through technological mediation, the body of the experimenter is able to enhance its perceptual capabilities, by reducing everyday experience in accordance with the constraints of quantifiable technological activity, and gain otherwise unobtainable information from the technological background. Through embodied technical knowledge and enhanced perceptual capabilities, human beings can enter into a cybernetic relation with technology to gain new experiences and powers. The dialectics between representational and material practices is embodied in disciplined labor.

The Dialectics of Technology and Objective Reality

Experimental apparatus do not always work in accordance with expectations. In fact, many scientists often joke that experiments rarely work in accordance with expectations and the only real law is Murphy's Law. The context of experimentation is one of frequent instability, anomalies, and malfunctions. As a student, the experimenter learns from other people what constitutes "working well," but they have to learn how to deal with instabilities, anomalies, and malfunctions by working with the apparatus in their attempts to stabilize its performance as a set of reproducible input-output relations, or functions. Much of representational and material practice involves the acquisition of a collection of experiences of tinkering with the technological configuration in order to maintain a stable output in line with expectations and intentions. Students may well learn that if a particular

---

[261] Merleau-Ponty (1987, 1999)
[262] Heidegger (1999); Rogers (2005)
[263] Ihde (1979, 1983, 1991)

characteristic signal is considered by the already established group members to be "noise" and consequently "undesirable," it should be "fixed" or "removed" by "changing a connecting cable," "adjusting the signal amplification," or whatever. The configuration of the apparatus is adapted in accordance with the expectations and intentions of scientists. Machines often do not conform to human expectations and intentions when scientists least expect it. Experimenters can encounter inconsistencies, errors, anomalies, and unexpected results, especially when the general consensus is that the experiment should work as expected and there is a vested interest in it working well. Experiments can confound the expectations and intentions of the scientific community (e.g. the Michelson-Morley Experiment and the failed measurement of "aether drift" of the Earth) and become pivotal in "paradigm shifts," or they are considered to be a dead-end (Morpurgo's apparatus to detect free quarks) and confirm the paradigm. The performance of the apparatus has a material aspect that limits and transforms social expectations and intentions, even though how that material aspect is encountered and understood remains socially mediated. It is a response to representational and material practices, which resists and shapes future efforts to arrange and rearrange materials to achieve effects, while human beings seek to realize teleological positings in material practices and modify their behavior in active response to machine performativity.

This material aspect can only be revealed and explored through the embodiment of the dialectic of representational and material practices, leading to structural transformations within the technological framework of science, which shapes future representational and material practices, and makes them available as cultural and intellectual resources. The work of experimentation is both socially organized and technologically mediated, wherein technological activity requires social activity but cannot be explained only in terms of representational practices, as specific kinds of material practices are required for specific scientific experiences and observations to be possible. Experimentation is a technological activity, but its outcomes cannot be explained only in terms of material practice due to the role of representational practice in planning and developing material practices. The dialectic empowers and limits technological activity. The dialectic cannot be said to exist independently of technological activity (given that it comes into presence as a consequence of it), but it cannot be reduced to being a set of socially constructed practices, as it requires technological activities to result in making new discoveries and bringing new powers into world. Of course technology depends on social structures, with all their historical and cultural conditions, for its existence and meaning, but the consequences of the dialectics of technology are beyond human control and change society and culture. Technology informs and constrains the social organization of labor and how scientists understand what it is that they are doing. How "the changing of the changeable" comes into presence within the apparatus is not something completely determined by human agency and control, even if its existence depends on human beings engaging with particular materials and arranging them in particular ways. It is for this reason that I have argued that it is necessary to understand scientific knowledge as the product of a dialectical development of the structures of the technological framework from within which experiments are emergent and situated through the ongoing refinement, adaptation, and transformation of the representational and material practices used to build, conduct, and interpret experiments. It is this aspect of experimentation that is taken by experimenters to be its connection with objective reality. It is this aspect that any interpretation of science must address.

How can we identify and characterize this objective reality? We need to look at cases when experiments persistently "fail" and machine performances do not meet the experimenters' expectations, match with their theoretical predictions, or fail to reproduce some expected result. Of course we need to acknowledge that human judgments are involved when this happens and these kinds of judgment do not occur in isolation from the culture from which science has emerged. The criteria under which such judgments are made are learned through education, training, and attending to the judgments of already-established practitioners. The experimenters will need to judge whether there is a problem with the design or construction of the apparatus, whether some mistake has been made in its operation, or whether some unknown "physical cause" is at work. If the expected result can be achieved, stabilized, and reproduced by rebuilding part or all of the apparatus, the problem will be determined to be a fault in the design or construction of the apparatus. The apparatus is modified to negate the source of failure and the planned experiment will be resumed. If the expected result can be achieved, stabilized, and reproduced by using a different technique (or by someone else performing the same technique), the problem will be determined to be a mistake made by an experimenter. Practices and the social organization of labor are modified to negate the source of failure and the planned experiment will be resumed. If the expected results cannot be achieved through modifying either the apparatus or practices, an unknown "physical cause" will be suspected. The experimenters will run through a list of possible transdictions that explain the failure and they will remove each one until the expected result is achieved, they give up trying, or "something new" is discovered. This involves exploring the possible sources of failure by making it instrumental in its own disclosure either by negating it or bringing it into presence as a phenomenon in its own right.

In most cases, it is only when this source of failure is an impediment to continuing the research project will it be considered worthy of further investigation. The extent to which Popper's "falsification principle" is a principle in scientific work is more of a mantra than a rigidly adhered to method. Scientists agree that any scientific theory or hypothesis should be falsifiable *in principle*, but working scientists rarely spend any time falsifying hypotheses or theories. Most scientific endeavors are attempts to make things work. A theory or hypothesis will only be rejected if it consistently fails to work and there is a workable alternative. Otherwise failures will be avoided or ignored and the work will continue as before. Anomalies are generally only seriously considered when the current model persistently fails and an alternative hypothesis is presented to account for them and suggest a new direction for experimental research. Without this "explanation" and suggested "new direction" the experimental result rarely is considered "publishable" and is often considered to be a "dead end." Challenging the model carries considerable career risks and requires the investment of time and resources, as well as the development and promotion of an alternative model, all of which may well lead nowhere. More often than not, scientists will focus their efforts on normal science and their work follows acceptable directions that lead to immediately "publishable results."

If any representational or material practice is considered to be "working well" within the context of the labor process, it will continued to be used while the experimenters deal with more interesting or problematical matters. Rather than spending their time "testing" hypotheses and theories, scientists spend their time adapting them to their work and looking for interesting results. It is this pragmatic acceptance of practices and theories within context that establishes stable technological objects according to their instrumentality. The stable technological object is made pragmatically meaningful through training and public demonstration in such a way as to "explain" the meaning of practices and theories by showing their results in the context of the work. The purposes of practices and theories are given a pragmatic basis of meaning in terms of their usefulness, which the student learns through mimetic attendance towards the uses, the intentions, and the expectations, implicit within the labor process. However, it is important to realize that a research laboratory is not comprised of homogeneous individuals and the group does not constitute a simple social agency. Scientists, like people from all walks of life, are idiosyncratic and have divergent interests; each person constitutes a complicated social agent in their own right, who may or may not cohere with all of the aims, expectations, and practices of the other complicated social agents within the group and the wider scientific community. Within any laboratory there will be a diversity of opinion about "how to perform the experiment," "how to interpret results," "how to solve this particular problem," or even if "there is a problem here." Of course the practices of any working experimenter must remain within the shared template of the general model if they are to be said to be performing the same experiment and communicating their experiences in commensurable terms, but from within the boundaries of the shared general model, the experimenter has considerable space for negotiation and choice about how to modify the apparatus and interpret its results. Practitioners may well share a commitment to make the experiments "work well" and to publish "new discoveries," but they may not necessarily agree about why the apparatus "fails" to "work well" or demonstrate "new discoveries."

The diversity of opinions on how to interpret, visualize, or respond to "failure" can be sufficiently incoherent and pluralistic as to leave a great deal of latitude regarding what constitutes "a solution." There is a dynamic social process of interpretation and negotiation involved when deciding how to respond to "failure." This dynamic social process is directed to transform a collection of diverse social agents within the group into a unified group, itself acting as a coherent social agent in interaction with other laboratories, conferences, and funding bodies, etc. There are many different values and criteria at play in dynamic social processes. Many of the choices required for the construction of "group decisions" are not always made on overtly "scientific" criteria.[264] Decisions are not made solely in terms of "internal logic" or "scientific rationality." Economic and political factors are also involved, as well as psychological and sociological factors. The group needs to gain and maintain prestige to attract funding or beat a rival laboratory, etc. Personal validation is also involved. "Failure" risks this and places the future of group in jeopardy. Consequently, laboratories will usually only publish "a failure" when they have already agreed upon "the solution." Any such "solution" must be made in relation to the expectations and conventions of the wider scientific community. "Solutions" are bounded, context dependent, and pluralistic selections because they are made in contexts of resource relationships, interests, and social connections that transcend the laboratory. Frequently "solutions" are published in contexts of opportunism and estimations of risk within a wider social context of the social organization of labor and the relations of production.

A stable labor process simultaneously constitutes a stable social order and a stable technological order, but, especially in the contexts of technological innovation and experimentation, "order" is often tentative and ephemeral. If "order" is understood in terms of "stability" and "reproducibility," the material aspect comes into presence as "instability" and "disorder." It is for this reason that I view "order" as a state of punctuated equilibrium. Labor adapts to the technological framework in order to (re-)establish a state of punctuated equilibrium through rearranging both the social organization and material configuration of technologies to (re-)achieve production and reproduction. The dialectic of technology is itself an ongoing and process of stratified stabilization as it adapts to the "instability" and "disorder" it creates by bringing heterogeneous objects together in novel contexts. Thus labor is informed and constrained by the stratified structure of the technological framework, as it responds and adapts to

---

[264] Knorr-Cetina (1981)

"instability" and "disorder" within that framework by producing and reproducing new representational and material practices to adapt to the incompleteness and heterogeneity of the framework itself. Taken as an ongoing historical process, the construction of the technological framework of science can be seen as an experimental process of building and exploring plateaus of "objective reality" within a flux of chaos, struggle, and change. These plateaus are the strata of interconnected machine-kinds and their families, their associated set of fundamental representations and theoretical interpretations, and the (reproducible) productive relations and forces that situate the technological framework within society, at any stage of its historical evolution.

Failures and contradictions arise because in complicated and innovative work, such as experimental science, technologies are *underdetermined*. They are used for purposes other than the use for which they were developed and they are transformed in response to the consequences of their use. If the meaning of any technology remains to be determined through a process of "trial and error," which is evaluated and re-evaluated through tentative judgments of the alethic modalities within an incomplete and emerging set of expectations and possibilities, as well as on the basis of socially established expectations and anticipations of limits and results. Experimenters do fly by the seat of their pants. The "changing of the changeable" within any apparatus is also underdetermined (otherwise there would be nothing to experiment upon) and this opens a creative space for choice in the selections and directions available to scientists. This space opens up in response to the material aspect of the experiment and allows considerable free-play and imagination in scientific work. It allows an experiment to begin and is only removed once a coherent consensus is made regarding the achievement of "stability" within the labor processes and the final technological configuration of the apparatus (which is subsequently reduced—"black boxed"—to being a technological object available for future use). The underdetermined aspect of the dialectics of technology is a condition for free creative labor and prevents the history of experimentation from being understood in terms of some notion of technological determinism.

It is impossible to determine the content of the structures of the technological framework from the outset because the content has been created in response to the problems caused by past efforts. The dialectic of technology moves from order to disorder, back to order, and then back to disorder, while each stratum of the technological is transformed in response to the problems that emerge from past efforts to create a new order, which, by doing so, introduced disorder into the already existing structures of the technological framework. In other words, it is the whole process of building the structures into the technological framework that is incomplete and experimental; no one is able to determine in advance exactly what form research and development will finally take when they are complete because the process of stabilization itself destabilizes the overall trajectories of research and development. New inventions solve a set of problems, but they create new problems, often in unexpected ways and contexts because they create new possibilities, powers, and experiences. Scientists may well have expectations about what the final form of their research should be, but these expectations are (re-)evaluated and transformed during the process of (re-)ordering the social and technological dimensions of the work in response to the disorder that their work generates.

Even when scientists (positivistically) remain within the (normal science) template of the methodology of their institutionalized field of research, the free-play in the construction of their plan of action and the heterogeneity of technological objects and social agents, which move between contexts via overlapping machine-families and the use of analogical reasoning, metaphors, and poetical imagination, is inherently unstable and unpredictable. The level of inherent instability and unpredictability is further increased when the technologies invented and developed within the closed system of the laboratory move into the open systems of the wider world. Scientists may well have developed lasers so they could explore the phenomenon of "the quantum coherence of light," but the meaning and significance of laser technology is transformed when it is further developed in the contexts of medical research, information technology, space flight, or military applications. All of the contexts of use and further development of a technology determine its meaning and significance—given that the objective reality of technology is the sum total of everything that it does and has been done with it. It becomes a transformative power in the world, generating new productive (or destructive) forces and relations of production (and consumption), which destabilize the existing technological and social orders, and transforms those orders and the horizon of possibilities open to us.

New transformative powers are reduced to technological objects and teleological positings, as "a means to an end," when they are stabilized within the technological framework as techniques, instruments, tools, or machines. They destabilize the existing socio-technological order (including its relations and forces of production) in underdetermined ways once they become available as a cultural resource, as standing-reserve for (unspecified) future use, which itself re-introduces the under-determinate nature of technology. The value-promiscuity of technology is itself a reflection of the heterogeneity, incompleteness, and pluralism of society and the technological framework within which science was emergent and structured as a transformative power. Any innovative process is one of converging heterogeneous technological objects and integrating these diverse transformative powers into a stable and coherent transformative power in a novel context to solve specific problems by inventing a hybrid or prototype, but the process itself permeates the boundaries between contexts and sends ripples of unstable and unpredictable consequences throughout society and the technological framework. This creates new problems and specializations, creating new representational and material challenges, and shows how the technological framework is itself inherently unstable and unpredictable. It is a societal experiment in how we can understand ourselves and the meaning of our own labor, the world in which we exist in relation to how we understand our productive possibilities, and the alethic modalities of the relations between the dialectics of technology and the objective reality it creates.

Realism and Civilization:

Is there no possibility of naturalism in scientific experience and observation? Where does this leave scientific realism? If experimental science is understood as a metaphysical interpretation of technology, and scientific theories, observations, and experiences are mediated by technology, and all of this is given cultural meaning as an objective means to explore Nature, how can we proceed with our understanding of scientific realism? Is it a psychologism? A paradigm? If the reader has been at all persuaded by my arguments above, we can agree that scientific realism entails shared metaphysical commitments, faith in scientists, and a scientific worldview and culture. Although often based on little more than a special kind of trust in one's own intuition, the social aspects of scientific realism as a philosophy for science certainly has some of the characteristics of a religion, even if the scientific worldview does not necessarily contain any shared theological commitments. Scientific realists can accept all of this and still claim that science is a special kind of cultural activity that connects-with and responds-to non-human Nature that pre-exists, transcends, and conditions all cultural activities, including the ability to know Nature through scientific practices and theories. Again, the faith in one's own perspective as having a special insight into the nature of reality that exists independently of human existence is something that is shared with religion. It is a way of viewing and living in the world that is given significance in terms of entities and forces that transcend and influence human existence. This way is given meaning and its confirmation in representational and material practices as something that brings human beings power and knowledge of an eternal and unchanging reality.

The cultural meaning of representations and techniques is embodied in practice within a community of practitioners, who are understood to be truth-seekers capable of connecting with objective reality through their practices, and challenging and testing the truth of each other's claims through reproducing each other's practices and their consequences. The relations between practitioners must remain free and open—allowed to develop in accordance with each and every practitioner's own curiosity and intellectual satisfaction—if there is to be trust within the scientific community that claims have been rigorously challenged and honestly tested. Such a society is indeed an *open society*, in the sense proposed by Popper (1945), for which the intellectual virtues of honesty, rigor, courage, perseverance, open-mindedness, and trustworthiness are developed as values and norms for scientific practice—forming a subculture embodied in representational and material practices through education and training—and form the intellectual and subcultural *ethics* of a community of practitioners, providing the rationale for motivations and intentions, and secured in relation to exemplars of scientific practice as teachers of expertize and insight—*as someone virtuous to imitate*.

The objectivity of the techniques and methods interpersonally developed within this community remain perpetually open to criticism while being intellectually secured (often unconsciously) by the practitioners for themselves in direct and equal measure to the extent that practitioners consider other practitioners to respect a shared ethics and be trustworthy in their reporting of their activities, experiences, and interpretations. Even when practitioners are in competition with each other, trust and ethics underscore the possibility of successful communication and adequate testing of the stability of methods and their results. Trust and ethics are also fundamental for scientific specialization to produce results that are meaningful for scientists who do not share specialized knowledge and techniques, and thereby allow science to be subdivided in accordance with a division of labor into specialized fields and sub-disciplines. Trust and ethics become the social glue that holds the scientific community together and gives science its meaning, value, and authority as a truth-seeking practice within contexts of justification and discovery, even when it is understood as a Kuhnian paradigm or disciplinary-matrix and a cultural activity with its historical conditions. In this respect, Michael Polanyi's philosophy of science, as described in his 1946 book *Science, Faith, and Society* offers us valuable insight into scientific realism and how it connects notions of objectivity and authority with a special kind of culture and its ethics.

First published after the Second World War, this book contains Polanyi's counter to Marxist theories of science developed in the Soviet Union during the assertion of Lysenko's theories of genetics and biology. Polanyi's argument was a defense of the independence of scientific research from social planning, understood in the context of events in the Soviet Union during Stalin's regime, while he acknowledged that, at bottom, this defense involved little more than a partisan defense of Western science against Soviet science, appealing to faith (or trust) in the former and skepticism (or distrust) in the latter.

As detailed historical studies of Lysenko and the suppression of geneticists in the Soviet Union have shown, the use of police-state tactics and terrorism within a highly bureaucratic and totalitarian society to impose ideological forms on science undermined the conditions under which the creativity, openness, and criticality necessary for science can flourish.[265]

Polanyi was a self-confessed realist philosopher of science who acknowledged that science is based on a faith in scientists, the social conventions and traditional practices of the scientific community, and the values of an open society. Polanyi argued that all realist claims about the objectivity of scientific knowledge of natural laws are based upon a personal commitment to a cultural belief in the existence of natural laws as a real feature of Nature that exists beyond human control, independently of scientific knowledge of them, and these laws cause an indeterminate range of consequences, some of which will be discoverable and others unknowable and unthinkable. The faith in the social conventions and traditional practices of the scientific community presupposes a belief that it is possible to discover natural laws through scientific activity and the meaning of all scientific research and training depends upon this shared belief in an underlying causal structure to reality. Without this belief, it would be impossible to sustain the idea that scientific knowledge of "the general nature of things" is universally applicable to explain the experiences of all human beings, in similar circumstances, in terms that (approximately) correspond to and describe the underlying realty that is independent from human experience. Polanyi called these terms "tokens of reality." Assumptions about the underlying causal structure of reality are implicated in the scientific understanding of (a) which questions are reasonable and interesting; (b) what would constitute evidence for the verification or refutation of any possible answer to these questions; and (c) what kind of concepts and relations should be applied to human experience in order to signify perception of the "tokens of reality." Scientific propositions are concerned with disclosing this reality, suggesting possible new experiences from which "the process of their discovery must involve an intuitive perception of the real structure of natural phenomena."[266]

However, Polanyi also acknowledged that this belief cannot be verified or refuted in relation to experience, given the available plurality of possible inferences, assumptions, or explanations that are required to relate any given experience to theoretical terms. Polanyi rejected the notion that descriptive exactness, predictive accuracy, or any unifying methodological operation has any role in the epistemology of science over and above the role played by ordinary perception and a scientifically trained intuitive grasp of the underlying causal structure of reality. There is nothing inherent to the experience of observing the movement of planets in the sky that confirms Copernicus' theory and it would be rejected if viewed from a strictly empiricist stance; nor is there anything in the experience of observing an object falling to the ground that confirms Newton's theory of universal gravitation. Concepts and fundamental representations are brought to particular experiences in order to unify these experiences as experiences of an underlying force, mechanism, or causal structure. Assumptions regarding the underlying structure of realty are tacitly learnt during scientific training and education, by imitating the discursive and material practices of established and exemplary scientists, and these constitute heuristic guides to action, rather than the epistemological basis for demarcation between science and non-science. Polanyi's epistemology of science was based on the psychological process of intuitively discerning aspects of reality that are not controlled by the observer but are involved in the shaping of perception, and human beings develop this intuitive

---

[265] Joravsky (1970); Roll-Hansen (2005); Rogers (2008)
[266] Polanyi (1946: 25)

capacity by learning scientific methods and the standards of scientific investigation, through imitating established and exemplary scientists, and these act as heuristic guides that shape scientific intuition, interpretation, and perception.

The decision regarding the selection and assimilation of evidence and facts are ultimately *matters of personal judgment* of the scientists involved in scientific inquiry and practical activity, and, for Polanyi, it is through the psychological and philosophical realist study of perception, judgment, and trust in relation to both the individual and society that the progressive and objective nature of science can be identified and understood. Hence he held the view that

"The scientist's task is not to observe any allegedly correct procedure but to get the right results. He has to establish contact, by whatever means, with the hidden reality of which he is predicating. His conscience must therefore give its ultimate assent always from a sense of having established that contact. And he will accept therefore the duty of committing himself on the strength of evidence which can, admittedly, never be complete; and trust that such a gamble, when based on the dictates of his scientific conscience, is in fact his complete function and his proper chance of making a contribution to science." (Polanyi, 1946: 40)

Polanyi rejected the notion that there is any logic of scientific discovery at all, but instead argued that it is an art, transmitted by examples of the practice that embodies it, without any precisely defined methodology, and, ultimately is based on personal judgment and conscience in being faithful to the scientific tradition. Learning how to practice science involves accepting this tradition and becoming a representative of it. Hence Polanyi's philosophy of science differed from Popper's in many crucial respects. Polanyi argued that the history of science shows innumerable examples where scientific propositions or hypotheses were not falsified by conflicting observations, but instead suggested a new mechanism to account for the discrepancy, such as Galileo's use of friction to explain the discrepancy between the motion of a ball on a plane and his theory of motion. The possibility of science as an ongoing activity directed towards the discovery of objective truth depends on the faith that the scientific tradition and its methods are progressive. Hence,

"To understand science is to penetrate to the reality described by science; it represents an intuition of reality, for which the established practice and doctrine of science serve as clues. Apprenticeship in science may be regarded as a much simplified repetition of the whole series of discoveries by which the existing body of science was originally established." (Polanyi, 1946: 45)

At most, Popper's epistemological principle of falsification acts as a heuristic guides for scientific practices *sometimes*, under some circumstances or within a particular research program, but they do not constitute the defining factors that demarcate science from non-science. Scientific discovery is a tacit process of making decisions and personal judgements, involving intuition, creativity, and heuristics, balanced by critical restraint, all of which are learned and valued from within a scientific tradition of historically developed and refined training and practice. According to Polanyi, the history of science shows that there is not any clear and enduring epistemological understanding of how discovery occurs. Science is not based on any enduring methods or methodology. Science is not a purely empirical pursuit either. Based on the evidence there is no reason to prefer the Copernican system over the Ptolemaic, given that both systems have about the same level of descriptive exactness and predictive accuracy. Both the interpretation of the facts and the facts themselves are based on historically conditioned and contingent interpretations involving general assumptions regarding the requirements of naturalistic explanations and assumptions to explain particular observations or experiments. The scientific process of discovery is the culmination of participatory and communicative acts directed towards understanding and demonstrating truth to other people about the causal structures of reality over and above any instrumental (practical) value or political expediency. Science is an inherently social process of communication, critically articulating and demonstrating its own truths to the satisfaction of the intellectual consciences of the members of the scientific community, in order to produce (and reproduce) an intelligible understanding or representation of reality—the scientific worldview—to members of that community.

However, this still leaves us with the question of how we could philosophically justify the claim that the intuitions of scientific practitioners corresponds to the underlying structure of reality. Polanyi asserted that the only explanation for this possibility is that we possess a "faculty to guess the nature of things in the outer world." It is this vague appeal to human physiology as the ground for the connection between intuition and perception that is an appeal shared by all realists, be they naïve, methodological, or critical. Polanyi emphasized the role of creativity in this and argued that

"The propositions of science thus appear in the nature of guesses. They are founded on the assumptions of science concerning the structure of the Universe and on the evidence of observations collected by the methods of science; they are subject to a process of verification in the light of future observations according to the rules of science; but their conjectural character remains inherent in them... discovery, far from representing any definite mental operation, is an extremely delicate and personal art which can be but little assisted by formulated precepts... All the efforts of the discoverer are but preparations for the main event of discovery, which eventually takes place—if at all—by a process of spontaneous mental organization uncontrolled by conscious effort." (1946: 38)

According to Polanyi, the choice between a naturalistic and magical interpretation rests upon a choice between traditions (or cultures). Choosing the naturalistic interpretation depends on a devotion to science as the best means of discovering objective truth; it depends upon a faith in the rationality and freedom of science as a human pursuit. Hence, for Polanyi, both scientific and mystical truths are based on faith which can only be upheld from within a community and the only difference between science and mysticism, on Polanyi's account, is between the types of community within which each approach is practiced. Polanyi anticipated Kuhn and Feyerabend's intellectual relativism. Whereas the truth of science cannot be demonstrated to someone who does not share a devotion to science, the devotion to science can become known in terms of the value of the free society it creates within which the continued pursuit of an open and free intellectual process, from within an ethical tradition or culture, where individuals are able to openly and freely interpret science within a community in accordance with their intellectual conscience and subject their interpretations to the scrutiny of their fellows. Hence Polanyi argued that the faith in science is premised upon a faith in the value of the open society upon which science both depends and provides exemplars of intellectual virtue and freedom.

Like Kuhn, Polanyi identified the disciplinary framework of science (and therefore its ethics) in terms of the standards and norms inherent to the scientists' intellectual values embodied in periodicals and books, the priorities of research institutes and funding bodies, as well as science departments in universities, —"a hierarchy of influence" as Polanyi termed it—and identified authority in science as being something that is more attached to persons (acting as exemplars) and transmitted through mimicry than it is to the dictates and rules of offices or institutions.[267] On Polanyi's account, which anticipated the sociology of science, science is itself a social network of experts and their areas of activity, established through ethical and interpersonal evaluations of credibility, trustworthiness, and respect between scientists. Authority is established in terms of reputation, experience, and expertise. The unity of science is dependent on scientists knowing who are the experts in neighboring fields, so to speak, rather than the maintenance of the same minimum epistemological standards in all the fields of scientific activity. Membership of this network is not established in relation to some centralized institution, nor is it established by conforming to some transcendent epistemological doctrine, but, rather, is a local decision that is made by already established scientists about who is considered to be a colleague, a fellow scientist, and a member of the scientific community. The scientific community exists in the interpersonal overlaps between self- and mutually- identified members of that community.

In this sense, the decision about whether any activity or proposition is scientific or not is a decision made by already established scientists regarding what is of use or interest to them as scientists, rather than whether it conforms to any abstract epistemological standards or unified method. Scientific authority is dispersed throughout the network—the scientific community—and the student submits to the authority of the exemplary scientists whom are taken to embody the scientific tradition in virtue of their credibility, trustworthiness, and the respect that others have for their expertise, knowledge, and skills. For Polanyi, a fundamental aspect of the scientific tradition is its demand that each generation is to be critical and reinterpret the nature of this tradition in order to better represent it. Dissent and criticism are major aspects of the scientific tradition and culture, alongside intellectual virtues such as honesty, discipline, independence, rigor, and originality. It is by embodying the scientific tradition that the student learns how to dissent, while also utilizing the traditional practices and interpretations, and the student learns how to competently criticize established authority, while also simultaneously maintaining a firm conviction in the soundness of the scientific tradition. By embodying this tradition through training and education, the student learns how to develop personal judgment and intuition, rather than rely on appeals to the authority of others; once this tradition has been fully embodied in practice, the student can reject authority, assume full responsibility before their own conscience, and become a scientist in their own right. It is this shared conviction that not only unites scientists in their faith in science, founding their relationships upon trust in each other's shared commitment to the same intellectual virtues, but actually forms the basis for their dissent from consensus. When scientists dissent from the current consensus, they actually appeal to the scientific tradition in order to convince other scientists that they are right to dissent because they hold it to be true that their dissent is more in line with the tradition than is the consensus.

It is in this sense that the heuristic premises of the scientific tradition are normative rather than epistemological in so far as they are intellectual virtues and ideals cultivated by the scientific community; the correctness of any established consensus is a matter of personal judgment, according to individual conscience, rather than something determined through the application of unified methodological principles. Science is in a state of permanent revolution in the sense that the *status quo* is constantly being challenged for its deviations from the ethics of the scientific tradition and it is perpetually being called upon to restore itself as each generation of scientists applies their personal judgment and

---

[267] Polanyi (1946: 48)

conscience to the task of reinterpreting and renewing the scientific tradition. In this way, each generation of scientists challenges the current consensus about how the scientific tradition is to be respected by showing the scientific community how they ought to respect it better.

When a government asserts the premises of inquiry, taking upon itself the decision about what constitutes moral or scientific truths, as well as the responsibility to conduct that inquiry and disseminate its results to the public, either in the form of information or legislation, then, even when it does so in the name of the public good, totalitarianism and the erosion of trust in science are the inevitable results. Hence, Polanyi argued that

"Whether a free nation endures, and in what form it survives, must ultimately rest with the outcome of individual decisions made in as much faith and insight as may be everyone's share. Any power authorized to overrule these decisions would of necessity destroy this freedom." (1946: 73)

Similarly when big business (multinational corporations) control the aims of science, subordinating the directions of scientific research and the dissemination of its results in accordance with their profitability and marketability, the intellectual freedom upon which science depends is suppressed and ultimately destroyed. It is necessary for science that it is an open process within a scientific community committed to open and free discussion between scientists sharing a common devotion, capacity, and obligation to pursue, discover, and communicate scientific truth; that members of the scientific community preserve their spirit of independence, exercise critical reason and dissent from the consensus when and where their conscience dictates that there are ways to better conform to the scientific tradition; and the scientific tradition preserves the intellectual freedom of conscience to decide how to best interpret that tradition.

Every generation of scientists takes on the responsibility of reinterpreting the scientific tradition in order to best serve that tradition, and should they neglect to attend to that responsibility then science would itself become meaningless as a human pursuit over and above achieving instrumental power or satisfying political ambitions. For this reason, Polanyi distinguished between *the General Authority of precepts and prepositions* and *the Specific Authority of doctrine and conclusions*.[268] General Authority is essential for the establishment of common standards and norms upon which science depends, whereas Specific Authority would destroy science completely. General Authority leaves the judgments and intuitions for interpreting scientific tradition to individual scientists within a community of scientists dedicated to science, whereas Specific Authority centralizes the decisions within a hierarchy—and ultimately a dictator—that acts on behalf of the community, as a law or ideal to which all individuals are to conform. For Polanyi, science constitutes Rousseau's ideal city-state within which

"Every succeeding generation is sovereign in reinterpreting the tradition of science… [that guarantees] …the independence of its active members in the service of values jointly upheld and mutually enforced by all." (Polanyi, 1946: 16-17)

The General Authority is akin to Rousseau's concept of "the General Will." It is formed through the reasoned and conscientious commitment of individuals to the scientific tradition, as the best expression of their intellectual efforts and aspirations as

---

[268] *Ibid* 57

scientists, as exemplars of that tradition. The scientific tradition is itself the basis for the social contract between scientists—as members of the scientific community—wherein each individual has the intellectual freedom to interpret how best to uphold that tradition in accordance with their own intellectual conscience and judgment. By relying on personal judgement and individual conscience, learned and refined during education and training, each scientist takes upon themselves the sovereign power to shape the substance of scientific activity and affect the interests of fellow scientists. When scientists take upon themselves this sovereign power, based upon a shared commitment to science as a whole, the scientists are able to formulate the General Authority as "the General Will" or "Scientific Consensus," which does not need any Specific Authority to arbitrate over these individual decisions. Whether and how the scientific tradition is to be respected becomes the responsibility of each generation of scientists.[269]

The social contract within the scientific community is that of a commitment of all scientists to the ideals and standards of the scientific tradition—a devotion to science as a whole—but, under the General Authority of the scientific tradition, all scientists have the freedom to interpret how the ideals and standards of that tradition ought to be respected, in accordance with their personal judgement on matters of intellectual conscience. Hence, Polanyi argued that

"[It] is impossible to safeguard against the mistakes of such decisions, because any authority established for such a purpose would destroy science. It is in the nature of science that it can live only if individual scientists are regarded as competent to state their views and the consensus of their opinions is regarded as competent to decide all questions for science as a whole… Their decisions are inherently sovereign because it is in the nature of science that no authority is conceivable which could competently overrule their verdict." (Polanyi, 1946: 60)

However, argued Polanyi, the competence of each generation of scientists to make such decisions for the whole of science is in direct proportion to their conscientious commitment to science as being the best means to discover objective truth because

"…if we believe in science, we will accept competent scientific opinion as on the whole valid, even though the final validation of any proposition will always involve a fractional amount of personal responsibility on our part." (1946: 61)

As Polanyi pointed out, this does not provide any reason for anyone outside the scientific community to share that commitment. This awareness is important because, as Polanyi argued, it is essential that discussion about the truth, meaning, and value of science must remain open and free within wider society, just as how to best interpret the scientific tradition should remain open and free within the scientific community. It is essential for the health of science that it remains open to challenge from rival and alternative traditions and cultural interpretations of Nature, such as animism, theology, mysticism, romanticism, vitalism, etc.

How should the outcome of such challenges be decided? Polanyi appealed to the intuition and conscience of citizens in a free and open society, guided by the principles of free discussion, fairness, and tolerance, transmitted by a liberal tradition of individual liberties and rights embodied in the institutions and practices of democracy.[270] The dissent and criticism of a "judicious public with a quick ear for insincerity of argument is

---

[269] *Ibid* 64-5
[270] *Ibid* 67

therefore an essential partner in the practice of free controversy."[271] As I have argued elsewhere, this depends on the existence of diversity and pluralism in public deliberation; the existence and activities of heterogeneous political, cultural, scientific, and humanitarian communities and organizations; the constitutional embodiment of fairness and tolerance in law and custom; and, the public dedication to ideals such as truth, justice, equality, freedom, and solidarity between all human beings.[272] An open society needs enlightened citizens that respect their intellectual culture and traditions by exemplifying them better than their predecessors did. The "democratic spirit" of any people is dependent on their respect for intellectual freedom of conscience; their commitment to truth and its honest expression and reception; the belief that truth and the virtues of honesty and intellectual courage can be learned and conveyed; and the common value and shared practice of open and inclusive communication, with an aim to mutual understanding, cooperation, and education, including in science and technology. The possibility of a democratic and scientific culture depends on the free and open cultivation and expression of the virtues of playfulness, cooperativeness, commitment, fairness, honesty, and trustworthiness, which bring together both visions of the open society and its ideals with the scientific tradition and the scientific worldview in the further intellectual and infrastructural development of science, technology, and society. This can only occur within a culture that values and develops these traditions in its practices and institutions, even at the risk of the revolutionary (and even evolutionary) potential and consequences of each generation taking upon itself the task of better exemplifying their culture's ideals and aspirations than their predecessors of power. The older generation must relinquish all fantasies of maintaining their traditions by imposing them from above. They must exemplify them in practice to teach them to the next generation, who will go on to exemplify them better, and by so doing better teach them to the next.

If Polanyi is right, it all boils down to a shared vision of what kind of society—what kind of civilization—that we exemplify in our representational and material practices and imagine ourselves improving upon by so exemplifying. If we aim to understand science and technology, we must understand the vision of the kind of society to which they are directed to reveal through constructing it—through producing and reproducing it—and refining and improving upon it from one generation to the next. This kind of society is *the technological society* and I have discussed its relation with science and technology at length and in detail elsewhere,[273] and I shall discuss it further and its relation with science and technology in the following chapter. By looking at its origin, its historical development, and its contemporary embodiment and trajectories, we can see how mechanical realism has been implicated in the technological society from its outset as a cultural dream and project to improve and civilize the world through the universal application of technology to all aspects of human life, including the human understanding of the world, its origin and nature, and how we understand ourselves as beings in the world. Popper was right that it ultimately comes down to a cosmological question of Man's place in the Universe, but what Popper did not understand was that the cosmology behind the scientific worldview is one in which Man's construction of an artificial world as a substitute and improvement upon the natural world—a dialectical transformation of Man's material conditions—is itself metaphysically postulated as a test of the scientific understanding of how the natural world came to exist and have the structure that it does.[274] Once we raise and explore the question of the origin of the human ability to make things and transform our material conditions, and how mechanical realism was one

---

[271] *Ibid* 68
[272] Rogers (2008)
[273] Rogers (2005; 2006; 2008)
[274] Rogers (2005; 2006)

cultural response to this question, we should questioned why we think that there is a single origin at all, and the extent which science is itself an experimental enterprise to answer this question. We must take a close look at the kind of civilization from within which science emerged and how science, in turn, has transformed that society. If we are to understand this grand societal experiment then we must take a close look at the technological society from which it emerged as an experiment.

6

THE TECHNOLOGICAL SOCIETY:

An early humanist critique of the substantive impact of science and technology on the human condition can be found in Jean-Jacque Rousseau's essay of 1750: *A Discourse on the Moral Effects of the Arts and Sciences.*[275] By adopting a Socratic defense of virtue and goodness, opposing intellectual conceit and vanity, and declaring "the light of reason" as the means for the understanding of the Universe and human nature, Rousseau considered the Scientific Revolution to be a restoration of the knowledge of the Ancient Greeks, after the barbaric Dark Ages. Through "a complete revolution to bring men back to common sense," he considered cultural splendor and magnificence to be "the effects of a taste acquired by liberal studies and improved by a conversation with the world." However, by satisfying the intellect and the body, the arts and the sciences empower governments to put themselves forward as the means to satisfy human needs and cultivate the talents upon which civilization is based. By increasing "artificial wants," the arts and sciences further bind human beings to the social order, modify human behavior, and provide artificial modes of expression for the passions. Human nature becomes expressed and restrained under the ornamentation of manners, such as politeness, affability of conversation, and social obligations to partake in the imperative to social benevolence. This leads to the refinement and abstraction of "our rude but natural" independence into "a servile and deceptive conformity." When government and law provides for the security and well-being of human beings, the arts and sciences "fling garlands of flowers over the chains which weigh them down" and quell the inborn desire for liberty, causing them to become civilized and "love their own slavery." According to Rousseau, the development of the arts and sciences corrupts human nature because they pacify it, lead to vanity, and place powerful instruments in the hands of those ignorant of truth, justice, beauty, and goodness. Without such philosophical knowledge, we cannot know how to make the right use of scientific knowledge, nor even if it has any practical value at all. By allowing the production of affluence and abundance, in the absence of philosophical knowledge, human beings are turned into an instrument of commerce and a resource for the State. Without a philosophical education to learn virtue and "the greatness of the soul," affluence and abundance lead to the increasing withdrawal of human beings into the private realm at the expense of the development and practice of public and civic virtue. By neglecting the development of a philosophical enquiry into morality, religion, and human well-being, education has been reduced to the acquisition of facts and skills, and human thinking has become focused on technical expertise at the expense of the development of a philosophical understanding of citizenship and happiness achieved through reason and "the voice of conscience."

The humanist tradition has continued Rousseau's deep seated concerns with the oppressive and distorting character of science and technology, and its reductive and erosive effect on human freedom in both thought and action. Science and technology have undoubtedly brought great benefits by providing techniques and devices to improve the material conditions of human existence, but it is the irrational conformity to the demands of technical systems that threaten to enslave humanity and erode the meaning of scientific knowledge. It was the positivistic subdivision of science into specialized technical and empirical disciplines that generated Edmund Husserl's despair regarding

---

[275] Rousseau (1987: 1-24)

the state of scientific knowledge, as articulated in his 1910 *Crisis of the European Sciences*.[276] Husserl was deeply concerned with the extent that meaningful knowledge was becoming increasingly unavailable to European culture due to technical specialization and increased levels of mathematical abstraction, which removed it from meaningful content in the lived world of human experience. For Husserl, theoretical knowledge should be intimately connected with the knowledge of how one should conduct one's life, and he was concerned with the fact that the empirical sciences have nothing to say on this question. In his view, what leads to a scientific culture are not facts and technical skills, but the formation of an enlightened and meaningful understanding of life. However, the positivistic presupposition of the importance of specialization and technique had undermined this understanding of scientific culture and alienated human beings from scientific knowledge. This philosophical crisis was a result of the degeneration of one true science—a unified, meaningful *Wissenschaft*, a comprehensive and unified knowledge of world that is meaningful for human beings—into the advanced positivistic disciplines, such as physics, chemistry, and biology, which, in his view, lacked scientific rigor, as well as meaningless for the question of how to live well, because they were unable to be unified into one true science through conscious self-reflection on knowledge and being conscious of the world and one's own mind. The cultural affirmation of the positivistic sciences was dependent upon historically conditioned judgments that had provided increased technological power but had lead humanity to a crisis.

In *Being and Time,* first published in 1929, Martin Heidegger developed Husserl's critique further and showed that scientific knowledge was founded upon a special and specialized mode of being-in-the-world.[277] Heidegger complained that modern science neglected to attend to the derivative character of theoretical knowledge from practical considerations, having concealed the central role of embodied human subjects in the constitution of meaningful knowledge from the lived-world of experience and practices. Heidegger criticized the positivistic assumption that technique was prior to meaning and that the modern sciences had a privileged access to reality. The positivistic interpretation of experimental science as being a neutral and privileged mode of testing theories by using measurement and calculation fails to recognize the role that interpretation has in bringing data, results, and events into experience. It also fails to address the extent that phenomena are scientifically understood in terms of a framework of meanings that are assigned and refined through an ongoing process of interpretations derived from historically situated technological activity. Heidegger's 1954 essay *The Question Concerning Technology* (Heidegger, 1977a) is an attempt to move beyond positivism and reveal the essence of technology and relate it to truth, and thereby explore its implications for human freedom and thinking, and for the human relation with the natural world. Heidegger was concerned with preparing a way in which we could question the essence of technology and develop a thinking relationship with it. For Heidegger, Western civilization is a progressively technological one committed to the quest for continually improved and increasingly efficient means to carelessly examined ends. Technology transforms ends into means and human beings are compelled to adapt to a technical substratum that has become so overwhelmingly immense that we are unable to cope with technology as a means, and, consequently, treat it as an end. What was once prized as a good in itself, for its own sake, is transformed into something that is only of instrumental value for the achievement of something else.

Heidegger was critical of the way that *techne* ("know-how") is treated as the ultimate virtue in modern society. In his 1939 essay, "On the Essence and Concept of

---

[276] Husserl (1970)
[277] Heidegger (1962)

Phusis in Aristotle's *Physics* Bk. I," Heidegger complained that modern science treats *phusis* as if it were a self-making artifact and science has been interpreted as if it were a kind of *techne*.[278] He posited that this interpretation of *phusis* is a consequence of the modern metaphysical conception of the essence of natural law as "a technique." In *Being and Time*, he was deeply concerned about the Galilean (and Cartesian) subject-object distinction, and its reinforcement of a cultural suppression of the question of Being in favor of a relentless search for theoretical knowledge about an abstraction. Heidegger characterized mathematical science as the objectification of Nature, which ultimately gives us an abstract "it" in mathematical terms that cannot account for the "earthiness" of the world. Science simply cannot grasp the essence of Being, as Being precedes the act of grasping. Hence, for Heidegger, mathematical projection appropriates natural beings and transforms them into objects without true individuality (or thinghood). Science conceals the essence of Being because it asserts that the world is solely comprised of the physical interactions between particles of inanimate matter and forces, which are entirely understood as abstract relations, and also because it presupposes a metaphysics that reduces every being to material for labor and technology. In the 1949 version of his *Letter on Humanism*, Heidegger complained modern science belonged to technology and that mathematical science had become the new metaphysics into which philosophy was becoming dissolved.[279] The power of technology obscured "the event of appropriation" (the origin of modern science), which is the anticipatory metaphysical act of mathematical projection over experience, and concealed how the ongoing activities of mathematical projection and experimentation gave science its meaning and objects. By the nineteenth century and the rise of the positivistic interpretation of the sciences, technique and truth had become bound together, and the way that science was historically grounded in technology via metaphysics was concealed (and simply excluded from discussion within the philosophy of science). By the mid-twentieth century, this metaphysics was unfolding in a new way, reducing all phenomena (including thinking) to data for the science of cybernetics, thereby reducing everything to mathematical 'information' and 'noise,' and 'feedback' loops between them, wherein the essence of Being and our understanding of thinking were reduced to terms accessible to computer science and robotics.

In *The Question Concerning Technology*, Heidegger examined how the instrumentalist and anthropological definitions of modern technology, while being correct as definitions, only tell part of the story and have made us blind to the essence of technology. The anthropological definition is that technology is 'a human activity' (subject to historical and cultural conditions and contingencies) and the instrumentalist definition is that technology is nothing more than 'a means to an end' (and therefore is understood entirely in terms of methods, procedures, and techniques). For Heidegger, the essence of technology is not something merely 'man made' or 'a means to an end,' and the traditional philosopher's assertion that technology is "neutral" is the worst possible misconception of technology because it immediately delivers us over to an unthinking relation with it. Of course Heidegger understood that technology is a human activity in the sense that human beings posit ends and decide on the best means to achieve them, and he understood that it is also an instrument in the sense that the manufacture and utilization of equipment, tools, and machines, as well as the needs and ends that they satisfy, all belong to what technology is. However, if we are to understand the essence of technology, we need to ask deeper questions: What do we mean by 'making'? What do we mean by 'means'? What is the instrument itself and from whence does it gain its

---

[278] Heidegger (1939: 20)
[279] Heidegger (1999: 259n)

instrumentality? Within what do means and ends belong, which categories of existence do they belong under, and what brings them together?

Heidegger's analysis of the essence of technology began with a distinction between modern technology and ancient craft practices. Heidegger based his analysis on an etymological distinction between modern technology and ancient handicrafts, alongside his discussion of human agency and the change in how these different modes of production disclosed reality and acted as modes of production. While modern technology only discloses things in terms of their usefulness or instrumentality, ancient handicrafts participated in "bringing-forth" (*poiesis*) beings into the world and these were considered to be goods- or ends-in-themselves, and thus were arts that participated in disclosing the truth and gaining knowledge through artistry (*techne*). They were intimately bound-up with truth—*aletheia*[280]—as a mode of disclosure of the real by bringing it forth through production, hence hence the *unconcealment* of the truth distinguishes truth-as-*aletheia* from the truths revealed by propositional linguistics or formal correctness in the (Roman) sense of *veritas*. Rather than being related to seeing, truth-as-*aletheia* was related to *poiesis* in the sense of Aristotle's fourfold causality[281] and understanding production as bringing-forth from out of concealment: that which was hidden is brought into Being. It was as if the artisan gave birth to the artefact. This revealed the intimate and causal relation between the craftsman and the artefact within which *poiesis* was revealed to the craftsman as the art of making something real and true. Alethic truth was bound up with modes of completion and perfection of the presence of the object, as it becomes brought forth into Being through the act of making it.

In contradistinction to ancient handicrafts, for Heidegger, modern technology was a mode of disclosure in which beings are set in place, framed, and ordered in such a way as to "put to nature the unreasonable demand that it supply energy that can be extracted or stored as such."[282] Everything is reduced to "energy"; not only perpetually available for something else, but understood only in terms of its instrumentality. Heidegger considered the essence of modern technology as being revealed as technology brings itself into presence as a mode of disclosure of the instrumentality of reality. Modern technology sets upon and challenges Nature to disclose itself, unlock and expose itself, as "energy" or material resources for future use. To make this point, Heidegger used the example of how modern technology challenged a tract of land to yield coal and ore, and land is disclosed as a coal-mining district and the soil as a mineral deposit (Heidegger, 1977a: 15). Air is set upon to yield nitrogen for the mechanized agricultural industry and the earth is set upon to yield uranium for the atomic weapons and power industries. Technology is

"…always itself directed from the beginning toward furthering something else, i.e. toward driving on to the maximum yield at the minimum expense. The coal that has been hauled out in some mining district has not been supplied in order that it may simply be present somewhere or other. It is stockpiled; that is on call, ready to deliver the sun's warmth that is stored in it. The sun's warmth is challenged forth as heat, which in turn is ordered to deliver steam whose pressure turns the wheels that keep a factory running." (Heidegger, 1977a: 15)

---

[280] Aletheia was the name of the goddess of truth in Parmenides's poem on the oneness of all being and the illusionary nature of change.
[281] The material cause (substance, material properties); formal cause (design, idea, shape); efficient cause (craftsman, human agency); final cause (end, purpose). See *Metaphysics* V 2 and *Physics* II 3.
[282] Heidegger (1977a: 14)

It is this availability for use in the future (without any consideration of the purpose or meaning of that future use) that Heidegger termed as *standing-reserve (Bestand)*.[283] Through modern technology everything is structured and directed according to its instrumentality. Objects lose their character as objects when they are disclosed as standing-reserve; they become transformed into technological objects available as means for further technological activity. It was for this reason that he considered the instrumental definition of technology as 'a means to an end' as correct, but it also concealed the truth about the essence of technology in a similar way as defining a dog as 'a four legged animal that barks…' It is correct but doesn't really tell us about the nature of a dog.

Modern technology can only disclose Nature as standing-reserve because it challenges human beings to exploit Nature in this way. This challenge has taken the form of an imperative which demands the participation of human beings in the ordering and disclosure of Nature as standing-reserve for the benefit of civilization and the human species. Human choice and consent is required, even if it is made and given in a rather unthinking manner, but those choices and consent are demanded of us all by society, in the name of contributing to society. By responding to these challenges, human beings are set-upon, gathered together, and ordered through the organization of labor and human relations into modes of disclosure of standing-reserve and putting it to use. Ultimately human labor and relations become standing-reserve as well. All existence becomes directed towards the advancement of the human species. Heidegger termed this way of disclosure as *Ge-stell*, which when translated as "enframing" retains the connotations of "frame" and "skeleton," to emphasize how modern technology structures human activity, and it is this way of disclosure that discloses objects only in terms their instrumentality or as a resource.[284] Under the sway of *Ge-stell*, everything is transformed into standing-reserve for future use—the whole of existence is placed at the service of the future. *Ge-stell* should not be understood as the content of technology (e.g. the particular arrangement of pistons, rods, and chassis in an automobile, and instructions of how to build and drive one) but it is rather the structure and framework of how human beings relate to existence through technology that falls under the term *Ge-stell*. It involves a decision or imperative placed by human beings on themselves for human beings to set-upon the world in this way. Heidegger termed the imperative to disclose the real as standing-reserve and put it to use, which simultaneously impels human beings to disclose the real in this way, as *Geschick*, which translates as *Destining*.

Destining should not be understood as the fulfillment of any kind of destiny or fate. It should be understood as heading towards a chosen destination. It is a commandment imposed by humans on humans—who chose to obey that commandment—rather than some cosmic plan or necessity. Rather than view it in the same way a train must follow the railroad tracks once the passengers climb aboard and the driver releases the brake, it should be viewed as the demand to get on the train in the first place. Continuing the train analogy, *Ge-stell* is the imperative to reach the destination by increasingly faster and cost effective means, and the destination is only another station along the way, another point to pick up more passengers. Destining is the decision to build railroads and make trains as a means of transportation, and *Ge-stell* is the imperative that we reorder the world in order to improve this means of transportation to get from point A to B by increasingly faster and more efficient means. Destining is not "a fate that compels… where 'fate' means the inevitability of an unalterable course" because human beings choose to participate, and human beings do not control what is disclosed.[285]

---

[283] Heidegger (1977a: 16)
[284] *Ibid* 19
[285] *Ibid* 25

Destining is an attitude towards technology and Nature that demands that human beings organize the material world to increase further the power to organize the material world, to increase further the power... and so on, and so on, and so on. This imperative turns back upon human beings and orders them in accordance with itself, turning labor into a mode of disclosing itself as standing-reserve. As railroad architects and engineers draw up their plans, gangs of laborers and machines are put to work, and tunnels are blown through mountains. Through *Ge-stell* human beings are only another resource available for management and the organization of work. Labor becomes revealed as an instrument at the disposal of labor to make itself into a resource for unspecified future labor. For Heidegger, it was this Destining of *Ge-stell* that "threatens to sweep man away into ordering as the supposed single way of revealing, and so thrust man into the danger of the surrender of his free essence."[286]

The twentieth century French social theorist and critic Jacque Ellul offered similar insights into the nature of modern technology and its relation to human existence. First published in French in 1954, Ellul's book *The Technological Society* is a narration of the tragedy of a civilization increasingly dominated by technology.[287] Like Heidegger, Ellul was concerned with the impact of technology on human relations and the question of how we can attain an authentic and free relation with it, and questioned the meaning of the dominance of technique for the human present and future. Ellul also considered technique to have its own reality, substance, and autonomous mode of being. Technology, which Ellul also termed as "the technical phenomenon," should be understood in terms of the sociological organization of human beings into the technical organization of society, which he termed *the technological society*. This society is in a punctuated state of social inequilibrium because it perpetually innovating new means to conduct and direct human activity to make that activity more productive and efficient. Ellul defined *la Technique* as "the totality of methods rationally arrived at and having absolute efficiency (for a given stage of development) in every field of human activity" and proposed that it should be studied as a sociological phenomenon.[288]

Technique refers to any complex of standardized means (or ensembles of means) for attaining any deliberate, stable, and rationalized productive behavior or intentions to achieve predetermined results. It is objective in the sense that "it is transmitted like a physical thing" through the organization of productive performances. Technique is the organized ensemble of practices that are used to secure any end whatsoever, and it can *in principle* only provide technical and quantitative solutions to technical problems.[289] Via his sociological examination of the role of technique in modern society and his historical analysis of the forces that have shaped its development, Ellul emphasized the erosion of moral values brought about by technology, and his concern was with the enslavement of human beings to an utterly meaningless society that was destructive to the human spirit and threatened our very souls.

Ellul rejected any metaphysical notion of technological determinism in his characterization of technique because things could always differ from the contingent actuality of the present. If technique is a "blind force" it is so because human beings have closed their eyes to the alternatives. Technique has the character of an imperative that only achieves its power because human beings respond to its demand. It is this conception of technique, in terms of an imperative to order the world in accordance with

---

[286] *Ibid*: 32
[287] Ellul (1964)
[288] *Ibid* xxxv
[289] *Ibid* 11-19; 79

its utility for some undefined future, which has a considerable parallel with Heidegger's conception of *Ge-stell*. Every intervention of technique is, in effect, a reduction of facts, forces, phenomena, means, and instruments to the logic of "efficiency" and "power." For what? More efficiency and power. Human beings become transformed into technical instruments that are defined in terms of their performance and productivity, as an integrated and articulated component in an ensemble of mechanisms and the materials they order and operate on. Technology sets upon and organizes human agency to organize the world and each other, and this imperative is globalized through the educational and technical dissemination of values, projects, techniques, and technicians. It is a global unification of a monolithic and totalitarian mode of social organization and cannot be anything else because the imperative towards "efficiency" and "power" intrinsically require the absorption and reduction of any plurality of means and ends into a single "one best technique"; the unification and universalization of technique is imposed in order to maximize "efficiency" through coordinated and standardized means, thereby treating the maximization of "efficiency" as an end in itself.[290] This involves the total organization of the human population and all of the natural world to achieve "efficiency" and "maximize results" in every area of human endeavor—without exception. This totalitarianism arises from the dominance of the technology society over the world as technology achieves its autonomy and objectivity by being available as "the best technique," no matter how short-lived any particular technique might be before it is replaced by a new and improved "the best technique."

The technical society—and therefore Western civilization—is entirely constructed in relation to technique to such an extent that only that which is technical can be considered to be part of objective or rational thinking. Everything must serve a technical end and anything non-technical is excluded as "inefficient," "subjective," or it is reduced to a technical form. The project of the construction of the technical society is a perpetual search for the "the best technique" to achieve any objective, as the perpetually expanding and irreversible role of technique is extended to all domains of life. The choice of technique is made in reference to the satisfactory stabilization of measurements, calculations, and productive practices, each determined in relation to an intelligible causal account and methodology. Such a choice cannot be divorced from the cultural background against which it is made and emergent from. Hence "efficiency" is a matter of consensus and a sociological phenomenon.[291] Once any technical choice has been made then technique becomes a technological object available for future work and is placed in competition with other technological objects for ordering in accordance with the technical imperative towards "efficiency." The winner is taken to be "the most efficient" and "the best technique," until it is replaced with another, and this will be the model for social organization, until that is replaced, and so on. All human beings are compelled, while under the sway of the technological imperative, to perform all tasks in "the most efficient" manner, which, of course, requires using "the best technique" as decided by consensus. Once any technique has become established as "the one best way" then it is no longer an object for deliberation and choice. The results of "the best technique" are indisputable until a "better" technique takes its place. Until it is replaced by another technique, "the one best way" achieves autonomy in practice because technical practitioners, also under the sway of the technical imperative, are obliged to use it. Failure to do so is seen as "irrational" or "incompetent."

Ellul did not presume or advocate technological determinism. He maintained that the situation could always differ from the contingent actuality. Technique can only

---

[290] *Ibid* 125
[291] See also Hughes (1983) and Bijker (1983).

become an autonomous "blind force" if we close our eyes to possible alternatives and repress our capacity to choose between them. It is our complicity in the technical rationalization of all human activities that leads towards the reduction of all aspects of life to the technological imperative and the "inner logic" of maximizing "efficiency." As he advised, in the introduction to *The Technological Society*, it is helpful to think of freedom "dialectically and say that man is indeed determined, but that it is open to him to overcome necessity, and that act is freedom. Freedom is not static, but dynamic, not a vested interest, but a prize continually to be won."[292] Criticisms of Ellul's philosophy of technology as being overly pessimistic have simply ignored the spiritualism that he advocated as an alternative.[293] Ellul recognized that modern society is not some monolithic structure constructed in accordance with some unitary essence, but involves tensions and struggles between competing tendencies and influences, and it is this tension and struggle that gives us some hope. The construction of the technological society is an inherent and dangerous tendency towards conformity to the technological imperative, but nothing has been determined in advance, and we can still win our freedom by choosing not to conform to its demands, and instead follow a more spiritual path.

Like Heidegger, Ellul warned us against considering the essence of technology as something technological—as a man-made means to an end. Ellul was concerned with the way that the dominance of the technological imperative has taken over all human activities and ordered everything according to mechanistic logic, techniques, and machine processes. Every material technique is subordinate to its immediate result and "efficiency" is determined by choosing the technique that is most immediate. The all-embracing totalitarianism of technique grew out of "the consciousness of the mechanized world," as Ellul put it, and it is this "consciousness" which manifests itself as the imperative to integrate everything into the mechanized world; it "will assimilate everything to the machine; the ideal for which technique strives is the mechanization of everything it encounters."[294] Technique has become the autonomous and integrated structuring agent for human society. It has led to the mechanization of every kind of human activity, via the integration of machines into every aspect of human life, and by applying the "know how" of mechanization to domains which were previously lacking mechanization. Civilization becomes synonymous with mechanization—and should not be thought of as necessarily in opposition to barbarism or brutality. Using the example of the Second World War, Ellul pointed out how the technical imperative to find and use the "most efficient" means pervades all human activity and "ranges from the act of shaving to the act of organizing the landing in Normandy, or to cremating thousands of deportees."[295] The technical operations of the railroads and camps of the Final Solution and the Holocaust were barbarism and brutality made "efficient," and their technical organization was implicated in "the banality of evil" wherein the participants were 'cogs in the wheels' and 'following orders' all the way to the top.[296]

Ellul characterized the modern conception of "rationality" in terms of a "technical rationality" which brings mechanics to bear on all that is spontaneous. It operates through systematization, division of labor, creation of standards, production of norms, and the reduction of method. It must be placed in contexts of political interests, organizations, and problems; every technology is a social ordering and organization of

---

[292] *Op cit*. xxxiii
[293] Ellul (1972; 1976b; 1991). Hanks (1984); Lovekin (1991) argued that the Christian spiritualism advocated by Ellul shows that he was something of an optimist.
[294] *Op cit*. 6-12
[295] *Ibid* 21
[296] Arendt (1963)

production processes directed to the satisfaction of social (political and economic) goals (or ends). A technological order is a social order *and vice versa*; non-human artifacts are participants in the shape and direction of society, and are part of the content and structure of society, which conditions and shapes the emergent goods (or ends) of society. Technological choices reiteratively shape and limit the social, economic, and political landscapes, which shapes and limits what the technological choices so happen to be; this gives the technological phenomenon its autonomy. The horizon of possibilities and the ways to reach it are reiteratively shaped by the social, political, and economic choices and problems that technology has promised to solve.[297] By bringing new things into the world, such as hydrogen bombs, antibiotics, contraceptives, radio, motorcars, etc., technology transforms the world, and new political, economic, and social possibilities arise because new modes of social organization become possible within this transformed world, which in turn opens up new problems and challenges for human beings to take upon themselves. These possibilities, when realized, shape the directions of technology, which shapes the content and structure of society, and so on. This is the dialectics of technology. Technological objects are non-linear complex objects that are emergent through interconnected and context-dependent feedback relations within the technological framework, which forms a background of established techniques and technological objects that are used to produce things or make changes in the organization of things and labor. Technology is no more neutral than the contexts in which it arises because the sum total of all the uses of any particular technology are the content and trajectory of its development, which is the technological society itself.

"Technical rationality" is a context-dependent and evolving form of decision-making directed in accordance with a bounded and evolving concept of "efficiency" as its basis for informed choice.[298] We should understand the technological system not simply as something that is designed to satisfy human purposes, but, instead, we need to understand it as the outcome of an evolutionary trajectory of choices made in context in relation to past efforts, and therefore emergent from and situated within a technological framework constructed from past decisions and efforts. The technological system evolves as its operations to adjust to the specific social, geographical, economic and political characteristics of its environment, and the contexts within which technology gains its use and meaning, and these in turn change that environment. As people overcome the technical problems posed by its development within its environment, which also includes other technologies and peoples, the technological framework is adjusted to incorporate new techniques and technological objects, which shifts the background against which future efforts emerge, and through either education or compulsion change human organization and thinking to incorporate those new techniques and objects. As the history of technology shows, attempts to implement the simplest idea can create vast numbers of unforeseen problems within any large-scale technological systems or widely used practice. Even seemingly straightforward technologies—say excavating and burning coal for electricity generation—can lead to highly complex and unforeseeable consequences, such as those of pollution and global warming, and radical changes in the organization of society (i.e. the Industrial Revolution). It is the social processes of solving these problems and responding to these consequences—adaptation—and not of implementing preconceived designs, which shapes technology, which in turn shapes society, and so on.

"Technical efficiency" is artificial, lacking spontaneity, and opposed to Nature—pitting human beings against Nature, both internal and external.[299] The technical

---

[297] See also Winner (1977); Walsh (1980); Wajcman (1984); Wallis & Baran (1990)
[298] See also Simon (1981); Mueller (1987)
[299] See also Marcuse (1971; 1991)

imperative is an attempt to create an artificial system that is supposedly more intelligible, controllable, and conducive to human well-being than the natural world. The artificiality of the technological society created through technique destroys and replaces the natural world. It does not even allow the natural world to restore itself or enter into a symbolic relationship with it, and, accordingly, these two orders are incommensurable. Just as hydroelectric installations take waterfalls and lead them into conduits, so the technical milieu absorbs the natural and turns it into something available for technology. We can see a great deal of similarity between how Ellul characterized technology and how Heidegger used the example of the hydroelectric plant to describe the way that the Rhine is disclosed as hydraulic pressure for an interlocking complex of turbines, electromagnets, power stations, and a network of cables, set-up to provide electricity as standing-reserve.[300] The river is damned up into the power plant and is transformed into a power supply. Even to the extent that it is still a river in the landscape, it only remains so as resource or an object for the tourist industry. We are rapidly approaching a time when there will no longer be any natural environment at all, except as a patchwork of national parks and conservation areas.

It is this project of replacing the natural world with an artificial one that Ellul considered to be *the societal gamble*: a gamble on the superiority of an artificial world over the natural world. It is the whole technical phenomenon—the technological society—which is itself a gamble. The technological imperative made technological innovation an autonomous method of perpetually searching for the best technique to perform any given task, and this created a universal mode of technical organization that was externally totalitarian and expansive in the name of "efficiency," while internally directed towards its own destabilization under the operation of the technological imperative to perpetually innovate more efficient means. Under the ideal of some perpetually deferred state of perfection, idealized in terms of absolute technological power over the natural world and human beings, the whole of society has been transformed into an experimental process of attempting to realize this ideal, as a grand social experiment in the progressive realization of Mankind's self-declared manifest destiny: the construction of the technological society—free from the capriciousness of Nature—within which human freedom is manifested as unlimited power over our material conditions.[301] In this sense, it is the whole technical construction of society as a technological society that is the societal gamble and experiment.

Western Civilization has become a product of the autonomous development of the technological society to the extent that only that which is technical is considered to be part of civilization. Anything non-technical is considered to be archaic, subjective, inefficient, or reduced to a technical form (e.g. consider how the study of fine art, poetry, or literature has become reduced to the analysis of techniques). The technological imperative is the perpetual search for the most efficient organization of society. It operates upon all facts, experiences, phenomena, and activities to reduce them to a set of techniques and their results in reference to technical standards, calculations, measurements, procedures, methods, and mechanistic causal accounts. Until it is replaced by a better technique, each reduction is represented as "the most efficient" technique available and the technical practitioner is obliged to use it to perform any given task in the most efficient manner. This shows how the technological imperative becomes a moral imperative. Genetic modification of plants and the use of herbicides are demanded by the moral imperative to feed the human population, and all opposition to genetic modification is represented as being in favor of famine and starvation. Even the means

---

[300] Heidegger (1977a: 16)
[301] See also Rogers (2006)

by which human beings reproduce human beings must be subjected to technique to improve upon the limits of human biology. If *la Technique* is the sum total of all such applications of the technological imperative towards the discovery and use of "the most efficient" means to achieve any given task then the technological society is the total reductive organization of all the possibilities of social being in accordance with the dictates of the technological imperative to improve upon Nature and replace it with a better world. This tends towards the totalitarian organization of humanity to maximize efficiency in every area of human endeavor by disciplining all human thought and practice in accordance with their integration within the technological framework.

It is this totalitarianism that empowers and sustains the autonomy of technological innovation as the highest good—the perpetual search for the most efficient technique or device to achieve any given task—and this finds its intellectual self-justification in technological determinism. The goodness of the project of constructing the technological society—as a societal aspiration or ideal—is premised upon the assumption that human freedom and well-being are contingent upon the liberation from the limitations and vulnerabilities of our organic state of being in the natural world, and a mechanical realist interpretation of reality shows that there is only one rational path of development at any given stage of the construction of the technological framework. It is a grand experiment—a societal gamble—in the emancipatory potential of science and technology to liberate humanity from toil, suffering, and ignorance by constructing a Faustian paradise on Earth, as a world of our own making, that seduces us with the promise of certainty, stability, and power. The construction of the technological society appropriates and absorbs the natural world, transforming Nature into something technological by removing spontaneity and ambiguity—or anything that cannot fit into the technological framework—and replacing it with reproducible and calculable relations of knowledge, necessity, and productive forces, and, thereby, promising to create a more controllable, intelligible, and predictable world for human beings.[302]

Science and the Technological Society:

Science is0 a technological activity that represents Nature in technological terms, and uses technology to test those representations by further implementing them in technological activity. Science does not exist in a vacuum. It exists in a society. This society 0is the technological society: a sociological organization and application of the discovery and implementation of efficient mechanisms and techniques in all aspects of human life, supposedly to improve the material conditions of human existence and liberate us from struggle, toil, and the capriciousness of Nature. The technological society has shaped the organization and application of science to the methodological problem how we investigate and understand the natural world. It has also shaped the social sciences and how we understand human relations. The technological society is a culture and mode of civilization that represents itself as the rational organization of human beings and the material world. Science promises to provide a better understanding of our material conditions and discover new "forces of production," which, in turn, allows new avenues of scientific research and technological innovation. Technology transforms, conditions, and constrains scientific experience and our expectations about the modal necessities, possibilities, and limits to our agency, which, of course, shapes how we understand the material conditions of our existence, and how we engage with Nature. Through the

---

[302] Rogers (2006)

metaphysical projection of mechanical realism, this understanding shapes how the scientific worldview is constructed—given meaning—and finds its corroboration in the construction of the technological society.

The technological society is premised on a representation of the human condition, given in terms of the scientific worldview, within which Nature provided the means for its own domination by Man. Technology is represented as a rational process accessible to universal and scientific reason and explanation of its instrumentality, given in terms of ideas of natural causality and necessity that are based on concepts of efficiency and productivity, themselves represented as governed by the exercise of natural mechanisms in accordance with natural law. The transformation from arts and crafts to modern science and technology arose from a transformation in the conception of artificial and natural powers as having the same fundamental origin and nature. After the sixteenth century, the artificial and natural were both understood and explained in terms of the same principles or laws; the only difference being that the former requires human intervention, whereas the latter does not. Nature was reduced to a sum total of universal mathematical laws, mechanisms, and materials, and nothing else. The emergence of this metaphysical and conceptual reduction allowed modern experimental science to become possible as a means of using technology to discover natural mechanisms, while technology became represented as an artificial process of utilizing the natural mechanisms discovered by experimental science to solve practical problems. Mechanical realism conceptually represented machines as the objective interface between Man and Nature, thereby reducing the theoretical understanding of Nature to a set of machine performances accessible to Man's reason and imagination. Man is situated within the scientific worldview as one who grasps objective reality by using technology both to change his material conditions of existence, and also through technological activity refine the human understanding of how Nature works.

For the mechanical realist, human freedom does not consist in an independence from natural laws and mechanisms, but, rather, in the knowledge of natural laws and mechanisms and the possibility of making these work towards the definite end of increasing technological power. Ideally, this knowledge is to provide the Promethean promise of technologically increasing the possibility of liberating us from the organically evolved alethic modes of our "natural state of animality" and the arbitrary capriciousness of that state when it is one under constant threat of hunger, poverty, disease, squalor, fear, pain, ignorance, and subjugation to forces beyond human control. The experimental sciences have been organized within the technological society according to their appropriateness and utility for the realization of the values and ideals of society by providing novel techniques and prototypes, and thereby expanding the instrumentality of the technological society in giving human beings new experiences and powers. Since the nineteenth century Industrial Revolution, the experimental sciences became bound up with the development and application of technology to all spheres of life in order to solve the practical problems of wider society and conquer Nature. Natural science promised the scientific systematization, unification, and clarification of everything, in accordance with the practical demands of the civic, commercial, and military ambitions of the patrons of scientific research.

The scientific worldview has been exported and globalized through the educational and technical dissemination of European values, projects, technologies, scientists, and technicians, but these values are themselves products of the technological successes and the discovery of new powers. This Western colonial unification of the world in terms of a technological mode of social organization is totalitarian because the technological imperative towards efficiency has absorbed, removed, and replaced any cultural plurality it encountered as it extended its grasp of the world in terms of a unification of all phenomena in order to maximize and rationalize power, coordination, and exploitation. This required the total organization of humanity to maximize efficiency in every area of human endeavor. This totalitarianism generated a monopoly over means—which itself was a monopoly over power—and therefore a monopoly over the jurisdiction over deciding between ends. Once mechanical realist metaphysics and the societal gamble had become culturally normalized (naturalized), the technological imperative to improve efficiency and productivity became a moral imperative. The drive to innovate, bringing novel inventions and new transformative powers into the world is inherently a moral drive to construct the technological society *as a better society*.

The German philosopher Friedrich Dessauer (1881-1963) was a proponent of the view that the technological imperative to transform the world for the better was a moral imperative that brought human beings into rational contact with things-in-themselves. The creative process of invention creates existence out of essence—therefore being a fundamentally transformative process—and the result is a working, practical solution to a problem. The autonomous, world-transforming consequences of modern technology demonstrate its transcendent moral value, and, for Dessauer, technology had become a new way for human beings to exist in the world that fitted into the Kantian understanding of a categorical imperative. Dessauer was a devout Catholic, who wrote several books on theology, defended technology in the strongest possible terms, and he also sought to open up dialogue with existentialists, social theorists, and theologians. He also opposed Hitler. For this last act, he was forced to flee Germany.[303]

Understanding how Dessauer understood the technological imperative as a moral imperative helps us understand how Ge-stell and la Technique took hold of society and human psychology. The technological imperative imposes a duty upon trained specialists to fulfil a social responsibility to discover and utilize the most efficient means at their disposal, and to create further means to liberate their fellow human beings of the arbitrary capriciousness of the vulnerability, disease, ignorance, and premature death that Nature imposes upon us in our "natural state of animality." The gathering and ordering of all plans and activity within a technological society in a way that corresponds to the technological imperative is seen to take us nearer to the realization of the success of the societal gamble—which is assumed to be a universal good—via the removal of "the evils of the natural world." Within the technological society, human freedom is enhanced by the knowledge of natural law and scientific truth is equated with the good and rational. Hume's famous 'Is-Ought Dichotomy' (or 'fact-value' distinction) is dissolved by the societal gamble. Scientific truth is a value and its pursuit is an obligation.[304] When freedom is associated with increased scientific knowledge and productive powers, this entails a cultural faith in the implicit goodness of the technological society. Consequently, the technological imperative is premised upon a metaphysical conception of its own possibility, alongside an idealization of society and what the human good life should be. The organization of the ongoing development and implementation of the experimental sciences has transformed the world in accordance with the posited anticipations of the form of truth, the ideal society, and the human good life that are implicit to this project. Thus the productive possibilities of the experimental sciences are situated within the whole project of the construction of the technological society according to the estimation of the powers associated with each science. The further development of the powers, potentials, and possibilities of any science are, in their turn, shaped by the way that they

---

[303] Cf. Mitcham (1994: 29-33); Tuchel (1982)
[304] See also Bhaskar (1993)

are embedded and situated within the organization of the work involved in the construction of the technological society. These potentials and possibilities should not be divorced from the organization of the agencies and labor processes that realize them. They are genuinely creative and transformative of the powers that they unleash upon the world through the technologization and mechanization of human relations.

The powers discovered by experimentation are the products of labor processes, each situated within a historical trajectory, projected towards the discovery and liberation of new powers over our material conditions. Thus the experimental sciences innovate and produce their own creative transformations of reality as sets, clusters, and ensembles of machines, techniques, and powers, in accordance with an ontological moral equation that scientific truth is good. Opposition to this view is considered harmful to society and resistance irrational and backward. The technical causal accounts that are abstracted and communicated as explanations of technological powers in terms of natural mechanisms are further represented as being successfully tested during the successful reproduction of those labor processes and the cognition of their future possibilities in material practice. Scientific knowledge is abstracted in hindsight as a result of extending the closed system of the experiment and removing all hindrances to its reproduction, while transforming all contact with the natural world into series of causal explanations juxtaposed with clusters of machine performances and their associated powers. The truth of scientific knowledge is not only viewed in terms of its instrumental value for enhancing possibilities for producing an intelligible explanation of power, but is perpetually deferred to the future because its realization is conceptually bound to its instrumental value in producing and explaining new powers, new possibilities for labor, and machine prototypes. Thus the discoveries of the sciences are inextricably bound to the efforts and decisions involved in the directions of the construction of the technological society, each directed in accordance with civic, commercial, and military aspirations; while, as an ideal, the technological society remains an incomplete, imperfect, and ongoing experiment that is endlessly challenged to test and confirm its objectivity by improving its "efficiency." It does this by innovating refinements of itself through the sciences, each embedded as an instrument in the ongoing, experimental construction of an artificial world to replace the natural world. The technological imperative to build this technological society is a societal gamble on the superiority of this artificial world as a substitute for the natural world.

Heidegger and Ellul on Science and Technology

For both Ellul and Heidegger, mathematical techniques were central to their definitions of science—in terms of quantification of phenomena and deriving predictions—and they considered only that which can be expressed numerically to be scientific and, hence, the scientific use of technique is that of reducing the possibilities of investigation to the calculation of numbers. But, how did Ellul and Heidegger relate science and technology?

Heidegger, Science and Technology

Heidegger was critical of the way that *techne* is treated as the ultimate intellectual virtue in the modern society. *Techne* is concerned with complete knowledge given in terms of "a true course of reasoning" and such knowledge, should it occur in experimental physics, would be the *end result* of experimentation. It promises to provide the abstract, general, and communicable knowledge of how to repeat the experiment. The technological objects produced in the experimental sciences are simultaneously produced for their own sake and for the instrumental value they have in satisfying the technological challenges of the wider society by providing more power as new source of standing-reserve. The imperative of *Ge-stell* is the transcendental precondition of modern technology that gathers together human beings and challenges us to reveal Nature in abstract and non-sensuous truths that are tested in terms of their availability for future use. It is a technological attitude towards truth and the world that compels human beings to set upon Nature in this way. Modern science cannot be understood simply in terms of "applied technology" because, through the use of mathematics, its represents itself as directed towards the realization and acquisition of *episteme*, and it thereby conceals the extent that it is a metaphysical art based on representational and material practices directed towards the technological acquisition of *techne* of Nature. Modern science, with its ideal being the acquisition of *techne*, is enframing and appropriating Nature, while simultaneously representing *episteme* in mechanistic terms. Modern science is the Destining of human beings towards the metaphysical realization and acquisition of the *techne* of Nature and the new powers that it promises to bring forth.

Heidegger accepted that modern physics "as experimental, is dependent upon technical apparatus and progress in the building of apparatus" and human beings are challenged and destined by *Ge-stell* to disclose Nature as the standing-reserve of energy.[305] This attitude, on the part of human beings, was first displayed in the rise of modern physics as an exact science, and physics was a "way of representing [that] pursues and entraps nature as a calculable coherence of forces" and even as pure theory "sets nature up to exhibit itself as a coherence of forces calculable in advance, it therefore orders its experiments precisely for the purpose of asking whether and how nature reports itself when set up in this way."[306] He maintained his view that "mathematical physics arose almost two centuries before technology" but claimed that because "physical theory prepares the way first not simply for technology but for the essence of modern technology" then modern physics "is the herald of *Ge-stell*, a herald whose origin is still unknown."[307] He argued that, despite the fact that "chronologically speaking" modern physics began in the seventeenth century and machine-powered technology began in the latter part of the eighteenth century, the essence of modern technology was "the historically earlier." Modern physics was itself challenged forth by *Ge-stell* in the demand that Nature is orderable as standing-reserve. This set-up physics as a means by which Nature was disclosed "in some way or other that is identifiable *through* calculation and that it remains orderable as a system of information."[308] Thus, for Heidegger, this is the reason why modern technology must employ exact physical science.

Heidegger did not describe how this happened and considered its origin to be mysterious. Thus he was unclear about how and why it was possible, and, consequently, in my view, he equivocated on the connection between science and technology. Furthermore, he was unclear about how the content of the projected ground plan of scientific research was refined and corrected as an ongoing activity in relation to technique. Heidegger was correct to characterize both science and technology as being bound together, but he did not explain how they were bound together. How was this possible? How are modern science and modern technology connected metaphysically? How does the power of modern science belong to *Ge-stell*? Heidegger was concerned with

---

[305] Heidegger (1977a: 14)
[306] *Ibid* 21
[307] *Ibid* 21-22
[308] *Ibid*: 23

how the object of modern metaphysical reflection was determined in advance in relation to a decision regarding *what is* and *the essence of truth*. What ontological interpretation of truth provided the basis for the foundation of modern science and technology? Under which metaphysics was the essence of modern science and the essence of modern technology brought together and connected via this ontological interpretation of truth? I agree with Heidegger that machine technology, as the most visible outgrowth of the essence of modern technology, was not simply "the application" of modern science, but, as an autonomous transformation of *praxis*, it made demands upon and shaped the form and trajectory of modern science. However, Heidegger did not describe how this happened and considered its origin to be mysterious. Furthermore, he was unclear about how the content of the projected ground plan of scientific research was refined and corrected as an ongoing technological activity.

Heidegger recognized that the metaphysics of modern science was mathematical projection, but did not provide us with a satisfactory answers to how this metaphysics became bound up with technology.[309] However, as I have argued above, a closer analysis of how and why science is performed reveals that the direction of research in experimental physics is driven by an imperative towards the novel and productive disclosure and implementation of mechanisms in novel kinds of machines. It is the disclosure of these mechanisms that binds truth with productivity, and metaphysically connects modern science and technology with knowledge. Heidegger's theoretical preconceptions about the objectification of Nature within the scientific worldview became an obstacle to a deeper inquiry into its metaphysical foundation, and (despite being highly critical of positivists) he relied on a positivistic conception of the object of experimental inquiry as being that which, in all but the most recent phases of modern science, was directly accessible to perception. As a consequence, he proposed a bi-partite and positivistic epistemology of law and phenomenon and was unable to reveal the ontic connection between science and technology. As I have already argued, the ontic connection disclosed by experimental physics is a tri-partite and realist ontology of law-mechanism-phenomenon.[310] The objective of experimental physics has always been to disclose the mechanism that connects phenomenal changes with the law that supposedly governs those changes. The abstraction of mathematical laws is only possible after this work has been done.

Technological innovation led the discovery of truth—allowing physics to be the herald of modern technology, as Heidegger put it[311]—because the experimental sciences are directed by the "how does it work?" question. By directing research towards the identifications of the "workings" of the causes of the phenomenon in question, modern experimental science necessarily equates "the physical" with "the mechanical," and the mathematical representation of the mechanism is the theory of the real. Thus modern experimental science requires a tri-partite ontology: (1) what is moved? = the object; (2) what moves it? = the mechanism; (3) what governs and limits that movement? = the law. By presupposing mechanical realism, modern physics operates upon a conception of the unity of its object (Nature), a unity of its means (the methodology), and, consequently, is able to present itself as a unified science aiming to disclose natural laws through technological activity, regardless of its specialization and any "paradigm shifts" it may go through. The ontology of the part of the world presented by modern physics as Universe—the Grand Machine—has extended itself in isomorphic parallel with technological innovation throughout the history of physics. The tri-partite ontology of

---

[309] Heidegger (1999c)
[310] Rogers (2005)
[311] Heidegger (1977a)

physics has remained invariant in its structure throughout this extension, and its content only varies according to which particular type of machine (with its associated mechanisms and laws) is under investigation. The objective structure it investigates is the structure of the technological framework within which it is situated and placed in metaphorical relation with Nature that is taken literally.

This tri-partite ontology reveals the extent that *episteme* has been transformed by modern science into *techne*. Scientific explanation is presented as a *techneic answer*, given in terms of general and abstract causal principles, which take the *epistemic form* of mathematically abstracted mechanisms. *Techne* and *episteme* have converged. *Techne* has become naturalized and *episteme* has become mechanized. The single point of distinction between these two is that the former has the experimenter as the efficient cause and the latter has "the inner workings of Nature" as the efficient cause. It is a mechanization of the Aristotelian *phusis* denied its teleology. By embodying the technological framework in representational and material practices, the scientists removes him- or herself from the account and the "efficient cause" of the experimenter and "the inner workings of Nature" become one and the same. The work of modern science is to generate a praxis that removes the experimenter from the account—revealing the objective structures of the technological framework as the objective structures of the natural world—and thereby project a mechanical conception of Nature over natural phenomena and experience. Whatever is "discovered" through technological activity is taken to be non-human Nature at work. This work can be done because non-human Nature has been constructed as the mathematically abstracted workings of non-human machines, but via mechanical realism these machines are interpreted as a means of disclosure of natural mechanisms and laws outside of the context of the machine that discloses these mechanisms at work. This is the methodological work of modern science since its outset in the sixteenth century to the present day.

Novel experiments involve a dialectical relation between novel techniques of making representations of the phenomenon under investigation and novel techniques of intervening and changing the behavior of the phenomenon, in order to map out the contours of interventions and machine performances in terms of representations of how the how the machine interacts with and discloses the phenomenon. The dialectics of technology has been fundamental to the trajectories and content of the natural sciences since the sixteenth century. This is far from being the straightforward and philosophically uninteresting process that the traditional philosophers of science have hitherto assumed it to be, when they have even been aware of it at all. The traditional notion that experimentation simply makes comparisons between theoretical predictions and observations is based on an abstract and false view of how experiments are actually designed, built, and performed, and it is also based on a whiggish history of science. It ignores the extent that experimentation is a historically situated labor process based on complex transformative relations between representational and material practices that are given cultural meaning and understood against a background of already developed theories, techniques, instruments and apparatus, which are related to each other within a technological framework that makes experimental science meaning and operational as a mode of disclosure. Traditional philosophers of science have ignored how the technological framework of the experiment has been culturally interpreted to present it as a possible and meaningful experiment into natural processes. They have ignored how both experiments and interventions have been constructed out of already existing representations and explanations of how all the instruments and apparatus involved in any experiment work together in relation to theoretical expectations about the phenomenon under investigation, and, how its results and observations are technologically mediated (informed and constrained), and, subsequently, related to an already existing cultural field of scientific knowledge, theories, and techniques. Once we pay attention to the role of the dialectics of technology within experimental natural science, especially in relation to transforming the criteria of testability and measurability, we can see how technological innovation transforms the criteria under which any theory is considered to be part of science, and, furthermore, how the established use of any technique determines what qualifies as legitimate or illegitimate scientific research.

Experimental science is an art aiming to achieve its own *techne* for its own sake as truth, as well as a manifestation of *Ge-stell* destined to perpetually use and test the truth of itself by gathering and ordering itself as standing-reserve for future innovation and experiment. It is also situated within a wider world that provides resources in exchange for novel prototypes. The technological objects produced by experimenters are simultaneously produced to discover their instrumentality (abstracted in terms of energy and mechanisms) and for the instrumental value they have in satisfying the challenges of ongoing research and the demands of the wider world. The Destining of *Ge-stell* is the transcendental precondition of modern technology that gathers together human beings, challenges us and sets us up to reveal reality in abstract and non-sensuous truths that are tested in terms of their instrumentality in ongoing technological activity. It is a technological attitude towards truth and the world that sets upon human beings and challenges us to set upon and challenge truth about change through making change happen.

In my view, Heidegger, by generalizing from Galileo's *Assayer*, Newton's *Principia*, and Heisenberg's positivistic interpretation of quantum mechanics had not attended to modern experimental physics closely enough, and, consequently, was compelled to consider the source of novelty and the essence of modern physics to be enigmatic. Heidegger argued that the methodology of science sets up Nature as an object of expectation over which "the ground-plan" is projected, and the objectification of "the real" in terms of "the certain" secures and guarantees some coherence of sequence and order. Mathematics participates in this methodology by setting up, as the goal of expectations, the harmonizing of all relations of order and "reckons" in advance with one fundamental equation for all possible order, and is not merely a reckoning by performing operations with numbers for the purpose of establishing quantifiable results. According to Heidegger, the "inconspicuous state of affairs" which conceals the essence of science can be revealed by taking particular sciences as examples and attending specifically to whatever is the case regarding the ordering, in any given instance, of the objectivity belonging to the object-area of those sciences.[312] For Heidegger, physics (in which he included macrophysics, atomic physics, astrophysics, and chemistry) observes Nature insofar as Nature exhibits itself as the objectivity of coherence of motion of material bodies. The elementary objects of classical mechanics were the motion of macro-objects in geometrical space; in nineteenth century physics these objects were fields and atoms; and in the twentieth century the interaction of elementary particles is the manifestation of the stratified "impenetrability of the corporeal." It is at this point that Heidegger's analysis hit its limit and could not go any deeper.

---

[312] Heidegger (1977c: 170-1)

For Heidegger, the essence of science was "rendered necessary" the moment that this setting-upon occurred, but that moment and its possibility remained mysterious.[313] However, once we take mechanical realism into account then we can reveal the possibility of this "moment" and its subsequent "necessity" within the cultural demands for power and certainty, alongside the practical successes of mechanics and mechanization. If we attend to how experimental physics is actually done, and what it is actually done to, we can attend to the specific ordering of the objectivity belonging to particular object-areas of physics without preconceiving, from the outset, the nature of the object of that pursuit. This object-area is comprised of technological objects and their associated technophenomena, collected and abstracted into machine-kinds and developed into machine-families—the structures of the technological framework itself—and connected to the scientific worldview, via exoframes and exoreferenced models, thereby associating machine-kinds with stratified fundamental representations made in terms of natural mechanisms and laws. The concept of "mechanism" merely grants objectivity to its status as revealing independent reality. The techneic laws abstracted from the alethic modalities of the contours of the interactions between technological objects and technophenomena are the determinations of the estimated possibilities, actualities, necessities, impossibilities, and contingencies of exoframed labor processes of experimentation itself. This means that it is the whole enterprise of experimental science that is an experiment into the possibilities and limits of labor.

Heidegger was correct to locate epochs of change, such as the change from classical physics to quantum physics, in the experience and determination of the objectivity of the appropriate object-sphere, while emphasizing that the essence of modern physics remains unchanged. However, for Heidegger, "in the most recent phase of atomic physics" the object vanished. Heidegger alluded to a change in the objectiveness of Nature into the constancy determined from out of *Ge-stell*.[314] Heidegger noted, in reference to the Wilson cloud chamber, the Geiger counter, and the balloon flights to detect mesons, that modern subatomic physics, despite its aim to make elementary particles exhibit themselves for sensory perception, can only provide indirect perception of the elementary particles via a multiplicity of technical intermediaries. It is this indirectness that Heidegger alluded to as being a fundamental change in the experience and determination of objectivity in the most recent phase of physics and indicative of the dominance of *Ge-stell*. My argument is that Heidegger would have seen the dominance of *Ge-stell* at work in physics from its beginning in the sixteenth century if he had taken mechanical realism into account. By presuming that the object of experimentation in pre-quantum physics was present to perception, Heidegger revealed his bi-partite structure of experimental physics in terms of law-object. Given that, for Heidegger, law is something that we project over the phenomena in terms of cause-effect relations in order to map out the changing of the changeable then the objects of investigation remains those aspects of Nature that are amenable to this project. It is an implicit consequence of Heidegger's analysis of science that the progressive criterion for selection of law could only be one of empirical adequacy of the law's description of the cause-effect relations to the changing of the changeable within the object-sphere. Thus, for this reason, Heidegger only marked out "the most recent phase of atomic physics" as something distinctly given over to *Ge-stell* because the objects of its object-sphere are unavailable to direct perception. Heidegger's interpretation of the aim of experimental physics as being directed towards the mathematical projection of laws describing the coherence of motion of its objects, and the constancy of the changing of the changeable, betrays his positivistic conception of science in general and physics in particular. The aim of science, on Heidegger's

---

[313] *Ibid* 169
[314] *Ibid* 173

account, is that of empirical description in terms of universal law. This should be unsurprising, given that Heidegger's two exemplars of scientific endeavor are Newton and Heisenberg, and that in Germany, at the time of Heidegger's writing, positivism was the dominant conception of science. I am not suggesting that Heidegger was a positivist (far from it!) but his conception of the goal of science was positivistic. Thus, for Heidegger, the object only disappears in subatomic physics because it cannot be directly revealed to sensory experience. Here we can readily see how Heidegger's own phenomenological pre-occupation with "that which presences" obscured his view of the phenomena of experimental physics. However, the aim of modern experimental physics has always been the disclosure of mechanisms, and its ontology has had a tri-partite structure (object, law, and mechanism) since the sixteenth century (Rogers, 2005). Particle physics is unconcerned with the detection of "particles," mesons for example, except as a means of investigating its models of elementary particle interactions. Physicists do not seek the "truth/top quark" for its own sake, but, rather, as a means of disclosing fundamental mechanisms.

Mechanisms have never been brought into presence, as objects, but are central to causal explanations for "that which presences." Since Galileo, physics has had pretensions towards a "deeper" ontological relation with the mapping of temporal successions of cause-effect sequences than merely confirming or refuting the law. These explanations are given as the underlying workings of that which allows "that which presences" to presence. "The real" disclosed by experimentation, via demonstrations given in terms of the temporal sequences of changes of the changeable, is brought into discourse as a causal account given in terms of the mechanisms that are proposed explain the phenomenal. The objectivity of "that which presences" is taken to disclose "the real" via the objectivity of the technological framework (the object-sphere) and the constancy (machine performances) of the changing of changeable (techno-phenomena) as means to the disclosure of "the real" as a mechanism of change in the world. The object-sphere, procedure, methodology, and mathematical projection are tied together via an exoframed model of the mechanisms in operation in the changes of the changeable. It is the projected exoframed model that allows mathematical projection, the object-sphere, methodology, procedure, law, and the ongoing activity of working towards securing representations to fit together as a coherent process of research. Working novel experimental science is not limited to sensory experience but, rather, explores what is disclosed by means of sensory experience. Its object-spheres, object-areas, procedures, laws, advancing methodologies, and mathematical projections, have always been means to this end, and, in this respect, it changes the changeable in the object-sphere to suggest causal mechanisms which are "tested" by implementing them in the ongoing technological procedures of experimentation. Thus the use of objects as standing-reserve available for future use and ordering has been central to the processes of "testing" in experimental physics since its outset in the sixteenth century and has been basic to all experimental science since. *Ge-stell* has been characteristic of modern experimental physics since its origins in the science of mechanics, which has become represented as the exemplar of "a hard science" that all other sciences, including the biological and social sciences, try to emulate. *Ge-stell* has been operational within the unfolding and ordering of the ontology of experimental science in all of its object-areas and is not restricted to the "most recent phase in atomic physics." Heidegger's positivistic conception of modern science concealed the operation of *Ge-stell* in experimental physics and, consequently, concealed the technological essence of modern experimental science.

Heidegger was correct to have considered objectivity to be essential to the setting up of modern sciences in general and modern physics in particular, as a methodologically and procedurally secured mathematical research projection of the ground plan of its object-sphere, which connects research with procedure. Modern experimental physics could have not begun nor operate without this set-up. Heidegger just misunderstood the exact nature of objectivity in modern physics. What Heidegger misunderstood is the objective of the pursuit of experimental physics. It is *techne*, as an ideal, that provides an asymptotic, imaginative link between practice, theory, and scientific rationality, knowledge becomes objective upon the creative establishment of the possibility of achieving a *techne* of how that knowledge was produced, and it is this that connects knowledge with methodology. Thus the process of scientific progress is one of aspirations towards *technical excellence* (understood as an intellectual virtue) that is to be achieved via the rigor of the techneic process of questioning and correcting both the content of theories, experimental and theoretical practices, and standards of justification. Gooding termed this process to be one of convergence.[315] This open-ended process is one that is perpetually directed towards the future and, as such, is one for which there are not any "logic of scientific discovery" to guide it because we cannot foresee the course of the dialectic of technology. It is one of bounded and evolving technical rationality and the creativity of the labor process itself.

Controversies regarding the artificiality of the results of techniques and preparations (what is a property of the object and what is a product of the preparation process) could only be achieved by reaching theoretical agreement as a consequence of appeals to technical rationality. This requires both theoretical accounts of the techniques utilized and of the object under investigation. The establishment of a clear distinction between these two accounts rests on rhetorical attempts to resolve any controversy one way or the other. Prior to publication, any experimenter needs to reflect upon the experiment (its purpose and execution) and anticipate any criticisms to her/his work; every experiment is situated within the background of a wider scientific community's standards and expectations in which it will achieve its significance and meaning. The working scientist will base her/his conception of what makes a reliable observation and/or technique upon recognition of the level and content of any possible criticisms to which her/his work may be subjected. It is this ability of the experimenter to anticipate how others will receive her/his actions and reasoning, which is a necessary condition for the critical self-reflection to be presented as "rational," while, in proposing any novel theory or experimental technique, it is necessary for the experimenters to break the accepted norms and standards of the current scientific community by engaging in a critically motivated reflection upon the correctness (or adequacy) of those accepted norms and standards. This opens up the possibility of contradiction and falsification through testing. Its resolution requires that the technological framework that the experimenters are working within is open and capable of being developed and changed. If the technological framework were fixed, final, and closed there would not be any potential for novel research because there would not be any possibility of establishing new techniques, theories, and objects for investigation. Novelty, controversy, inconsistencies, and new specializations arise from the innovative interconnection (the relating and conjoining) of machine-kinds. Again, incompleteness is the source of both contradictions and the possibility of their resolution in terms of something new.

Experimental work is an unending process of producing a series of refinements for which successive refinements reveal and transgress the limits of previous efforts by transforming the technological framework within which the work is situated. This process is perpetually one in which the experimenters must make technically rational judgments about alethic modalities of the experiment, which emerge from a background of past

---

[315] Gooding (1990)

efforts and perceived limits, and interrelate technological objects and technophenomena. It is a process of exploring the boundaries of the whole process of refining and developing the technological framework of experimental science. Thus experimental science must be an ongoing activity that is constantly open to change. But it does not pursue the objectivity of its object-spheres or its object-areas, or even "Nature." Instead, it pursues its own development towards its own completeness. The object-spheres and object-areas (the machine performances of machine-kinds interpreted via associated fundamental representations) within distinct specializations in experimental sciences are means to this end through the ongoing stratification of the technological framework. Particular specializations are simply delimited by specific kinds of machine, each situated within the stratified technological framework of experimental science, wherein its structural development and exploration of ontological depth go hand in hand.

"The empirical" within experimental science is limited to machine performativity and the technophenomena this reveals, but the epistemology of science is secured against the ontological interpretation of the interrelation of technological objects and technophenomena to disclose the mechanisms that are supposedly in operation during the "changing of the changeable." These mechanisms could only be conceived as natural by presupposing mechanical realism. On this account, the function of an object-sphere is to disclose the fundamental mechanisms that explain how interventions within an object-sphere generate the machine performances actively produced and investigated by experimentation. Modern physics is only concerned with the motion of matter insofar as it discloses those fundamental mechanisms at work in a machine such as a pendulum or neutrino detector, just as experimental genetics is only concerned with laboratory rats insofar as they can be used to disclose the connection between fundamental genetic sequencing and flagging techniques and interpret these in relation to biochemistry and animal physiology. It is only by recognizing the metaphysical centrality of the concept of mechanical realism within the rationality of science that we can understand the purpose of the performative art of experimentation and the nature of the reality that scientists aim to disclose. Physicists, chemists, geneticists, or neuroscientists are no different in this respect—and, since its positivistic turn, psychology and the social sciences have attempted to emulate. Thus we can see how experimental science has been performed for its own sake—as an art directed towards the realization of its own *techne*—and bound up with *Ge-stell* in the ongoing, endless development of the technological framework and its connection with the scientific worldview, made in terms of the mechanistic theory of "the real" and "tested" by transforming technophenomena into technological objects available for future use.

Modern science cannot be understood simply in terms of "applied technology" because, through the use of mathematics, it represents itself as directed towards the realization and acquisition of *episteme*, and it thereby conceals the extent that it is a metaphysical art directed towards the technological acquisition of *techne* of Nature, wherein Nature is itself understood in terms of its instrumentality. Modern science is directed towards the acquisition of the *techne* by enframing and directing the trajectory of the human understanding of Nature through technological activity, while simultaneously representing *episteme* as something technological and testing that representation by bringing new powers into the world through technological innovation. Modern science is a destining of human beings towards the metaphysical realisation and acquisition of the *techne* of Nature and the new powers that knowledge promises to bring forth. Directed towards the acquisition of the *techne*, science frames and directs the trajectory of the human understanding of the natural world within the technological framework of ongoing research, while simultaneously representing *episteme* as something technological via conceptions of natural mechanisms and laws. Modern physics is a destining of human beings towards the metaphysical realization and acquisition of the *techne* of Nature and the new powers that it promises to bring forth. It is through the deepening of this destining in the science of cybernetics, in terms of its transparency through embodiment and the illusion of the "steersman" metaphor, which brings with it an increasingly unquestioning relationship with modern technology. This destining finds its first and most explicit scientific expression in Norbert Weiner's (1965) studies into the possibility of the science of cybernetics and the study if control systems, and Kenneth Sayer's (1979) reduction of the philosophy of mind to the science of cybernetics, and the conception of the natural order emergent within chaos and thereby reducible to information or energy understood within the terms of thermodynamics.[316] This brings with destining a danger to our chances of developing a free relation with technology. It also has the most profound impact upon our understanding of the world.

Ellul, Science and Technology:

Ellul's analysis of the relation between science and technology had similarities with Heidegger's, but rather than assume an empiricist's model of science (as Heidegger did), Ellul considered science to be more of a rationalist's enterprise. He accepted that it is the creation of general explanatory theories that makes science distinct from technique. However, Ellul was critical of the view that modern science is pure theory and technology is applied science. He argued that this view is "radically false" because it is only true of the nineteenth century physical sciences and is not true of science and technology in general.[317] In his view, technique preceded science—and "even primitive man was acquainted with certain techniques"—but technology only began to develop and extend itself when science appeared. Science requires the application of technique as a necessary condition of its existence because, without technique, science would be merely hypothesis and speculation—an interpretation of observation and experience. Technique gives science its objectivity. How? For Ellul, modern science became bound up with technique during the Industrial Revolution's development of the machine and the application of technique to all spheres of life. How did precision and explanation solve the problems of technique and bring together theory and technology? Unfortunately, Ellul did not explain this to any greater depth than arguing that there is an increasing interaction between scientific research and technical preparation to such an extent that science is incapable of progressing without the technical means to do so. For example, according to Ellul, Faraday was unable to precisely formulate his theories about the constitution of matter because of a lack of high-vacuum techniques. While this is true, Ellul needed to explain why this should be the case. If technique and Nature are independent, as Ellul maintained—"not even entering into a symbolic relationship with it"—then why is technique necessary for natural science? Why were high-vacuum techniques necessary for the precise formulation of Faraday's theories? Surely, if technique does not have anything at all to do with Nature then there would be not any necessity for any particular technique for the scientific synthesis to proceed. What was the "matter" that Faraday wished to explain? Ellul did not address these questions either.

Ellul argued that, even though the experimental sciences of the seventeenth and eighteenth centuries did not provide the motivation for the application of the technological imperative to all areas of human endeavor, they prepared the way for the

---

[316] See also Prigogine and Stengers (1984); Deleuze and Guattari (1994); Du-Puy (2000)
[317] Ellul (1964: 7-28)

nineteenth century's positivistic conceptions of societal progress being represented in terms of the conditions for the construction of the technological society.[318] The application of mathematical rigor to the design of any machine was not only constrained by the demands of practical applicability, but preconditioned the prejudice that progress involved experimenting with the development of the designs most amenable to mathematical calculation and practical utility. While science provided general explanatory theories from the outset, it required the application of technique as a necessary condition of its existence. Since the nineteenth century, it has become increasing the instrument of the rationalization of the application of techniques to all areas of human activity. In the twentieth century, science became dominated by the technological imperative to the extent that the development of atomic theory and the atomic bomb were necessarily connected (Ellul, 1964: 86). The work of technique, the mechanization of all human spheres of action, was a systematization, unification, and clarification of everything. In turn, technique required science to progress because it had to wait for science to provide the mathematical solutions to the problems posed by the search for "the best technique" and the development of its confirmation and explanation. How did science provide these mathematical solutions? How were the problems encountered? Why were they problems? How did scientists know where do start? Ellul did not address these questions. He did insist, however, that the border between technological and scientific activities is not sharply defined and that technique provides preparatory work for scientific synthesis. Science has become the instrument of technique because scientific discoveries are increasingly implemented in everyday life before the consequences of that implementation have been considered.

Modern scientific research increasingly requires large teams of researchers, enormous amounts of money, and the aid of increasing complex machines. One need only compare the Hubble Telescope to that of Galileo's to see the validity of this point. The Hubble Telescope is not merely the device but also involves the organized technical activities and machines of NASA and the Flight Operations Team at Goddard who collect and transfer the data from the orbital telescope to scientists at the Space Science Telescope Unit, who use computers and models to translate the data into scientifically meaningful units (wavelength, brightness, position, distance, focus, etc.) and images for astronomers around the world. The construction of the telescope itself is far beyond the optics of Galileo's telescope. It also involves a wide range of instruments and techniques in photography, spectrography, and thermometry to translate date into visual imagery of infrared and ultraviolet radiation, and see temperature as well as light sources, and also interpret and calibrate guidance sensors to determine position and orientation. The work of large scientific research is increasingly technical work and "pure science" is becoming increasingly "applied technology." To illustrate this point, Ellul used the example of way that the steam engine was the product of technical trial and error sequences of invention and improvements and scientific explanations came much later. However, he did not provide us with an account of how those explanations were forthcoming. Nor did he show us how they were related to invention. Ellul did observe that science has been becoming increasingly governed by technique since the nineteenth century to such an extent that the smashing of the atom and the smashing of Hiroshima (and Nagasaki) are manifestations of the same imperative.[319] However, Ellul did not provide us with any explanation of how atomic theory and the development of the atomic bomb were manifestations of the same imperative. He did not explain that it was only through the construction and detonation of the first uranium fission device that atomic theory gained its objectivity. It is only in the wake of the explosion of the Trinity bomb in New Mexico

---

[318] *Ibid* 38-45
[319] *Ibid* 86; see also Chaloupka (1992)

could atomic theory become a theory of the real. The Second World War merely gave it a context for the application of the imperative to build it and unleash its power; once destruction of Japanese cities was deemed a necessary goal by the Allied command to win the war, the use of atomic bombs on a Japanese city was inevitable once the test bomb demonstrated its destructive power by bringing it forth into the world. The imperative of technique demands it—only efficiency in destruction will ultimately save lives, on the promise of ending the war earlier—and receives its theoretical confirmation in its execution. Indeed, as Robert Oppenheimer famously remarked, "I am Shiva, creator and destroyer of worlds."

There was considerable equivocation and inconsistency, on Ellul's part, in his description of the relations between science and technology. He maintained the view that mechanical progress "is limited by the physical world" and that the drive for efficiency is the mobilization of "the forces of nature" and an "intervention into the inorganic world."[320] However, he also insisted upon the fact that the only thing that matters technically is "efficiency" and this is distinct from Nature—in opposition to it. The law of technique can only be discovered by the total mobilization of human beings, body and soul, and implies "the exploitation of all human psychic forces" and, yet, technique "has its own specific laws which are not the laws of organic or inorganic matter... Man is still ignorant of these laws."[321] Does this equivocation reveal an inherent contradiction in Ellul's argument? How could science and technology be related in this way and have nothing in common with the natural world? Did Ellul take "the physical world" and "the inorganic world" to be distinct from "the natural world"? Unfortunately, he did not discuss the relations between these "worlds" further. In my view, the equivocation in Ellul's thesis was a consequence of his rationalist account of mathematics and characterization of science as mathematical and explanatory, on one side, and the absence of any account of the technique(s) which provided a clear link between mathematics and technology, on the other. What was the scientific synthesis? How did science produce explanations? Ellul argued that the precision of any machine is only possible because of the elaboration of its design with mathematical rigor in accordance with its use, but he did not provide any account of how this elaboration of its design could be performed. He argued that the precision of any machine is only possible because of the elaboration of its design with mathematical rigor in accordance with its use to the extent that practical activity rejected gratuitous aesthetic preoccupations in favor of the idea that the most readily adaptable technique for use and mathematical calculation is taken to be the best. This meant that practical activity rejected gratuitous aesthetic preoccupations in favor of the idea that the line best adapted to use is always the most efficient. It was supposedly necessity and the certainty of mathematical calculation that characterized the technical world but Ellul did not provide any account of how mathematics became bound up with the technical imperative and conceptions of "efficiency." His position was that modern science depends upon technique, while he simultaneously maintained the traditional view that *somehow* technology is "applied science" because it is mathematical. He accepted that the technological imperative somehow developed out of the science of mechanics, but its origin was "mysterious and enigmatic" and possibly bound up with magical rituals.[322] For Ellul, the "modern worship of technique derives from man's ancestral worship of the mysterious and marvelous character of his own handiwork."[323]

---

[320] *Ibid* 88, 103
[321] *Ibid* 324, 429
[322] *Ibid* 5, 21
[323] *Ibid* 24

Yet, as I argued in chapters 3 and 4, if we take a closer look at the cultural transformation in the status of mechanics in the end of fifteenth century, we can see how the importance of the practical advances of technology secured their mathematical representations as having objective correspondence through their use to solve practical problems as geometrical demonstrations of the objectivity of those representations of mechanisms in terms of the instrumentality of the associated mechanical devices. Mechanical realism had underwritten metaphysics and natural philosophy from the Renaissance onwards, with the application of Euclidean geometry to the problems of producing mathematical representations of natural mechanisms at work in terms of the template of the six simple machines as the six most efficient mechanisms for the achievement of each of a fundamental set of six basic motions. The representational and explanatory link between mathematics and mechanics had been established before the end of the fifteenth century, and increased technological power and advantage was taken to be the corroboration of the truth of the new sciences. Metaphysics reduced the methods of science in accordance with the limits and powers of technique, relating it with the scientific worldview, and reducing the idea of the physical world to instrumentality itself, while being integrated into an ongoing cultural project of using science and technology to improve the human condition and build a good society by mechanizing it. Through science, technology is represented and explained as an intervention into the physical world by mobilizing "the forces of Nature" to change the physical world, and the ability to do is taken to be a demonstration of technology as a means of confirming science as the theory of the real. This was as true for Galileo and Descartes, as it is today. From the outset, the name of science was invoked along with promised new knowledge, powers, and experiences, and the ongoing construction of the technological society has been taken to prove the validity of science ever since. We can see a clear technical intention in the writings of Francis Bacon, the reports of the English Royal Society in the sixteenth century, and its association with the English Navy, and the efforts of the sixteenth century Italian mechanists and engineers in building fortifications, weapons, and useful devices to do mechanical work. The practical value of the new sciences secured the epistemological value of the whole enterprise of the new sciences from their very beginnings. Not only were craft practices and the innovation of novel tools and instruments central to the possibility of the experimental work and ideas of the new sciences, but they also confirmed them and granted them their objectivity.

The seduction of technique was ever present in the sixteenth century drive for the achievement of commercial, political, and military advantage in the competitive contexts of European nationalistic and mercantile ambitions. It was more than Bacon's dream or the stuff of fiction; the technical imperative came with an expectation of success as well. The destining of science and technology came hand in hand and their practical value secured the scientific worldview as an explanatory representation of the world, which was readily available as a justification of the technological imperative and societal gamble in terms of their necessity and rationality. Yet it is their necessity and rationality that is at stake and remains unknown. Everything still depends on whether the technological society will ultimately be good for human beings, and whether it has ever been good for human beings, which are philosophical and political questions rather than scientific ones.[324] This is why science and technology must be viewed as experimental at a cultural level, while the question of 'what is a good society?' remains outside of remit of science. The value of science and the goodness of the technological society was assumed from the outset, as an *a priori*, while seeking its confirmation in the production and movement of goods *a posteriori*. The technological society is itself an experiment in the possibility of human beings building a better world, in which they know that world and

---

[324] Rogers (2006, 2008)

how to live in it through reason and knowledge of the material and necessary conditions of human existence; thereby understanding what is natural or contingent in human behavior and thinking, and through science and technology liberate human beings from those material conditions and necessities by using natural forces to change the structure of the world. Whether it will do this or enslave us further to our material conditions remains open to question. The rationality of the societal gamble is itself at stake.

Labor, Science, and the Technological Society

The technological society is built through organized and disciplined labor processes, utilizing techniques and technological objects, and these comprise the sociological structure of the technological society as it reproduces and extends itself. Experimental science is also structured and shaped by organized and disciplined labor, and, like all labor, is premised on human interests, needs, and ends, and its expectations and assumptions about the object of inquiry are based on the purpose of labor. This insight opens up the possibility of a Marxist interpretation of science.[325] Science is inherently a means to satisfy objectified purposes, ambitions, and challenges, and these human interests are not only required to set-up any experiment in the first place, but are also fundamental to how the object of inquiry is investigated and understood. The teleological positings of science are situated within the technological framework of the research program and generate the directions of that program and the anticipations of its outcomes, which makes the outcome of human relations and practices situated within the technological framework. The decisions and choices adopted in the execution of any experiment are made with the explicit aim of satisfying specific purposes and challenges that have been posited by or for others; all which allows the experimenter to determine the conditions under which the experiment can be said to have been successful or a failure, by explaining or describing to others how to perform the experiment and repeat its outcomes, and interpret those outcomes in terms of a fundamental set of representations of machine performances.

In his critical response to the technological determinism inherent to Marxism, Robert Heilbroner proposed what he termed as a "soft" version of technological determinism that holds there to be an "inner logic" to technological innovation, but acknowledges that technological development is also directed by socioeconomic factors, which, although strongly influenced by technology, are not completely determined by it, given that cultural factors also shape socioeconomic factors. Heilbroner's "soft" version of technological determinism describes technological development in terms of having specific stages, each with their own stratified technological conditions. The hydroelectric power plant, for example, requires the development of the steam mill, which required the development of the hand mill, which required the development of basic gears, which required the wheel, etc. Each strata of technological development contains the previous strata as prerequisites, regardless of whether any particular culture develops them or not, but should any culture develop any particular stratum of technological development, it must have already developed its prerequisite strata. Each and every technological innovation exists within a stratified technological framework that is both a condition for its existence and shapes its usage and further development. Implicitly assuming mechanical realism, Heilbroner explained the "inner logic" of the limits and directions of this trajectory—its alethic modalities—as being conditioned by the discovery of natural laws and mechanisms. The historical discovery of new "forces of production" is made

---

[325] Lukács (1978)

possible by the discovery of natural laws and mechanisms, but whether they are discovered or implemented is conditional on cultural and political influences, which cannot be reduced to either economic or technological factors. Heilbroner was aware that different cultures and societies have different goals and values, which lead them to develop different technologies in different ways, due to the considerable cultural and social pressures on the directions of technological development, but he argued that both the possibility initiating a course of development and its continued evolution follow a determinate pattern rather than a random or contingent course. Heilbroner's position presupposed that technological development is based on the sequential and objective discovery and application of natural laws and mechanisms of Nature and, therefore, predictable and continuous, once initiated and sustained within any culture. Comparing Heron's and Watt's steam engines, Heilbroner argued that technological innovation is not simply a matter of producing a prototype, but also requires the successful integration that prototype within wider society as an effective and productive machine. This involves the establishment and development of a cultural infrastructure of supporting technologies (e.g. iron foundries and machine tools), as well as the discovery and dissemination of the technical knowledge of how to produce and reproduce the machine. It is this need for *technological congruence* that constrains the sequencing of technological development with the interrelation and connection between previously divergent and unrelated technologies that transforms a prototype, alongside the construction of its supporting infrastructure, into a new technology. Technological innovation requires a transformation of social organization and cooperation between different industries, technical specialists, and technologies, all operating within a stratum of organized and specialized knowledge, labor, and resources.

New technologies only become "forces of production" when they are successfully integrated within an already existing stratum of forces and relations of production, which, of course, destabilizes previously stabilized forces and relations of production. Once integrated into a new infrastructural stratum, creating new forces and relations of production, the social, economic, and political structures of society are transformed, thereby changing social relations and societal organization, bringing with it new goals and a horizon of possibilities, and new trajectories of research and development, as well as bringing new problems and risks. The "inner logic" of technological development imposes constraints and directions upon the development of society by transforming the composition and division of labor and the hierarchical organization of work, creating new societal needs and the means to satisfy them, as well as creating new forces of production for the further development of technology. Technology both reflects and shapes social relations, which in turn reflect and shape technological activity, which responds to and is limited by cultural and political values, directions, and influences. Any technological development must be congruent with the already existing technological infrastructure of society and also compatible with cultural attitudes and influences. Hence, even though "the technical" is a subcategory of "the social," it cannot be reduced to it.

Bruce Bimber (1994) termed Heilbroner's theory of technology as a nomological variety of technological determinism because it presupposes that the conditions of technological development are governed by natural laws. As I have argued above, this nomological variety of technological determinism lies at the core of the whole scientific enterprise since the sixteenth century and the societal gamble on constructing the technological society as the cultural confirmation of its own truth. The discovery of natural laws has been superimposed over the stratified structures and trajectories of technological development, as abstractions of the labor processes and struggles involved in their invention, stabilization, integration, and use within the already existing overlaps between technological objects that constitute each stratum of technological development. The cultural acceptance of mechanical realism has concealed how science operates by transforming Nature into something that can be manipulated through technological activity, which is metaphysically represented as utilizing natural laws and mechanisms without transforming them, while representing natural phenomena in technological terms and placing them at the disposal of technical apparatus and techniques of measurement as the test of their truth. The conceptual contradiction between the artificial and the natural has been avoided by allowing the artificial to be metaphysically represented as being a natural consequence of natural human perceptive, cognitive, and reasoning abilities, as evolved within a material world. This representational step has allowed technological activity to be confirmed as the rational implementation of natural mechanisms in material practices; simultaneously an act of discovery and practical application. In the case of experimental science, knowledge and experience are both simultaneously represented as *techneic* and *epistemic*—idealized as characteristic of an objective asymptote of interwoven chains of causal accounts—each representing technical procedures in terms of natural mechanisms and laws to explain the efficacy of techniques in the design, construction, operation, and interpretation of the performances of the apparatus in response to human interventions. Yet, measurements cannot be simply attributed to the properties of the object measured, but must also be linked to the techniques used to make those measurements. It is the cultural belief in technological determinism—hard or soft—that allows technology to become invisible or transparent as nothing more than a means to an end. Whereas, if we suspend belief, we can see how scientific narrative underwritten by mechanical realism and technological determinism has concealed the substitution of technophenomena for natural phenomena during experimentation. Scientific knowledge is an inherently metaphysical understanding of Nature, which represents natural entities and processes within a scientific worldview as being the phenomenal manifestation of natural laws, which allow and limit the realization and exercise of natural mechanisms that cause phenomenal changes in the world of experience, while the theoretical entities and predictions derived from representations of those laws and mechanisms are tested within the context of technological activity. It is only to the extent that an event or phenomenon can be situated, reproduced, and manipulated within a technological framework can it be said to be available to scientific investigation and, hence, taken to be part of objective reality. Only that which can be situated within the technological framework of experimentation and represented in terms of functions and consequences can be taken to be real.

Within this scientific worldview, human beings are represented simultaneously as natural beings—with their own organic evolution—and an efficient cause that is capable of grasping the objective reality of their own material conditions by intervening into and manipulating those conditions by discovering and implementing "natural mechanisms" in material practices. The cognitive products of this effort are taken to be disclosures of the same underlying reality that causes the phenomenal natural world, human nature, and the "inner logic" of technological activity. The historical development of technology can be thus represented as the disclosure of deeper ontological strata, representing innovation as a mode of discovery, which can be unified as a progressive trajectory through modifications and refinements of the scientific worldview, while the progressiveness of science can be verified in terms of its practical successes and instrumental value in its own development and refinement. Experimental science aims to disclose our productive possibilities and explain these in relation to a scientific worldview, rather than merely compare theory with experience (as the traditional philosophers of science would have us believe). It abstracts these productive possibilities into sets of alethic modalities, represented as corresponding to the objective reality via mechanical models, which are tested in terms of their instrumentality for future research, as well as for the practical problems of the wider world. The theoretical understanding of the process through which one set of technological objects are transformed into another set of technological objects is taken to be the theoretical representation of the objective natural mechanisms that permitted those transformations to take place.

The ontology of the natural world has been substituted with a metaphysical interpretation of a collection of the performances of different kinds of machines and the epistemology of science has been reduced to a methodology of innovating, relating, and justifying procedures and techniques, by demonstrating their instrumentality in bringing forth novel technological power, and/or refining and developing the technological framework from which scientific research emerges. The scientific understanding of human limits, power, and capacity for control is identified, refined, and developed within this framework in order to present an understanding of ourselves and the world that offers both as being objectively within our cognitive and manipulative grasp. Only those aspects of the world that can be situated within the technological framework are understood as objective properties of the world, while the human abilities to confirm or refute theories, experiment upon and explore reality, develop and interpret new experiences, etc., are all situated, related, organized, and transformed within the technological framework. The refutation of scientific hypotheses and the collection of empirical data are only scientifically possible by remaining within the technological framework and, therefore, if this is the case, the scientific worldview and the technological determinism upon which it is based are totalitarian and internally self-evident paradigms, which converge and reinforce each other through the technological imperative and the discovery of new powers.

All genuine novelty in experimental science is the product of a technical complex of heterogeneous components combined into an ensemble, alongside novel conceptions and representations of those ensembles and their components, which are combined and integrated with other ensembles in order to produce and reproduce functionality, and thereby relate that ensemble as a new technological object to other technological objects and techniques through the stratified structures of the technological framework. All technological objects are complex and cannot be understood without addressing their connections and interactions with other technological objects and the purposes that they were constructed to satisfy. Converging and integrating heterogeneous components into a stable and communicable set of techniques and their associated machine performances produces all novel technological objects. The labor processes of experimentation are the refining processes of stabilization of techniques and machine performances, and their associated technological powers, as labor moves between plateaus of stability punctuated by transition periods of destabilization and restabilization. This is what Kuhn termed as normal science, which encounters anomalies and contradictions as it approaches its completion, thereafter followed by crisis and the emergence of revolutionary science to explain those nominalizations resolve and those contradictions. When labor is performed for its own sake then it is an art directed towards its own self-perfection, but in the experimental sciences the achievement of perfection or completion lies on an unreachable asymptote, projected at the end of all experimentation and the completion of the technological framework. As such, experimental science always remains incomplete, with the meaning of its results deferred to the future and directed to providing instrumental powers for the attainment of something else. The experimental sciences are organized and disciplined labor processes that produce and reproduce themselves for their own sake, allowing science to participate in discovery by discovering its own possibilities, while finding its ongoing corroboration in the production of energy and power. They straddle the distinction between *techne* and *Ge-stell*.

Through organized and disciplined labor, the speculative mind has transformed into a rational agent, pragmatically defined in terms of performance and function; the experimenter is as an integrated and articulated component within an ensemble of functioning agents, each brought together and given meaning within the technological framework as an experiment. Implicated in the cultural construction of a totalitarian society obsessively driven towards the maximization of efficiency, which, at any stage of achievement, objectively adopts the best available technique to achieve any given goal, excluding all goals for which there is not a technological solution, and to the extent that any and every alternative is unthinkable. Every best technique is made in reference to the satisfactory stabilization of measurements, calculations, and productive practices, determined in relation to an intelligible causal account of machine performances, each emergent from the technological framework of expectations and limits on the alethic modalities of human labor. When the performance of the machine is ambiguous, the experimenter acting as a rational agent becomes framed by technology, which ferociously tests the machine in accordance with the technological imperative towards efficiency, to achieve a reproducible and determinate outcome, further refined and corroborate in its reproduction, as every agent is placed in competition with the others to find the most efficient winner, as the one best way to achieve any given goal, until it is replaced by another. Having achieved its autonomy in practice, due to the technically rational obligation to use it, it remains indisputable as the efficient means and its use is no longer a matter for deliberation. It was for this reason why Heidegger and Ellul were both concerned with the way that the dominance of the technological imperative towards efficiency has taken over all human activities and ordered them according to their utility, integrating and assimilating everything into the technological society, applying mechanisms, techniques, machine processes, and mechanistic logic to all areas of human activity. Technique has become autonomous as an integrated substance and totality of human agency within the world. Just as hydroelectric installations take waterfalls and lead them into conduits, transforming them into electricity available for future work, so the technological society absorbs and appropriates human labor, and therefore itself, in the perpetual construction of itself.

It was for this reason that, in his criticism of Marxism, Ellul claimed that capitalism was only one aspect of the deep disorder of the nineteenth century and that capitalists were not the dominating motivating force behind the development of technique, instead being merely opportunistically aware of how to extract profit from technological developments.[326] The ills of technique, such as the flooding, pollution, and erosion caused by deforestation for industrial agriculture and the use of chemical fertilizers on the poor quality soil that remains, are products of the relentless development of technique to make agriculture more efficient, which brings obvious advantages as well as unforeseen consequences, and the capitalistic exploitation of cash crop markets is simply an opportunistic use of this development.[327] Hence, according to Ellul, the dominance of technique upon capitalist economies imposes an impersonal centralism upon the economy through the mechanisms of the stock market, in which economic planning is reduced to an anarchic "the order of the day" for the economy as a whole, while the technical imperative lies at the core of the cycles of creation and destruction that generate new opportunities for increasing profits and reducing wage costs, at the

---

[326] Ellul (1964: 5, 63). Ellul cites Lewis Mumford's *Technics and Civilization* (1963) as support for this claim, but as I shall explain below, Mumford's view on the relationship between technology and capitalism was more complicated and critical than this.
[327] Ellul (1964: 104-5)

expense of waste and unemployment.[328] This impersonalized centralism does not result "from the machinations of evil statesmen" or "wicked financiers and industrialists" and it cannot be controlled by "public opinion." Even the State, powerful that it is, has become dominated by the development of technique, and, by the end of the nineteenth century, capitalism had already become eclipsed by the efficiency of the technical method and the construction of a technological society. Capitalism opposes the construction of the technological society when the technological imperative towards the achievement of maximum efficiency opposes the drive to increasing profit.[329] Hence, Ellul recognized that

"Capitalism checks technical progress that produces no profits; or that it promotes technical progress only in order to reserve for itself a monopoly… The pursuit of technical automatism would condemn capitalist enterprises to failure. The reaction of capitalism is well known: the patients of new machines are acquired and the machines are never put into operation." (Ellul, 1964: 81)

Technical progress is antithetical to capitalistic individualism, which Ellul described as a form of anarchism, and he considered technique to be "the most important factor in the destruction of capitalism, much more than the revolt of the masses."[330] Ellul recognized that laissez-faire capitalism opposes the universalization and centralism of technique to such an extent that he interpreted the Marxist critique of capitalism as being essentially one that advocated the liberation of the development of the technological society from the restraints of laissez-faire capitalism because, for Marxists, the construction of the technological society is a good, which will free the proletariat and is the condition for the realization of communism.[331] Ellul considered the Marxist vision of a communist system to be one orientated to technical progress, in all areas, giving free play to all technical automatism in every field of human activity, in order to maximize efficiency and productivity (both of which are conditions for the post-scarcity society), and, therefore, he considered that the Marxist vision of communism would necessarily lead to a totalitarian technological society over and above the dictatorship of the proletariat. This view was also predominant in his later work. In *The Technological Bluff* (1990), Ellul argued that the unthinking production and consumption of gadgets has nothing to do with efficiency. Once we take all the resultants of a technique into account, including waste, pollution, external costs, excess production, and in-built obsolescence, then much of contemporary technological innovation has nothing to do with efficiency. Due to its reproductive power, modern technology has permitted the development of a universal economy of impersonal products, but the economic irony of the capitalist "market forces" model is that it is based upon an acceptance of the scarcity of resources, while it is profoundly wasteful and inefficient, exacerbating the problem of scarcity rather than alleviating it; showing how its foundation on consumerism is based on the creation of "need" rather than its satisfaction. As Marxists would explain, this reveals a fundamental contradiction between capitalism's drive to profit and the technological society's drive to efficiency. Capitalism places limits upon the development of science, once it has provided means to satisfy its commercial, military, and civic ambitions, because (1) further research is expensive and requires long term and sustained investment, without any guarantee of profits; (2) increased productivity and efficiency can also place limits on the return on previous investments in the means of production by making those means obsolete. The oil industry buys up patents for alternative and sustainable electricity

---

[328] *Ibid* 184-97
[329] *Ibid* 200-1
[330] *Ibid* 198; 236-7
[331] *Ibid* 144

generating and transportation technologies, as well as new battery technology, and sits on them in order to protect its investment in its infrastructure. Hence, Marxists would argue that the liberation of the technological society from capitalism is necessary to free scientific research and the development of the technological society from the distortions and limitations caused by capitalism. What can we make of this? If capitalism is subordinate to the technological imperative, as Ellul claimed, and this imperative is antithetical to it, then how can capitalism resist and pervert it in this way? If the development of technique is the most destructive force over individualistic and anarchistic capitalism, then why did the totalitarian, communist technological society, driven to maximize efficiency in all areas of human activity, fail to emerge as the dominant form of society?

Technics and Civilization:

Lewis Mumford argued that capitalism and modern technology conditioned and reacted to each other, but they must be clearly distinguished from each other.[332] To this end, Mumford characterize the development of the technological society into three phases: the *eotechnic*, *paleotechnic*, and *neotechnic* phases.

Mumford termed the technological complexes of the industrial revolution as the *paleotechnic phase*. He considered the railroad system to be the most efficient form of technics of the industrial complex produced during the paleotechnic phase. Inspired by the societal gamble, the scientists and inventors of the nineteenth century maintained the aim of achieving a more humane and rational society. The nineteenth century extended the Enlightenment effort to reorganize and rationalize scientifically every aspect of society, but due to the dominance of the paleotechnic phase, this was extension was directed in accordance with the capitalist drive for profits rather than the technological imperative. At the onset of the modern era, the entrepreneurial spirit of capitalism empowered the technological society in order to develop the industrial capacities for precision engineering and mass production, with the goal of maximizing the return on investment. The technological imperatives of maximizing efficiency and productivity and the capitalist economic imperative of maximizing profits were in phase, investors collaborated in supporting technological innovation and invention, which in turn fed their entrepreneurial zeal and provided them with increased wealth, but, when the capitalist economic imperative dominated technical rationality, while lagging behind the technological imperatives, investors withheld and suppressed technological innovation in order to maximize the profits on previous investments.[333] For example, in the eighteenth century, Benjamin Franklin proposed a method to utilize unburnt carbon in coal smoke by recycling and burning it a second time in the furnace. This method was never used. Steam power is highly inefficient. About ninety percent of the heat produced escapes in the steam and smoke. The steam engine was also extremely noisy. James Watt's efforts to improve the efficiency of the steam engine and reduce its noise were also dismissed by industrialists as being too expensive to implement. This is also apparent today in the dominance of the coal, oil, and automobile industries and the suppression of alternative modes of electricity generation (such as wind, water, and solar) and public transportation (as well as electric cars). Such industries have sufficient resources and influence over governments to prevent the investment in and development of sustainable and alternative technologies, at least until their current investments have maximized their profitability

---

[332] Mumford (1963)
[333] *Ibid* 167-8

and these industries can control and profit from those alternative technological innovations.[334]

For Mumford, the paleotechnic phase was truly dreadful. It caused war, slavery, oppression, squalor, fear, disability, illness, and premature death on a massive scale. It was even more capricious than the Nature it supposedly countered, while it opposed, dominated, and perverted the technological imperative upon which its wealth was made possible. It was the product of the dominance of the capitalist economic imperative to increase efficiency (understood in terms of maximizing profits by reducing costs) over the technological imperative to increase efficiency (understood in terms of optimizing productivity by reducing effort required to achieve the work). Due to the high level of skill of premodern craftsmen, their exchange-value was based purely on the use-value, rarity, or beauty of their products, whereas labor was a commodity in the personhood of the worker. In modern industrial society, directed towards the mechanization of skill and mass production, the craftsman continued to be a producer of luxury items and artworks for the wealthy, whereas the exchange-value of the worker became measured in terms of wages and time. Hence, even though the nineteenth century was essentially an intensification of the unity between natural science and technology, efforts to invent and implement more efficient means of burning coal and transforming steam power into mechanical power were ignored by industrialists; the industrial revolution was actually extremely inefficient because it squandered huge amounts of iron, coal, and human lives by continuing to implement and use inefficient technologies simply because they were more profitable in the short term. The environmental pollution and a deterioration of human health during the nineteenth century were an inevitable (and, arguably, widely known) consequence of this inefficiency. However, iron, coal, and human beings were cheap and abundant. Thus, they were quite expendable. The dictates of the capitalist drive to maximize profits had sway over the technological imperative of maximizing technical efficiency. The damaging effects that this policy had on human health and living conditions (including the illnesses and diseases caused by smog, pollution of air, soil, and rivers, overwork, low levels of sunlight, poor hygiene, overcrowding, the use of child labor, poor education, and the growth of slums) are historically documented as being commonplace during the nineteenth century. The destruction of varied human potential and social diversity—as well as forests and rivers—in order to maximize the profits and short-term gains of the industrial capitalists, offset the social value of the availability of cheap manufactured goods and the bulk transportation of commodities. Human beings were treated with the same disregard and exploitation as the natural world. They were simply a resource—a source of commodities—to be ordered, used, and discarded when exhausted, injured, or killed. The well-documented growth of child labor and the destruction of the craft base shows how skill levels and wages were reduced, while the length and intensity of working days increased, regardless of the social consequences, in order to reduce costs and maximize profits. The impoverishment and degradation of the factory workers was essential for the development of the whole basis of the industrial revolution as a capitalist revolution. Through the use of land monopoly laws and the destruction of traditional education for ordinary people, the industrial capitalists were able to nurture and propagate the foundations of industrial discipline, and establish for themselves monopoly control over production. The factory system was the culmination of this industrial discipline and social engineering. Human beings were transformed into functions and components and, after all alternatives had been destroyed (such as the crafts and rural agricultural base as modes of sustainable ways of life) and no other labor opportunities were present, these human beings were bound to machines to survive. Meanwhile, the paleotechnic society maintained and exploited the high levels of poverty,

---

[334] See Lovins (1979).

ignorance, and fear among the general population. The result of the Industrial Revolution was the creation and construction of an increasingly poor and restricted society for the majority of people, while a small minority grew extremely wealthy and powerful from the profits.

The paleotechnic phase was driven by an emergent capitalist economic imperative to maximize profits (by reducing costs), which transformed labor into a commodity, and utilized the machine as a mechanism through which the maximum surplus value could be extracted from human life. Out of sheer avarice and an overwhelming lust for power, combined with indifference (perhaps even contempt) for anything that could not be mechanized and utilized, the whole of social reality was transformed into a means of the production of wealth for a social elite—the owners of the means of production—while representing the sacrifices of the workers as being in inevitable consequences of social progress, even if any such progress was illusionary for those that were making the sacrifices. As Mumford put it,

"The starvation of diminution of life was universal: a certain dullness and irresponsiveness, in short, a state of partial anesthesia, became a condition of survival. At the very height of England's industrial squalor, when houses of the working classes were frequently built besides open sewers and when rows of them were being built back to back—at that very moment complacent scholars writing in the middle-class libraries could dwell upon the "filth" and "dirt" and "ignorance" of the Middle Ages, as compared with the enlightenment and cleanliness of their own." (Mumford, 1963: 181)

The industrial and commercial ambitions of the industrial capitalists required the representation of progress in terms of modern science, inventions, profits, power, machinery, luxuries, and comfort. These ideals were exported to other societies by means of trade and warfare, regardless of whether they actually benefited those societies, and were socially justified by allowing some of the benefits to the exploited, "underprivileged class"—as they were euphemistically called—provided that this was done prudently enough to keep the underprivileged class diligently at work in a state of passive and respectful submission.[335] The mechanization of labor, developed by engineers and inventors in order to increase the profits of their paymasters, soon became beyond the control and comprehension of the factory worker. The operation upon society of the emergent capitalist economic imperative reduced wages, lengthened working hours, deprived the workers with decent rest, recreation, and education, robbed children of a proper childhood and genuine opportunities for growth and self-determination, destroyed the centrality of family life, and left the worker ill, crippled, and impoverished in his old age (should s/he live that long).

"The paleotechnic period, we have noted, was marked by the reckless waste of resources. Hot in the pursuit of immediate profits, the new exploiters gave no heed to the environment around them, nor to further consequences of their actions on the morrow. 'What had posterity done for them?' In their haste, they over-reached themselves: they threw money into the rivers, let it escape in smoke in the air, handicapped themselves with their own litter and filth, prematurely exhausted the agricultural lands upon which they depended for food and fabrics." (Mumford, 1963: 255)

This inhumanity and irrationality was considered to be morally justifiable because of the unquestioned social acceptance that technological innovation based on the

---

[335] *Op cit.* 285

experimental sciences was the route that humanity could overcome the savagery, brutality, squalor, and ignorance of its organic state and achieve the Enlightenment ideal of building a world within which science and technology would inevitably lead to a rational and humane society. Despite all the suffering that the Industrial Revolution was evidently causing, it was accepted as self-evidently true that progress was confirmed by each and every scientific discovery and technological innovation. This belief in progress was the unquestioningly acceptance of the societal gamble. It did not matter whether any new scientific discovery or technological innovation actually brought real benefits to society, or even whether it was evidently quite disastrous, because it was represented and justified as a stepping stone towards the realization of the ideal technological society within which all human needs would be satisfied. In short, each and every scientific discovery and technological innovation was an *a priori* benefit to society simply because it had been invented, regardless of any immediate benefits or harms it brought with it.

On this basis, it simply did not matter that the industrial workers lived lives of drudgery as impoverished machine operators. It did not matter that the real quality of life of the workers of the Industrial Revolution was more impoverished than the medieval peasant. It did not matter that during the so-called Great War (1914-18) that the new weapons of the industrial age (bombs, poison gas, artillery, flame throwers, chemical explosives, and the machine gun) killed more people and caused greater devastation than a century of warfare did in the medieval period. The facts of social reality simply did not matter. Progress was represented as the self-evident consequence of these terrible facts, which were represented as being the side effects of progress, as if they were simply an awful part of the process of human scientific and technological evolution. This was the historic mission of the construction of the technological society and the self-evident truth of that ideological abstract took positivistic precedence over the facts of concrete reality. As Mumford succinctly put it,

"Life was judged by the extent to which it ministered to progress, progress was not judged by the extent to which it ministered to life. The last possibility would have been fatal to admit: it would have transported the problem from the cosmic plane to a human one." (Mumford, 1963: 85)

Due to its reliance on the societal gamble as its justification, the paleotechnic phase was a continuation of the sixteenth century, humanistic preparatory stage of the Scientific Revolution, which Mumford termed as the *eotechnic phase*, but was empowered by the continuation of premodern ambitions for the increased wealth and power of a social elite (often over each other as well as the workers). The scientific worldview, within which natural processes were represented in terms of mechanical efficiency, allowed the mechanical clock to represent the rational cosmos while ordering the working day upon the regimentation of lived-world temporality (in terms of the standardization, universalization, unification, and quantification of time). Time became a commodity. It became equated with money either in terms of return on investment or wages for labor. The use of the clock to mechanize and regiment the working day reduced human existence to sequences and timeserving activities.[336] Taylorism became possible. The mechanization of labor involved a dehumanization of the human being into sets and sequences of motions and the associated results—made formally "scientific" through the work of Frederick Taylor and the development of "scientific management" techniques—requiring a quantifiable and measurable form of social regimentation as its condition, appropriate to a technological framework within which that labor could be deskilled,

---

[336] See also Stiegler (1998)

controlled, and intensified in order to increase productivity while reducing wages. The capitalist economic imperative of maximizing profits concentrated masses of workers in large factories, mechanizing and standardizing the processes of work, and reducing the standard of living for the workers to being that of bare sustenance. The individual worker was forced to accept the conditions of mechanized labor, or starve, commit suicide, or become a criminal.

Hence, even though Mumford was critical of Marx's technological determinism, he agreed that capitalism had perverted technics by placing it under the direction of the capitalist economic imperative that acts upon the machines at its disposal and transforms their operation in accordance with the lust for profit.[337] For Marx, the mode of production is the dominant historical factor, and, hence, the hand-mill was bound up with selfdom in a feudal system and the steam mill with the wage-labor of industrial capitalism. According to Mumford (and Marx), the capitalist economic imperative opposes and perverts the emancipatory nature of technology, for the advantage of a few capitalists at the expense of many workers, and thus, if we agree with Mumford (and Marx) that the purpose of any machine is to liberate human beings from labor, capitalism placed the operation of the machine in contradiction to its essence by intensifying labor to reduce costs and increase profits. However, according to Marx, it is because capitalists (and their apologists) consider the essence of technology to be identical with its function within the capitalist production process—as a means to increase profits—that there cannot be any antagonism or contradiction between the nature of technology and the capitalist employment of it. While Marx's analysis shows the contradiction between the capitalist economic imperative and the technological imperative, for the capitalist, the technological society is the means to achieve the capitalist economic imperative and therefore there cannot be any such contradiction. As Marx put it,

"The contradictions and antagonisms inseparable from the capitalist employment of machinery, do not exist, say [capitalist economists], since they do not arise out of the machinery, as such, but out of its capitalist employment! Since therefore machinery, considered alone shortens that hours of labor, but, when in the service of capital, lengthens them; since in itself it lightens labor, but when employed by capital, heightens the intensity of labor; since in itself it increases the wealth of the producers, but in the hands of capital, makes them paupers – for all these reasons and others besides, says the bourgeoisie economist without more ado, it is clear as noonday that all these contradictions are mere semblance of reality, and that, as a matter of fact, they neither have an actual nor a theoretical existence." (Marx, 1967: 441)

Marx's scientific vision of a socialist utopia was profoundly influential on Mumford characterization of the neotechnic phase of the construction of the technological society in accordance with the scientific and technological imperative towards increased efficiency. Mumford described this as being a definite counter-march against the paleotechnic methods because the role of modern science in the rational development of efficient methods, rather than trial and error, is the crucial aspect that distinguishes between the neotechnic from the paleotechnic phase. The paleotechnic phase of the construction and development of the technological society occurs when the capitalist economic imperative dominates the technological imperative and the construction of social reality, whereas the neotechnic phase occurs when the technological imperative operates within the technological society without hindrance from the capitalist economic imperative (or any other influence, such as familial blood

---

[337] See also Mackenzie (1984); Winner (1977: 65-77)

lines, religion, tradition, etc.). Thus Mumford argued that the neotechnic phase began in the early nineteenth century with the scientific improvement of the efficiency of the water turbine, as well as the subsequent work on electromagnetism, thermodynamics, atomic power, and quantum physics. However, as Mumford pointed out, the neotechnic phase had yet to develop its own form and organization by the end of the 1920s, and, consequently, had not yet displaced the paleotechnic phase. Hence, according to Mumford, the modern world is yet to embrace the neotechnic phase and its social institutions lack the degree of adaptive and cooperative intelligence necessary for the fulfilment of the neotechnic phase of the technological society, even with lean technology and a scientific rationalization of supply, demand, and distribution. The neotechnic phase it itself based on the scientific worldview, bounded technical rationality, and the societal gamble, directed, according to the technological imperative, towards making the world universally better for all human beings by liberating us from scarcity and struggle. It does so only in so far as human beings are agents within the construction of the technological society, and hence a sociological phenomenon given over to neoliberal interpretation.

Mumford argued that, in order to develop genuine technical rationality, transforming the productive and technological basis of society from the paleotechnic into the neotechnic phase, doing away with the wasteful and retarding effects of profiteering, we must devalue the capitalist economic imperative and integrate the scientific processes of technological innovation within a rational re-evaluation of human needs, to assimilate and co-ordinate technology in accordance with human needs and abilities. Manifest both as Heidegger's *Ge-stell* and Ellul's *la Technique*, the technological imperative of the neotechnic phase extends itself to all aspects of life in order to improve them; it aims at producing better births and survival chances through technics (rather than simply increasing the number of births), providing better opportunities for health care, education, and nutrition, while removing illnesses, poverty, and the squalor of the paleotechnic phase.[338] While the pecuniary interests of the paleotechnic phase dominated the mechanization of labor in order to rationalize the production process (reduce costs), for the worker, labor has been irrationalized once it has been aesthetically reduced to a series of arbitrary procedures (that on an assembly line are dissociated from the final product) and arbitrarily quantified in terms of time spent for wages received. It is this reduction of labor to a system of exchange-value (labor for wages) that Marx termed as alienation of the workers from the products of their labor.

The irrational productive process is abstract and contrary to the interests and motivations of the workers, and, hence, workers need to be coerced (through threats of unemployment, homelessness, and starvation) in order to participate. Given that coerced workers are either indifferent or antagonistic to the production process—only being concerned with wages—then this state of affairs inevitably leads to a reduction of worker efficiency, when measured in terms of productivity, even if it permits a reduction of labor costs and is represented as more efficient in terms of the imperative to maximize profits. The reduction of the skill base of the general population to justify the reduction of labor costs and increase the expendability of each worker, as part of the overall strategy of coercing labor, reduces the total social capacity for technological innovation and development by reducing the level of participation in the envisioning of the trajectories of innovation and development to those envisioned by the owners of the means of production. Whether at the level of the individual factory or the whole of society, there is a contradiction between the capitalist economic imperative and the technological imperative. An essential characteristic of the neotechnic phase is that it aims at creating a whole aesthetics of the technological society—itself based on a vision of the rational as

---

[338] Mumford (1963: 263)

being that which aspires to realize the universal good—in order to improve the quality of life for all, by making the means of living life and overcoming problems more efficient in accordance with "the application" of science. During the neotechnic phase, labor relations are organized in accordance with the technological imperative to maximize efficiency—these relations are objectified and reified in terms of natural laws abstracted from the technological processes from which labor obtains its meaning as a rational activity. Such as society is one in which the operations of technology and labor are so seamlessly convergent towards automation and effortlessness that, as the asymptote of the completion of the technological society is approached, technology becomes increasingly totalitarian and ubiquitous, invisible and naturalized. As Mumford put it,

"Whereas the growth and multiplication of machines was a definite characteristic of the paleotechnic period, one may already say pretty confidently that the refinement, the diminution, and the partial elimination of the machine is a characteristic of the emerging neotechnic economy. The shrinkage of the machine to the provinces where its services are unique and indispensable is a necessary consequence of our better understanding of the machine itself and the world in which it functions." (Mumford, 1963: 258)

Due to the rise of consumerism and the middle-class in the industrialized world of the twentieth century, the class struggle within the so-called developed world has become transformed into a struggle against globalization within the so-called underdeveloped world. Globalization is a continuation of the methods of the paleotechnic phase given neoliberal representation in its development of neotechnics in the so-called developed world. Through the operations of globalization, the paleotechnic phase is as dominant in the modern world as it was when Mumford noted that society was still undergoing a period of transition between the paleotechnic and neotechnic phase of modern civilization. What we have witnessed is a transition from the exploitation of the proletariat in the developed world to the exploitation of the proletariat in the underdeveloped world. Not only has this allowed for capital flight and union busting in the developed world, thereby securing more power in the hands of the owners of the means of production and distribution, which has been a fundamental to the transformation from national industrial capitalism to international finance capitalism, but it has concealed the operations and consequences of paleotechnic industries (such as alienation, urban sprawls, pollution, poverty, exploitation, etc.) behind globalization, thereby allowing the developed world to present itself to itself under the thin veneer of neotechnic forms, such as lean supply-on-demand systems, online shopping, recycled packaging, sustainable electricity generation, and automated delivery systems, alongside the perpetual warfare industry, mass incarceration, and sweatshops and slave-labor factories, industrial farming, mass consumption and pollution, and the exploitation and devastation of the natural world.

Once the exploitation concealed behind globalization has been exposed, we can see that the paleotechnic phase has not only continued, but it has been intensified and extended worldwide into militaristic systems of oppression and imperialism designed to preserve the status quo of the use of paleotechnic methods in the Third World under the guise of the neotechnic aspirations of the developing world, and its ongoing construction in the developed world. This allows society to be represented as more progressive, efficient, and advanced, all on the basis of a highly inefficient and destructive globalized economic system, while it has increasingly directed the technological imperatives to construct a powerful war-machine and global system of dominance directed in accordance with the dictates of the capitalist economic imperative. This has been done for the benefit of an economic elite, which, through the private ownership of mass media and information technologies, has increasingly dominated the political superstructure of society by becoming the gatekeepers of the political process. The gap between the rich and poor has grown throughout the world, as increasing numbers of people face unemployment and homelessness in the developed world, or take low-paid work in the service sector, while workers in the developing and undeveloped worlds are compelled to work in industrialized factories, farms, and mines for their bare sustenance. Governments in Africa, Asia, and South America are used by multinational corporations to suppress environmental protections, unionization, and democracy movements in order to maintain low labor costs and supply cheap manufactured goods and technical services to the markets of the Europe and North America. As Mumford put it,

"In the persistence of paleotechnic practices the original anti-vital bias of the machine is evident: Bellicose, money-centered, life-curbing, we continue to worship the twin deities, Mammon and Moloch, to say nothing of the more abysmally savage tribal gods… The neotechnic refinement of the machine, without a coordinate development of higher social purposes, has only magnified the possibilities of depravity and barbarism." (Mumford, 1963: pp. 264-6)

Once we acknowledge that the paleotechnic phase dominates the politics and industry of the modern world, we can explain the relentless exploitation and oppression occurring around the world, leading to dictatorships, nationalistic and class-based struggles, and perpetual warfare. We can also recognize that the natural world is still being exploited to the point of total degradation without any regard for the long term consequences for life on Earth. As Mumford put it, "the paleotechnic remains a barbarizing influence. To deny this would be to cling to a fool's paradise."[339] If the dominance of the methods, tactics, and strategies of the paleotechnic phase are not supplanted then "the very basis of technics itself may be undermined, and our relapse into barbarism will go at a speed directly proportional to the complication and refinement of our present technological inheritance."[340]

While the economic and technological imperatives can be analyzed as distinct imperatives within society, each of which differently shapes the trajectories of the technological innovation and scientific research, and often in contradictory or conflicting ways, the history of the development of the modern world shows that economics cannot be divorced from science and technology, and that the scientific conception of the natural world, developed from the sixteenth century onwards was a preparatory stage for the development of both the capitalist economic imperative and the technological imperative. The form and content of modern society, with all its contradictions and conflicts, are emergent from the interaction between the capitalist economic imperative and the technological imperative, which become manifest as the outcomes of the dialectic between the paleotechnic and neotechnic phases of the construction of the technological society. This comes down to contradictions and conflicts in the visions of what a good society should be, and how the methods and policies to deal with those contradictions and conflicts are dealt with. When society allows those contradictions and conflicts into the open, to be dealt with and resolved by people in their own terms, we have an open and pluralistic society; when those contradictions and conflicts are suppressed and ignored, or only dealt with and resolved by an elite, we have a closed and totalitarian society. Mumford failed to recognize that the eotechnic phase—characterized by the Francis Bacon's dream of new discoveries and powers to liberate human beings from natural forces and material constraints—has been directed and developed in accordance

---

[339] *Ibid*: 213
[340] *Ibid* 211

with civic, commercial, and military ambitions from the outset. Modern science and technology were never developed in isolation from the obsession with increased wealth and power, and, therefore, the scientific worldview and the changing derivations of technical efficiency and aesthetics from its refinements have been implicated in the development of paleotechnic phase. While the neotechnic phase does prioritize the overall scientific development of society rather than using technology only to increase the wealth of a social elite, it does not have the pure connection with an eotechnic phase that Mumford considered the paleotechnic phase to have perverted. The paleotechnic phase was a development of the eotechnic phase, thereby serving the extension of civic, mercantile, and military power, and, the neotechnic phase is an eotechnic idealization of the paleotechnic phase that emerges when the thrust of development follows the technological imperative rather than the capitalist economic imperative. The neotechnic phase remains an idealization of the societal gamble—the construction of a perfect, technological society—but due to a lack of concrete independence from the content of the paleotechnic development of industry, science, and technology, it remains an imaginary vision of a future society—a projected form—that justifies the paleotechnic phase as being a necessary historical stage in its development. It is this kind of technological determinism that underwrites both Marxism and the most adamant apologists for capitalism alike.

Science, Technology, Capitalism and Marxism:

Marx proposed a vision of a socialist society within which, through technology, everyone would be able to satisfy all material needs, while being sufficiently liberated from work in order to realize the creative and free essence of human labor, education, and arts. This vision was based upon Marx's profound faith in the rationality of the societal gamble, and his conception of historical materialism presupposed technological determinism. It was this conception that would inevitably put the development of the technological society in contradiction with capitalism. It was inherent to Marx's philosophy that the relations and forces of production could oppose each other, and once the proletariat controlled the means of production—the machines at their disposal—the emancipatory power of the machines would emancipate the proletariat. In other words, left to its own devices, developed under the technological imperative, the technological society would be a liberating and egalitarian society that would use the machines at its disposal to reduce the working day, lighten the intensity of labor, increase the wealth of the workers, and enhance the technical efficiency of society (resulting in less waste, pollution, overproduction, and unemployment). For Marx, emancipation is an historical event (not a mental act) that is brought about by changes in material conditions through the development of science and technology. It was a central feature of Marx's philosophy that labor is the mediator between (historical, changing) society and the (ahistorical, unchanging) natural world (which forms a material substratum for society). He presupposed the validity of the precepts of mechanical realism and represented technology and labor as the material intermediaries between human beings and natural forces, and thereby he understood science as having revolutionary potential by bringing new productive forces into the world. For Marx, social reality is transformed by labor and new forces of production, but it is the possibility and nature of that transformation that is taken to be objective—thereby giving human relations concrete meaning as productive relations—and, hence, direct political theory towards the rise of class-consciousness as a species-consciousness of the revolutionary potential of the technological society to be an emancipatory society—liberating human beings from scarcity, struggle, and toil—and is the culmination of history (and human evolution).

Marx's theory assumed scientific realism in its formulation and characterization of dialectics in order to inherently presuppose the rationality of the societal gamble, as it shapes the form and content of the technological society. Dialectics cannot reveal anything final and absolute, but, rather, reveals the transitory and contradictory character of everything. Marx propose the dialectical method as being the only method by which it is possible to have knowledge of the totality, based on knowledge of the relations between all aspects of society and the society as a whole, as a historically progressive and evolving social reality based on struggle between Man and Nature, and between economic classes, to produce and reproduce the material infrastructure of society. Its form and content are changed and developed during historically situated struggles to overcome social contradictions—not just in thought, but also in practice—and these struggles shape science and technology, as well as being shaped by them. While Marx did accept that relations of production (which could oppose the forces of production, as discovered by contemporary sciences) were the (contingent) products of history and not (necessary) natural laws, he did accept the objectivity of natural laws and their effect on our organic and material conditions for existence. For Marx, human beings distinguish ourselves from other animals as soon as we begin to produce our means of subsistence, a step which is conditioned by our organic being, and, we are indirectly producing our mode of existence by producing our means of subsistence. The possibility of dialectics depends on our ability to expound the real process of production, starting out from the material conditions of life itself, and to comprehend the forms of social interaction connected with this and created by this mode of production.

For Marx, the whole world is represented as a natural, historical, and intellectual process that is in constant motion, change, transformation, and development, and dialectics is the attempt to trace out the internal connection that makes a continuous whole of all this movement and development. Dialectical materialism was represented as the basis of all history and the true mode of analysis required to explain all the different theoretical and social products and modes of thought, and the whole of history—as a process—could be depicted in its totality and therefore the interaction between various sides, aspects, and elements can be understood in relation to the whole. Marxism is based on the conviction that the dialectical method is the scientific means to achieve progressive truth about the material causes of social reality, only if it is developed and deepened by its advocates in relation to the overarching project of the conscious construction of a better (more rational, egalitarian) society through scientifically guided productive activity understood within historical and political context. The form and content of dialectical materialism presupposed the societal gamble, and therefore presupposes a faith in the final goodness of the technological society. This faith will be justified at the end of history—which is synonymous with the completion of the technological society. It is this conviction that locates Marx's theory—as the ideological expression of a method—within modernism, given that the objectivity of Nature is postulated as the ontological foundation for the historical and material development of society, providing humanity with the real possibility of change and mastery over the forces of production. It is directed in accordance with the scientifically founded construction of the totality of social reality: the construction of the universal socialist society emergent from industrialization and modernization of all aspects of human life. It is this goal that makes dialectical materialism bound-up with the project of constructing and perfecting the technological society—as a scientific and emancipatory post-scarcity society. History itself is represented as a unified process of the struggle for the development of the technological society; the completion and perfection of the technological society would be the end of history. Human evolution is represented via the scientific worldview within which human beings have the form of modern man, free creative labor is represented as the human essence, and the construction of the technological society is presented as realization of human emancipation from our material conditions. Marx considered natural science, as a theoretical and intellectual relationship between human beings and the natural world, to be internally bound together with practical activity, industry, and with the organization of labor, in order to disclose new forces of production and thereby transform the relations of production. But, according to Marx, due to the division of labor prevailing in class society, natural science is divorced from the material process of production and transformed into an abstraction of reality that was permeated with the ideology of capitalism. Marx believed in the objectivity of the technological framework, while being skeptical of the scientific worldview.

For Marx, the human, scientific understanding of Nature has always been based upon practical activity, while human beings are a part of Nature for whom their identity emerges from the scientific understanding of Nature and the technological environment that human beings are able to construct, within the limits proscribed by Nature. The material practices of previous generations create and transform the aim and means of scientific research, the instruments of observation, and the theoretical understanding of Nature, and the theories and practices of the natural sciences are transformed by the productive and social conditions from which they emerge, due to its historically developing relations with commerce, industry, and human sensual activity. Accordingly, Marx considered Newtonian mechanics as one of the conditions for the development of industrial capitalism. But the relations between science and society are historical in character and dependent on developments of production, which transform the social conditions (such as the class-based relations of production) that can assist or hinder the scientific development of production. While Marx considered natural science to be a transformative, revolutionary force, he was very much aware that the character and application of scientific methods and theories were historically shaped by their social and productive conditions. Thus, the conception of natural science and the scientific worldview are transformed with each and every major scientific discovery, and any real understanding of Nature is only possible on the basis of a developing understanding of the history of the development of human society, which arises from and forms a specific part of the natural world. Conversely, the understanding of political economy also calls for the study of natural science, as a condition of technical and economic development; as an essential condition for the development of productive power and efficiency. Marx presupposed the objective material reality explored by natural sciences, but took great effort to examine the dialectical materialistic process through which representations of that objective material reality are produced. Thus, even though Marx accepted that the natural sciences discover natural mechanisms through a dialectical process of relating and developing theoretical and material practices, it is inherent to his theory that the conception of mechanism (and not merely refinements of the representations of mechanisms) must change through this process, yet there is no evidence from his own writings that Marx was aware of this. Thus, even though specific theories and practices are transitory parts of a dialectical process, it is through the engagement with reality through material practices that provides science with its objectivity and meaning. The significance of the scientific establishment of objectivity through technological activity—and its implications for how Nature has been represented and understood—is simply not addressed and explored in Marx's writings on science and the nature of the natural world revealed by science. This is also apparent in Engels's *Dialectics of Nature* (1883). Marx's (and Engels's) conception of natural science was based upon an unreflective acceptance of mechanical realism.

With all their qualitative differences, for Marx, natural and social sciences are unified in that they both study the relations between theory and practice within a material world from which they both historically emerge and change. The ethical and aesthetic evolution of the consciousness of society is emergent as a unity between theory and practice, within the framework of the construction and perfection of the technological society developed in accordance with the technological imperative to maximize efficiency and productivity. Due to the underlying technological determinism and essentialism in Marx's writings, he simply assumed that once society and production have evolved to a sufficiently scientific level of development, science will be necessarily transformed from a condition of the exploitation of the proletariat into a condition for the emancipation of humanity as a whole. For Marx, the achievements of modern science have brought triumph to materialism, because they explain the objective laws, the material bases, the interconnection, the conditions, and causality of the material world. According to Marx, these achievements enlarge the theoretical basis of the practical activity of Man directed towards the mastery of the forces of Nature. The conditions for the emergence and liberation of the consciousness of the proletariat—as being both a class and the basis of all productive power—are identical with the conditions for the unfettered construction of the technological society. Hence it is implicit to Marx's own theory of revolution that the evolution of the class-consciousness of the proletariat is identical with the conscious development of the technological society—as a rational society—and the class struggle is the outcome and dialectical product of the contradiction between the capitalist economic imperative and the technological imperative within the societal construction of the technological society.

The collapse of feudalism and the emergence of capitalism was a consequence of the rising power of the Renaissance merchants.[341] The acquisitive ambition of mercantile commerce predates the development of industrial capitalism, which in its turn was only able to emerge because it was able to acquire modern technology to provide it with forces of mass production. Industrial capitalism emerged from the intersection between feudalism, mercantile commerce, and modern technology. By the beginning of the nineteenth century, capitalism was the driving force in the dissemination of the sciences and the construction of the technological society. The incentive to mechanize all aspects of production lay in the greater profits that could be found through the enhanced productive power and efficiency of modern technology. With the abolition of the feudal estates, capitalism could emerge as an economic imperative to accumulate capital and maximize profits, thereby imposing this imperative over the technological development of society. However, if the technological society is driven only by the technological imperative then profitability takes a backseat to technical efficiency. According to Marx, once the proletariat achieves class-consciousness—identifying itself as the productive power of society—then the demise of capitalism is inevitable once its basis for power turns against it. The workers' revolution is to be the spontaneous emergence of the awareness of the power of the proletariat—consciously self-identified as the advanced productive force of the technological society—and the obsolescence of the bourgeoisie and all class relations of ownership and control. According to Marxist theory, the emerging class-consciousness of the proletariat promises to liberate the technological imperative from the imposition of the limitations of the capitalist economic imperative. Hence, if we adopt Marx's technological essentialism, it is the emergent consequence of the aspirations of the eotechnic phase that have been perverted and jeopardized by the

---

[341] Marx (1967). See also Landes (1959).

dominance of capitalism, and this dominance resists the completion of the technological society as a post-scarcity society.

Looked at in this light, the Marxist post-revolutionary transition from a capitalist to socialist society is nothing less than the total transformation of society into the technological society. This transition involves the transformation of the whole of society into workers and consumers; thereby dissolving the proletariat as a class by transforming the whole of society into the proletariat. By dissolving all human relations into the conscious totality of the technological society, by reducing human agency to the organization of labor and the economic and technical relations of production and consumption, the Marxist idea of the socialist revolution is itself premised on the liberation of the technological society to become a totalitarian society—not to be misunderstood as a dictatorship in the common use of the term, but rather as a society in which any alternative is seen as irrational, reactionary, or counter-revolutionary. In this sense, "the dictatorship of the proletariat" is itself none other than a totalitarian society governed by the technological imperative. Thus the proletariat and techniques are the historical objects that provide the content of the technological society, and the technological imperative drives its dialectical construction. When this has achieved a class-consciousness of itself as a technological society, this content and form become the ontology upon which that society and the understanding of its teleological positings (vision, norms, and ideals) are founded and pursued.

It is implicit to Marxism (due to its uncritical acceptance of the rationality of the societal gamble) that the development of socialism is identical with the total integration of all relations into the teleological positing of the construction of the technological society—as a totalitarian, scientific society—free from the contradictions between the capitalist economic imperative to maximize profits (through economic leverage and by lowering labor costs) and the technological imperatives to maximize efficiency (itself understood as the maximization of productivity for the least labor). It is not only the case, within Marxist theory, that the essence of human being is free creative labor, but is therefore absolutely imperative for the possibility of socialism that the teleology of human being and the technological activity converge to the point of identity. In the technological society, intentionality and functionality are the same because human agency is identical to technological activity. Thus the abolishment of alienation is the conscious submission of the individual to the totality of the ontology of society, the form and content of which is the totality of the integrated productive and creative relations between human beings, and its vision of itself as an egalitarian, post-scarcity society. Labor becomes a social category in which material dialectics becomes defined in terms of experimentation, innovation, and struggle for the emancipation of labor as a free, creative act that is performed for its utility within the social totality. The technological society totalizes all being and becoming through the social category of labor to order everything within the teleological construction of itself. Thus, through material dialectics, labor becomes the subject-object of experience to such an extent that dialectics is manifest as the drive to the liberation, formulation, and perfection of itself as a universal method, as an expression of the liberation, formulation, and perfection of society, as a technological society.

Hence we can see that the Marxist definition of history as being the history of an unceasing overthrow of the objective realities (struggle against our material conditions) that shape human life is itself based on an interpretation of modern history that is premised upon the societal gamble on the rationality of the technological society, as being the means by which that overthrow could be possible and therefore the foundation upon which a post-scarcity society could be constructed. It is also based upon a conflation between historical and technological determinism in which the future of humanity is seen to be synonymous with technological progress in all areas of human activity and relations. Socialism, as envisioned by Marxists, is the final culmination of the construction of the technological society in which both the revolution and "dictatorship of the proletariat" involve the extension of the societal gamble and the technological imperative into the totality of all human relations, with automation and absolute leisure being the asymptotic ideals on the envisioned horizon of the end of history: the completion of a post-scarcity society and such a society would bring liberation from struggle and inequality. The overcoming of alienation—the complete integration of all human subjects into the objective structures of the technological society—will dissolve individualism by making it obsolete by reducing society to a form in which individualism either has no (rational) meaning or is simply a form of personal retardation (a remnant from the old, pre-revolutionary, class-based society).

Due to the dictates of the capitalist economic imperative upon the concrete and practical implementation of the technological imperative, capitalist economic aspirations have a transformative effect on both the form and content of society, while being empowered by its application. All at stages of its construction, since the seventeenth century, the technological society has been ontologically based on the development of industrial capitalism, and it is for this reason that the neotechnic phase did not come to dominate over the paleotechnic phase, which continues today as relentlessly as it did during the nineteenth century. Indeed, while the dominant tendency is towards the construction of the technological society, dominating the institutional arrangements of the State, transforming them into technocratic systems of public administration and regulation, but the inefficiencies of human individualism and a "laissez-faire spirit" tends to oppose this dominant tendency. However, as Ellul noted, we cannot rely on any allusions to the "free-market" and the "laissez-faire" pretensions of liberal capitalism because capitalism has become eclipsed by the efficiency of technical methods and the rationalization of systems of production, distribution, advertising, and consumption.[342] Liberal capitalism has become transformed into industrial capitalism, and even when it distorts the technological imperative, say by maintaining the use of inefficient technologies by suppressing patents or building in obsolescence, with all the concomitant waste and unsustainable consumptive patterns, this is only a corruption of how the technological imperative is implemented, rather than a genuine and democratic alternative or opposition to it.[343] The patterns of production and consumption generated by industrial capitalism (with its waste, pollution, and social costs) are inefficient and exacerbate the problems of scarcity, but they are still constrained and directed in accordance with the technological imperative towards the construction of the technological society. As a result, Ellul interpreted the Marxist critique of the inefficiencies of capitalism and its contradictions with the technological imperative as being based on the presupposition that the technological society is essentially good and emancipatory, and, therefore, requires the liberation of the development of the technological society from the inefficiency and inequalities of liberal capitalism and the intense exploitation and waste caused by industrial capitalism. Hence Ellul considered the Marxist vision of communism to be one orientated towards technical progress, in all areas, giving free play to all technical automatism in every field of human activity, in order to achieve maximize efficiency, and, thereby, its fundamental result would be the construction of a totalitarian technological society.

---

[342] Ellul (1964: 184)
[343] See also Ellul (1990); Rogers (2008)

Marx's conception of "the dictatorship of the proletariat" in a post-capitalist society was not that of a coercive state apparatus, governing as something separate and above the proletariat. It was supposed to the democratic expression of the consensus of the proletariat—as producers of societal wealth—as a free and equal association of disciplined industrial workers who would collectively determine on the basis of technically rational decisions the division of labor, the forces and relations of production, and how to satisfy social needs.[344] Hence, Marx's theory does not provide any detailed discussion of how dissent and critical discussion should relate to the deliberation and decision-making processes of the majority; nor does it provide any theoretical basis for defending minorities (including ethnic, religious, and gender difference) and the rights of individual citizens. It was not simply the case that Marx considered these to be bourgeois ideas, but rather he considered them irrelevant, given that in "the dictatorship of the proletariat" only the question of how best to satisfy the demands of technical rationality and increase productivity would form the rational basis for deliberation and decision-making. Such a society would be totalitarian because dissent and criticism of the technological imperative would be irrational and have nothing to contribute to progressive societal development; divergence from the technical imperative being represented as the product of backwardness and ignorance of the objective conditions for societal progress. Within any such society, it would become quite impossible to place any concerns into the public realm that did not deal with technological development or the relations of production and consumption, such as ecological or moral concerns, that did not impact on the material conditions of human existence. Only science and technology would provide the conditions for the rational development of socialism into a communist society, and, in this respect, Marx's vision of post-capitalist socialism was very different from the "actually existing socialism" of the Soviet Union.

According to Marx's theory, Russia was too industrially backward for a socialist revolution and the construction of the post-capitalist state. The Russian situation actually led Marx to reconsider his theory of the conditions for a socialist revolution and the construction of a post-capitalist state to also include predominantly agricultural economies. Inspired by Georgy Plekhanov, Marx considered the Russian *mir* (traditional village based commune) to be the basic social unit for communal living, collective agriculture, and the fulcrum for social regeneration in Russia.[345] With the aid of modern technology, such as agricultural machinery and chemical fertilizers, the *mir* could provide the foundational social unit for a socialist revolution in Russia, without needing to develop an advanced and disciplined industrial proletariat through capitalism first. A similar theory was proposed in 1850 by the nineteenth century anarchist Alexander Herzen.[346] In some respects, this path was taken when Lenin and the Bolsheviks adopted the New Economic Policy in 1921 and permitted the development of an agricultural cooperative economy based around the *mir* and a mixed urban economy until Lenin's death in 1924. It was only after its architects, such as Nikolai Bukharin, had been undermined by Trotsky and the Left Opposition and the poor harvest in 1929, that the conditions were right for Stalin to impose centralized collectivization.

The Soviet Union was the product of a series of distortions and suppressions of Marx's theory, which created an authoritarian and centralized bureaucratic administration of state-controlled capitalism that had very little resemblance to Marx's vision of communism. It also distorted and suppressed the technical and scientific conditions upon which Marx's vision of "the dictatorship of the proletariat" and the transition from

---

[344] Marx (1938; 1967, chap. 1, sec. 4; 1992)
[345] Shanin (1983)
[346] Herzen (1956)

capitalism to communism depended.[347] The authoritarian tendency was apparent in Lenin and the Bolsheviks from the outset, which despite all of Lenin's rhetoric in *State and Revolution*, opposed any form of workers' control over the means of production. This was pointed out at the time.[348] If one compares Lenin's *State and Revolution*, with his previous writings in the *April Theses*, and his subsequent writings in *The Immediate Tasks of the Soviet Government*, published after the Bolshevik seizure of power, one can see that, apart from in *State and Revolution*, Lenin had little tolerance for worker's control, largely considering it to be anarchism. It is arguably the case that Lenin's "libertarian" and "democratic" views expressed in *State and Revolution* were designed to win over the workers' soviets to the Bolsheviks prior to the seizure of power. The fundamental ideology of the Bolsheviks after seizing power opposed workers' control did not deviate from that expressed in Lenin's *What Is To Be Done?*, published in 1901, wherein he advocated nationalized and state-controlled industrial monopolies for the benefit of the people. From the moment they seized power in October 1917, Lenin and the Bolsheviks (acting as a minority with the support of the Red Army) imposed their vision of socialism upon the majority in a way that required a series of radical distortions of, and deviations from, Marx's theory.[349] The Bolshevik government—the Sovnarkom—refused to negotiate with and consult with the Soviets from the outset.

The Petrograd factory workers were a minority, even in Petrograd, and many of them disagreed with the Bolsheviks (many of whom also disagreed with Lenin's call for a seizure of power and one-party rule). The workers' soviets and unions only sought workers' control over the factories—or simply wanted better wages and working conditions—and did not seek control over the State.[350] The Petrograd soviets only supported the Bolsheviks' seizure of power after it was a *fait accompli*. During her travels around post-revolutionary Russia, Louise Bryant observed that there were few supporters for the Bolsheviks among the peasantry (which comprised about 80% of the population), who were largely disappointed by the Kerensky Provisional Government's failure to implement the promised land-reforms and predominantly supported the Left Socialist Revolutionary Party rather than the Bolsheviks.[351] Even John Reed, who at the time of writing *Ten Days that Shook the World* (first published in 1919) was both sympathetic and optimistic about the Bolsheviks as a democratic movement for genuine social revolution and justice, observed that the Bolshevik seizure of power was widely resisted and opposed among workers and peasants.[352] The Bolshevik seizure of power was not the result of a workers' uprising, but the workers' support for it was based on the need for a stable and practical government to deal with the political and economic circumstances created by the Kerensky Provisional Government's failures, including its violent suppression of the Bolsheviks, radical socialists, and the workers' soviets since April; its refusal to fulfil its promises to create a democratic Constitutional Assembly wherein the workers' soviets would share power with other democratic groups, such as political parties, municipal councils, trade unions, and the delegates from peasants' cooperatives; and, the deprivations caused by Russia's involvement in the First World War.

Due to the absence of a disciplined proletariat and advanced means of production, the application of Marx's theory of socialism was contingent upon

---

[347] Joravsky (1961, 1970); Zaleski (1969); Bailes (1978); Graham (1988); Gorokhov (1992), Roll-Hansen (2005)

[348] Luxemburg (1918); Rhys Williams (1921); Goldman (1923)

[349] Carr (1966); Althusser (1971); Lewin (1975a); Liebman (1975); Bellis (1979); Fitzpatrick (1994); Gooding (2002)

[350] Mandel (1984)

[351] Bryant (1919)

[352] Reed (1966)

industrialization, as was also the capacity of the Soviet Union to defend itself from military invasions or threats by capitalist countries; this entailed the centralized planning of industry and agriculture in order to force the pace of industrialization to brutal extremes.[353] The growth of centralized, bureaucratic administration and a militaristic, institutionalized intolerance of dissent were quite inevitable under such circumstances and, therefore, Stalinism was not quite the aberration of Leninism that some Marxists would prefer to claim. Marcuse's critical analysis of the Soviet Union highlighted several contradictory tendencies and a fundamental ambivalence inherent to the attempt to realise Marxism—as an emancipatory and progressive aspiration for human freedom, happiness, and social justice—through suppressive and oppressive means that failed to realise any of the Marxist ideals in practice.[354] Marcuse saw the policies, apparatus, and operations involved in the construction of the Soviet Union as being essentially a series of responses to the West. He identified "objective tendencies" that potentially could result in top-down liberating reforms as the Soviet Union developed into a technological society, and claimed that these "objective tendencies" were apparent in Khrushchev's reforms, but Marcuse argued that these tendencies were suppressed by the continued Stalinist 'hard-liner' objective of strengthening the State in response to the aggressive and interventionist policies of the Kennedy Administration.

Marxist theoretical efforts to synthesize class-consciousness for the technological society—in order to connect the political form and technical content of society in the person of the proletariat—ended up with the industrialization of the Soviet Union and China, which was little more than an intensification of the paleotechnic phase. This was no historical accident—supposedly due to Lenin and Mao's contingent problem of how to implement Marxism in a predominantly peasant-based society—but is inherent to Marx's theory of society. Marxist theory was proposed as the ideological means (of "consciousness raising") to make the proletariat conscious of the extent that they are a unified, exploited class and that capitalism is by no means a natural and inevitable system for society. But, due to the technological determinism inherent to Marxism, the capitalist accumulation of capital and the development of industry based on mass production technologies were represented as a necessary historical stage towards the development of socialism. Thus, in peasant-based societies, the post-revolutionary society was forced (via Party decrees, five-year plans, agricultural collectivization, bureaucratic centralization, Gulags, torture chambers, re-education centers, and a police-state apparatus, etc.) to conform to the paleotechnic phase of industrialization, with their suffering interpreted as sacrifices for the construction of a future post-scarcity society within which people struggle no more—the End of History, in both the sense of terminus and goal. The Soviet Union constructed the technological society with greater intensity and scale than the nineteenth century Industrial Revolution had done in Europe. This was not only seen as "catching up" to the West, but was taken to be fundamental to the transition to socialism. Ironically, Marxists (and Leninists and Maoists) have represented capitalism as a necessary link in the chain of human development, while justifying itself—and its own systems of exploitation—in terms of a form of socialism that has the content of monopolistic state-capitalism based on nationalized industries, wherein the workers are employees of the State, and the State controls wages, prices, and the supply of products.

Even though Marx challenged the idea that the form of society had a "natural" basis, he uncritically accepted the mechanical realist representation of the foundation of technology and presupposed that technology has a neutral and autonomous essence. Thus, while Marx criticized the "naturalistic" interpretation of the social sciences, he

---

[353] Lewin (1975); Davis (1980); Sutela (1991)
[354] Marcuse (1958). See also Kellner (1984)

accepted that the experimental sciences were, in fact, natural sciences that bring new forces of production into the world and provide us with objective facts about the material substratum upon which society is built. Marx's interpretation of Nature represents the operations of technology as being the process of exploiting the complex of interacting mechanisms at our disposal, treating Nature as something at our disposal, as a resource, and nothing else besides, available to labor as the means to change the conditions of human existence. Due to Marx's unquestioning acceptance of mechanical realism and the societal gamble, his vision of a post-revolutionary society presupposed the construction of the technological society as the route to communism, and therefore entailed reproducing the very structures of domination and inequality that he tried to expose and overcome. Hence, in his 1859 *A Contribution to the Critique of Political Economy*, Marx wrote:

"In the social production of their life, men enter into definite relations that are indispensable and independent of their will, relations of production which correspond to a definite stage of development of their material productive forces. The sum total of these relations of production constitutes the economic structure of society, the real foundation, on which rises a legal and political superstructure and to which correspond definite forms of social consciousness. The mode of production of material life conditions the social, political and intellectual life process in general. It is not the consciousness of men that determines their being, but, on the contrary, their social being that determines their consciousness. At a certain stage of their development, the material productive forces of society come in conflict with the existing relations of production, or—what is but a legal expression for the same thing—with the property relations within which they have been at work hitherto. From forms of development of the productive forces these relations turn into their fetters. Then begins an epoch of social revolution. With the change of the economic foundation the entire immense superstructure is more or less rapidly transformed. In considering such transformations a distinction should always be made between the material transformation of the economic conditions of production, which can be determined with the precision of natural science, and the legal, political, religious, aesthetic or philosophic—in short, ideological forms in which men become conscious of this conflict and fight it out. Just as our opinion of an individual is not based on what he thinks of himself, so can we not judge of such a period of transformation by its own consciousness; on the contrary this consciousness must be explained rather from the contradictions of material life, from the existing conflict between the social productive forces and the relations of production. No social order ever perishes before all the productive forces for which there is room in it have developed; and new, higher relations of production never appear before the material conditions of their existence have matured in the womb of the old society itself. Therefore mankind always sets itself only such tasks as it can solve; since looking at the matter more closely, it will always be found that the task itself arises only when the material conditions for its solution already exist or are at least in the process of formation." (Marx, 1904: 1)

Marx presupposed that the ontological foundation of technical rationality is transformed into a social reality from a natural reality (as revealed by the natural and experimental sciences) and, thus, presupposed that the material foundation of the technological society is based upon the rational utilization of natural forces—hence, Marxism fails to recognize and analyze the substantive capitalistic basis for the structures and trajectories of technical rationality, except in terms of its bearing on economic and productive reality, which is represented as a perversion or distortion of the emancipatory essence of technology. By understanding how scientific realism underwrote Marx's political theory of the transition from capitalism to communism, we can see that, in the *Dialectics of Nature*, Engels did not extend Marx's theory too far beyond its self-prescribed limits as a method for understanding history and changing society, when he assembled arguments to show how the dialectical method is confirmed by the natural sciences as a method to understand and manipulate human and natural processes alike, and therefore would form the intellectual foundation of a scientific society. Lenin did not stretch Marx and Engels too far when he said in 1920 that

"We must show the peasants that the organization of industry on the basis of modern, advanced technology, on electrification, which will provide a link between town and country, will put an end to the division between town and country, will make it possible to raise the level of culture in the countryside and to overcome, even in the most remote corners of land, backwardness, ignorance, poverty, disease, and barbarism." (Lenin, 1972: Vol. 30, p. 335)

In fact, technological determinism so underwrote the Bolshevik vision of how to transform "backward" Russia, that no-one questioned his commitment to Marxism when Lenin declared that: "Communism is Soviet power plus the electrification of the entire country."[355]

Advancement, for the Bolsheviks, was clearly a matter of reproducing the technologies of the West and surpassing them along the route towards creating a post-scarcity scientific and technological society within which struggle, toil, and inequality would be things of the past.[356] This belief was central to both the *Pravda* propaganda and serious philosophical writings of Nikolai Bukharin.[357] Leon Trotsky stated that human beings should use science and technology to rearrange mountains and rivers, to improve on Nature and transform the world.[358] And, for Stalin, scientific realism and technological determinism were taken for granted in his 'official' interpretation of the dialectical method and historical materialism, and the transition from capitalism to communism.[359] Soviet Marxism was emergent as a political form of technological determinism—with the assumption being that control over the means of production would change the mode of production—postulated as an effort to unify the consciousness of the productive sector of society (the proletariat) in order to advance the conscious and rational construction the technological society.

Marx was fully aware that the activities of capitalists are not just simply parts of an economic process of maximizing surplus value, but are perpetually driven to reproduce the class structure.[360] The capitalist economy is a consciously driven social process that is directed to maintaining the autonomy of the capitalists as a social elite, directed to perpetuate the proletariat as the instrument of capital, as well reproduce as the proletariat acceptance of capitalist propaganda regarding the intellectual, moral, and natural superiority of the capitalists. An analysis of the class relations between the owners of production and the exploited workers, who actually produce the wealth and productive power, is central to any rational and moral development of science and technology to the extent that the resolution of this inequality is a precondition of rationally and morally using science and technology to solve human problems in a way that equally benefits all humanity. Yet, by treating the societal gamble as if it was a manifest destiny for humanity as a species, Marx failed to recognize that the ontological basis the technological society

---

[355] Lenin (1972: Vol.31, p. 516)
[356] Joravsky (1961); Graham (1998)
[357] Bukharin (1921; 1925; 1931)
[358] Trotsky (1960: 251-3)
[359] Stalin (1938)
[360] Marx (1967: 578)

is artificial and contingent upon the same exploitative and totalitarian foundations as industrial capitalism. The implementation of these technologies simply reproduced the same hierarchical structures of management (except the economic elite was replaced by a political elite), along with the intensification of labor, pollution, squalor, ill-health, and the same problems of industrialization experienced by workers in capitalist societies. Yet, supposedly, once the domination of the technological society by the capitalist economic imperative was removed, the proletariat would ontologically remain identical with the technological society. The "dictatorship of the proletariat" would be nothing less than the unfettered development of the technological society in accordance with the technological imperative. So, if this true, why is the project of constructing socialism through industrialization doomed to repeat the same inequalities and dominating structures as the capitalist societies that it was supposed to replace?

Regardless of any genuinely democratic aspects of "the Heroic Period," the Soviet Union was doomed to repeat and intensify the oppressiveness of the paleotechnic aspect of the technological society, once, after the Civil War, under the dictates of Stalin, it moved into the centralized program of agricultural collectivization to fund industrialization by exporting the collectivized grain to the West. The "dictatorship of the proletariat" became a repetition and intensification of the capitalist industrial-military complex, except in the Soviet Union it became a system of state-capitalism, which repeated and intensified all the injustice, horrors, brutality, and cruelties of nineteenth century industrialization. This occurred alongside Stalin's betrayal of the ideals of the revolution, via his extension of Lenin's *Cheka*, the Red Terror, and special prisons and the construction of a corrupt, terroristic and autocratic police-state that silenced or murdered all dissenting voices. The bureaucracy, special prisons, purges, secret police, Gulags, forced marches, and famines were not only the means by which Stalin maintained and intensified his dictatorship, but they were consequences of the forced implementation of centralized industrialization on a predominantly agricultural, peasant-based society in accordance with the dictates of the technological imperative. This was repeated by Mao and the Chinese Communist Party, with exactly the same results, when centralized industrialization was imposed on the predominantly agricultural China, and arguably similar but smaller-scale efforts at industrialization in Cuba also were a repeat of the same paleotechnic expansion as an expression of socialist reform. Operating under the same abstract and reified relations between human beings, within which the meaning and value of any human relation is circumscribed in accordance to estimations of "efficiency," themselves defined in terms of the successful implementation of natural mechanisms discovered by the experimental sciences, the totalitarian ideal of the technological society is for it to order all of its mechanisms, components, and resources (including human beings and the natural world) and coherently integrate them into a complete and unified automaton capable of satisfying all human needs.

Thus the dialectical relationship between necessity and freedom, in opposition to and defined by an objective natural state of being, would remain represented as something quantifiable and mechanical in terms of increased productivity, efficiency, and power. Productive forces and labor converge as they dialectically transform one into a manifestation of the other, and this convergence establishes the ontological basis of the technological society. However, the operation of the capitalist economic imperative (regardless of whether it operates in a system of private capitalism or state-capitalism) disrupts that convergence. This disruption is concealed by representations of the dialectic of technology as being governed by a "natural" master-slave dialectic within which the efficient is defined in terms of reduced costs and increased productivity, for the benefit of an elite acting on behalf of humanity, rather than the scientific application of natural mechanisms to transform and improve the aesthetics of labor. Thus efficiency in any capitalist system (regardless of whether it is laissez-faire or state-controlled) must be defined by the streamlined reduction of labor to an increasingly inexpensive commodity, rather than the empowerment of labor as an essential human activity.

In many respects, with the hindsight of history, due to the technological determinism that underwrote both capitalist and Marxist economic theories, some degree of fatalism, equivocation, and recapitulation regarding the power of capitalism should be expected since the collapse of the Soviet Union. However, the view that capitalism is the unlimited and invincible victor of the twentieth century is one that blindly fails to recognize and attend to the environmental unsustainability and social destructiveness of unchecked capitalism. The failure of socialism to establish itself as a genuine alternative to capitalism does not undermine Marx's (and Mumford's and Ellul's) criticisms of the inefficiencies and injustices of the capitalist society that the socialist vision was supposed to replace. As Langdon Winner pointed out: "The fact that various communist and socialist alternatives have now fallen into disfavor has not eliminated the fundamental problems that gave rise to them."[361] Even though the Industrial Revolution was premised upon the promise to produce goods for masses of people, which previously were available only for monarchs, the feudal nobility, and wealthy merchants, while improvements in technology promised to eliminate the barriers to universally shared wealth, it is evident from the position of historical hindsight that industrialization produced greater inequality and concentration of wealth and power in the hands of a social elite, regardless of whether that elite was an economic or political elite.

Herein lies source of the contradiction between capitalism and the technological society. Industrial capitalism is founded upon the technological society, but the capitalist economic imperative to maximize profits and the technological imperative to maximize efficiency only ran in phase during the late eighteenth and early nineteenth century. During this period, the economic imperative empowered the technological imperative, thereby creating mass production and transportation technologies, which in turn provided the means to reduce costs and increase profits. However, from the mid nineteenth century onwards, due to the scientific development of the technological society, the technological imperatives began to outpace the capitalist economic imperatives in terms of both efficiency and productivity. Once it became possible to provide longer lasting and precise manufactured goods at an increasingly lower cost—perhaps finding its culmination in Tesla's promise of free electricity for all humanity and J.P. Morgan's use of his influence to suppress and silence Tesla—the capitalist economic imperative required artificial reductions of quality, in-built obsolescence, inflations of price, patent control and suppression, waste, and the use of antiquated industrial processes in order to maximize profits.[362] The divergence between these imperatives became the contradiction at the heart of the modern world which traded-off universal technological progress for increased profits for a few. It is the failure of the capitalist economic system to integrate and equate the economic imperative with the technological imperative, and thereby realizing that harmonious neotechnic totality in terms of a moral imperative, which has been its ideological failure. The consequent failure of capitalism to satisfy the societal gamble by constructing a world that is the best of all possible worlds has resulted in an arbitrary system of oppression and exploitation for the benefit of a few. It is due to this failure that ecological collapse, perpetual warfare, and the possibility of a

---

[361] Winner (1992: 3)
[362] Seifer (2001)

descent into a new and terrible form of barbarism looms over the world in the wake of an unstoppable and unsustainable juggernaut.

One-Dimensional Man and the Societal Gamble:

Herbert Marcuse was profoundly influenced by Heidegger's philosophy, as well as the theories of Marx and Freud. Marcuse was concerned with the extent that the dominance of the technological imperative is leading to the construction of an irrational society based on structures of domination and conformity rather than free relations developed through critical reasoning. Marcuse presupposed the ideals and vision of a free and rational society articulated in *Reason and Revolution* (first published in 1941) and *Eros and Civilization* (first published in 1955).[363] Marcuse was concerned with the extent that instrumental reason has dominated human thinking by repressing any aspirations, ideas, or values that cannot be submitted to instrumental reason. He argued that the human capacity for individuality, spontaneity, and critical reasoning are suppressed by the technological imperative's drive to pacify existence and create an advanced state of social cohesion and conformity. He was concerned with restoring genuine individuality, spontaneity, and critical reasoning in the face of the totalitarian nature of the technological society. In *One-Dimensional Man* (first published in 1964), Marcuse presented a social critique of the structures of contemporary industrial society as a dominating society that represses human individuality and any ideals or values that cannot be defined in terms of the narrow bounds of technical rationality.[364] In response to his critics, Marcuse modified and developed his description and analysis of one-dimensional society in his later works *An Essay on Liberation* (1969) and *Counterrevolution and Revolt* (1972), but his basic position against the technological society and other social forms of domination and conformity remained the same until his death in 1979.

Marcuse defined technology in broad terms and considered the technological imperative to be a profound threat to human individuality and freedom. He described modern society as a technological society within which production, labor, consumption, leisure, culture, and thought, are integrated into the dominating structures of that society in order to produce an advanced state of conformity. All aspects of human life, including culture, thought, leisure, labor, relations of production and consumption, the identification and satisfaction of individual and social needs, and the interactions between the public and the private realms, are reduced to forms that can be integrated into the technological society in accordance with the constraints and demands of the technological imperative. Only those thoughts and activities that conform to these constraints and demands are included and empowered; opposing or alternative thoughts and actions are repressed or suppressed as being "irrational," "utopian," "naïve," or "impractical." Like Ellul, he considered this technological society to be a profound threat to human individuality and liberation. Also, like Ellul, he considered technology in terms of a sociological definition that did not define it merely as the sum-total of tools, instruments, and material practices, and which cannot be isolated from its social and political contexts of emergence. Instead, he defined it in terms of a social system that, as a totality, determines the productive activity, needs, and aspirations of the individual, as well as the socially needed occupations, skills, attitudes, and the operations of servicing and extending machines. Technology cannot be isolated from the uses that it is put to. It mediates the thought and action of the individual in the technological society, negating

---

[363] Marcuse (1974; 1999)
[364] Marcuse (1991)

the distinction between public and private life, between individual and social needs, and institutes an effective form of social control and cohesion through conformity. Thoughts and actions that conform to the operations of the technological society are brought together and empowered, whereas, oppositional actions are prevented and corresponding thoughts are left speculative and impotent. This threatens to construct a totalitarian society within which the operations of a system of domination pervades and mediates each and every thought and action.

In advanced industrial societies, the political realm is reduced to a technocratic system of institutional arrangements to control economic and societal development, where ideology is little more than a propaganda tool for legitimating this system, which is itself subordinated to the latest techniques of propaganda. The technological imperative reduces every policy and legislation to a series of technical refinements and innovations of the technocratic system, which provides empowering mechanisms only to those who conform to it, and further reinforce its power and legitimacy by doing so. Every qualitative social change is either reduced to a quantitative refinement of the system or remains impotent and imaginary. Hence, as Marcuse argued, the totalitarianism of the technological society is not necessarily enforced by any tyrant or police state, but by the overwhelming and anonymous technological power and efficiency of that society. The use of coercion and police state tactics are only necessary when the technocratic system is incomplete, or breaks down, and people resist the technological imperative to construct the technological society. In capitalist societies, individuals are increasingly integrated into an economic system that demands the total accommodation and submission of all human beings into that system by empowering those human beings who conform to the system and disempowering those who refuse. Unemployment, homelessness, and starvation become mechanisms for social control. This leads to a selection mechanism in favor of those individuals who conform to the expectations, norms, and practices of the current relations of production and consumption. This also reinforces the autonomy and legitimacy of the technocratic structures of planning and management, by normalizing whatever "the best" technical strategy and method happens to be, and eliminates any genuine possibility of radical social change or the development of practical alternatives. Once alternatives become impossible then society becomes "one-dimensional," according to Marcuse, and closes itself off from the possibility of the liberation of human beings and the rational development of the means to alleviate scarcity and need.

Technical progress may well be defined in terms of the historical development and differentiation of the means for achieving "the scientific conquest of Nature," understood as "the pacification of existence," but, rather than achieving "freedom from toil and domination" this has led to "the scientific conquest of man."[365] Once human beings have conformed to the technological imperative, we become resources and objects within the technological framework available for administrative and managerial calculation, manipulation, and control; selected and ordered in accordance to with the latest techniques and technologies of administration and management. Once our individuality is suppressed, we become alienated from our essence—our free and creative subjectivity—as beings capable of transforming our own existence. We cease to strive for our potential to live authentic (self-determined) economic and social lives. We not only cease being able to transform the conditions of our existence, but, also, we cease to comprehend it, and, therefore, the technological imperative drives us further into powerlessness, compulsion, conformity, and incomprehensibility—all of which intensify the irrationality of human relations with technology and each other—and leads to the suppression and repression of the potential for genuine social change and human

---

[365] Marcuse (1978; 1989; 1991). See also, Rogers (2006; 2008)

emancipation. The domination of human thinking by instrumental reason threatens to lead to the construction of a totalitarian technological society, which reduces human thinking to calculation, as the measure of objectivity, and the expression of consumer preferences, as the simulacrum of freedom.

Writing in the 1950s and 60s, Marcuse was critical of the "one-dimensional" trajectory of both industrial capitalist and "Soviet Marxist" state-controlled industrial societies. Although highly critical of the Soviet Union, he refused to accept capitalist propaganda regarding its moral superiority over communism.[366] His critique of the technological society applies to industrial society in general, regardless of its political superstructure. Technical rationality is necessarily a form of domination and control, irrespective of its ideological justification. The technological society is a totalitarian system of domination—in both capitalist and communist countries—within which the operation of the industrial and technical system pervades each and every thought and action. Thus rationality is reduced to technical rationality in industrial societies. Marcuse described how technical rationality was implicated in the construction of a system of totalitarian social control and domination.[367] As Marcuse argued, the totalitarianism of the technological society is not enforced by any tyrant or police state, but by the overwhelming and anonymous technological power and efficiency of that society. Political transformations are themselves reduced to technological refinements and innovations, and, thus, each and every qualitative social change and political act is directed into the technological development of the technological society. Within the capitalist driven construction of the technological society, individual human beings are increasingly integrated into an economic system that demands the total accommodation and submission of all human beings into that system. Alternatives are not tolerated, and as a matter of survival, increasingly, individual human beings are compelled to conform to the norms and practices of society. The structures of bureaucracy, planning, and management within industrial capitalist society have created a totally administered society without opposition—integrating private corporations with governmental institutions via lobbying and public contracts, and the privatization and deregulation of the public sector. This threatens individuality because it demands conformity and destroys all possibilities of radical social change. The technological society orders and controls human beings by integrating them into a system that organizes all thought and action in such a ways as to only allow change and reason to operate within its framework. Once alternatives become impossible and unthinkable then society becomes "one-dimensional."

Marcuse recognized that technical progress is defined by the movement towards ameliorating the human condition as its goal, and this goal was rooted in using the scientific conquest of Nature for the scientific conquest of human beings. This goal is able to defeat all protest against technology in the name of the historical prospects of "freedom from toil and domination" in order to abolish labor and struggle (between human beings, as well as against Nature) to achieve "the pacification of existence." He argued that once the technological society became capable of realizing the liberation of human beings from toil and struggle, it closed itself off from this possibility, and became based on a principle of containment of technical progress in order to preserve the capitalist vested interests in maintaining control through owning the means to alleviate immediate scarcity and need. Hence, Marcuse argued that industrial capitalist society is driven by

---

[366] Marcuse (1958)
[367] Marcuse (1991: 226-7)

"…a trend towards the consummation of technological rationality, and intense efforts to contain this trend within the established institutions. Here is the internal contradiction of this civilization: the irrational element in its rationality." (Marcuse, 1991: 17)

Without putting it in these terms, Marcuse argued that there is an internal contradiction between the capitalist economic imperative and the technological imperative. Like Mumford, Marcuse equated the technological imperative (when unfettered or uncontained by the capitalist economic imperative) with the Marxist vision of an industrial socialist society. According to Marcuse, through political revolution, the political apparatus of capitalism and the contradiction of private ownership of the means of production and social needs are to be destroyed, technical rationality is to be freed from the irrational constraints and contradictions of capitalism, and will thus be able to sustain and consummate itself in the new socialist society. Once the control of the technological society will be placed in the hands of the workers, engineers, technicians, and scientists, who will take control as a matter of survival, then there would be a qualitative change in the nature of the technological society from satisfying the needs of a small social elite to satisfying the needs of the technological society itself. Hence, Marcuse argued that the claim that the capitalist economic system is rational is itself based on a conception of the "economic prosperity" of industrial capitalism that ignores the waste, violence, destruction, exploitation, and repression upon which that system is sustained. The self-proclaimed rationality of the advocates and apologists for the capitalist economic system is itself irrational and simply involves ignoring its irrational elements and dismissing these factors as "externalities." He was highly critical of the way that "democracy" in capitalist society is based on media manipulation and conformity; how its wealth is generated through a dehumanizing, alienating, slave labor system; the ideological fetishism involved in consumerism; and the insanity of the highly dangerous military-industrial complex upon which advanced capitalism depends.

Marcuse presupposed that human beings possess the ability to make a true distinction between real and illusionary needs. Once this distinction is made then it becomes possible to show that the contemporary media representations of "individualism" and "freedom," as well as the postulated means to achieve them (i.e. consumerism and the acquisition of capital), are actually dominating forms of propaganda from which the individual needs to be liberated in order to preserve his/her individuality. Even though the economic and political structures of contemporary "democracies" are represented as being the condition of progress and liberty, they are in reality systematized mechanisms of domination that coerce individuals to conform to societal norms and practices, strengthening them, and perpetuating the very system that enslaves us. Once the individual's economic activity is limited to sell his/her labor as a commodity in accordance with the dictates of a system over which s/he has no influence or control, the individual becomes a functionary of that system. Once the individual's political activity is limited to vote for the choice of representative from between members of the same social elite, the "democratic" participation of the individual is reduced to being a functionary for the ratification of the illusion that the system is democratic. Once the media is under the control of a social elite, which can control and suppress information and the voices of genuine opposition, while being an instrument for "shaping public opinion," the media becomes nothing more than a relentless propaganda machine to establish conformity to the norms and practices of key sectors of society.[368] Once all the individual members of society are completely integrated into the system—as a system of conforming functionaries—then society is able to absorb and dissolve all opposition. It is thus able to direct all thought and action indefinitely, perpetually stabilizing capitalism, on the basis

---

[368] See also Rogers (2011; 2012)

of being the champion of human freedom, happiness, and individuality, regardless of its destruction of the environment, dependence on a globalized war-machine dedicated perpetually to create conflict and fear, while eroding all individuality and resistance, in order to sustain an increasingly wealthy social elite at the expense of the majority of the world's population. Hence, the "stability" of the capitalist system is only possible because of it is able to conceal its contradictions and reality by presenting and disseminating the illusion of its own rationality, necessity, and beneficence, by creating the problems for which it postulates itself as the solution. However, it is the reality of these contradictions that undermine the system's stability and expose the extent that the stability of capitalism is an illusion propagated through mass media deception and domination.

Technical rationality frames all aspects of human life by imposing the technological imperative upon all forms of though and action, in such a dominating mode of instrumentalism that it ultimately erodes human freedom and genuine individuality; it uncritically conforms to existing norms and practices, and, therefore, cannot provide the basis for critical and rational evaluation of societal norms and practices, except in the narrow sense of examining the instrumental efficiency of practices in relation to norms. As an alternative to this narrow mode of rationality, Marcuse attempted to develop a critical and dialectical mode of thinking, which was to negate existing forms and categories of thought in favor of realizing higher potentialities, norms, and categories of reasoning. This involved the development of the human ability to abstract universal concepts and categories, instead of merely conforming to the concepts and categories given by society, in order to create a critical standpoint from which the conditions that suppress the potential for self-determination could be negated. For Marcuse, scientific thought is unable to place concepts in opposition to the facts of immediate experience, given its empirical pretensions, and, hence, it is unable to sustain any dialectical tension between "is" and "ought," despite the fact that science establishes its truth against experience. Science must conform to experience and therefore cannot make a judgment that condemns the established reality. A new scientific concept can only be opposed to the previous scientific concept to the extent that it conforms to experience better, but any conceptual opposition to experience and the current state of affairs is taken by science to be a sign of falsehood. Marcuse advocated the critical development of historical analysis of these universal concepts as the means to critique and transcend the empirical and its validation of ordinary discourse and behavior. Critical and dialectical social philosophy was to theoretically facilitate the processes of genuine social resistance and change by critically establishing the basis for individual transformation in relation to the ideals and vision of liberation and happiness, against which current tendencies and conditions could be critically evaluated and rejected in relation to the potential implied by the application of our universal concepts, such as "freedom" and "happiness," and the way that empirical actuality does not achieve these potentials.

According to Marcuse, the aim of critical social theory is one of providing the theoretical basis for radical social change. The theoretical task is to show the historical contingency of contemporary norms and practices, while also showing that there are real possibilities for alternative norms and practices, based on a rigorously historical analysis of social tendencies in theory and practice. However, Marcuse presupposed that it is possible to make distinctions between existence and essence, actuality and potentiality, and, appearance and reality, by appealing to universal concepts. Marcuse was inspired by the Renaissance struggle against superstition, irrationality, and dogma, as expressed in both art and philosophy.[369] He considered it to be the creative source for individual

---

[369] Marcuse considered the dialectical method as demonstrated in the dialogues of Plato to have been one of the most influential sources for Renaissance metaphysics.

rationality and liberation, upon which the real creation of an advanced and rational civilization is to be founded. Technical rationality undermines and suppresses this creative source. By positing an idealization of human essence and rationality, according to Marcuse, norms and potentials for human happiness and freedom can be identified and articulated, which could then be opposed to and negate the existing state of affairs and all its illusionary and irrational values and norms. This is an imaginary act of postulating visions of the human good and a good society against which the status quo can be critiqued.

Marcuse presupposed that the essence of the human subject is that of a free, creative, and self-determining being, which contains possibilities to be realized and qualities, such as values, aesthetic characteristics, and aspirations, standing in metaphysical opposition to an object-world, available for the cultivation and enhancement of human life. Reason links the subject and the object-world, placing the two in dialectical opposition, as a metaphysical duality, which allows reason to intuit the essence of reality to the subject in mediated contradiction to appearances. This essence is increasingly suppressed by the technological society, within which aspirations, values, and needs, are represented as being objective norms for which there must be technical means to satisfy them, otherwise they are rejected as idealistic, impractical, and irrational. Marcuse argued that Renaissance metaphysics has been superseded by technology that replaces the metaphysical and sublimating concepts of subjectivity and objectivity with practical and desublimating concepts of instrumentality and efficacy, which are simply understood through a common principle of ordering all thought and action in accordance with the arrangement and sequencing of means and ends. He argued that the dialectical relation between subject and object, itself a precondition for the subject's ability to negate the current state of existence and realize potentials and possibilities that do not exist yet, is eroded within the technological society because it assimilates the subject within the object-world. All relations between the subject and the object-world have become mediated and defined by the technological society—situated within the technological framework—and, as a result, the human ability to be critical of the status quo, identify real needs, and engage in authentic, original activities to satisfy them has been eroded. Thus, "one-dimensional man" has his/her intentions and means to achieve them presented to him/her as technological objects. The human capacity for authentic thought and action is eroded and forgotten, and, finally, s/he has no alternative but to submit to the imperatives of the technological society and conform to its norms and practices.

Once human beings have conformed to the technological imperative then we become objects for manipulation, administration, management, and control. Hence, we become alienated from our essence, our free and creative subjectivity, and are no longer capable of realizing our potential to live authentic, self-determined, social and economic lives. We lose our individuality as soon as we submit to becoming another object within the technological framework of the administration of society. We not only lose our ability to transform our existence, but we lose our ability to even comprehend it, and, therefore, the technological society atrophies the possibilities of genuine social change and human emancipation. Once the subject has been assimilated as a technological object, it becomes subjected to the same technical norms and practices as any other object, and, hence, loses the recognition of the distinctly rational and creative human ability to engage in transforming norms and practices in order to discover and realize more liberating possibilities. However, while I agree with the thrust of Marcuse's critique of the technological society, he failed to realize the extent that technical progress is rooted in a moral and metaphysical project—the societal gamble—that is profoundly embedded in the psychology and ideology of the Enlightenment ideals. Once we recognize the foundation of the technological society upon the metaphysics of mechanical realism, the scientific worldview, and the societal gamble, we can recognize that the technological society is not in opposition to Renaissance metaphysics, but, in fact, is a consequence of it. Marcuse was aware that modern natural science "develops under the *technological a priori* which projects nature as potential instrumentality, stuff of control and organization."[370] He recognized that this *technological a priori* involved a transformation of concepts of society and Nature, which is understood as the *a priori* intuition of the Universe as a technological reality within which science projects and responds. Matter is understood, in terms of pure science, not simply in terms of its practical use for a specific task, but as pure instrumentality in itself. Here we see the considerable overlap between Marcuse's critique of the technological society and Heidegger's critique of modern technology. Hence, for Marcuse, even though the rationality of pure science is not immediately directed to specific practical ends, it is itself a form of technical operationalism that is developed under an instrumentalist horizon.[371]

"The principles of modern science were *a priori* structured in such a way that they could serve as conceptual instruments for a Universe of self-propelling, productive control; theoretical operationalism came to correspond to practical operationalism. The scientific method which led to the ever-more-effective domination of nature thus came to provide the pure concepts as well as the instrumentalities for the ever-more-effective domination of man by man *through* the domination of nature. Theoretical reason, remaining pure and neutral, entered into the service of practical reason. The merger proved beneficial to both. Today, domination perpetuates and extends itself not only through technology but *as* technology, and the latter provides the great legitimation of the expanding political power that absorbs all spheres of culture." (Marcuse, 1991: 158)

It was for this reason that Marcuse considered scientific rationality to be a mode of political rationality that was bound up with technology, as a mode of social control and domination, due to its function as a

"…rationalization of the unfreedom of man and demonstrates the 'technical' impossibility of being autonomous, of determining one's own life. For this unfreedom appears neither as irrational nor as political, but rather as submission to the technical apparatus which enlarges the comforts of life and increases the productivity of labor." (Marcuse, 1991: 158)

However, due to his neglect to attend to the question of how it was possible for Nature to be ontologically represented as the *technological a priori* and for *technical operationalism* to be epistemologically represented as natural philosophy since the Renaissance, Marcuse failed to recognize the shared metaphysical precepts that made modern science and technology both possible as unified aspects of each other. No doubt due to the influence of Husserl and Heidegger, Marcuse treated scientific thought as if it were positivistic in the sense of being defined in terms of (1) the validation of cognitive thought by the facts of experience; (2) the orientation of philosophical thought to the physical sciences as the model of scientific thought; (3) the belief that progress depends upon this orientation; (4) the rejection of all metaphysics; and, (5) the object world is understood in terms of its instrumentality.[372] Marcuse accepted Husserl's interpretation of Galileo's physics having emerged from and referred back to a pre-scientific world of practice and practical arts.[373]

---

[370] Marcuse (1991: 153)
[371] *Ibid* 156-7
[372] *Ibid* 172
[373] Husserl (1970)

In my view, for reasons I have already given, Husserl was correct but failed to address the question of how it was possible for Galileo to connect epistemologically the practical arts with natural philosophy. Husserl failed to show how Galileo was able to connect geometry with practical activity in such a way as to correlate ideational truth with empirical reality through mechanical devices. As I have already argued, the characterization of scientific thought as positivistic has neglected to attend to the mechanical realist metaphysical basis of experimentation that permitted the epistemological and ontological aspects of science and technology to be represented as being consequences of the same natural laws.

The technological society is founded upon the metaphysical postulation of the emerging ontological goodness of itself because it is represented via the societal gamble as the scientific means by which human liberation and progress become possible. It is itself a negation of the human organic and natural state of animality, with all the ignobility, suffering, and limitation that state of being entails. It is a dialectical confrontation with Nature, postulated as movement to construct a replacement and improvement upon Nature, driven towards its own conception of perfection as absolute human power and freedom from all limitation and constraint. Due to his neglect of this metaphysical foundation, Marcuse did not recognize that the technological society is also based on an equation between truth and value. In fact, for the experimental sciences, a valueless truth is an untestable contradiction in terms, but the value of truth is understood in terms of its value for the discovery of its instrumentality, rather than simply having immediate practical use in a single context of application. Marcuse failed to reveal the cultural meanings that permit science and technology to be represented within the modern world as the source of human liberation and enlightenment. Without addressing how this representation was possible, his critique of the technological society, although quite correct about its outward appearance, failed to expose and critically analyze the metaphysical foundations of that society upon the societal gamble. Marcuse was only able to address the social consequences of the societal gamble, rather than examine the historical conditions that made possible its representation as a rational societal project. However, once we recognize this metaphysical foundation, we can recognize that the technological society is itself incomplete and is an idealization of the final goal of all struggle, labor, science, and technological innovation. Hence, it is not a rejection of the classical dialectic between Being and Non-Being at all, as Marcuse claimed,[374] and, hence, while it is indeed true that the technological society has not answered "the question of Being," it is actually an experimental attempt to answer this question by creating itself as the answer to this question. It is fundamentally transformative of Being and how we conceive of it.

The technological society is an idealization of the future of the whole process of innovating itself from the past, concrete, technological ontology created in confrontation with Nature, and thus is bound up with its own Becoming. It is at this level that the reified, abstract potential of the creation of the technological society is idealized as the Being of Becoming, to be realized by acting in accordance with natural law and implementing natural mechanisms in the ongoing construction and perfection of society. The technological society is postulated upon its conception as the completion of history, the destiny of the modern world, and a negation of the past and the natural world. The process of creating the technological society as a replacement for the natural world is itself the modern dialectic between Being and Non-Being that is made immediate and concrete in its perpetual Becoming. Thus its confirmation of its own truth is perpetually deferred to some asymptotic ideal of a complete understanding of the *techne* of the human

---

[374] *Op cit.* 136, n. 4

condition that is projected onto the horizon of the becoming of the future. Marcuse made the error of presuming that the technological society was premised on irrational instrumentalism or will to power—a headlong and mindless pursuit of innovating new means to carelessly examined ends—but this is only true if we limit our analysis of the technological society to its outward appearance. Once we reveal the metaphysical foundation of the technological society, we can recognize its moral and ideological essence as an idealistic, dialectical construction of the means to confront Nature and technologically construct "a better future."

However, according to Marcuse, human beings are to realize Nature's inherent possibilities in accordance with representations of Nature's potential. This shows the extent that Marcuse's own philosophy of science was very much indebted to the societal gamble and mechanical realism, once reason had been purged of the one-dimensionality of the technological imperative, and instead adopts a dialectical relation between actuality and potential. One finds Marcuse's clearest statement of his faith in the societal gamble in his *An Essay On Liberation*:

"Freedom indeed depends largely on technical progress, on the advancement of science. But this fact easily obscures the essential precondition: in order to become vehicles of freedom, science and technology have to change their present direction and goals; they would have to be reconstructed in accord with a new sensibility—the demands of the life instincts. Then one could speak of a technology of liberation, product of the scientific imagination free to project the forms of a human Universe without exploitation and toil." (Marcuse, 1972: 19)

The technological society, driven by the technological and capitalist economic imperatives, is inherently premised upon a dialectical movement from the past into the future (transforming the past into the increased potentiality of wealth and power to be realized in the future) that is represented as the emergence of increased liberation from material limitations. It is a negation of the past and a sublimation of the future. Through labor and invention, acting upon the past as its stock of resources, the present becomes instrumental in the perpetuation of that dialectic, which is always represented as offering realizable future rewards. Hence, it is futile for critical thinkers like Marcuse to oppose the technological society with a vision of human liberation and individuality because it is already based upon the metaphysical postulation of its own teleology of human liberation and progress, hence, it will not be negated by such a critique because it will be able to absorb any such critique and represent it as an affirmation of itself. All such critiques can be represented as impatient demands for the liberation and empowerment of all humanity, and this is exactly the same ideal upon which the technological society is premised. All such critiques can be represented as having their heart in the right place, so to speak, but lacking practicality and realism.

The technological society is a societal aspiration, rather than a concrete reality. Modern society is a complicated and heterogeneous mix of diverse and often contradictory tendencies, aspects, characteristics, parts, dimensions, levels, classes, ideologies, and cultures. The capitalist economic and technological imperatives co-exist within the complex reality of the modern society as societal demands and challenges, which operate upon the direction of the development of modern society. Often these conflict with one another and generate social incoherence in the ongoing transformation of social relations, structures, and institutions, most evident when its innovation and development is directed to overcoming the disasters and problems caused by its prior inventions and industrial processes. The modern representation of human beings as rational agents—understood in economic and technological terms—is a consequence of technological and scientific progress, as theoretical abstractions, which entail ideological commitments and metaphysical assumptions. Often these theoretical abstractions merely functioned as an instrument of propaganda or self-justification, while little effort, if any, was actually made to implement them in practice. As Mumford put it: "Before the paleotechnic period was well underway their images were already tarnished: free competition was curbed from the start by the trade agreements and anti-union collaborations of the very industrialists who shouted most loudly for it."[375]

Since the collapse of the Soviet Union, industrial capitalism has ceased to be an ideology directed to order and prioritize the development and trajectories of technological innovation and infrastructural construction in accordance with a vision of the best of all possible worlds. Now the Cold War is over, ideology no longer has a technical function. Instead, capitalism has become a mechanized system of increasing profits and therefore maintaining and reproducing the social inequalities that interconnect technologies with the economic systems of production, exchange, and consumption. Otherwise there would be no supply and demand. Lean supply on demand technologies have become self-identified as the necessary outcome of efficiency. The interconnected dominance of capital and technique in modern society have pervaded to all areas of the bodily passions, including sex, sport, and war, and when combined with mass media, all these aspects become transformed into spectacles of mass entertainment and commentary. The media domination of mass democracy has simulated public debate by transforming it into the presentation and dissemination of opinions and moral positions as if these were objects (or brands) available for consumption. These become 'off-the-peg' objects for adoption and discussion within a system of mass democracy, within which, capital, technology, and mass media form a totalitarian machine, and culture and propaganda become one and the same.[376] This totalitarian unity represents the production and consumption of standardized goods, services, and entertainment as being the modern form of the natural order, expressing and reflecting our individuality through choices in modes of 'off-the-peg' conformity—thereby reducing human individuality and identity to a series of consumer preferences. Of course, this representation is one that is designed to naturalize the position of the social elite that owns and controls the access to capital, technology, and mass media, to increase the perception of the artificiality and impossibility of any real social change in the social order. According to such a perception, opposing capitalism would be an irrational act of opposing natural laws, on the basis of an imagined, alternative society that was doomed to fail because it contradicted fundamental human nature and the natural order of things. Yet, industrial capitalism is no longer based upon a liberal capitalist ideology—in either its anarchist or libertarian forms—directed to improving society by creating the conditions for individual endeavor and prosperity, in accordance with an Enlightenment vision of humanity, but has become an unreflective, unquestioning reproduction of itself, as a good-in-itself—without any coherent vision of the society that it is constructing. Industrial capitalism has become irrational, except as a political and economic means of sustaining and reproducing the power base of a social elite, and thereby forcing the social elite to conform to it most of all. It is the whole societal gamble that has become an irrational project of the unthinking and uncritical extension and enhancement of the power of the technological society.

---

[375] Mumford (1963: 269)
[376] Horkheimer & Adorno (2002: 94-136); Ellul (1964, 1972); Rogers (2011; 2012; 2016)

The Mass Society and Device Paradigm:

José Ortega y Gasset's *The Revolt of the Masses* (first published in 1930) has often been cited as a critique of direct democracy and majority rule, in favor of republicanism and competitive elitism.[377] However, this is something of a misrepresentation. It is actually a critique of "mass society," within which the organization, standards and directions of societal development are informed and constrained in accordance that which is common to all. It is a critique of the development of society in accordance with the anticipated needs of "everyman," which is, of course, no one. It is a critique of consumerism and conformity. The conditions for human individuality and excellence have been suppressed and eroded by "mass society" to provide the conditions suitable for "mass man." The "mass man" seeks only improved levels of consumption and security, avoids effort and struggle, and, for Ortega, is the product of societal inertia and conformity. Both "mass society" and "mass man" are the results of abundance and comfort produced through relations of mass production and consumption—the products of scientific and technological developments—rather than being the sum total of a multitude or the expression of some herd instinct. The "mass man" is an artificial construct, who is made possible by the technology, but the "mass man" refuses to take responsibility for societal development, insisting that this is the responsibility of the State as the organizer of technical and administrative methods to improve society. Meanwhile the "mass man" asserts that, as "an individual," s/he has an absolute right to security, comfort, employment, income, and privacy, and it is the responsibility of the State to provide the conditions for these rights through the organization of technical and administrative methods. "Mass society" is the ongoing reduction of liberal democracy into a technocratic system of regulated relations of production and patterns of consumption, wherein "mass man" is an entirely economic and consumptive mode of being. Ortega argued that industrial society based on liberal democracy has become a technocratic "mass society" and there is a widespread disregard for any critical reflection on epistemological and moral standards.

In this respect, Ortega's conception of the "mass man" and "mass society" preempted Albert Borgmann's conception of *the device paradigm*. In his book *Technology and the Character of Contemporary Life*, Borgmann proposed and developed the idea of the device paradigm to describe and analyze the way that the structures of modern society have stabilized into particular patterns of technological activity.[378] Borgmann was critical of Ellul for being overly pessimistic and deterministic about modern technology, and, following Heidegger, he was also critical of instrumentalist and anthropological philosophies of technology. Borgmann defined modern technology—in a manner consistent with Heidegger and Ellul—as the "characteristic and constraining pattern of the entire fabric of our lives" in mass society.[379] His critique is based on the argument that the device paradigm—in the form of stable patterns of production and consumption—erodes the traditions within which our practices gain their meaning. He acknowledged that science and technology have provided powerful devices and techniques to improve human health, agriculture, mobility, etc., which have overcome the limitations of our organic state of being by providing the means to relieve the burdens of disease, hunger, and vulnerability, but he was deeply concerned with the way that technological activity has also eroded the meaning and character of human existence by eroding the conditions under which we discover human meaning and character.

---

[377] Ortega y Gasset (1993)
[378] Borgmann (1984; 2000). See also Higgs, Light, & Strong (2000)
[379] Borgmann (1984: 3)

Borgmann described these conditions in terms of a culture of *focal practices and things* that require our attention, care, effort, and patience, which place challenges and demands upon us and lead to the development of virtues, discipline, celebration, shared experiences, all of which he considered to be part of human well-being and excellence. By attending to focal practices and things (which resonates with Heidegger's understanding of "gathering"), we reawaken our connections with those practices and things in our everyday lives, focusing human relations in a way that engages us with each other and provides continuity within traditions. Focal practices and things are ends-in-themselves and are intimately bound-up with our social being and identity as human beings, gathering together particular experiences, human relations, and cultural meanings. Focal practices and things are the specific practices and things of our lives that are of great importance for our sense of well-being and our reflective care for the good life. Examples of focal practices and things would include social activities, such as the traditional family meal, learning and playing music, conversations between friends or family, and ceremonies, but they can also include individual activities such as long distance running, walking in natural landscapes, and crafting tools or objects. Focal practices and things relate the particularities of our experiences to the generalities of traditions, unifying them through cultural meanings and providing connections between means and ends, individuals and their community, as well as enjoyment and celebration of them as ends-in-themselves. It is this lived relation between the particular and the general that gives our lives a sense of continuity, meaning, and value. Borgmann termed this as a "telling continuity" to refer to the way that it this lived relation that situates the particularities of experience within the larger continuity of one's life, community, history, and a sense of place in the world. It is through the act of focusing upon these practices and their meaning that the focal thing "gathers the relations of its context and radiates into its surroundings and informs them."[380] Hence, for Borgmann,

"A focal practice, generally, is the resolute and singular dedication to a focal thing. It sponsors discipline and skill which are exercised in a unity of achievement and enjoyment of mind, body, and the world, of myself and others, and in social union." (Borgmann, 1984: 219)

By describing how they relate our everyday practices to traditions and also demand attention to detail, shared experience, commonality of purpose, and the development of virtue (understood in an Aristotelian sense), Borgmann described how focal practices and things have a "commanding presence" within a tradition that teaches us their meaning. While his idea of focal practices and things was clearly inspired by Heidegger, it was also inspired by the writings of Emerson, Thoreau, Melville, and Aldo Leopold, as well as Native American oral traditions. Deeply rooted in this American narrative tradition, Borgmann was concerned with how modern technology—as *Gestell*—has concealed Being within a framework of instrumentality for future use, and, he was also concerned with the extent that we have complacently become quite unthinking about technology, which has in many respects become an ensemble of unconsidered means to achieve ill-considered ends.

Even though technology ties us to an active and engaged relationship with the world through labor, when that relationship becomes a commodity understood only in terms of wages, exchange, and consumption, we lose our capacity for a deeper, bodily, mental, and social relation with the world. The technological society—driven by the capitalist economic imperative to maximize profits—has trivialized life and turned human

---

[380] *Ibid* 197

beings into passive and isolated commodities and consumers. According to Borgmann, the rational revaluation of technological innovation must be based upon a community engagement with the articulation and critical assessment of the concrete proposals of technology in contrast to the current conditions of their fullness, in accordance with considerations of sufficiency for a practical appraisal of the good life, which must be evaluated in terms of traditions, focal practices and things, and *communities of celebration*.[381] Borgmann argued that devices replace focal practices. The microwavable meal replaces the traditional cooked family meal, the television replaces conversation and other entertainments, and the electronic music system replaces live performances by musicians. Thus it transforms our social relations with the material conditions of our existence and how we interpret their meaning and value. Devices erode our traditions by making them obsolete or quaint (parochial), which replaces the commanding presence and telling continuity of focal practices with the thrills and fetish of using the latest disposable device to achieve culturally given ends, which may well not be achievable without such devices. This pattern of production and consumption erodes our need for focal practices and things because "the peril of technology lies not in this or that manifestation, but in the pervasiveness and consistency of its pattern".[382] It was for this reason that Borgmann acknowledged the *affluence* of the device paradigm and distinguished it from the *wealth* of engagement through focal practices and things, and called for restraint of the device paradigm in favor of a return to focal practices and things as ends-in themselves.[383]

While modern technology has evidently increased human power by providing an ensemble of devices and techniques, it has also reduced our commitment to developing our social relations and human excellence by eroding our need for focal practices and things. How has modern technology eroded our need for focal practices and things? For Borgmann, modern technology is the total ensemble of disposable devices and techniques designed to alleviate effort. According to Borgmann, these disposable devices and techniques are glamorous in their appeal. They are disposable in the sense that, situated in everyday life as mere means to ends, they are replaceable and, thereby, have only an instrumental relation to the ends to which they are directed towards satisfying. Any device or technique may well be thrilling or glamorous in terms of the powers or opportunities it affords, as well as its novelty and desirability by other people, but it also can be replaced with a new, improved device or technique and, therefore, it does not have any intrinsic value or meaning. The human relation with devices and techniques is reduced to one of invention or use; an economic relation of either production or consumption. While devices such as the computer or the telephone may well allow us to be remotely connected to distant people and places, they remain mere means to an end as a medium for communication. Devices, such as the automobile, for example, are mere means to ends, which are exchanged, via the medium of money, as instruments to satisfy our present needs (travel to work, take the kids to school, attract members of the opposite sex, attract the envy of members of the same sex, etc.) as well as afford us future opportunities. Even though we may well be seduced by the shape, color, and power of automobiles, the vehicle itself remains instrumental, given that it can always be replaced with a better or the latest model, or an alternative means of transportation. Of course cars can be transformed into focal things, such as vintage cars, lovingly restored, carefully maintained, driven around town or in rallies as a pleasure for its own sake, and understood in terms of the history of its design, development, and usage, as well as made meaningful through personal experiences. It is possible for a mass produced motor car to become a focal thing, just as it is possible for a hand-made and restored vintage car to remain a

---

[381] See also Borgmann (1990)
[382] *Op cit.* 208
[383] *Ibid* 223-31

device. Whether an automobile becomes a focal thing, an end-in-itself, or a device, a means to an end, is determined through our relationship with it rather than any essential category of how it came into being (i.e. through advanced industrial engineering and relations of production) or its intrinsic properties (speed, elegance, power, and beauty). As a focal thing, a particular automobile has inherent value as an end-in-itself, a work of art, made meaningful in terms of everyday focal practices because people give it that value. As a device, motor cars are used as mere means until they are discarded as junk, instrumentally valuable for parts or recyclable materials because people only value them as means. While it is evident that people do develop fetishistic relations with devices, such as cars (or boats, clothes, televisions, or music systems, etc.), these devices remain disposable as things in the sense that the relationship is with the glamour of the device, the sense of ownership of the latest model, and also the functionality (or prestige) it affords.

The increasing complexity of technology is concealed using "user friendly" push-button or point-and-click interfaces, which increases the separation between production and consumption, further disconnecting means and ends. The internal construction of the device (e.g. a mobile phone) is concealed completely within its casing, as a set of functions to deliver a commodity or service (convenient communication). This not only increases the disposability of the device, but it increases the dependency of the consumer upon the producer. Due the increased complexity of technology, alongside their "user-friendliness," devices deskill their users by alleviating the burden of having to learn how the device works. Ironically, modern technology disconnects the consumer from "the technical" and allowing the consumer "to use up an isolated entity, without preparation, resonance, or consequence."[384] Technology is built into society and into societal structures that constitute "the inconspicuous pattern by which we normally orientate ourselves." (Borgmann, 1984: 105). It is in this sense that the technological infrastructure of society becomes invisible—except as an available ensemble of devices, as 'off-the-peg' societally given means to achieve societally given ends—and the structures of conformity become invisible to us, as users of devices, and present to us the illusion of freedom of choice (masquerading as a consumer-choice between brands or models of device).

"The promise of technology," as Borgmann termed it, is the two-fold promise of liberating us from the ills and limitations of the natural world and, also, embracing a vision of the good life that technology offers to make possible.[385] Hence, Borgmann was aware of the societal gamble and the implicit faith in science and technology as providing the potential to construct a better, more certain, and rational artificial world, the technological society, as a replacement for the natural world. It preconditioned the promise emergent from the Enlightenment tradition that science and technology will liberate us from superstition, fear, toil, suffering, and perhaps even death, allowing human beings the freedom for creativity and to become masters of our own destiny. However, we have strayed considerably from this vision. We have become enslaved within patterns of production and consumption. The Enlightenment vision of a rational society populated by free and equal human beings has become reduced to a positivistic measure of our "standard of living" calculated in relation to our level of consumption. A world of intrinsic value has been replaced by a system of devices, commodities, and exchange-values. The enjoyment of a meaningful present has been replaced with the thrill of purchasing and using the latest devices as the measure of well-being. The device paradigm, as the overall pattern of use of the ensemble of devices, transforms the natural

---

[384] *Ibid* 53
[385] *Ibid* 35-48

world into standing-reserve, structured in terms of its instrumentality within patterns of production and consumption.

Yet, as Borgmann observed, increasing levels of production and consumption do not lead to satisfaction, despite the promise of technology to bring liberation and enrichment. We are left alienated from deeper cultural and personal meanings, disengaged from our deeper human aspirations, or simply mindlessly locked into patterns of production and consumption. Borgmann argued that we need to provide genuine alternatives to this unreflective consumptive relation with technology by reawakening focal practices and things, once again becoming mindful and respectful of their commanding presences and telling continuity within our communities of celebration. According to Borgmann, we need to take time to develop interpersonal relations with each other through these reawakened focal practices and things, paying close attention to how these practices and things are meaningful within our individual and community lives, aiding the development of our attention, skills, capacity for effort, carefulness, and understanding of our history and traditions. This requires a commitment to our everyday activities as being ends-in-themselves. It requires a return to the enjoyment of a shared present, for example, over a shared cooked meal at the family table, or while sitting around a wood burning stove, or during a conversation, or while walking along cliffs, through woods, or across hills, or any of the everyday activities shared with family, friends, or neighbors. It is only through the collective devotion to the meaning of our lives that we will be able to recover these meanings, reawakening the possibility of human excellence and the good life through the effort required to learn and develop reflective and mindful attitudes to our lives.

Technology replaces other activities. For example, television replaces family conversation or public entertainments (such as going to a cinema, a theater, or a concert). In many respects, the device paradigm has distracted us from "great embodiments of meaning," as Borgmann put it.[386] However, as I have argued elsewhere, the erosion of traditions and cultural meanings has not been caused by the development of modern technology in society, but, instead, technology acts as a substitute for traditional practices and, thereby, conceals the erosion that has already taken place.[387] Otherwise it would be difficult to explain why focal practices and things, with their "commanding presence," would be so readily eroded by the device paradigm. We would have to explain why human beings choose the thrilling over the meaningful, if, as Borgmann claimed, the meaningful is an essential aspect of human identity. Of course one can appeal to the convenience and ease of using modern devices in relation to the difficulties involved in learning traditional practices—it is much easier to download music from the Internet than learn to play a musical instrument—but, in which case, one would still need to explain how focal practices and things lost their "commanding presence." If one only learns the intrinsic value of focal practices and things through the attention, care, and effort of learning traditional practices through focusing, this still does not explain why human beings stopped valuing them as goods or ends-in-themselves. How did this happen? In my view, it is only when sitting around the fireplace or hearth, to use Borgmann's example, had already lost its value as a focal practice that the transition to electric heating became desirable—delivering heat as a commodity—once the focal thing, such as the wood-burning stove, lost its commanding presence. Otherwise Borgmann's argument would entail that although human beings first began considering wood-burning stoves only in terms of their instrumental value, as a device to deliver warmth, they subsequently learned their intrinsic value from the pleasure and meaning that their preparation and sitting

---

[386] *Ibid* 188
[387] Rogers (2006; 2008)

around them brings, but, if this is the case, then there is no reason to presuppose that new focal practices and things will not emerge from sitting around in an electrically heated room. Borgmann's rejection of this possibility seems to draw something of an arbitrary line between kinds of technology based simply on his own subjective estimation of the sufficiency of any technology. Hence, according to Borgmann, playing a musical instrument is a focal practice but playing a DVD is an act of consumption, even though in both cases the focus would be a group of people coming together over the shared enjoyment of the music. Likewise, Borgmann arbitrarily considered a wood-burning stove to be a focal thing, but an electric heater to not be. After all, could we not equally argue that the wood-burning stove eroded the communal gathering of a village or tribe around a shared fire, its secrets only known by revered elders? Did not the wood-burning stove erode the magical and communal ceremony of fire-making? Did not the tinder-box erode the "commanding presence" of making fire using a friction bow and dried twigs and moss? Taking this argument to its pre-Promethean extreme, one might even argue that the knowledge of how to make and use fire eroded the "commanding presence" of the natural world and the traditional practices our ancestors had developed to survive in it.

It seems Borgmann's primary objection to modern technology is to the effortlessness of it as a means to satisfy our desires. Hence the focal thing is not so much the wilderness experience while backpacking or the splendid trout caught by fly fishing, to use his examples, but it is the effort involved in having those experiences. He considered effort to be a good because it because of its value in building character through making particular kinds of experience possible through effort. The problem is not modern technology at all, but is our indifference to the development of our own characters, if, like Borgmann, we consider the development of good character to be an intrinsic good. But, what does this mean? By unburdening us from making the effort necessary to achieve any given ends, by providing the end as a commodity to be purchased, the device disengages us from caring about the conditions of our existence, except as ends to be satisfied by the employment of mass produced means. Our existence becomes mediated by the device paradigm and is dependent upon it; the quality of that existence is to be purchased through participating in an economic system of exchanges of money for labor. If this is the case then the problem of the device paradigm is that—tragically—its success in satisfying our material needs has eroded the conditions under which we develop a meaningful and existential relationship with the conditions of our existence. Conformity to the device paradigm results in such an easy life that we are not compelled to engage in the practices that develop our characters as beings embodied in a physical world. We become shallow consumers, without any regard for understanding the conditions of our existence. If, as Borgmann claims, the meaning of rationality is bound together with a deep philosophical, moral, and creative vision of the good life, then we can consider ourselves to be rational to the extent that we are capable to reflect upon and understand the conditions of our existence, as well as possess the skills and means to control or change them in relation to moral norms and ideals for human excellence of character. His argument presupposes that the device paradigm does in fact, in itself, prevent us from being rational by disengaging us from the material and moral conditions under which we develop character through both the physical and intellectual effort required to envision and realize human well-being.

However, it is on this pivotal point that my theory of the problem of technology differs from that of Borgmann's. In my view, the problem of technology is not that it *caused* us to become shallow and irrational consumers, but that it *conceals* from us the extent that we have become shallow and irrational consumers by removing the conditions under which we would confront that fact. By supplying the means to satisfy our needs, at the press of a button or the flick of a switch, it conceals from us the nature of the causes of the erosion of our desire to maintain the traditional practices which previously satisfied our needs. For example, in my view, television and microwave meals are not responsible for the demise of the traditional cooked family meal, but, *by acting as a substitute* for this traditional practice—by providing the microwavable TV dinner as an alternative means to prepare and eat food while being entertained—technology has concealed the erosion of traditional family relations that had already occurred. It conceals from us the fact that something has been lost. We do not notice our lack of conversation while we eat, nor our unwillingness to prepare and cook our own food from raw ingredients. Our silence is drowned out by the television. Our indifference to cooking is hidden by the ease of buying our food prepackaged, putting it in a microwave, turning a dial, pressing a button, and waiting a few minutes. We do not notice that the family has ceased to be a community of celebration. We do not notice that we have become alienated from each other and how our food is produced. We do not need to practice patience. Everything is fine. We eat what we want, at our convenience, without all the difficulty and effort of attending the family dinner. What modern technology does is conceal from us the reasons why our traditional practices became unsatisfactory in the first instance by providing us with an alternative that makes those traditional practices obsolete as a practical activity. This covers up the cracks in our society by replacing already waning traditional practices with a technological device that satisfies the same need (say eating food) without questioning why the traditional practices were waning.

Once we take this into account, we can see that underwriting Borgmann's call for a return to traditional focal practices and things as the means to develop character is a set of moral prescriptions and norms for social being. This somewhat conveniently neglects the inequalities that were involved in many of these focal practices. For example, it normalizes the sex-based division of labor that was involved in the traditional cooked family meal, and, by placing the blame for its demise on modern technology, Borgmann has ignored the social changes that had taken place which meant that, thereafter, women no longer needed to accept their traditional role as unpaid housekeepers and cooks. It conceals the extent that the traditional community of celebration was based on fundamental gender inequalities. What modern technology did was simply compensate for this social change by allowing the family to eat without depending on women to cook and serve the meals, while also continuing to shield men from the burden of learning how to cook and serve food to their families. It would be quite impossible to return to this traditional practice without re-imposing the inequality inherent to the traditional division of labor within the family, but what modern technology has done is prevent a critical examination of the reasons why men are traditionally reluctant to take a share in the preparation and serving of meals, and it does this by making the practice seem obsolete and the question seem moot. Yet, many of the traditional attitudes and prejudices have remained intact, even though technology has provided new means as a replacement for the traditional practices. Once we acknowledge how this inequality has become concealed by reducing all criteria of cultural norms and values to those of practicality and the satisfaction of desires, thereby allowing a 'technical fix' to all human problems, we can recognize how the technological imperative has made society more shallow and conformist by treating all human relations as those of achieving means to ends, without developing any critical and philosophical perspective on the ends presented to us as goods.

Having said that, it also needs to be said that we should not take the somewhat reactionary path of reifying traditional practices, as if these are self-evident ends-in-themselves. Contrary to Borgmann's appeal to tradition, as if it is an unquestionable good in itself, we should be aware that recovering the focal practice of the family meal should involve consciously, mindfully transforming it to accommodate social changes by equitably sharing the burden of learning how to cook and prepare meals throughout the whole family, regardless of gender. This would not be a return to traditional practices at all, but would be a focal accommodation to the fundamental changes that have occurred within the family structure, while recovering those focal practices from the device paradigm that concealed the reason why there was a need for such social changes in the first place. Rather than blame technology for all our ills, putting our faith in a somewhat reactionary call for a return to traditional practices, we need to carefully examine why our traditional practices eroded in the first place. In my view, the erosion of our traditions occurred because they were inadequate or inappropriate due to their inherent incoherence, inequalities, and inconsistencies, and, as a result, they could not withstand societal changes. Rather than put our faith in tradition, we need to examine critically our personal and cultural goals and goods alongside understanding whether and how societal structures empower, distort, or suppress them. In some respects this will involve critically reflecting on how the device paradigm has failed to live up to the promise of technology, how our faith in the societal gamble is misplaced, and in other respects it will involve critically reflecting on how our traditions failed to realize their ideals and values, either by failing to live up to them or by being riddled with internal contradictions and falsehoods.

Putting aside the essentialism and romanticism inherent to Borgmann's philosophy of technology, as well as the normative moral conservatism which underwrites his choice of focal practices and things, my criticism of his argument is the technological determinism which underwrites his causal connection between the device paradigm and the erosion of traditional focal practices and things. In my view, his argument is akin to claiming that the bullfighter's red cape caused the death of the bull because it distracted him from the bullfighter and his sword. My argument is that the pernicious aspect of technology is not that it causes mindless and unsatisfying lives, as Borgmann claimed, but, much worse than this, it conceals the irrationality and arbitrariness of the relations and structures of modern society by providing a successful system of providing means to distribute commodities for those who conform to the system and are able to integrate themselves within it. The device paradigm does not undermine a philosophical and spiritual engagement in life, but conceals the fact that such an engagement is absent. Even if we agree with Borgmann's claim that modern society lacks an explicit vision of the good life over and above a certain level of consumption, we would still need to show that the device paradigm has also undermined such a vision. My position is that it is not evident that the device paradigm has done this, but, due to its success in providing commodities and satisfying material needs, it has hidden the need for such a vision from public awareness and deliberation. We no longer question the ideals and vision of societal development; these are given to us as 'off-the-peg' ends for which society provides the devices to satisfy them, or it is working on inventing the device to satisfy them. For example, the ongoing development of sophisticated means to extend the human lifespan treats human longevity as an end-in-itself, without addressing the questions of the meaning and quality of human life. By covering our irrationality and arbitrariness with technical rationality, we have become unconscious of our irrationality and arbitrariness and no longer confront them and the fears that generate them; instead our focus has become reduced to the search for the best means to solve technical problems to satisfy human needs and wants. The pervasive conformity underwriting the development of the structures and content of society, constructed without any reflection upon the reason for these structures and content, nor on their consequences beyond their immediate instrumentality within the system, all remains unchallenged.

Moreover, even if it is true that people in modern society are in large part significantly unhappier than people in primitive or traditional societies, as Borgmann (and Ellul) claimed, we still have to show that it was the development of the device paradigm which decreased human happiness. Putting aside the considerable difficulties involved in evaluating the levels of happiness or unhappiness of people in different societies, cultures, and historical eras, it seems evident that, even if we could make such a comparison, we would be still confronted with the difficult (perhaps impossible) task of demonstrating the existence of a causal relation in order to argue that the device paradigm was the dominant causal factor responsible for the unhappiness of people within modern society. In my view, it is not at all evident that people in modern society are generally unhappier than people in traditional societies; I would argue that they have different understandings of what happiness means, and they give it different significance and value, but, even if it is true that they are unhappier, it is arguable that the device paradigm did not cause their unhappiness but distracted them from the true causes of their unhappiness by allowing them to satisfy their material wants and needs without having to discover and confront the sources of their unhappiness. For example, rather than address one's alienation and loneliness, one goes shopping, plays a computer game, watches TV, or takes a pill.

If there is some truth to my argument then the increased development of the device paradigm will not make us unhappier, but our conformity to it will further distract us from confronting and changing the causes of our unhappiness. It does this by continually substituting the thrills and glamour of new devices and higher levels of consumption for the need for philosophical reflection and critique of our vision and choices regarding the development of society. The device paradigm does not suppress our capacity for reflection and critique, but simply distracts us from the conscious recognition of their necessity to prevent the irrational development of the technological society. Without such reflection and critique, the technological development of society substantively structures and conditions the possibilities and directions of our modes of engagement with the conditions of our existence by increasingly systematizing human relations into systems of media and exchange.[388] This will further superimpose technical systems over our intentions, expectations, limits, and interactions, until it has reduced them to those capable of being satisfied by technical systems. At which point all human relations will be reduced to the device paradigm and Sophocles' warnings about the tragedy of the human condition will become a nightmarish reality, wherein, like all but one of the denizens of Aldous Huxley's *Brave New World*, or Yevgeny Zamyatin's *We*, our descendants will simply not care for anything other than their position and functionality within the system, until such time as the technological society has consumed the natural world completely.

If we do not want human beings to become powerless slaves to technological innovation then we have to subject technology to articulate critical thinking. Philosophy, in a general sense, needs to reaffirm itself as the primary mode of critical reflection upon the conditions, values, meaning, and ideals of our technological society. We need to philosophically scrutinize the vision of the world that we are working towards creating. Is it really desirable? Would it really be good for us to live in that world? However, we are neither titans nor gods. It is impossible to accurately predict the forms that technologies and societies will take in the future. We cannot know whether our current actions will be good or bad for us in the long run. As Nietzsche and Chuang Tzu remind us, evil comes out of good actions, and vice versa. We are in a state of innocence regarding the future, but we are not beasts that are only concerned about the present. We think about the future and can try to find the best course of action to try to achieve our ideals

---

[388] See also Habermas (1987)

and visions of how we should live. Of course, I do not intend any connotation of human supremacism by this remark. If a beast does not care about his or her future, only about his or her present, it is quite possible that there are many animals on Earth (or elsewhere) that also think about their future and try to find the best course of action in the absence of knowledge. In which case, they would also not be "beasts" either. In the absence of certain knowledge, every action is a gamble on its own goodness. It is for this reason that we should embrace social plurality and diversity because, once we accept our existential innocence about the future and the consequences of our actions, we also must accept that we simply do not know which course of action is for the best. We are all guessing. If all human beings are equally innocent regarding the future and the goodness of our actions, then we should adopt an egalitarian stance about human goods, values, and purposes.

Society has to accommodate different visions of the ideal society, which may stand in conflict with one another. The philosophical question 'what is the good life?' has a plurality of answers. It is thus better for society, as a whole, if its citizens try as many varied courses of action as possible and to seek to satisfy as many different goods, values, and purposes as possible. We need to explore alternatives if we are to maximize our chances of discovering good courses of action and to minimize our chances of making a terrible mistake that damages the whole of society. It is simply the case of not putting all our proverbial eggs in one basket. A pluralistic society has a greater chance of hitting on a good course of action, through its trial and error processes, involving as many people as possible in as many different ways of life as possible, than an authoritarian and dictatorial society which collectively follows the same path that was pre-emptively set down by a single individual or social elite. It is simply a question of the advantage of diversification over specialization in an unpredictable and changing world. Once a pluralistic society, comprised of diverse communities and individuals exploring as many different ways of life as possible, has developed widespread possibilities for communication and debate among its citizens, then we are able to learn from each other's trials and errors, and debate and discuss how best to live our lives in the light of our collective experience. Such deliberation and reflection also needs to be pluralistic and diverse, incorporating alternative visions for society, as well as divergent, critical evaluations of the nature of rationality, all of which should be available to aid our decision-making processes about how try to live life well. Such a society is inherently a democratic society. Of course, in a democratic society, evolving in complicated and pluralistic ways, some individuals and communities are going to make some terrible decisions and, due to the capriciousness of Nature and the unpredictability of technological innovation, unfortunate events and consequences can interrupt the most carefully crafted plan. However, even if some individuals or communities fall afoul of their mistakes, natural disasters, or unforeseen consequences, a pluralistic society has a better chance of surviving and learning from mistakes and misfortunes because other individuals and communities will be trying different things for different reasons and, therefore, the impact of those mistakes and misfortunes will be varied. By recording and studying its detailed history, as well as engaging in critical discourse with people from other communities, we will be able to learn from our whole society about our possibilities and limitations, about our experiments, about our failures and successes, and about how we can celebrate our differences as being our greatest source of creative power and our greatest asset for constructing a sustainable society. It will remain an ongoing, genuine, and mindful effort towards a sustainable and desirable life for as many people as is humanly possible, and that is perhaps as close to progress as it is possible to be, even if we cannot all agree on what a sustainable and desirable life would be.

Technology and science need to be integrated into society—not simply through the trial and error laissez-faire of use, consequence, and accommodation, but through an open, critical, and democratic evaluation of the vision for society as a whole. Such a public and pluralistic evaluation of the benefit and impact of technology and science should be made in relation to evaluations of human well-being and potential. When the direction of technological development and innovation is "governed" in accordance with the short-term interests of a social elite, without much regard for human well-being or potential, we find ourselves dominated by institutionalized technological practices that are socially damaging, unjust, inefficient, and environmentally disastrous. When the social evaluation of technology and science are restricted to whether they satisfy the short-term interests of an oligarchy, the development of the technological society will remain that of an inflexible, undemocratic, and totalitarian means of acquiring more wealth and power for a social elite. However, the democratic development of the technological society does not simply depend upon the egalitarian distribution of technology and scientific knowledge—and political power as a result. It requires these, but it also needs flexible, democratic participation in the evaluation, implementation, and re-evaluation of criteria under which technology and science should be evaluated. Public participation should not be limited to merely making a choice between representatives and their proposed "solutions" because this will continue to limit decision-making to the narrow criteria of bounded technical rationality, political expediency, and cost benefit analyses. Public participation needs to be involved at all levels of decision-making, bringing a diverse and pluralistic stock of imagination, knowledge, experience, and values, in order to broaden the criteria of evaluation needed for a genuinely open, societal exploration of visions of the ideal society and the human good life. As Borgmann argued, the democratic and rational revaluation of technological innovation must be based upon a community engagement with the articulation and critical assessment of the concrete proposals of technology in contrast to evaluations of current knowledge and values, made in accordance with considerations of sufficiency for a practical appraisal of the good life.[389]

Decentralization and Democracy:

In *The Road to Serfdom*, first published in 1944, Friedrich Hayek argued that a centralized system of planning would inevitably lead to totalitarianism and tyranny and, therefore, socialism was a utopian delusion.[390] By collectivizing power and putting it at the disposal of an authoritarian committee or leader, within a system that demands that all individuals concentrate all their efforts for the benefit of the whole society, the decision making resources of the whole society would be concentrated in a small group of people. It is quite inconceivable that any small group of people would have sufficient intelligence and knowledge to amass a sufficient degree of foresight in order to adequately plan the construction and development of a society. It is also quite inconceivable that a small group of people would be able to anticipate and plan for every consequence of every possible event that could occur in the future. Centralized planning increases the societal vulnerability to collapse or disaster and, therefore, it will inevitably result (sooner or later) in economic failure or the inability to cope with some unforeseen event. It would also inevitably lead to a complete loss of individual freedom. According to Hayek, only the worst elements of society—those that utilize deception to appeal to the lowest common denominator and basest instincts—are capable of gathering the massive support that is needed to govern the majority that have no strong convictions of their own ("the docile

---
[389] Borgmann (1984:114-24; 1990)
[390] Hayek (2001)

and gullible," as Hayek put it), but will accept any system of values providing that it is "drummed into their ears sufficiently loudly and frequently," and it is based on ideas that take advantage of human weakness, such as hatred for an enemy or fear (of enemy infiltration or terrorism, for example). Thus, as Hayek argued, the majority will be comprised of those that are most readily influenced by propaganda because they are possessed by vague and ill-formed ideas and they are most readily aroused by their emotions and passions. Hayek argued that when elected officials embark on a course of economic planning for the whole of society, often described in vague terms using populist rhetoric, then there will be an agreement on the need for central planning and an absence of any agreement regarding how to implement the plan. This, of course, is doomed to failure from the outset because its only criterion for success is that the majority has agreed upon the policy.

Moreover, Hayek argued that, due to the diversity and plurality of the possible courses of action in society, it is impossible for democratic assemblies to function as planning agencies. Compromises only result in the construction of an unworkable plan or everyone being dissatisfied with the results. Ultimately, democratic assemblies need to delegate the task to experts or charismatic leader and this will result in power residing in the hands of a single person or a few individuals. Hence, Hayek argued that the best way to limit and decentralize power is to distribute it throughout all the individuals of that society and allow competition between those individuals to provide solutions to our problems. He argued that it is only by basing economics on decentralized competition can we hope to preserve individual freedom and optimize societal creativity, by utilizing individual knowledge and creativity. Hayek advocated liberal capitalism as the best form of social organization because, providing that there is some basic insurance against the common hazards of life and a clearly thought out legal framework, individual competition to acquire personal wealth provides the best basis for human efforts to be effectively coordinated without any intervening authority. Putting aside that Hayek did not make it clear how this level of insurance was to be decided and distributed, nor who was to do all the thinking through of this supposedly clearly thought out legal framework, my main contention with Hayek's thesis is with the fundamental idea that liberal capitalism provides the best ideological basis for people to organize their efforts. In my view, the problem with liberal capitalism is its unquestioned premise that a principle of laissez-faire competition will lead to decentralization. This is not necessarily the case.[391] Unfettered economic competition always favors those with the most wealth, resources, and power because it allows them the greatest access to technology, and to enjoy the benefits of an economy of scale and capital flight to regions or countries with cheap labor. When empowered by the technological society, liberal capitalism will inevitably lead to an industrial capitalist economy based upon corporatism and globalization, which results in increasing amounts of wealth and power in the hands of a small social elite that is capable of moving its capital to countries which provide it with the cheapest labor, while being able to provide the means to produce more goods, at lower costs, and transporting them across the world.

Industrial capitalism is far removed from Benjamin Franklin and Thomas Jefferson's ideological visions of an idyllic nation comprised of liberal individuals engaged in competitive capitalism in order to increase their personal wealth and, thereby, the wealth of the nation. There are also obvious moral criticisms about the hypocrisy of the foundation of political and economic rationality of the American form of liberal capitalism—using the prosperity of the United States of America as evidence for the success of free-market individualism—when its wealth is historically based upon the

---
[391] Rogers (2006; 2008; 2011; 2012)

slavery of Africans and the theft of the land of the Native Americans. Liberal capitalism may well have been a reasonable and achievable ideal when it was based on the craft based workshops and small farm based agricultural markets of postcolonial Massachusetts and Virginia, but when empowered by the technological society, liberal capitalism leads to the concentration of wealth and power in the hands of a few individuals because small businesses and farms are unable to produce goods cheaper than large businesses or farms and, therefore, an economy based on competitive individualism will inevitably end up with a few individuals controlling all commerce, industry, agriculture, and mining. An economy based upon competition allows individuals with greater access to resources to be able to increasingly dominate the economy, which eventually results in an oligopoly planning the world economy and the directions of technological innovation. It may well be the case that more people in the so-called developed countries have access to cheaper goods, but the consequence of industrial capitalism is that the majority of individuals end up having the terms of their labor and material conditions of their life decided by the few.

Once corporations acquire equal resources and power as nations, the only difference between the centralized state and a corporation is regarding whether the members of some executive committee for a state bureaucracy or the board members of a corporation makes the decisions that have effects on millions of people who have no say whatsoever in the economic and political development of their lives. Either way, the concentration of power and the social stock of available imagination and creativity are reduced to that of a few individuals dictating the criteria for evaluating the development of the technological society. Once this is coupled with a political system within which those that can afford access to mass media and expensive political support are the only ones that have a chance of election to legislative bodies then the process by which the legal framework and levels of social insurance are established are decided by the wealthiest members of society. When empowered by the technological society, liberal capitalism transformed into industrial capitalism and is as likely as any socialist state to result in a totalitarian system of centralized planning that is organized solely on the basis of preserving the *status quo*. The best conditions for propaganda are exactly those produced through industrial capitalism; where a massive media infrastructure is owned by a social elite with the vested interest in maintaining the *status quo* and is the main source of information and connection between individuals and government, which also has a vested interest in maintaining the *status quo*.[392] Also, as Horkheimer and Adorno argued, modern media—"the culture industry"—has degenerated into the production of social conformity and mass entertainment.[393] Under these conditions, the conformity of individuals to the *status quo* is the most likely result. Hence, contrary to Hayek, once we take the rise of the technological society in account, we can see that liberal capitalism is just as likely to result in totalitarianism and ineptitude as any other system, and, consequently, is also the road to selfdom for the majority of people.

The contradiction between the capitalist economic imperative and the technological imperative is most evident in modern industrial societies that are considered advanced and prosperous, while, within these countries, a significant proportion of the population lacks access to education and health care; modern housing and hygiene systems; civic security and legal support; nutritional and balanced foods; and opportunities for creativity and leisure; all the while, a minority enjoys all the fruits of technological empowerment and controls the directions of its further development. In order to maintain a sufficient proportion of the population for low-paid and low-skilled employment—a standing-reserve of workers—while also keeping labor costs as low as

---

[392] Ellul (1965: 90-116, 232-50); Herman & Chomsky (1988); Chomsky (2002); Rogers (2008; 2011; 2012).
[393] Horkheimer & Adorno (2002)

possible, industrial capitalist societies must deny a significant proportion of society any real access to civilized and democratic life, and produce and reproduce a culture of mass dependency on the system of relations of production and consumption, as dictated by the owners and controllers of capital. Any failure to keep labor costs down results in lower profits for investors, which would result in higher opportunity costs, and, inevitably, a flight of capital to investments capable of delivering higher returns. Industrial capitalist societies must maintain a repressed level of knowledge and skills among the working-class; limit the possibilities for mass participation in the political evaluation of the directions of the technological development of society; and prevent any alternative economic systems from being successful. As a consequence, the technological imperative has become subordinate to the capitalist imperative to maximize profits, creating an irrational consumer society dependent on globalized relations of exploitation and perpetual warfare, insensitive to its pollution, waste, corruption, and cruelty, even at the risk of destroying life on Earth in the process of maximizing profits. Despite all its promised competitive advantages of the self-regulating, laissez-faire market, industrial capitalism has led to the concentration of capital in oligopolies and monopolies, which, of course, creates an antidemocratic basis for the technological society and leads to all of the problems of centralization that Hayek considered to be the inevitable consequence of socialism. Moreover, under the technological imperative, we should rationalize consumption—by consuming more efficiently and wasting less—rather than squandering precious resources to supply the insatiable and increasingly shallow wants generated by the device paradigm. Of course we need to address the extent that profiteering opportunists of the advertising and public relations industry have hoodwinked us into participating in irrational and meaningless purchasing behavior, by equating the good life with increased consumption, and pandering to our fears and other pathologies. In the political sphere, this has resulted in the domination of corporations over mass media, political campaigns, candidates, and elections, along with legislation, regulation, enforcement, and the whole political process.[394]

We need to recognize that the good life depends upon much more than increasing our levels of consumption. If we wish to rationalize the productive and consumptive base of our society then we need to determine the extent that localized industry and agriculture can satisfy our needs. How we evaluate the costs and benefits of globalization? Do we really find it acceptable to ignore the oppression of the rights of foreign workers simply because the factories that utilize cheap labor in other countries can produce cheaper manufactured goods for us to purchase? According to the apologists for capitalism, globalization results in a freer and more democratic society everywhere, but this simply ignores the consequences of unsustainable levels of consumption and the capitalist imperative to maximize profits. We need to seriously question to what extent that an economic system that defines "competitiveness" and "efficiency" in terms of reduced wages and longer working hours for the majority of people—as well as moving people away from economic self-sufficiency and causing unemployment, waste, and pollution—is actually leading us in the direction of a free and democratic society, or whether it is a nothing more than a thinly veiled system of coercion and exploitation for the benefit of a few at the expense of the many. However, once we submit to the technological imperative, we cannot accept the practical benefits of technology without also accepting its moral imperative of enhancing the material and aesthetic conditions for the whole of society by removing inefficiency, waste, and pollution. The technological imperative is the engine that drives the transition from the paleotechnic to the neotechnic phases of the development of the technological society. The social denial of this moral imperative in favor of a discourse limited to talk of costs, capital flight, and the loss of jobs—all of

---

[394] Rogers (2011; 2012)

which have been used to justify the unfettered dominance of the capitalist economic imperative—leads society to fragment and degenerate into an irrational *adhocracy* of inequalities and exploitation—alongside the development of systems of public welfare and administration—within which a small elite benefits by manipulating and sabotaging the educational and psychological development of the rest of society, while controlling mass media and the political superstructure.

The technological imperative is one that must eliminate social distinctions because universal and collective goals are those of the empowerment of labor and the use of science to liberate human beings from the limitations of our material conditions. In order to maximize the efficiency and productivity of society, as a totality, empowerment and liberation must be distributed throughout the whole of society. Technological efficiency, capacity for innovation, and sustainability of any society—as well as its ability to respond to unforeseen natural disasters and other unpredicted events—are all dependent on the internal level of diversity, plurality, and cooperative responsiveness of society, which are all maximized when knowledge, skills, access to resources, and capacity for decision making are distributed throughout the society, and tested by their application in the further development of the technological society. Thus, like Ellul, Mumford considered that the dominance of the technological imperative would inevitably lead to communism.[395] However, Mumford argued that this conception of communism needs to be post-Marxist because he rejected the paleotechnic values upon which he considered Marxism to be based, which it certainly was in the Soviet Union. After he expressed admiration for "soviet courage and discipline," Mumford rejected the idea that communism needs to adopt the methods or take the political and institutional form proscribed by Marx, Lenin, Stalin, and the Soviet Union. He stated that his notion of communism—as a universal system for distributing the essential means of life—owed more to Plato than Marx.[396] He also pointed out that within many modern societies schools, libraries, universities, museums, swimming pools, hospitals, sports facilities, and public parks are supported by and available to the community at large. Emergency services, such as police, fire, and ambulance, are already provided on the basis of need rather than ability to pay and, hence, and likewise could be considered as basic communism. Also, even though it takes the form of state-administered welfare, a basic communism exists in most modern countries as far as provisions for the unemployed and the elderly are concerned.

The capitalist insertion of their pecuniary interests into the technological development of society necessarily limits (and even retards) the empowerment and liberation of society because it must disempower the majority of society disproportionately in order to protect the social privileges of the minority. Indeed, industrial capitalism provides large numbers of people with a wide variety of low priced, mass produced goods, providing that there are sufficiently large numbers of people coerced to work for low wages, but the true cost of that "benefit" is that an even larger and increasing number of people are unable to develop sustainable and democratic communities because the control of their economic basis is far removed from them. It becomes impossible for citizens in a community to plan the development of their lives if the technological and economic infrastructure of their community can be removed or transformed by people outside of their community, simply because it is more cost efficient to operate industry or agriculture elsewhere. It simply does not matter whether the decision is made by a board of directors or a committee of commissars, the outcome is the same: the majority have no control over the economic and political development

---

[395] Mumford (1963: 355-6)
[396] *Ibid* 403

of their communities and lives. If the implementation and development of the technological society is dictated in accordance with the short-term economic and political interests of a minority, it is impossible for any community within that society to develop a democratic and rational basis for its own organization and development because it cannot plan for its future. This limits the bounds for strategic decision-making and informed public discourse, hence, leads to increasingly limited criteria for the intellectual and political development of society, which, of course, reduces the diversity, pluralism, creativity, and motivation of its citizens—which results in economic conformity and political apathy—and, therefore, reduces the total capacity of society as a whole to adapt and respond to unforeseen changes or events.[397]

However, as I argued above, the development of society solely in accordance with the technological imperative would also to lead to a totalitarian system. Leaving decision-making to a technical elite would result in a technocracy that would dominate all aspects of life. The Soviet Union further compounded this by placing this technocracy under the dictates of a political elite. The centralized and dictatorial Soviet Five-Year Plans were a good example of how a reduced stock of decision-makers were a structural handicap and antithetical to the whole social process of intelligent, flexible, and adaptive organization of technological innovation within a complex and unpredictable world, thereby damaging the ability of society to respond to drought or any natural disaster, as well as to provide for human needs in a manner that respects the freedom and dignity of all. After Stalin imposed centralized plan after centralized plan to transform Russia from an agricultural feudal nation into a modern industrialized state—a militaristic system of state-capitalism—the authoritarian and ideologically-driven projects implemented by Soviet engineers and architects, often working under the fear of arrest for sabotage or counterrevolutionary ideas, imposed their technologies upon local communities and workers without any knowledge about the conditions and complexities of the local situation and circumstances, which resulted in inefficiency, waste, famine, pollution, and environmental catastrophes.[398] Democracy is a condition for genuine scientific and technical progress because centralized planning fails to foresee all the details in advance, and, as a consequence, generates an anarchy of *diktats* and all the inefficiencies and self-deception that follows from this.[399] Limiting the criteria for decision-making to those of an elite, be it a technical, political, or economic elite, reduces the stock of imagination, skill, and experience required for the rational, intelligent, and adaptive development of society within a complex, open-ended, and changing world. Even if the capitalist economic imperative were to be rejected in favor of the technological imperative, it would result in the assimilation of all human activity, including science and politics, in the collective construction of increased technological power and control over the material world, while reducing the capacity of society to adapt to changes in the world, including changes in itself as an ongoing, heterogeneous, and experimental society. The development of increasingly powerful technologies, without any moral imperative or constraint, apart from the acquisition of more technological power, would create an unstable, unsustainable, and irrational society because it lacks any vision of human well-being and the good life, apart from one that reaps the benefits of the technological imperative. Under the direction of the capitalist economic imperative, such a society inevitably hits its limit with the development of the device paradigm, as it promotes conformity, shallowness, and unsustainable consumption.

---

[397] Rogers (2006; 2008)
[398] Gorokhov (1992)
[399] Rogers (2008)

Even if the bounds of technical rationality were evaluated in accordance with a democratic imperative—maximizing the level of inclusion and participation in the evaluative and decision-making process—we would need to recognize that in any heterogeneous society, the democratic process is unlikely to result in any single vision or set of criteria for evaluating and deciding the future development of society. Given a sufficient level of diversity, any democratic society must acknowledge and respect a certain level of pluralism. This means that, given any problem or matter of shared concern, there are a plurality of possible solutions or resolutions, each endorsed by a different group of people. Furthermore, there is an absence of any consensus about the nature of the problems and concerns we face, and, even when we agree on problems and concerns, there are differences in how we prioritize them. In the absence of any universal agreement on how we envision the good life and the purpose of human existence, which also entails a vision of a good society within a natural environment, the criteria for the evaluation of technical rationality are at stake, and the technological society takes on the character of an experiment.[400] Hence, the democratization of the technological society requires that there must be pluralistic outgrowths in the development of the technological framework itself, allowing different ontologies to be explored; this not only allows a plurality of different directions and levels of technological development, thereby allowing community autonomy in the development of its technological infrastructure, practices, and goals, but also fostering competing fundamental paradigms of how technology works and how it should be developed and implemented in both theory and practice. It also allows the possibility that some communities could 'retreat' from technology use, at least in the sense of having autonomy in the decision regarding the necessity or sufficiency of any given technological object or direction of development. Communities such as the Amish would be examples of this.[401] The development of the technological society must be subjected to pluralism—allowing different communities to adopt different directions and levels of technological development—to optimize societal diversity and flexibility. Neither liberal capitalism nor centralized socialism are an adequate ideology for guiding the democratization of the technological society because their criteria for success will be determined by the monism of the interests of an elite—be it economic or political—rather than the needs of the local communities within which the technologies will be implemented and developed. Both of these ideologies lead to the public exclusion from decision-making or only allows the public to participate after all the important decisions have been made. They also tend to reduce the evaluation of proposals to very limited criteria and suppress any critical discussion about alternatives, either by imprisoning dissidents or discrediting them. In both the USSR and the capitalist West, technocratic administration supported the power of the elite and embodied ideological commitments rather than a practical response to societal problems.[402]

Elites tend to choose to implement the technologies that provide the results that they want and ignore all other effects and their implications for society. Either way, we end up with having technologies and policies imposed upon us by a boardroom or committee. The restriction or limitation of decision-making to specific classes of society, whether an economic or political class, is a societal mistake because it reduces the breadth and latitude of the tactic knowledge, skills, experience, and imagination that is brought to bear in making decisions regarding societal development. This is made even worse when the elite conspires to limit the access to education for the majority of people—using it as a means of propaganda in favor of the *status quo*—and suppresses all forms of genuine political opposition; this actively destabilizes the cohesion of communities in order to

---

[400] Rogers (2006; 2008)
[401] James (2013); Umble (2000)
[402] Thomas (1994); Cunningham (1987)

create a society of disempowered and obedient individuals. It radically reduces the societal capacity for creativity, flexibility, and adaptability by not only reducing the societal stock of tacit knowledge, skills, and imagination of the population, but also disabling the capacity of the population to self-organize at either a local or societal level. Put simply, it is a matter of numbers. Increasing the number of citizens that are involved in the experimental development and organization of society increases the quantity and diversity of ideas, skills, experience, and creativity that is brought to the task. My argument is that we need to recognize that optimizing the capacity for social participation and cooperation is the best way to organize the development of society, but, once we realize that pluralism and diversity optimize the flexibility, adaptability, and creativity of our society then it is evident that the decentralized democratization of our society is the best political system for people to decide how to develop society.[403]

In order to avoid the limits and totalitarianism of centralization, the economic and political structures of society should be decentralized into local communities governed by the people who live in them. Providing that these communities are able to communicate and interact with other communities, throughout the whole of society, this optimizes the creativity of society by maximizing participation and embracing social pluralism, and maximizes the chance for the development of sustainable communities and a healthy society. How a community elects to run its economy and technological development should be a matter of local decision, based on its own resources, knowledge, and experiences. The ability of its citizens to negotiate and cooperate with their neighbors to obtain what they need or when local decisions are of concern to any neighboring communities would lead to a ground-up approach to decision making that remains located within the communities for which the matter in question is of concern. Once a community focuses on its own development then there are plenty of opportunities for citizens to look beyond the concept of labor as a commodity to be exchanged for wages, to pay the rent and bills, and, instead, see it as a shared and cooperative social activity to build a better community in accordance to whatever vision of a good community that its citizens come together and decide upon. Even in a culture that values individuality and private property, once the local community becomes the local focus of all economic and political relations, while individual rights and the democratic process are legally protected by the national or federal government, then it is possible for society to optimize its pluralism and diversity along the boundaries of differences by localizing and limiting decisions about both means and ends to communities and associations between communities, and enhance the available stock of experience and creativity of society as a whole. It is simply a matter of hedging one's bets by trying all the solutions that people seriously propose. This will not only empower people to experiment and organize themselves when making decisions about the development of their communities, allowing different communities and individuals to try different solutions, as decided by the community members effected by those decisions, but it will also provide society, as a whole, with a greater capacity to adapt to unforeseen events and rapidly recovery from disasters. It also facilitates the limitation of the societal impact of any policy or decision to the locality within which it is implemented and developed. It at least acts as a damping factor on the consequences of mistakes by limiting them to a locality.

---

[403] Rogers (2006; 2008)

Democratizing the Technological Society:

The real issue is not technology and progress *per se*, but the variety of possible technologies and paths of progress, among which we must choose. Modern technology embodies culture and the values of the technological society and especially of its elite, which rest their claims to hegemony on technical mastery. Andrew Feenberg adopts this view and was critical of predictions of the technological society as condemned to authoritarian management, mindless work, and the unrelenting consumption, and he also rejected the view that technical rationality and humanist values are "contending for the soul of modern man."[404] He considered these views to be clichés. Feenberg rejected the dichotomy between technological rationality and humanist values implicit in the debate between instrumentalists—who hold that technology is a neutral means to rationally satisfy ends or goals established in other social spheres—and substantivists—who hold that technology is an autonomous phenomenon, overriding all traditional or non-technical values, shaping both humanity and the natural world, and disseminating and transforming ends and goals, as both an environment and way of life. He took this distinction from Borgmann, and cited Nicholas Rescher and Emmanuel Mesthene as proponents of the instrumentalist theory of technology, and cited Ellul, Heidegger, Habermas, and Borgmann as proponents of the substantive theory of technology.[405] Instead, as his point of departure, Feenberg took the Marxist line that the degradation of labor, education, and the environment is rooted not in technology *per se*, but in the antidemocratic values that govern technological development. Maintaining the distinction, on one side, between the essence of technology (which he left open) and, on the other, the perversion or distortion of technology (which he considered to be antidemocratic and in favor of the elite), Feenberg called for the development of critical theory to generate a cultural critique of technology, which could rationally evaluate the larger context of technology, thereby articulating, examining, and judging the values and relationships that have become central to the exercise and organization of political power through technology.

Feenberg argued that the political directions of technological development are based on ontological decisions about what it means to be human and what kind of civilization we wish to construct. He argued that the exclusion of the vast majority from participation in this decision is the underlying cause of many of our problems, and a good society should enlarge the personal freedom of its members while enabling them to participate effectively in a widening range of public activities, therefore, a profound democratic transformation of modern industrial society will resolve these problems. Technological rationality stands at the intersection between ideology and technological power, where the two come together to control human beings and resources in a way that installs the values and interests of the elite in the very design of rational procedures and machines, even before these are assigned a goal, tacitly regimented as rules and procedures, devices and artefacts; as "technical codes" that routinely reproduces the pursuit of power and advantage. Hence, Feenberg argued that technology should be situated as an ambivalent process between alternatives within a scene of political struggle and debate. Consequently, he postulated that there are at least two different paths of technological development available to us, for example, whether we use computers to control or liberate communication, whether we build our cities around public or private transport, or whether we construct factories as an assembly line or a workers' cooperative. Hence, there are always at least two possible civilizations that we can choose between when choosing between technologies. He argued that if this choice is made through grass

---

[404] Feenberg (1991)
[405] Borgmann (1984: 9). Rescher (1969); Mesthene (1970)

roots, public participation in the implementation of technology then new paths of technological development and the construction of civilization will become available to us. However, even though I agree with Feenberg that the paths of development of the technological society depends on human choices, we also need to look at how the dialectics of technology determines—informs and constrains—those human choices within a society that has already embraced the societal gamble. We need to look at how the dialectics of technology is at the core of its natural sciences and how shapes how we understand the natural world.

Before discussing Feenberg's critical theory of technology further, I shall make some general comments about the conditions under which the technological society could become a democratic society. We need to take a further step back before we can rationally examine and evaluate technologies, in order to choose between them, and examine the rationality of our vision for society, human well-being, and the good life. We need to place these metaphysical conceptions of being human and the natural world under critical scrutiny. Once we recognize that each technological development is an experiment, we can take some consolation in that fact that, in a pluralistic society, different communities will be developing different technologies in different ways, as well as having different goals and ideals, and we would be able to compare the development of these communities and learn from the ongoing successes and mistakes. Hence, it is essential that we develop critically democratic evaluations of technical rationality and how we understand the alethic modalities of technological activity.

As Richard Sclove argued, the aim of democratic evaluation of technology is not to predict the future, but it is to evaluate the technologies that we already have, and ask ourselves whether these technologies enhance or reduce our capacity for democratic participation.[406] The more that we participate in this process then the better we will become at evaluating, reforming, and replacing existing technologies, rather than irrationally innovating new technologies and implementing them for the sake of more power. If this process is performed in an open and cautious way then we will be able to recognize adverse structural effects quickly, modify and replace them before they become irreversible or disastrous, and make this information widely known to others. Sclove argued that strongly democratic communities should exercise *reasonable self-restraint* when making decisions in order to avoid "translocal harms," such as pollution, for example.[407] While I agree with Sclove that reasonable self-restraint would help community governance and its relations with other communities by avoiding "translocal harms," it is not prudent for a model of strong democracy to overly rely on this.[408] After all, it seems to me that we would not really be in the predicament of needing to radically reform our political system of decision-making if we were universally capable of exercising reasonable self-restraint, and it also raises the important question of how we decide what reasonable self-restraint is, and who decides this. After all, it is quite possible that a democratic community could decide that homosexuality, feminism, or interracial marriage cause "translocal harms" and should not be tolerated. All manner of oppression and intolerance can be disguised under the mask of preventing "translocal harms," and history provides us with countless examples of terrible acts of mass murder and oppression all in the name of preventing "translocal harms." Instead, what we need to do is to raise and answer the deeper constitutional questions of how the boundaries of political jurisdiction should shift in order to encompass a broader notion of community when local decisions are perceived potentially to lead to "translocal harms," but also protects minorities and

---

[406] Sclove (1995)
[407] *Ibid* chap. 7
[408] Rogers (2008)

individuals from the dictates of the majority. What we need is an inclusive concept of community enshrined in constitutional law, which is something that preserves a republican form of government by protecting enumerated rights over and above the contingencies of the democratic decision-making process, while asserting the notion that everyone equal under the law, alongside the idea that the legislature itself has limits that cannot be legitimately transgressed, no matter how popular transgressing those limits may be. When the consequences of local decisions are likely to extend beyond the local community then a broader notion of democratic community should be constitutionally available in order to facilitate the participation and representation between members of all involved and concerned citizens, while also respecting the equal rights of all citizens, regardless of whatever the majority decision may so happen to be.

It is an implication of such ideas that the boundaries of "the community" are capable of being shifted to include all those concerned with the wider consequences of local decisions in response to wider perceptions of possible "translocal harms." Communities become fluid entities that wax and wane depending on the overlapping interests and concerns of its members. The boundaries of political jurisdiction of the democratic process should be flexible enough to adjust in accordance with the perceived need of wider participation of those concerned with any local decision and its potentially regional or national consequences, while also respecting the rights of all people both within and without any particular community. It is essential that human rights transcend the dictates of local communities; otherwise democracy is likely to degenerate into autocratic feudalism, demagoguery, or 'mob-rule' extremism. It is also essential that participation is as inclusive as possible, otherwise democracy will simply reproduce inequalities, classes, elitism, and only represent the interests and concerns of a minority. The exclusion of concerned and involved citizens from the decision-making process should make any resulting decision undemocratic, and, therefore, a violation of fundamental constitutional principles and illegal. The removal or disregard of individual rights would also make any resulting decision unconstitutional, and, therefore, undemocratic and illegal. It is at that point that excluded citizens could call on the intervention of the regional, national, or federal government, or the aid of other communities.

A strong democracy requires a broader notion of community than that of simple geographical proximity. It requires a network of relations and interests among people who know and communicate with each other. It requires shared life-world experiences and projects, thereby generating shared problems and concerns. Thus, any individual may well be a member of several communities, not limited to geographical location, and discover that different communities have overlapping concerns and ideas about the future development of society. This is even more apparent since the invention of the Internet, which allows people to form communities remotely. If the democratic process constitutionally demands that shifting boundaries of jurisdiction are required in order to include all concerned citizens in any decision, while respecting the rights of all people, then there is no need to rely on reasonable self-restraint, even if we acknowledge that it would be beneficial and it is harder to proceed without it. Reasonable self-restraint can be learned through democratic participation, rather than being a condition for it. It is important for the health and sustainability of a society that the search for commonality does not degenerate into social intolerance, conformity, bigotry, and xenophobia. Intercommunity exchanges and affirmation of the societal value of diversity and pluralism are essential for the health of society, therefore, as Sclove argued, even when consensus is hard to achieve, a genuine commitment to strong democracy increases the level of mutual respect and tolerance for other people.[409] When all participants feel respected by the community then there is a greater chance of achieving mutually beneficial agreements, actually listening to the ideas of other people, and also citizens are more likely to accept decisions that they disagree with.

In a strong democracy, the function of government would be to protect and empower the democratic process within local communities, rather than make decisions for people. The mediation and content of the democratic process must be left to the local citizens of those communities, as would also the implementation and administration of any decisions. The exact nature of relationship between local communities and central government must remain a constitutional agreement to accept a particular form of political process and arrangement of institutions, thereby protecting this arrangement from the caprice of "democratic excesses." Whether government is comprised of a stratified or federalized system of democratic forums and institutions; how participation and representation are to be combined and balanced; whether democratic participation makes the nation-state and idea of national government obsolete; all involve complex questions that are, unfortunately, beyond the scope of this book. However, the minimum requirements of the relationship must be that the processes of governance are transparent, all citizens have equal rights and representation, and citizens have access to and can participate in government at all levels. It is also important that citizens from different communities also can join together in NGOs that can facilitate collective action, and, hence, act as a check and balance against abuses of power and ineptitude on the part of any government. These NGOs can also help the democratic process and facilitate the exchange of knowledge, skills, ideas, and resources between different communities, while also operating with the support of local communities.

One of the oft-repeated criticisms of strong democracy is that nothing would ever get done if everyone was involved in deciding what was to be done. However, this is not really a good objection. It not only presumes that the qualification for strong democracy is that everybody decides everything, which is itself an unrealistic demand, but it also presumes that it is always better to do something quickly rather than wait until there is a well-thought out course of action, which is itself an unintelligent and foolish demand. Of course, at least initially, genuinely democratic decision-making will take much longer than that of authoritarian dictatorship. Perhaps that is unavoidable, but it is also clearly the case that any decision is more likely to be a well-thought out decision if it is made and agreed upon by all the people likely to suffer the consequences of that decision, rather than on the whim of a single individual, committee, cartel, or board of directors. It may well take a long time for a community to come together and agree on important decisions, but the reward of this effort is that the decisions that it makes are likely to suit people for a long time and perhaps, with luck, only need subtle modifications or further elaborations to adapt to future changes in circumstances. This is actually much more efficient and sustainable than having to follow the arbitrary dictates of a tyrant, even if s/he can make them in the blink of an eye, because arbitrary dictates will probably need to be completely different tomorrow or they will likely fail to achieve their goal. This is even more evident when we consider the historical cases of all the squanderous and murderous efforts made by tyrants and corrupt governments to cover up their bad decisions and suppress all criticism—tyranny and fallibility seem inexorably bound together. On the other hand, by making participation open and accessible to all those that wish to participate, on a community-based and voluntary basis, sharing common goals of communicating successfully and cooperating with others, it is also evident that a great deal of action can

---

[409] *Op cit.* 160

occur in a manner that leads to societal stability and consensus, without violence and bloodshed.

Democracy does not imply that any single elected committee or body of representatives needs must debate and ratify every decision (as if the decision about whether to build a nuclear reactor to supply a region with electricity has the same social weight as the question of whether I should have mashed potatoes or yams for my dinner). All that strong democracy minimally requires is that there are social processes, through which people can bring their concerns to public attention and deliberation, socially initiate change in policy or practices, and organize participation in setting the agenda and a plan of action. As long as there is an enforced legal framework that protects the democratic and individual rights of citizens, allowing all those concerned to participate in decision-making, openly communicating and sharing information, then people will be able to learn about proposals and participate, if we so wish, in the decision about whether and how *any* proposal should be implemented and developed, *not all*. This legal right to participate would allow citizens to be able to shape legislative and electoral agendas and procedures, when we are affected by those agendas and procedures, in order to develop and adapt them to the democratic needs of different communities and our different circumstances. It is important that the democratic process itself can be adapted to the plurality and diversity of local needs, concerns, experience, and ideals, and, providing that citizens are protected from abuse and disenfranchisement, then this degree of flexibility is itself highly desirable for the development of strongly democratic communities. It is important to realize that the public participation in the design and development of communities is not simply a matter of electing to proceed with one design proposal over another. Public education about how to participate in the development of urban areas, say housing, is not about just about developing the infrastructure of our communities, but is about the local commitment to developing democratic processes as practical, public, and inclusive modes of decision-making.[410] The immediate benefit achieved from public participation in the whole design process is the improved ability to participate through the acquisition of and experiences of how to participate. The more ordinary citizens are involved in the design of our communities then the more intimate we will become about the processes of how decisions are made within a complex society, and how social structures have been constructed to prevent and hinder public participation. If the democratization of community development is to become a reality then it is imperative that we learn and develop strategies for diverting economic and political resources and powers into our communities. As well as putting pressure on the current political order to allow greater public involvement and representation, this involves actively participating in the development of our communities, using our current knowledge, skills, and resources, without waiting for the permission to do so. This involves deciding how to use the available technologies and resources through democratic forums and processes that we democratically agree upon at a community level. It also involves teaching each other basic skills that allow us to become more self-sufficient and capable of reasonable self-restraint.

Another oft-repeated objection against strong democracy is that it would inevitably involve protectionism and the regulation of regional, national, and international trade and, as a consequence, it is too idealistic to work in the real world of globalized trade and large corporations. This objection is that the protection of local economies runs against the international economic system (institutionalized through the General Agreement on Tariffs and Trade (GATT), the World Trade Organization (WTO), the World Bank, as well as other international organizations committed to globalization) and genuinely democratic trade agreements would effectively disband GATT and the WTO

---

[410] See Ward (1983)

in favor of decentralized local trade agreements and an international commitment on the part of local communities to provide sufficient democratic representation and participation to effectively democratize world trade; the organization and protection of labor and workers' rights; egalitarian access to communications, technology, and transportation; the protection of indigenous peoples and environments; the dissemination of technology and scientific knowledge and skills; and the establishment of pluralistic democratic forums for the representatives from every nation to discuss ideas, concerns, and grievances (without being dominated by the most powerful and wealthy nations). It is claimed that this would be impractical or impossible.

My response to this objection is a straightforward question. If it is impossible to democratize world trade and politics because of the established vested interests of powerful nations and corporations, then to what extent can we describe Western countries as being advocates and promoters of global democracy?—or even comprised of genuinely democratic countries at all? It is also frequently argued that any form of workers' rights increases the chance of capital flight to countries that do not provide workers' rights. However, the more economically diverse and self-reliant that a community becomes then the more it is able to accommodate the consequences of capital flight, because the community does not depend on a single industry (controlled by investors and directors from outside of that community) and is thereby able to make demands about the working conditions and rights of its citizens. Hence, one of the important democratic functions of government is to protect the right of every community to govern its own economic development by legally protecting and encouraging sustainable economic relations between communities that are on an economically equal footing, as far as levels of wages and workers' rights are concerned; hence promoting fairer trade and protecting local economies, and also offering more opportunities and incentives for local citizens to start their own businesses and workshops to satisfy local needs. This would prevent large corporations from taking economic advantage of their antidemocratic policies (such as using coerced cheap labor in underdeveloped countries to control markets in developed countries). Of course, in order to safeguard sustainable economic relations between communities from abuse and corruption, it would be necessary to establish regional, national, and international procedures for monitoring trade, without the need of any centralized authority; democratically addressing grievances, and punishing offenders. It involves collectively boycotting developments, businesses, and industries that are imposed upon our communities, unless we are able to participate in whether and how they are to be designed, implemented, and developed. And, of course, this will result in difficulties, inconveniences, struggles, and failures, but if we wish to live in a democracy then we must come together and take control of how our communities are developed. This requires great efforts on our part. Otherwise the decision-making process and power will remain in the hands of an elite that is relentless and tireless in its efforts to come together and take control of how our communities are developed.

The ability of communities to learn how to protect themselves by preventing exploitative and antidemocratic trade' through education, persuasion, voluntary enrolment in cooperative and collective action. Large businesses and corporations need to continue to invest in developed and democratic communities, even with their higher labor costs and levels of workers' protection, because if investors wish to take advantage of the markets of those communities then they have to invest in local commerce, industry, and agriculture. It is on the basis of our experiences of living in a community that we learn how to nurture collective practices, mutual respect, and develop a sense of commonality and friendship with our neighbors. These are basic experiences for the development of translocal democratic organizations and institutions to establish collaborative relations, such as economic relations, with people from other communities and countries. It is essential that the vast majority of people learn how to communicate and cooperate, if we are to achieve enduring international peace, justice, environmental protection and sustainability, and mutually beneficial trade agreements. Moreover, the localization of trade allows citizens to know all the participants, where the goods come from, and how they were produced, which develops a rich life-world relation with our material conditions.

This also provides us with a response to the apologists for globalization. For example, Milton Friedman in his defense of "free market globalization" claimed that the globalization of the market makes prejudice meaningless and obsolete. Friedman, quipped,

"The purchaser of bread does not know whether it was made from wheat grown by a white man or a Negro, by a Christian or a Jew." (Friedman, 1962: 109)

Indeed, all other things being equal, this is largely correct. But, in my view, the problem is that when the process by which wheat is grown and made into bread is completely anonymous and abstract, the purchaser of bread simply does not care where it came from or how it was made. Labor becomes simply one commodity among others. When bread comes into existence only at the moment that it is picked off the supermarket shelf, globalization generates an indifference to the material and social conditions for the production of our food. It allows us to be content to be ignorant about the conditions upon which our lives are made possible. This is an understandable state of denial in the face of undeniable moral truths, because once we are aware of the terrible conditions under which most of the people who grow our wheat in the Third World live and work, as well as the pittance that they are paid for their backbreaking efforts, then the bread becomes increasingly hard to swallow. Our only alternative to blissful ignorance is to develop callousness towards others, as a precondition for our pleasure in our food. Globalization turns both the natural world and people into standing-reserve, and, as a consequence, we should consider it to be a manifestation of *Ge-stell*, and, with all its talk of market discipline and efficiency, it is bound-up with the societal gamble on the technological society, driven by the capitalist economic imperative.

Sclove noted that a study of several thousand American workers concluded that they spend more mental effort and resourcefulness in getting to work than in doing their jobs.[411] The physical organization of the life-world, especially its public spaces and the working environment, reflects the power relations of society and can act as an obstacle to democratic participation. It is important that we democratically participate in the design and construction of the public spaces and working environments of our communities. The power relations inbuilt into the organization of labor and the working environments within which most of us spend much of our adult lives naturalize structures and hierarchies of dominance and conformity within their technological infrastructure and operational procedures. It is essential that we are able to transform these power relations through democratic participation in the organization of our work environment, whether it is in agriculture, industry, or commerce, and this can only be achieved if we are able to democratically participate in the formulation of strategy, management, and policy in accordance with our needs and the needs of our communities. Of course investors will want to receive a good return on their invested monies, but that does not justify the subordination and abuse of workers and our communities. It is also quite

---

[411] Sclove (1995: 85)

reasonable for technologists to aim at the maximization of efficiency and productivity, but that does not justify the imposition of technological innovations on communities and all the changes that such experiments inevitably induce. In a genuine democracy, the capitalist economic and the technological imperatives must be subordinated to the public good, as decided by the public. In other words, democracy counters the totalitarianism implicit in the reductionism at the core of the societal gamble and the technological and capitalist imperatives, but it does not need to reject them as *a priori* antidemocratic evils. It may well be the case that particular workers and communities accept and agree that the capitalist economic imperative or the technological imperative is in the public good—for those workers and communities—but that is a matter for local democratic consensus, rather than being treated as if these imperatives were *a priori* universal truths that can never be rationally questioned and criticized.

The democratic process requires such a rational process of questioning and criticizing the ends and means of the economic, political, and technological development of our communities. We should structure our own labor in accordance with our perception and consensus about the local needs of our communities and how we wish to balance the time spent on work, leisure, and political participation. This will allow more citizens to participate more in the political process, and that will help to create greater social equality and responsibility. Sclove suggested that work-release time to enable citizens to have the time to perform political participation could well be established in a way that is analogous to jury service.[412] However citizens choose to do this, either through job sharing on an individual basis or careful organization of the whole community, through increased childcare support, through permitting flexibility about the retirement age, through encouraging better paid part-time employment over poorly paid full-time employment by improving the efficiency and productivity of businesses, farms, and workshops, or more imaginative ideas than these, are matters of local decision and experimentation. Unless people are able to live and work convivially, creatively, and with a level of responsibility that challenges our abilities and aspirations, as well as equally sharing social burdens and tiresome tasks, then we will be unable to truly achieve the societal conditions that are necessary for egalitarian and participatory democracy. Democratic changes in how work is performed will only occur if the labor process is rationalized in order to optimize the intrinsic creativity and importance of the work for each and every worker, as well as the level of participation in the decisions regarding the tactical and strategic management of the work. Of course there needs to be a balance between maximizing the aesthetics and productivity of the labor process, but boring and dangerous labor should be automated, whenever possible, and shared, when automation is not possible. Advances in the technological processes of production should be used to shorten the average working day (rather than reduce labor costs) and permit the worker more time for leisure, education, family, and political participation. Increases in productivity should be used for the benefit of workers rather than just to increase the wealth of the elite.

We should be able to enjoy our working life as much as possible, after our labors have satisfied our material needs and the needs of our community within which we live, and, therefore, we should be able to have considerable plurality and latitude in our choice of labor. Democratic participation in factories, workshops, offices, farms, as well as other organizations and institutions, would inevitably increase the labor and running costs (lowering the profitability), but because it would also inevitability increase pride and a sense of community; it would reduce the incidents and costs of illness, sabotage, depression, alcoholism, violence, drug addiction, crime, strikes, and unemployment.

---

[412] *Ibid* 204

Hence, while it would reduce the dividends of shareholders, it would also reduce the tax bill too, because it would require less social services to deal with the problems caused by overwork, work related stress, and poverty. Moreover, through democratic participation, the costs of local governance would be reduced, as the need for a bureaucracy is decreased—possible if citizens share the administrative burdens of the community; as people become increasingly aware of the needs of the communities within which they live, and develop a greater stock of skills and resources for the satisfaction of community needs. Why should a community pay taxes to a bureaucracy to tender an irrigation project to an outside contractor, when, after a few public meetings, the citizens of that community could easily come together to design and build one for themselves for a fraction of the cost? It only makes sense to do this, if we lack the skills to do it ourselves. By developing a plurality of skills and cooperating with one another to improve the community's infrastructure, citizens would be able to enjoy higher wages and lower taxes, a much higher quality of life, and live in an increasingly sustainable and increasingly self-reliant way—with resilience cultivated through continued democratic participation. Moreover, citizens would be able to decide the local price of the community produce, and, as a consequence, higher wages would not necessarily lead to inflation, as if it were a natural law that one follows the other.

Once a community's local economy and public purse are under the governance of its citizens, then it becomes possible for local people to invest collectively in technological projects that we agree have social and political benefits for our communities, in accordance with our vision for the development of our communities. However local democratic procedures and forums are decided by local citizens, it is essential that public participation is not limited to the evaluation and ratification of contracting projects, but is also involved in the overall implementation, development, operation, management, and monitoring of the technological infrastructure of that community. Through participating in the everyday operations of a wide range of technological infrastructures and practical activities, we increase our technical knowledge, skills, and appreciation of the wider implications of implementing and developing technologies. Moreover, we would gain increased social skills and critical appreciation of difficulties, implications, and responsibilities of democratic participation. Local people who know and interact with each other should satisfy local needs, as much as is practically possible, because this keeps power and resources in the community and also develops a sense of self-reliance, resilience, communality, and social responsibility. It is better if we are able to build and repair the technological infrastructure ourselves, using the skills, knowledge, and resources of our own communities, but when outside help is needed it is essential that local people have overall control of the project.

Of course, as Sclove pointed out, apart from in a few Amish communities, we lack an example of a societal tradition of subjecting technologies to democratic scrutiny, evaluation, and control.[413] Our tradition is that bureaucrats, politicians, and corporations make these decisions for us, which, of course, they represent as the technically rational decision in accordance with technical and economic factors determined by impersonal natural laws and market forces, while they make the decision which best suits their interests or those of their paymasters. The "technically rational decision" is represented as progressive, even if it remains untested and experimental, and is often used to oppress local communities that have traditional, sustainable, and well tested technological practices that are sufficient for their local needs. This kind of dictatorial 'top-down'

---

[413] *Ibid* 104

approach and suppression of traditional practices can be clearly seen in the treatment of Native Americans in the United States of America.[414]

Hence, it is often argued that democratic participation would slow down the process of designing and constructing technological projects. However, this is not a well thought through objection. It is much better to explore every conceivable aspect of a proposed project before it even reaches the drawing board, rather than having to deal with the expensive and disastrous consequences of an inappropriate or ill-conceived project after it has been constructed. It is much better to take advantage of local knowledge and participation during the planning process than to ignore or suppress it. It is quite simply more cost effective to design and build a sustainable, locally appropriate technology, and it requires local knowledge and participation in order to have a better chance of actually doing that.[415] While there are many philosophical, political, and practical problems that would need to be addressed in order to optimize the chances of developing high levels of democracy and public participation, all of which are beyond the scope of this book, such as the relations between the communities, the constitutional structure of democracy and the role of representation, democratic rights and obligations, commercial and military secrecy, media and propaganda, and many other such problems, my argument here shall be restricted to responding to the claim that the process of technological evolution is just too complex for ordinary citizens to make the correct choices about which technologies should be developed and implemented because ordinary citizens are simply not competent to participate in the nuts and bolts of the design process.

Even though this objection makes an important point about the limitations of public participation in the design of technologies, it presumes that design choices and specifications are defined by the internal workings and technical operations of the technology—their "inner logic"—in relation with their intended results, whereas the democratization of the technological society involves a much more holistic and pluralistic view and does not involve the removal of the input of technicians, economists, bureaucrats, and professionals. There is nothing inherent to democracy that requires us to ignore expert advice or to treat all opinions as equally valid. It is simply the case that the decision-making agenda should not be pre-emptively constrained to technical choices and judgments as an *a priori*. There are many different ways that the public can be involved in the design of technologies, at every level, and the greater the degree of participation that ordinary citizens have then the more they will appreciate the technical problems and limitations involved in making choices and judgments when designing technologies. Early public participation in the research and design of any prospective technology allows for accommodative adaptations throughout the whole process, which leads to the implementation and development of more flexible and socially responsible technologies, greater public receptivity to new R&D ideas (rather than suspicion that yet another technology is going to be hurriedly imposed on an unwilling public), greater likelihood of general satisfaction with results, and a more fair and inclusive process of implementation and development. Democracy provides opportunities to discover the social contingency of any technological endeavor, which would otherwise be obscured; motivation for the public development of participatory competence in technological politics; a broadening of critical reflection on public needs, ideals, and concerns; a deeper examination of the possible consequences of any new technology; and, an increased awareness of the antidemocratic structuration of our technological infrastructure.[416] As Sclove pointed out,

---

[414] Jennings (1976)
[415] Harrison (1987); Appfel-Marglin & Marglin (1996)
[416] *Op cit.* 183

there are numerous cases of technological projects that successfully involved public participation, which failed only due to outright opposition from powerful institutions.[417] The ability of ordinary citizens to participate in deciding how to construct the technological society is not so much a question of technical competence, but, instead, is a question of how citizens can overcome bureaucratic, economic, and political resistance to public participation. Modern technological and political systems tend to promote centrally coordinated, technocratic administrations and hierarchies—technocracy—that use "technical experts" as an instrument for the preservation of the status quo.[418] The most difficult challenge for strong democracy is not how to decide which technologies to implement and develop, but is how to overcome the power of antidemocratic elites, ideologues, and institutions that see public participation as being a threat to their hegemony.[419] It is a problem of overcoming the status quo of vested interest, power, and privilege within the political economy, and, thereby, transforming the status quo into a more democratic form of society.

How can this transition occur? Much of the prevailing technology and architecture of our society are designed and deployed in accordance with the purposes of the elite, which materially and ideologically opposes democratization, and, therefore, we should not expect an effortless transition from authoritarianism to democracy.[420] As long as "the technical" and "the social" are represented as being distinct aspects of modern society, thereby allowing "the technical" to operate with autonomy within its own realm, then the technological society will remain incompatible with democracy. It is essential to examine critically this distinction in order to show how "the technical" has been placed over and above "the social" in the political decision-making process in order to limit the criteria through which any technological processes are publicly evaluated—often limited to an assessment costs and risks—and established prior to any public consultation to direct and limit democratic participation in the ongoing development of the technological society. However, examining technological development solely on the basis of narrow evaluations of costs and risks ignores the larger moral and political dimensions. If people are going to have a genuinely democratic role in the construction of the technological society, we must be able to participate in deciding societal goals, setting the agenda, strategic planning, design, education, communication, public contracting, policy formulation, media deliberation and criticism, seeking alternatives, and the actual processes of the implementation and development. The democratization of the technological society requires the development of democratic procedures for the evaluation of science and technology; democratic participation in the deciding the directions of science and technology; and, the ongoing identification, implementation, and development of knowledge and technologies that help the democratic process.[421]

However, a crucial problem for the democratization of the technological society occurs when "technical experts" are required in order to understand the technical details needed to make informed choices between possible directions of technological innovation and scientific research. How can citizens participate in scientific and technical decisions if "technical expertise" is needed to make those decisions? How can "the social" inform "the technical" in these contexts? How can we democratize the technological society without having to technologize the lay public and either turn everyone into

---

[417] *Ibid* 193
[418] Winner (1977)
[419] Dickson (1974, 1984); Winner (1977, 1986, 1992); Day (1988); Frost (1992); Postman (1992); Sclove (1995); Myklebust (1997)
[420] *Op cit.* 81
[421] Rogers (2008).

"technical experts" or reduce public participation to a "rubber-stamping" process of legitimation? Can "technical experts" act as translators, thereby allowing the lay public to make decisions, in a way that is compatible with democracy and technology? As Raphael Sassower pointed out, even though "technical experts" should act as translators between the public and the technologists and the scientific community, the public needs to be aware that such translators have their own agenda; overly simplify the situation in order to be able to communicate with the public; conceal problems in order to gain public acceptance for a proposed project or calm public fears; and, overestimate their degree of certainty because of their perception that the public requires certainty.[422] Sassower argued for an examination of our expectations regarding the feedback loop between the public and the "technical expert." He was concerned that the scientific enterprise would be damaged or distorted if it was forced to cater to the anxieties and fears of the public and "constantly pandered to the public and its expectations" for certainty and assurances; it would be disastrous for the public if "technical experts" were forced to conceal errors, failures, and doubts.

Marx Wartofsky argued that this is a political problem concerning the process through which the norms and definition of rationality are to be achieved.[423] For Wartofsky, the democratization of technology requires the technological and scientific education of the public, alongside with the emergence of a public awareness of the undemocratic distribution of knowledge and power within the construction of the technological society. Rational democratization of technology requires increased access to technological and scientific training, as well as a critical, political engagement with the inequalities of access to technological power. I agree with Wartofsky's claims about the need to address politically inequalities in the distribution of technological power within society, requiring greater levels of technological and scientific education for the public, but, in my view, it is also an essential requirement for the democratization of the technological society that scientists and technologists are also educated in alternative forms of rationality and the wider criteria involved in the development of the quality of life. It is not just the case that the public needs to become "technical experts," so to be able to rationally participate in the decision-making process, but also "technical experts" need to become aware of the non-technological criteria involved in rationally evaluating human goods and societal progress. As C.P. Snow argued, it is essential that we get beyond the "two cultures" of the science v. humanities ethos in the education system.[424] Rather than uncritically following the societal gamble on constructing a technological society "to escape the human condition," we need to remain philosophically related to it through a thinking relation with labor.[425] We need to overcome the increasing specialization of technical and mathematical language, which, as Arendt argued, allows scientists and technologists to "move in a world where speech has lost its power," and wherein technology and philosophical thinking have "parted company for good."[426] We need to place philosophical reflection, critical thinking, rigorous deliberation, and political debate at the fore of public participation in deciding the directions of science and technology, and how the technological society is to be constructed.

Scientists and technologists need to be well read in the humanities, while, at the same time, dedicated to promoting a high level of humanitarian culture in society. Once we recognize that technology and science cannot be developed in isolation, without being

---

[422] Sassower (2004: 72-3)
[423] Wartofsky (1997)
[424] Snow (1964)
[425] Arendt (1958)
[426] *Ibid* 3

developed in accordance with the centralized plans of administrative bureaucracies and social elites, the development of technologies and sciences requires broad public participation, debate, and initiative as its basis. In other words, it is essential that "technical experts" become better members of the lay public and that we dissolve the distinction between "the technical" and "the social." But, this does not mean that science and technology must become subordinate to ideology or political expediency. Political power over technological power must be developed in accordance with a pluralistic debate about the norms and definition of rationality over and above bounded technical rationality, and this places political power in the crosshairs of public scrutiny and deliberation. It is the task of coming together and deciding how public participation is to be conducted that is the fundamental realization of the potential for democracy in a technological society. It is not only impossible to adequately define the norms and definition of rationality from the outset, due to the pluralism and heterogeneity inherent to modern society, but it is actually counter-productive because it would be antidemocratic to do so. Rationality is at stake and open for public deliberation. Democratic debate about technology and science is not necessarily just about how we control and shape the development and implementation of science and technology in society, but it is also going to include a critical and philosophical deliberation about the overall goals, ideals, and values of democracy and human existence itself.

Of course ordinary people would benefit from an increased access to a scientific education and technical literacy. This would help citizens make informed choices. However, the most important criteria for the decision-making process—such as the knowledge of how to achieve the human good life and universal happiness—are those for which there is an absence of "technical expertise" or even any universal consensus, for that matter. Once we recognize that the rational development of technology and science are bound together with the rational evaluation of human well-being, the good life, and the ideal society, then pluralism, diversity, and democratic participation need to be recognized as being of paramount importance for technology and scientific research and development. As Sclove pointed out,

"However, if the most important knowledge about a technology involves not its internal principles of operation but its structural bearing on democracy, then presumably the latter kind of knowledge should constitute the very core of technological literacy. Yet experts, even the elite, typically know little about this first-order issue—not even that it is an issue. Must one not reluctantly include among the technologically illiterate—in that term's socially most meaningful sense—the majority of technical experts?" (Sclove, 1995: 53)

Moreover, once we recognize that the consequences of the implementation and development of technologies in the world cannot be determined in advance and predicted, with anything even close to "absolute certainty," then we should no longer accept that idea that the "technical experts" have any greater capacity to make decisions about the technological development of society than any other members of the public. At best, their specialization only qualifies them to testify on a very narrow aspect of any proposed scientific research or technical project. Once we understand that every technological object is transformed once it begins to interact with other technological objects within the technological framework, as well with as other complex features of the world, we must also understand that "technical experts," as specialists in narrow technical fields, cannot claim to have any expertise regarding the societal and environmental consequences of any new technologies. Most "technical experts" are completely technologically illiterate about areas outside their specialization, and, as the complexity of technology increases and its non-linear interactions within the world become increasingly ambiguous, "technical experts" become increasingly unable to demonstrate any certainty or foresight over and above that of the ordinary citizen. In this regard, the "technical experts" are just as much members of the public as the rest of us. This is also true of professional politicians and bureaucrats—"political experts"—who are no more or less competent than any other citizen when it comes to questions of norms, values, and ideals, and their expertize is limited to a very narrow sphere of activity. If both "technical experts" and "political experts" are on a par with ordinary citizens when it comes to the question of the social and environmental consequences of new technologies, then it is quite foolish to rely on "technical experts" and "political experts" to make decisions for us. "Experts" will simply make guesses and decisions that suit their own agenda and interests, or, even if they remain "impartial," their decisions are informed and constrained by the technological framework itself. Unless they can demonstrate some objective and rational knowledge about what is good for humanity as a whole—knowledge of the forms of goodness and truth—we would be wise to presume that there are not experts on that question. It was for this that the development of democratic participation in the technological society does not depend on making each member of the public into a "technical expert," but, instead, depends on reforming and maintaining the potential for enlightened debate concerning the vital interests of the public, free from manipulation by mass media propaganda.[427] The task of public education is not simply that of imparting facts, knowledge, and skills required for the acquisition of "technical expertise," but it is about awakening intelligence and skill in public affairs and democratic participation.[428]

Furthermore, if technology is inherently at odds with the practical possibility of developing democracy—given the complexity of technology—then, perhaps, it is our obsession with technological innovation that we should be questioning, rather than our democratic ideals, and we should be very wary of leaving technological implementation and development in the hands of "technical experts" and "political experts." As Sclove succinctly noted,

"After all, it was not panels of laypeople who designed the Three Mile Island and Chernobyl nuclear plants; who created the conditions culminating in tragedy at Union Carbide's Bhopal, India, pesticide factory; who bear responsibility for the explosion of the U.S. space shuttle *Challenger*; or who enabled the *Exxon Valdez* oil spill." (Sclove, 1995: 49)

It may well be technologically imprudent—as well as undemocratic and complacent—to leave technological decisions in the hands of the "technical experts" and "political experts." If this is the case, the implementation and development of technology would be better and more stable if it was subordinate to the democratic process, and we should not allow the "experts" to dominate the agenda regarding science and technology. Technological decisions are contingent and value-laden decisions, the consequences of which cannot be known in advance, and "experts" have a vested interest in arguing that the most important questions are those that the "expert" can answer. Maintaining "expertize" has economic value. Hence "experts" tend to view technology as a means to solve the specific problems that it is posed as a solution for and they tend to limit their analysis to quantifiable criteria, such as construction time, job creation, revenue, cost, risk, investment return, and durability. However, there are many other criteria that need to be considered, such as long-term environmental and social impacts, transformation of human character and the life-world, and also moral and religious concerns. It is paramount importance for the health of society that these qualitative and complicated

---

[427] Hickman (1997); Rogers (2008, 2012)
[428] Rogers (2008)

criteria shape and limit the bounds of technically rational decisions. The belief that all technological change is progressive and beneficial is naïve and it is highly arrogant to impose new technologies on a community, especially given that every technological change is a social experiment. It is not only fair that communities should be allowed to consider whether they wish to be experimented upon, but there is also a better chance of success for any experimental introduction of new technologies into a community, if the community participates in the implementation and development of those new technologies. All decisions are made in the absence of any certainty regarding the social and environmental consequences of any technical action, whether constructed in accordance with a narrow technical agenda or a broad democratic one, and, consequently, the best reason for democratic participation is that broader criteria increase the chances of anticipating problems and suggesting remedies or alternatives. The complexity of science and technology does not provide a good reason to exclude democratic participation. It is a good reason to encourage it.

7

BACK TO NATURE

In Plato's dialogue, *Protagoras*, the chief interlocutor, from whom the name of the dialogue is taken, narrates the following tale: When the gods decided that the time had come to populate the earth with living beings, they entrusted Prometheus and Epimetheus with the task of producing them and providing them with suitable qualities. Epimetheus took over the concrete work, while Prometheus reserved for himself the right of supervision. After having wisely distributed among the different living species the characteristics that would enable them to survive and reproduce harmoniously, Epimetheus discovered–when the moment came to produce human beings–that he had already exhausted the natural qualities. So he was obliged to make a being that was naked, weak, devoid of any special feature, and inferior to the animals. In order to remedy this oversight, Prometheus stole fire and the arts (that is, the principles of making) from Hephaestus, and he stole from Athena the arts of the intellect (that is, the principles of science and wisdom). These qualities were diversely distributed among human beings and they, by using them, were able to secure their superiority over the animals by producing artefacts and building cities. However, humans showed themselves incapable of living in communities, as they split into factions and fought with one another. At this juncture, Zeus, much concerned about the destiny of humans, punished Prometheus and charged Hermes to bring to humans the political virtues of justice and modesty. After these virtues were bestowed, humans were able to live a just and harmonious life in their cities. Obviously, this tale is a myth, but, perhaps, it is timely that we reflect upon how we have largely failed to exercise these virtues and, especially when harnessing our technological powers to our political ambitions, we do not act wisely. War and conflict, poverty and inequality, injustice and deceit, and waste and corruption have been the results. The current mode of human civilization is unsustainable, as it lurches from crisis to crisis, and we are at the brink of global catastrophe. In order to direct our efforts to building and living in harmonious and sustainable communities, exploring the diverse possibilities and alternatives that our imagination and ingenuity suggest, and putting our technologies to work in helping us all live a good life according to shared visions of what it means to live a good life within a community and on a shared world. If the gift of Prometheus is forethought then is it not time for us to use this gift to put forth and deliberate our visions for society? Is it not time to subordinate the development of science and the construction of the technological society to the virtues of justice and humility, and—if I may dare say it—wisdom?

Even in the ancient world of Plato, and later for Aristotle, the image of a human being as vulnerable without tools, arts, and reason, has been an enduring image. It has become the image of technological man—driven by his fear of the natural world and the uncertainties and wildness therein. As I have argued above, this image has conditioned how science and technology have been developed and understood. Materialism and instrumentalism—viewing Nature as nothing other than a material substratum for human use—are the consequences of this cultural conditioning. These doctrines have shaped modern thinking, with its underlying rationalism and modernism, and constructed the modern view of both human nature and the objective world in materialist and instrumentalist terms. The societal gamble and faith in the technological society is reflected in that image and enlarges it into a manifest destiny for human kind. This image is all-pervasive. It can be seen in all modern ideologies and political theories, ranging from liberalism to communism, or from anarchism to fascism, or from ideological capitalism and the invisible hand of the free-market to Marxism and the command-economy.

In Marx's writings, Nature satisfies human basic needs, such as food, water, air, etc., and, in turn, human beings use technology to change the environment in which human beings exist. By doing this, humans bring new possibilities, problems, and needs into the world. He wrote in *Grundisse*:

"The earth is the great workshop, the arsenal which furnishes both means and material of labor, as well as the seat, the base of the community… The earth is the original instrument of labor as well as its workshop and repository of raw materials." (Marx, 1993: 472, 485).

In the *Economic and Philosophical Manuscripts of 1844*, Marx reduced Nature to being the "inorganic body of Man" that is "the instrument of his life activity" and "his direct means of life."[429] Yet human beings are also part of Nature and, as such, are natural beings—possessed with natural abilities, powers, tendencies, limits, needs, and instincts—that, like animals and plants, are conditioned by Nature. The reality of being human is that of being an embodied and sensuous being that is both conditioned by Nature and capable of using our natural attributes and properties to innovate technology and change our natural environment to produce a technological environment. Human beings and history cannot be understood without understanding the natural conditions upon which *Homo sapiens* depend to exist and thrive, and also the physical organization of human physiology and its relation to other beings within the natural world. It is Nature that provides human beings with the ability to transform our material conditions and produce the means to sustain and reproduce our existence, but, by doing so, we change our conditions and material reality.[430]

While Marx argued that social and economic relations of production were the products of history and not outcomes of the laws of Nature—relations which shape how we represent and understand Nature—he also unquestioningly accepted the objectivity of the laws of Nature, as revealed by the natural sciences, and their effect on our organic and material conditions for existence. Human beings are distinguished from other animals because we produce our means of existence—a step which is conditioned by our organic being but not determined by it—and, by producing our means of existence, we are directly producing (and changing) the conditions of our existence. It is this fundamental distinction that underwrites Marx's conception of history. Human beings have a history because we have technology, but it is important to realize that, for Marx, technology does not just provide the means by which we manipulate and control our material conditions; it also provides the means by which we consciously realize our essence as human beings *qua social beings*.[431] As Marx put it, in the 1859 Preface to *A Contribution to a Critique of Political Economy*:

"The mode of production of material life conditions the social, political and intellectual life process in general. It is not the consciousness of men that determines their being, but, on the contrary, their social being determines their consciousness." (Marx, 1904)

---

[429] Marx (1970: 112, 180-2)
[430] Marx (2011: 7)
[431] Blackledge & Kirkpatrick (2002)

Free creative labor is central to discovery in science, but, according to Marx, in bourgeois society, labor has been intellectually devalued and commoditized in terms of its exchange-value. Hence, instead of discovering the contours and possibilities of labor—as an intimate and sensuous relation with representational and material practices—science is culturally represented in bourgeois society as a means of discovering structures of the material world that exist independently of labor, which mirror the values and norms of society, and naturalize them as a result. Nature is appropriated by industrial capitalism purely as a "raw resource" available for exploitation for the benefit of a few; an act which is subsequently objectified as being part of the natural order of things, represented in terms of the natural struggle and competition for scarce resources. Marx termed this as making the Earth an object of "huckstering."[432] Yet, within Marx's theory, the harms and damage imposed on Nature by this capitalist "huckstering" are actually only harms because they have consequences for future generations of human beings. Pollution and waste are inevitable products of capitalism, with its culture of boom and bust, overproduction and overconsumption. Marketeering and advertising creating false wants and needs, profiteering from debt; creating and maintaining scarcity; leading to cycles of recession and unemployment. Pollution goes hand in hand with poverty, the degradation of natural landscapes, poor health (including mental health), social apathy and alienation. While it can be argued that these harms are the results of governmental corruption and neglect of the public good, Marx argued that they an inherent consequence of the capitalist system itself—ultimately resulting in a deliberate eugenics policy to remove "the surplus population" and those that are deemed as useless. The waste generated by capitalism maintains the system of scarcity upon which it depends for its profits (e.g. polluting tap water creates the need for bottled water, which creates the need for plastic and glass bottles, etc., which then creates waste disposal and recycling businesses).

Marx considered technological activity to be the condition for objective scientific knowledge and assumed that it has an emancipatory essence to liberate human beings from labor and scarcity, which has been perverted by capitalism by reducing wages and intensifying labor in order to increase profits by lowering labor costs and increasing productivity. For Marx, there was an "absolute contradiction" between the technical demands of industrialisation and industrial capitalism.[433] The contradiction between the capitalist economic imperative and the technological imperative, wherein capitalism distorts or perverts the emancipatory essence of technology, leads to structural contradictions between "the forces of production" and "the relations of production," despite being both manifestations of the same societal gamble on the goodness and rationality of constructing a technological society as an improvement over the natural world.[434]

Making and using tools are an evolutionary force, alongside the making and use of fire—the first force of production. It is this step that supposedly differentiates *Homo sapiens* from our hominid ancestors and other animals. The tool is an extension of the human hand, just as "the instrument" is an extension of the human senses, which have been shaped through natural evolution. This way of thinking underwrote Engels *Dialectics of Nature* (posthumously published in 1883). This provides human beings with the means to "impress his stamp on Nature" and thereby reproduce the differentiation between human beings and other animals.[435] It is this differentiation, made possible by the natural

---

[432] Marx (1970: 210)
[433] Marx (1967: 533-4)
[434] Rogers (2006)
[435] Engels (1934: 46-8)

evolution of the brain, senses, and hands, which separates humans from Nature by allowing us to make our own environment. It is through tool manufacture and use that allows *Homo sapiens* to depart from natural history, which ends at the evolution of tool use and manufacture, and, thereafter, make our own history. The "most essential and immediate basis of human thought" is "the alteration of Nature by men."[436] Critical of the reductionism of Darwin's "struggle for existence" and Hobbes's *bellum omnium contra omnes*,[437] Engels considered this reductionism to be a cultural projection of bourgeois society over Nature, and he highlighted the distinction between human beings as *producers* and other animals as *collectors*. Assuming this distinction, Engels showed that human history cannot be reduced to and completely explained by natural history, even though our origins and source of our abilities can be found in the natural world. Human society forms a distinct stratum over and above the natural substratum, and, therefore, human history and the development of the means of production are socially produced and reproduced, and transformative of the world.

Deforestation, agriculture, the artificial selection and breeding of plants and animals, use of artificial fertilizers and pesticides, and the industrialization of food production, are all examples of the means by which human beings change the natural environment and "master" Nature. For Engels, although he did not put it quite in these terms, human freedom is to be achieved upon completion of the technological society. This is a freedom *from* our natural conditions and limitations, which dominate human beings until we develop, produce and reproduce, the technological means to dominate our natural conditions, transform them, and master the construction of society, and the production and reproduction of the means to satisfy human needs. This societal ambition presupposes that human beings will become "the real conscious master of Nature." This is to be achieved by placing "natural forces" under human control to transform them into "productive forces" via labor and knowledge. Labor is the "objective mode of existence" and any scientific knowledge must obtain its objectivity in relation to labor. This is a crucial step, according to Engels, for human beings to move from "the realm of necessity" to "the realm of freedom" through constructing the technological society. Affirming the societal gamble, Engels considered the construction of the technological society to be "a world-emancipating act" and "the task of scientific socialism." Through labor, the distinction between human beings and other animals is produced and reproduced, thereby further justifying the exploitation of animals and treating them as nothing more than a resource to the satisfaction of human needs and wants. The slaughterhouses and factory farms of the industrialized world stand on the foundation of representing Nature, which includes other animals, as a resource and nothing else. However, as Engels pointed out, these "conquests of Nature" bring unforeseen consequences, such as soil erosion and landslides, the consumption of organic material and increased dependency on artificial fertilizers, the loss of groundwater and increased dependency on pumping and irrigation, and new diseases and health problems, such as alcoholism, obesity, heart disease, and diabetes.

Both Marx and Engels considered Francis Bacon to be the founder of materialism and experimental science.[438] According to Marx and Engels, physics—the exemplary empirical science—was the best means to achieve a rational understanding of Nature in terms of matter and motion. It is significant, in this respect, to note that Marx's doctoral thesis *The Difference Between the Democritean and Epicurean Philosophy of Nature* was on ancient atomism and its implications for materialist conception of Nature. It was on the basis of

---

[436] Engels (1934: 404-5)
[437] War of all against all.
[438] Marx (1975: 172)

this conception that Marx turned Hegel's idealism on its head. For Marx and Engels, science and technology are brought together through production and changing our material conditions. While "natural forces" are discovered by human beings through scientific activity, "productive forces" are the products of human labor and technological activity using the scientific knowledge of "natural forces."[439]

"Thus at every step we are reminded that we by no means rule over Nature like a conqueror over a foreign people, like someone standing outside Nature—but that we, with flesh, blood, and brain, belong to Nature, and exist in its midst, and that all of our mastery of it consists in the fact that we have the advantage over all other creatures of being able to know and correctly apply its laws." (Engels, 1934: 239)

The feedback relation between science and technology is a transformative and dialectical relation between "natural forces" and "productive forces" applied to the construction of society, which is the basis for human beings to fulfil our essence *qua* human beings. This places science in both the contexts of discovery and production, thereby problematizing any clear distinction between pure and applied science, and binds it to human emancipation and self-discovery. As I have already argued, this distinction is further problematized when we pay close attention to the way that "productive forces" are used in experimentation to discover "natural forces" at work in the performance of apparatus.

In the writings of Marx and Engels, the ideal society—the communist society— would be a technological society. Technological determinism underscored Marx's view of history and the conditions for the transition to a communist society. In the writings of both Marx and Engels, it is necessary that human beings are liberated from toil by science and technology, while being able to satisfy all the material conditions for human existence. Technology provides the medium for the construction of the dialectical relationship between the historical development and natural evolution of human beings as social beings. Natural science is a theoretical relationship between human beings and Nature that is internally bound together with practical activity, industry, and with the development of "the forces of production." However, economic class has led to a division of labor which abstracts science from this internal bond and suppresses a conscious recognition of it, while developing the scientific understanding of objectivity in accordance with the values and ideology of capitalism. According to Marx, once we recover a consciousness of this internal bond, we can understand how the scientific understanding of Nature has been developed through practical activity, industry, and the development of "the forces of production," alongside a scientific understanding of how human beings have evolved from our organic conditions by transforming the material conditions of our existence. Marx stated that "…technology discloses man's mode of dealing with Nature, the process of production by which he sustains his life, and thereby also lays bare the mode of formation of his social relations, and the mental conceptions that flow from them."[440]

By liberating human beings from scarcity, technology has the potential to liberate the creativity of human beings by allowing time for education, cultural activities, and leisure, but, technology has been historically conditioned by scarcity and exploitation. When technologies have been built to oppress one class of human beings for the benefit of another, the actuality and potential for oppression is enhanced. This effects the future development of technologies and leads to an increasingly sophisticated and powerful trajectory of the research and development of technical means of oppression and control

---

[439] Marx (1967: 386-8)
[440] *Ibid* 352, n.2

(i.e. surveillance technologies, propaganda, mind-control, etc.). According to this view, if technology is free from the distortions and perversions of a class-based and hierarchical society, it will necessarily liberate human beings from scarcity. Under the technological imperative, inherently enhancing the potential for human emancipation, the aim of technological innovation is to substitute automated machine performances for all human labor. The historical struggle to construct the technological society as a post-scarcity society promises to culminate in the creation of a seamless and integrated infrastructure of automated relations of production and distribution subordinate to human will. According to this optimistic view, the fundamental relationship between human beings and the material world will be transformed from labor into design.

It is the technological determinism inherent to Marx's thinking that was the major influence on the school of critical realism.[441] Anarchist theorists have also assumed technological determinism in their accounts of the current state of affairs and how a free, egalitarian, and rational society can be achieved. In the writings of Pierre-Joseph Proudhon and Michael Bakunin, a libertarian, egalitarian, and rational society was inherently a post-scarcity society.[442] Scarcity generates the fundamental struggle between human beings and the natural world, which had been distorted throughout history into a struggle between human beings, therefore for Proudhon and Bakunin, science and technology were the means for human liberation from struggle, inequality, and superstition. Proudhon and Bakunin, like Marx and Engels, represented science and technology as the objective and rational means for human liberation from superstition, toil, and suffering by dominating the natural world, discovering new forces of production, and, driven by the technological imperative to improve productivity, empowering the future construction of a technological society as a post-scarcity society. Contemporary developments in anarchism (with the exception of anarcho-primitivism) continue this traditional assumption of the emancipatory potential of science and technology. This assumption underscored the social libertarian writings of Murray Bookchin, for example.[443] Although there are important and fundamental differences in their conception of revolution, proponents of both anarchism and Marxism have called for a revolutionary seizure of the means of production in order to recover the emancipatory potential of science and technology from the distortions created by a class-based, hierarchical society. Anarchism and Marxism differ on their conception of the role of the State in this revolution, which for the anarchists is an inherently oppressive organ of power, while, for the Marxists, the State ("the dictatorship of the proletariat") is historically necessary for the post-revolutionary transition from capitalism to communism. The idea of "historical necessity" and the demands of the technological imperative, as revealed by scientific knowledge, have been represented as one and the same under the Marxist conception of socialism and the construction of the post-scarcity society, wherein the State would "wither away" and its functions absorbed by civic society. Via advanced administrative methods, scientific knowledge, and technological automation, human beings will administer things rather than each other. At least in this respect, Marxism (along with anarchism) are descendants of the Scientific Revolution and the Enlightenment. Adherence to the demands of technical rationality (intellectually justified in terms of technological determinism) became represented as a moral duty towards the future of humanity in general. It was for this reason that technical modernization in industry and agriculture were as important for the anarchists in the CNT-FAI during the Spanish Civil War, as they were for the Bolsheviks during the Russian Revolution, as well

---

[441] Bhaskar (1986; 1989; 1993)
[442] Proudhon (2011); Bakunin (1990)
[443] Bookchin (1971; 1996; 2005); see also Biehl (1998)

as for Soviet propaganda in the post-revolutionary period. Hence the famous Bolshevik slogan: "Communism is socialism plus electrification."

According to Marx, it is the conscious engagement with reality through material practices and sensuous activity that provides science with its objectivity; the understanding of this engagement is itself developed and differentiated within a dialectical, historical process of transforming "the relations of production." This is done in order to resolve the contractions generated by integrating new "forces of production" into these relations and the subsequent necessity for their further development and differentiation. Science and technology are thereby conditioned by the historical stage of development and differentiation of "the relations of production," within which they emerge, but they also have an intrinsic potential for the progressive transformation of society into the technological society. The aims and means of scientific research have been transformed through the material practices of previous generations, creating new instruments of observations, new experiences and refinements, which allow new theoretical representations and conjectures to be possible. The theories and practices of science have been developed within the bounds of "the relations of production" at any given stage of the development of "the forces of production" and, thereby, cannot be abstracted from these bounds without distorting our understanding of them and hindering the development of scientific objectivity. Once society has achieved a sufficiently advanced level of technological development, science will transform itself from being a means for the exploitation of the proletariat into a condition for the emancipation of humanity as a whole. This will not happen by magic. Marx's idea is that the exploitative "relations of production" will be inadequate and inefficient for the integration of the new "forces of production" into society, and in a society driven by the technological imperative this means that "the relations of production" must change. At which point, science will rationalize society by providing the means of human mastery over "the forces of production," alongside the conscious awareness of human essence as free and creative, which will demand that the inadequacies and inefficiencies of "the relations of production" are resolved in a way that satisfies human mastery, freedom, and creativity.

Once the proletariat controls the means of production—the machines at their disposal—the emancipatory essence of technology would liberate them to control the conditions of their existence. Upon completion, the technological society and the proletariat would be unified as one totality—into a global, self-creating, self-conscious social being—wherein struggle and scarcity would be overcome through the unfettered development of "the forces of production." The emancipated proletariat would become the consciousness of the technological society. At which point, the proletariat would dissolve as a class into the conscious totality of the technological society and Marx's vision of communism would have been achieved. Marx's vision of a communist society was that of a post-scarcity technological society, wherein societal development would be directly controlled by scientists, technicians, and producers, comprising the vast majority of the population of such a society. After abolishing the capitalist perversion of its essence for the benefit of the minority, the technological imperative would construct the technological society for the benefit of the vast majority. The technological imperative is the driving force for societal change.

The conditions for the emergence of a disciplined and class-conscious proletariat, as the agent for revolutionary change, are identical with the conditions for the unfettered construction of the technological society. It is the demand to overcome the limits placed upon the technological imperative by capitalism (in accordance with its demand for an adequate return on investments) that will lead to the overthrow of capitalism by a disciplined and class-conscious proletariat. According to Marx, the revolutionary seizure of the means of production could and should only occur once science and technology, within industrial capitalism, have reached a barrier to their further development. This barrier is itself the result of the limits of capitalist economics, and a technically advanced, disciplined, organized, and class-conscious proletariat becomes aware that it constitutes the productive sector of industrial and urbanized society. According to Marx, capitalism must have fully matured and reached its own developmental limit before the seizure of the means of production by the proletariat becomes historically necessary to overcome the limitations that capitalism imposes on the construction of the technological society. Once the proletariat achieves an advanced state of discipline and class-consciousness, identifying itself as the organized agent of productivity, then the demise of capitalism is inevitable once its technical limit has been reached and the proletariat consciously takes upon itself its historical task to construct and complete the technological society. Thus it was intrinsic to Marx's conception of "revolution" that it could only succeed once the proletariat have become capable of liberating the technological imperative and trajectories of scientific research from the limits of capitalist economics. In this sense, the emerging class-consciousness of an advanced proletariat should not be understood as a fraternal sense of moral outrage at the privileges of the bourgeoisie, but, rather it is the conscious realization of the necessity of the liberation of the technological imperative from the limitations of "the relations of production" inherent to capitalist economics. The revolution occurs when bourgeois society has become obsolete and an obstacle to progress towards the construction of the technological society. Hence, for Marx, revolution would inevitably result from the contradictions between the ongoing technological innovation of "the forces of production" and the inefficiencies of the "relations of production" generated by capitalism. Providing that society has achieved a state of advanced industrial capitalism, within which the vast majority of the population would comprise a disciplined and class-conscious proletariat, the revolutionary seizure of the means of production and the transition to "the dictatorship of the proletariat" would be the rational, technical means to construct a post-scarcity society, the end of struggle and inequality, the absorption of the administrative functions of the State into civic society, and the achievement of communism. Human beings would be liberated by technology from material needs, toil, and struggle, which would offer the freedom for the pursuit of education, leisure, and creative activities, providing that they obey the demands of the technological imperative. This is what Marx meant by "the end of history."

Contrary to many of his critics, Marx's vision was not that of a police state controlled by a central committee, wherein dissenters were arrested and sent to labor camps, torture chambers, or death by firing squad. His vision was of a technological society wherein the technological imperative was inherently a moral imperative and scientific basis for the construction of an egalitarian and communist society on the back of a post-scarcity society and the obsolescence of labor and its exploitation. Providing that the technological imperative is liberated from the constraints and perversions of capitalism, the post-scarcity society is the inevitable outcome of unfettered technological innovation based on scientific knowledge of the material conditions of human existence and our species-specific interests. While it is arguably the case that Marx's authoritarian tendencies were apparent during the 1872 split in the International Workingmen's Association ("the First International"), the problem with Marx's theory is not his authoritarianism, which may well have more inherent to Marx's character than his theory, but that the application of his theory would necessarily result in a totalitarian society due to the presumption of the rationality of the technological imperative as the only progressive and emancipatory force for social development. It is in this sense of totalitarianism (under the guise of "necessity") that we can see how the technological determinism and faith in the societal gamble inherent to Marx's own theory was applied as an ideology to the agricultural economies and peasant-based societies of Russia, China, and Cuba, which led to a reproduction and intensification of the same kind of squalor, poverty, pollution, and exploitation caused by the Industrial Revolution in Europe and America, thereby producing and reproducing systems of state-capitalism and the same contradictions between the relations and forces of production. These contractions in "the post-revolutionary society," due to the premature attempt at the transition to socialism, required an authoritarian, centralized party-based superstructure and militaristic police-state to sustain it, while simultaneously suppressing it.

Hence, despite the fundamental deviation from Marx's theory, resistance or opposition to Marxist ideology was to become represented within early Soviet Union as the product of backwardness and ignorance of historical and technological necessities—albeit theoretically couched in Marx's language of dialectics and historical materialism. Bourgeois idealism was simply blinded by self-justifying abstracts that had persistently failed to grasp the realities of human needs and the historical development of the means of their satisfaction. It was only after the Bolsheviks had seized power that the language of "counter-revolution" and "sabotage" came to dominate; this language was used justify the use of violent and oppressive measures to impose ideology, itself largely functioning as a propaganda instrument to justify industrialization and militarism. Traitors and wreckers were everywhere. As well as suppress dissent among technical experts and scientists, the setting-up of a bureaucratic police-state to consolidate power in the hands of the Party, which infamously eradicated opposition and dissent, and provide slave-labor in the special prisons and camps of the Gulag[444], was fundamental to the construction of the technological society in Russia. The cult of infallible leadership promoted by Lenin and Stalin—and to a lesser extent continued by Khrushchev—was itself a human face on the historical and technological determinism that underwrote Soviet ideology and policies. It is not a historical accident that the Cold War took the technological form of an arms race (extended into a space race as well as nuclear arsenal development) and an economic war of attrition, alongside the *Realpolitik* of war by proxy in the Third World and a battle of international espionage, propaganda, and counter-propaganda. Intensified and distorted under the Bolshevik nationalization of industry and the Stalinist agrarian collectivization, the post-revolutionary transition to socialism became identified by the Party with the technological imperative to maximize productivity for the State (cosmetically termed as "the Soviets" or "the People"). Any resistance or opposition to the State—any failure to achieve planned levels of productivity and specified technological stages of development—was represented as nothing less than counter-revolutionary sabotage against "the People."

At the heart of this project is not only dictatorship over people, but also is the ambition to have dictatorship over Nature. This is evident in the history of Soviet agriculture, from Stalin's forced collectivization to Lysenkoism and the Virgin Lands project under Khrushchev. The "dictatorship of the proletariat"[445]—itself an expression of the technological imperative—fundamentally presupposes that the End of History is

---

[444] Solzhenitsyn (1973)

[445] Arguably, writing in the nineteenth century and educated in the classics, Marx used the term "dictatorship" in the Roman sense of the term. This Roman sense meant a temporary dictator who could save the Republic in terms of crisis and who had strictly limited constitutional powers, rather than the twentieth century use of the term to describe the autocratic leadership of the Soviet Union, Fascist Italy and Spain, and Nazi Germany. If this is the case, "the dictatorship of the proletariat" referred to a temporary period wherein productive workers would decide societal development for themselves and on behalf of the unproductive.

the end of all struggle. This is not simply an end of all struggle between economic classes, but is also an end to the struggle between human beings and Nature. The inherent vision to this utopian view of the technological society is ultimately totalitarian. As Hannah Arendt argued, totalitarianism involves domination of every sphere of human existence by means of the elimination of all human spontaneity and freedom.[446] A totalitarian society (especially during its advanced stages) can be benign and peaceful, without any conflicts or resistance, because opposition or dissent simply would be unthinkable or futile. While considerable violence would be necessary to create a totalitarian society (perhaps starting out as a dictatorship), once such a society has been constructed and its operations are all-pervasive and seamless, violence or the threat of violence is no longer necessary. The totalitarian society systematically controls members of society by repressing and channelling the possibilities of thought and action. "The dictatorship of the proletariat" would be the pursuit of the technological imperative as being *the only path* to construct a rational, egalitarian, post-scarcity society. It is therefore important to distinguish between a totalitarian society and a society under dictatorship. A dictatorship is a form of government (usually personified as a supreme leader) that imposes its authority through the threat or use of violence, whereas a totalitarian society is one wherein alternatives or acts of resistance are impossible. As both an individual and mass participation in accordance with an ideological vision of all the possible stages and destinations of historical development or human destiny, the totalitarian society aims to transcend and direct all human potentialities by eliminating and suppressing all alternatives to itself.

Hence a distinction should be made between the "critical" and "scientific" aspects of Marx's theory.[447] My criticisms of the inherent technological determinism within Marxism are directed towards the "scientific" aspects of Marx's theory, which have been dominant in the development of Marxism in theory and practice. As I discussed elsewhere (Rogers, 2008), if we take the "critical" aspects of Marx's theory into account, while remaining critical of the "scientific" aspects of his theory, as Andrew Feenberg has done, we move towards a substantive and critical theory of technology. Marx's assumption that technology is the primary historical agent for social change presupposed technological determinism, and it is this "scientific" aspect of Marx's theory that underpinned a totalitarian vision of society. It is in this "scientific" sense that Marx's vision of communism necessarily presupposed the societal gamble on the goodness of the technological society, and was necessarily totalitarian in its adherence to the dictates of the technological imperative. These dictates lie at the heart of the human attempt at "the conquest of Nature."

This dictate can be seen in how contemporary critical theories of technology represent human freedom. As descendants of analytical Marxism, they attempt to explain how historically developed and structured technologies constrain and inform human choices and practices—how they shape our visions of our possibilities and choices. Although these social and critical theories reject technological determinism, which holds that society is the product of technological development and the trajectory of this development is governed by historical or natural law, they instead emphasize the contingency of human decisions and relations in the development of technology within society.[448] Yet, they only reject technological determinism in its most positive aspects, as they too assume the technological framework from within which these contingent human choices are made. They also subsume Nature—itself viewed as a social construct—under

---

[446] Arendt (1958)
[447] Gouldner (1982)
[448] Rogers (2008)

an overarching concept of society. Hence, they take categories of social relations as their starting point when they ask: how can the social sciences objectively understand society when they presuppose positivism and its methods? However, they too assume a positivistic conception of "the empirical" that underscores their own interpretations. Hence, while they understand that the failure of social science has been the outcome of positivistic pretensions to be "an empirical science" by imitating the mathematical projections of the natural sciences like physics, chemistry, and biology, they do not address or understand the metaphysical precepts on which these sciences are based. As such, they cannot look any deeper that "cultural relativism" or "political power" to find the answer to explain the particularities of human choices, and the contingencies of the development of the sciences and technology.

The Domination of Nature

Human beings have killed and been killed, have dominated and been dominated by, other humans, and this historically preceded "the domination of Nature" by technologically empowered human beings. Examples of this date all the way back to the interactions between *Homo sapiens* and Neanderthals, which led to the extinction of the latter—or the story of Cain and Abel, and the world's first murder—depending of which story of human origins we so happen to believe. The ancient civilizations of Mesopotamia, Egypt, India, and China all show countless examples of domination and exploitation of human beings by other human beings. Which ancient civilization did not exist by slavery, warfare, plunder, and conquest? Civilization has been built upon slave labor in the quest for power over all men, and the domination of women by men is historically and causally prior to the domination of Nature by the human species as a whole.[449] As well as the domination of women by men, consider the domination of children by adults, alongside the domination and subjugation of all other peoples by white European men in the name of civilization; the domination of people by monarchs, tyrants, and despots; criminals, bandits, and raiders; wars and genocide; torture, plunder, and rape; alongside cruel punishments, genocide, and acts of terrorism over the whole population. Countless examples date back to the tactics of Genghis Khan and the ancient practice of burning cities to the ground if they did not immediately surrender to their new master—and, of course, killing the whole population but a few to spread the word—to see how terrorism has been part of warfare since ancient times. Ancient history is a horror story. The technological imperative led to the human domination of Nature—at least as a cultural ambition or aspiration—but the centrality of domination and exploitation to economic relations pre-existed capitalist societies (i.e. ancient systems of slavery in Babylon, Athens, and Rome, and the medieval feudal system and right of conquest), so the critical theorists' critiques of modernity, science and technology, and the Enlightenment only tells us half the story. The slaughter of the Great War was only possible because of modern weapons and munitions, but the Hundred Years War almost matched its carnage, even if at a much slower rate.

Since the end of the last Ice Age, over 10,000 years ago, during what is now called the Holocene extinction, human beings were responsible for the extinction of many species of what are termed as *mega-fauna* (mammoths, giant sloths, sabre toothed cats, giant birds, etc.) through hunting, agriculture, and deforestation. With the advanced power of modern technology, the human capacity for destruction of the natural world has reached hitherto unimaginable proportions, and it is the neglect to attend to the violent, exploitative, and totalitarian aspect of prehistoric and ancient human relations

---

[449] See Griffin (1978) for an interesting read.

towards Nature as a material substratum or resource that has distorted our thinking about Nature. We should not limit our critique of technology to examining its modern ideological distortions, as if the domination of Nature is only a modern phenomenon. Without attending the ancient roots of anthropocentrism and domination, we are doomed to continue the same patterns of fear, brutality, and destruction. This has resulted in a repetition of the same reductionism of Nature that has been presumed by the traditional philosophers of science and by Marxists and critical theorists alike, and they have failed to provide us with an adequate philosophy of ecology, which situates within it human beings as social beings, and is itself premised on sustainable ecological relations within the natural world.[450] The idea of Man's domination of Nature is an ancient one. After all, we can find this idea written in *the Old Testament*:

Genesis 1:26 "…and let them have dominion over the fish of the sea, and over the fowl of the air, and over the cattle, and over all the earth, and over every creeping thing that creepeth over the earth."

The concept of the domination of Nature has been central to Critical Theory (especially of the Frankfurt School) and a great deal of New Left thinking since the early 1970s.[451] They presuppose that human liberation the opposite of the domination of Nature. Science, technology and capitalism have been analyzed in terms of their dehumanizing tendencies, destructiveness, and totalitarianism; they are shown to be out of control, enslaving human beings, and destroying the natural world. Nature is reduced to a resource, and nothing else. They argue that this reductive destructiveness stems from a materialistic interpretation of Nature. This reduction is common to capitalists and Marxists alike. Yet critical theorists also presuppose that technology is a fundamental aspect of the human essence—with language being the other fundamental aspect.[452] They also presuppose that Nature is a material substratum for human society and freedom. Yet, if this presumption about Nature is correct, it is unclear as to why the domination of Nature is a moral concern or problem at all. The critical theorists follow the Marxist assumption that Nature is nothing more than a material substratum of matter and forces, yet they do not explain why it is morally wrong to exploit matter and forces, if that is all that Nature is. If Nature is nothing more than matter in motion and the material conditions for our existence, its domination would have no moral meaning or significance whatsoever, just as rearranging Lego pieces in different combinations has no moral meaning or significance. It is not clear that a material substratum can be dominated at all.

This failure and its source within the cultural project to construct the technological society was identified by Norbert Wiener (1964) as an outcome of the rise of cybernetics. The technological society is viewed as a technologically structured sociological phenomenon, much as it was in the writings of Jacque Ellul and Lewis Mumford, and it was explored in depth by Herbert Marcuse in relation to totalitarianism (2001a), critical theory (2001b), and political aesthetics (2001c). This has resulted in a duality between the technological society and the natural world. While technology is examined by social theorists in terms of its social contingencies and conditions, the tendency due to this duality either reduces Nature to a social construct or it delegates responsibility to scientists to provide us with the facts about Nature. Critical social theory and social constructionism both presuppose this. These theories (or 'schools of thought') have taken Marxism as the starting point for their analyses and criticisms of the

---

[450] See also Bookchin (1996: 188-9)
[451] Horkheimer and Adorno (2002); Horkheimer (1974); Schmidt (1971); Leiss (1972); Parsons (1977); Marchant, (1994); Coulson (2003)
[452] Sarles (1985)

technological society, and it is this starting point that has led to the contradictions and limits inherent to these theories, because they presuppose that Nature is itself nothing other than a social construct or a material substratum available to human exploitation.

We need only to consider the modern application of the results of laboratory genetic modification to agriculture and medicine to see how this attempt at the "conquest of Nature" has been intensified (somewhat recklessly) and has opened a Pandora's Box of unforeseen consequences and possible disasters, possibly of a global scale, all in the name of ending world hunger and disease (and making a profit along the way). In many respects, the lure of genetic modification as a panacea for "the imperfections" of Nature repeats the ancient and medieval lure of alchemy.[453] The accusation that scientists today are "playing God," was an accusation that was raised against alchemists during the medieval because of their intention to change the order of the natural world by using artificial means.[454] The utopian speculations of the Medieval and Renaissance periods were strikingly similar to nineteenth and early twentieth century debates about eugenics and our contemporary debates about the ethical and religious issues surrounding modern genetic engineering and *in vitro* replication of life.

"… in the modern world of bioengineering and genetic wizardry, the ever-growing possibility of ectogenesis holds no less a grip on our own visual sensibility, even if its explicit association with alchemy has been lost… predicted results of ectogenesis, cloning, the farming of women, and genetic engineering were prefigured by premodern fears that included the production of a diabolical master race, the reduction of women to the status of a hollow incubator, and the prenatal modification of intelligence and gender—all issues that our ancestors found fascinating and at times abhorrent, just as many of us do today. The wellsprings of these dreams or nightmares run deeper than any modern bioethicist or free-market promoter of biotechnology can possibly imagine." (Newman, 2004: 301-2)

Obviously, with the partisan application of hindsight, we could respond that alchemy could not fulfill its promises and it was a pseudo-science, whereas modern genetics is a proper science. Of course I acknowledge this, but it seems to me, on one side, that this point is somewhat whiggish because the early alchemists did not know what the limits of their art were and which of their claims were fantastic until they attempted them, and the purpose of their efforts was to discover these things by trying to realize their claims *experimentally*; on the other side, it is also somewhat presumptive because it assumes that genetic engineering will be able to realize the equally fantastic claims made about its potential. It may well be the case that genetic engineering will not only fail to fulfill its promises and, hence, could more similar to alchemy than some people today would like to think, but it also may well be so ill-conceived as to be disastrous for long-term and sustainable agriculture, reproduction, ecology, and human health. Contemporary advocates of genetic modification[455] similarly subject us to fantastic claims and promises about the perfection of the world and the human species, such as the end of world hunger, immunity to all disease, the end of ageing, the prevention of birth deformities, flawless beauty, high intelligence, perfect memory, complete birth control, total cellular regeneration, human immortality, etc. Current developments in biotechnology have even led feminist proponents of *gynogenesis* to advocate its use to produce artificially a completely female humanity on the premise that such a society

---

[453] Newman (2004, chap.4)
[454] *Ibid*: 33
[455] Naam (2005); Easterbrook (1996)

would be a considerable advance over the natural state of affairs.[456] These examples show that our current debates about biotechnology and genetic modification should be situated within a long tradition of debating the relation between the artificial and the natural in terms of human arts and the ability to manipulate natural beings for the human purpose of creating a better world.

Critical theorists follow a Marxist line as their point of departure for their critiques of capitalism, and, as a result, assume that the bourgeois revolution was complete and the capitalist economic imperative—blind to all else but maximizing profits—dominated as the determining factor of social organization since the Enlightenment. The assumption is that the primitive accumulation of capital—plunder, conquest, and slavery—began at the start of the modern era with the European colonization of Asia, Africa, and the New World. As Marx put it,

"The discovery of gold and silver in America, the extirpation, enslavement and entombment and mines of the aboriginal population, the beginning of the conquest and looting of the East Indies, the turning of Africa into a warren for the commercial hunting of black skins, signalized the rosy dawn of the era of capitalist production. These endemic proceedings of the chief momenta of primitive accumulation."[457]

Exploiting the surplus-value of labor and extracting profit is a fundamental mechanism of capitalism, regardless of whether it is laissez-fare "free market" capitalism, advanced finance and corporate capitalism, or state-capitalism. Surplus-value (profit) was defined by Marx as the exchange-value (sale price) minus costs of "forces of production" (the sum of labor, resources, and technology). Capital accumulation results in an increase of "forces of production," increased profits and production towards maximum until overproduction, oversupply, and redundancy results, followed by loss of profits, increase of debt, and decrease of production. This crisis for capital results in recession, bankruptcy, credit and bank crises, and unemployment. Layoffs, closures, selloffs, scraping, and asset stripping decrease the "forces of production" and decrease the money supply, while simultaneous allowing for the further capital accumulation in fewer hands. In *Das Kapital*, Marx talked at length about these cycles of boom and bust, and their waste and inefficiency. The need for cheap labor and natural sources has led to unequal exchange, exploitation, resistance, and the use of violence and the threat of violence to control populations. At a political level, this results in cycles of domination and the unsustainable, destructive, and totalitarian control over religion, language, culture, politics, the movement of peoples, and the technological infrastructure. Primitive accumulation destabilizes colonized/conquered populations and their traditional practices and organizations, further exploiting and degrading the local population and natural environment, and destroying local economies and creation of culture of dependency on the colonizers/conquerors in order to transform the local population into wage-labor for mines, plantations, factories, and sweat shops. All this has resulted in the natural world becoming reduced to natural resources for exploitation.

Yet, we should interject and ask whether the bourgeois revolution was complete. Did capitalism replace all pre-capitalist traditional practices and all their inherent inequalities—or was industrial society developed only in accordance with capital and the ownership thereof? Or did the Industrial Revolution leave intact gender discrimination, religion, family blood lines, racial discrimination, and many traditional beliefs and practices (such as religious superstitions and rituals, family relations, etc.) in forms that

---

[456] Sourbut (1996)
[457] Marx, *Capital*: pt. 8, chap. 31, p. 914. See also Galeano (1997).

cannot be reduced to capitalism and the products of mass manipulation and the culture industry? Although problematized and challenged by the bourgeoisie revolution and the rise of liberalism, pre-capitalist social structures, inequalities, and hierarchies have remained determining factors in social organization even today. These factors have, at least in part, shaped how science and technology have been practiced, developed, and implemented within society. They have influenced and determined the modern relation with the natural world. If the real causes of natural ecological degradation and resource depletion are overconsumption (due to irrational consumerism as well as overpopulation) and militarism, alongside destructive industrial and agricultural practices, then, although capitalism remains a major contributing factor, we need to look also at pre-capitalist psychological factors, such as egoism and greed, as well as fear and ignorance, as major contributing factors that cannot be reduced to either the capitalist economic imperative or the technological imperative, even though both of these imperatives are continuations and intensifications of these psychological factors.[458] With the technological imperative thrown into the cultural and psychological mix, modern society grew with inherent and irresolvable tensions, contradictions, struggles, and fragmentary, irrational modes of domination of groups of human beings by other groups of human beings, alongside treating Nature and nothing other than a resource for human exploitation, but further empowered and reinforced by science and technology. The turning point is that, through science and technology, the ability to dominate and reduce everything to its instrumental value has become intensified and extended across the whole planet and every facet of life.

The cultural construction of the technological society is a "self-fulfilling prophecy"—a mythology—as it is justifies itself by declaring itself as the solution to all problems, and thereby reproducing the societal gamble on the projected and future success of technology despite its ongoing incompleteness and the endless creation of new possibilities and problems. The technological imperative the cultural drive to do this for its own sake, but the enduring motivation for the construction of the technological society is based on myths of its limitless power to be realized in an imagined future technological utopia. The capitalist economic imperative puts this utopia at the service of the economic elite, as their instrument. Globalization has resulted in the whole planet as being represented as nothing other than a collection of natural resources, labor, and markets for exploitation as capital. It has resulted in the further ecological and economic degradation of the Third World.[459] "Market efficiency" has become the justification for the extension of colonialism and imperialism through the capitalist economic imperative and the technological imperative. "Free trade agreements" and "reduction in trade tariffs" are outward manifestations of the total transformation of the whole planet Earth, including its human population, into a system of resources and labor (including child and slave labor) and consumer and investor markets, which cycles of extraction, production, and consumption providing profits for the owners of the means of extraction, production, and distribution. Globalization is a form of economic totalitarianism, given political expression through neoliberalism, and an ideologically-driven means of control over developing and Third World countries through debt and the dominance of the policies of the World Bank and International Monetary Fund. This not only has resulted in entrenching the dependency of the Third World on cash crops and selling off natural resources, but has led to a planetary ecological crisis: desertification, extinctions, deforestation, overfishing, pollution, and planetary resource depletion. This is intensified by a growing population and the drive in the Developing World to increase levels of

---

[458] Rogers (2006)
[459] Miller (1978)

consumption and technology, the success of which is measured in exclusively terms of increased GDP.

Mass consumerism and overpopulation have resulted in incredible pressures on the ecology of the Earth. This era has been termed as the *Anthropocene,* with many dozens of species becoming extinct each day from human activities, and the further destruction of rainforests, coral reefs, ocean populations, man-made climate change and global warming are simply the tip of the rapidly melting iceberg. Increasingly, we are seeing a global food and water crisis; well past Peak Oil, we are now witnessing the birth of Peak Water. Increased pressure on remaining ecosystems, in search of land for agriculture and the extraction of natural resources, has further degraded the stability of the whole ecosystem. This has resulted in further climate change, degradation of soil, and the overall degradation of the natural world. Of course, as Malthus warned, Nature counters with famine, drought, soil erosion, landslides, floods, hurricanes, fires, disease, invasive species, disease, and cancer, and human beings respond with war, plunder, police states, and political domination over all the ways to sustain life. This situation is rapidly becoming irreversible and further climate change is inevitable. Arguably, with a world population of 7 billion people, which is likely to become 10 billion people by 2050, the ecological carrying capacity of the Earth and its ability to sustain life will be severely and irreversibly exceeded. At that point, planetary Armageddon is inevitable. As vital resources to sustain life become scarce, human pressures further increase, and this results in the fragmentation and disintegration of the economic system and the Earth's ecology. The intensification of measures imposed by political economy to prop up economic system and the rule of the economic elite—such as the use of military and police forces, alongside all-pervasive propaganda—further depletes resources and increasingly more and more of the economy is diverted into reinforcing and maintaining the status quo. This not only threatens the rise of global fascism and corruption but also results in the intensification of the conditions of collapse of the technological society into barbarism and warfare, as the infrastructure of society is cannibalized to provide the resources and means to develop a totalitarian police state and the war machine, and deal with the consequences of the collapse of civilization.[460] The intensified human struggle for land and resources combined with refugee crises, mass immigration, warfare, criminality, terrorism, and barbarism has resulted in the further intensification of the relations of domination and control, the rise of fascism and dictatorships, the collapse of civilization, global war and civil war, all of which intensifies the existential crisis we face as a species. World War III is already happening in the Third World and increasingly the economic basis of society is directed to the production of weapons and the extraction of natural resources to fund the war economy. Bombs, tanks, planes, firearms, and bullets all irreversibly destroy and deplete resources and technological objects, and further toxifies (chemical, biological, and radioactive) soil, air, and water. In Iraq and Afghanistan, as well as most of Africa and much of South America, descent into barbarism and violence becomes the political norm, and any survivors are forced to scavenge on wasteland or landfill sites, or flee to the horror of refugee camps. Under these circumstances 'back to Nature' results in an increasingly savage death rate among what are effectively a growing population of hunter-scavengers (and sometimes literally cannibals), while other species (bacterial diseases, rodents, insects, arthropods, etc.) form a new (semi-)stable ecology after period of instability or punctuation (which could finish off the human species completely). A new state of equilibrium will emerge and the planet will "shake us off, like a bad case of fleas," as the American comedian George Carlin quipped.

---

[460] See Marcuse (2001a)

Destructive cycles of transforming "productive forces" (including labor, science, technology, and natural resources) into the means of waging war (i.e. ordinance, weapons, equipment, etc.) and countering social disorder (police, courts, prisons, etc.), as these means are irreversibly used and destroyed in their usage, further intensifies planetary resource depletion and the decimation of the human population.[461] Total planetary ecological collapse and mass extinction of species (including total or near total human extinction) is the inevitable result—it is as if the fire of Hephaestus burns the world before burning itself out. The demographic transition through increases in production and consumption has been one of reducing both mortality and birth rates—leading to fewer children being born, but living longer to bear children of their own—has resulted in stable populations in developed countries as a result of the advancement of technology (such as medicine and birth control) and an increase in the level of consumption. However, this has happened at the cost of the intensified exploitation of the Third World, leading to intensified crises in those countries, further destabilization, the rise of puppet dictatorships, the use of violence and terrorism to suppress resistance and independence movements, and an increase in migration to the developed world from the Third World. The transformation of Third World agriculture into the producers of cash crops for the developed countries' consumption has and will continue to result in rising food prices, the inability to feed the people of the Third World, famine, poverty, increase of infant mortality rate, increase in birth rate, intensification of deforestation and slash and burn agriculture, further environmental degradation, etc. The establishment of cycles of dependency and destabilization through use of puppet regimes, debt, and foreign aid to impose developed countries policies on Third World countries to maintain access to natural resources and cheap labor, leaning to an increased intensity of resistance (popular or otherwise) and foreign military interventions. Civil war, nationalism, and militarism necessarily leads to the further destruction of resources and labor through their consumption by military forces and the industrial-military-complex to ensure a steady supply of natural resources from the Third World to the First World, alongside adequate access to the cheap labor of the Third World. The situation is rapidly becoming one of parasitism and instability; unequal and non-symbiotic relations all premised on the control over the means to commit violence. The failure of capitalist economics rests, ultimately, on its unsustainability, and the need to use violence and the threat of violence to maintain the economic structures upon which it depends.

Overpopulation, overconsumption, waste, intensified extraction of natural resources, food crises, famine, resource depletion, police-states, violence, and war will inevitably result in a Malthusian conclusion to the global crisis. The belief that this crisis generated by these economic relations and policies can be dealt with by contraception and raising the standard of living in the Third World through globalization is a myth.[462] Turning this myth into policy—in accordance with the ideology of neo-liberalism—has resulted in increased pressure on and exploitation of the natural world, as well as the human population within the Third World. Within the so-called First World or Developed World, the situation is one of "internal colonization" wherein the population and natural resources of "advanced" countries are increasingly put at the service of multi-national corporations.[463] Any justification of inequality of access to the means to satisfy human needs and survival masks a belief in the inequality of a human right to life. Such a belief is premised on the assertion that some people are worth more than others. Once such a belief is inserted into political and economic policies, hegemony, elitism, eugenics, racism, ageism, sexism, and speciesism are inevitable consequences. The technological

---

[461] Georgescu-Roegen (1971); Bukharin (1929)
[462] Mamdani (1972)
[463] Silvard (1987)

society itself becomes an instrument for the reinforcement and extension of inequalities to preserve and reproduce the status quo of privilege and power.

While critical theorists and Marxists alike have made damning critiques of the relations between science and capitalism, their concept of Nature also presupposes that the natural world is nothing more than a material substratum governed by natural forces, laws, and mechanisms. These are available as a resource for labor to satisfy human needs, and nothing else. This raises all sorts of problems for the Critical Theorists' and Marxists forays into ecology and environmental philosophy. It is difficult to see why alienation—understood as human separation or 'aloneness'—from Nature is a moral problem at all, if Nature is nothing more than the sum total of all matter in motion. It is not at all clear why alienation from Nature (when understood only as a material substratum for human use) is of any moral concern at all, or even regrettable. If Nature is understood only as matter in motion, subjected to external forces, it is unclear why subjecting it to instrumental reason, treating it only as a means to ends, is something that needs critique, except insofar as a human being is treated as nothing more than another object in the world, and, therefore, available as a resource, like any other. Hence their argument against the domination of Nature rests on a concern with the repression (or discipline) of the internal nature of human beings, understood in terms of spontaneity and a desire for liberation—the origins of which are not explained—and it is this repression that makes human beings more like objective and external Nature, which leads human beings to treat other human beings as simply one resource among others, and, by doing so, reduces human beings to objects for instrumental reason. To turn the problem of alienation from Nature on its head and make it a problem of alienation from human *inner nature* simply changes the problem of alienation into an affirmation of spontaneity, without any clear explanation of why spontaneity is equated with liberation, or why it should be affirmed. There is also no explanation why spontaneity is desirable as a good-in-itself—or how it is related to Nature. This is taken by critical theorists to be a self-evident truth.

Horkheimer and Adorno acknowledged, from the outset of their analysis of the Enlightenment, that, as a cultural project, it was set up and set forth to liberate human beings from superstition and fear of the natural world, and establish "the sovereignty of Man over Nature."[464] The consequence of this cultural project was the dissolution of myths, the rise of the scientific worldview, and "the disenchantment of the world."[465] Like Marx and Engels, they considered Francis Bacon to be "the father of experimental philosophy," which they claimed was premised on instrumental reason, exploitation, an equation between knowledge and power, and the aspiration to dominate Nature. Knowledge is reduced to outcomes, acts of manipulation and control, where computation (or calculation) and measures of utility became the tests of knowledge, with quantification being the ultimate determinant of truth. Mathematics has dominated human thinking since the Enlightenment precisely because it allows the quantification, abstraction, and calculation of Nature, and all those aspects of experience that cannot be thus reduced and quantified are relegated to the subjective realm or are considered to be transitive cultural phenomena. By reifying the quantitative aspects of experience, science and technology have distanced and alienated human beings from the natural world. It is the quantification of Nature and its manipulation that, for Horkheimer and Adorno, placed science, technology, the natural world, and human beings and the disposal of commodification and capitalism. Technological activity has dominated human thinking from the Enlightenment onwards to the extent that it has become totalitarian. The scientist has dominion over things in so far as s/he makes them. S/he transforms Nature

---

[464] Horkheimer &Adorno (2002: 3)
[465] *Ibid* 28

into an abstract totality of laws, mechanisms, and objects available for manipulation and control.

Horkheimer and Adorno lamented the loss of "tribal spiritualism' and "animistic nature,"[466] but what does this loss amount to? It is not so much an affirmation of "tribal spiritualism" or "animistic nature"—whatever these mean—but, when unpacked, this amounts to a lament about the loss of human freedom in the technological society. Their concern was that, *by imitating Nature*, which is represented as an objective material substratum available as a resource for use, human beings became controlling, dominating, and exploitative of each other, thereby all humans become controlled, dominated, and exploited by others, as a means to ends, just like any other object in the world. This is like Thomas Jefferson's criticism of slavery in his *Notes on the State of Virginia* (1785) on the grounds that it makes "lazy brutes" of the slave-owners, and "unfit to be citizens of the Republic."[467] Their moral concern against the exploitation of Nature is its effect on the nature of human beings *qua* exploiters. This not only makes Horkheimer and Adorno's concerns about human alienation from Nature mystifying—as it is hard to see how imitating Nature and becoming like Nature is the cause of our separation from Nature—but their criticisms of the scientific worldview and technological society ultimately unpack as being based on a somewhat naïve historical belief that systems of domination of human beings did not exist prior to the end of the 15th century. Yet, slavery and serfdom, as well as all manner of inequality and exploitation, and every conceivable shade of cruelty and violence, all existed in pre-capitalist societies, and certainly long before the Scientific Revolution and the Enlightenment; the inequality and exploitation of humans by humans cannot be attributed to modernity, science and technology, except to highlight them as continuances or intensifications of already existing cultural practices and intentions (i.e. the exploitation of the labor of others).

Horkheimer and Adorno were unable to provide us with any intelligible, alternative representation of Nature, or any alternative relationship with Nature, and, as a result, their moral objections against the domination of Nature should be translated into moral objections against the repression and denial of human "inner nature" and how this allows the domination of human beings by other human beings. Hence "the revolt of Nature" was, for Horkheimer, "the rebellion of human nature" against systems of domination, civilization, and "the repression of instincts".[468] Following the Marxist orthodoxy, history is reduced to struggles and conflicts—violent social explosions—and the rise of class awareness that capitalism is wasteful, irrational, exploitative, and destructive. Capitalism responds to this rebellion by using propaganda and the culture industry to make alternatives seem absurd, flawed, frightening, irrational, dangerous, or quite simply impossible. Nature is represented as a place for the poor and destitute, which only offers hardship, scarcity, disease, injury, starvation, and death. Yet capitalism also justifies itself in naturalistic terms and represents itself as Nature made efficient. Should the rebellion against capitalism continue and even intensify, the agents of capitalism readily resort to the tactics of fascism and the use of violent, oppressive, police-state and militaristic measures to defend the natural order of things. It is no coincidence that fascism has always appealed to a natural order to justify its extreme measures. The domination of Nature and the domination of human beings by other human beings go hand in hand, once instrumental reason is put in service of preserving the status quo, but this should be unsurprising, given that Horkheimer's moral objection against the domination of Nature rests on the claim that instrumental reason turns upon human

---

[466] *Ibid* 54-7
[467] Jefferson (1998).
[468] Horkheimer (1974: chap. 3)

beings, dominates us, and makes us less free. Horkheimer's moral objection to the domination of Nature and the destruction of ecological systems—"breaking down the mechanisms of self-renewal"—is that the consequences for human beings of this "immoral act" are unpredictable and frightening for our survival as a species. Thus "the domination of Nature" is itself represented in terms of human irrationality and self-destructiveness; ourselves dominating and turning against us, making us less free, rather than committing any immoral act against non-human nature. Ironically, Horkheimer's objections to instrumental reason are themselves couched in terms of instrumental reason. We end up with instrumental reasons explaining why it is bad for us to use instrumental reason to dominate Nature.

Marcuse in his essay "Ecology and Revolution" argued that capitalism (and the industrial military complex) is destroying the ecology of the natural world, which he considered to be "the sources and resources of life itself."[469] Like Horkheimer and Adorno, his objection to the destruction of the natural world panned out to be an objection to the inequality of distribution of resources and the suppression of human inner nature, rather than pointing to any specific immoral act against Nature. The immorality of the destruction of the natural world remained, implicitly, an immoral act against humanity. Marcuse's commitment to ecology was little more than an aspect of his commitment to an aesthetic of the environment within which human equality would flourish, except when capitalism, and other systems of repression and exploitation, imposes themselves on human nature and pursues ecocidal activities for the sake of profit and domination of one class of people over another. Thus Marcuse's objection to the domination of Nature by science and technology was actually an objection to how science and technology have provided the means to dominate human beings. Marcuse's attempt to turn Nature into a subject, equal to the human-subject, was an attempt to appeal to the liberation of Nature as a condition for our own liberation of our "internal nature," but this attempt was a continuation of the reductionism and anthropocentrism that attempts to understand all things via analogies with ourselves. As a result, despite objecting to the technological exploitation of the natural world, Marcuse's theory was an anthropocentric projection of Man over Nature that failed to provide any intelligible, alternative conception of Nature, other than something human beings imitate or exploit as means when they repress themselves and exploit each other. Thus, like Horkheimer and Adorno, Marcuse was unable to provide any alternative to an anthropocentric approach to the natural world because he was unable to provide any intelligible account of how Nature is "a revolutionary force" beyond its use as an inspiration and justifications for the anti-capitalist sentiments of the New Left of the 1970s, which largely rested on appeals to human freedom and equality.[470] Torn between instrumentalism and romanticism, the critical theorists have been unable to provide any new conception of Nature to the extent that for Jürgen Habermas it was unthinkable that Nature could be anything other than a material substratum and source of forces of production, and the ecology movement should be rejected as based on little more than romanticism.[471] Critical theorists have come full circle by reaffirming the distinction between human beings and Nature (and all other beings) as premised on language and technology, ultimately affirming the anthropological and instrumentalist definitions of both, and reaffirming the classical philosophical and anthropological definitions of human beings.

---

[469] Marcuse (1971).
[470] See Woddis (1972) for an interesting discussion of Herbert Marcuse and "the New Left." See also Singer (2000) for a more general and critical discussion of the concept of equality in the thinking of "the New Left" in relation to Darwin's theory of evolution. Also, cf. Marcuse (2001c).
[471] Habermas (1979)

This has left the door open for the social constructionists. In his book, *Against Nature*, Steven Vogel argued that the concept of Nature has been problematical for social and critical theorists, leading them to deal with it inconsistently throughout their writings due to the influence of Marx and Marxism on social and critical theory.[472] Critical Theory takes Western Marxism as its point of departure for its critiques of the epistemology and methodology of science, while it also attempts to deal with the philosophical problems that arise after rejecting the possibility of developing Critical Theory as a "scientific method." Vogel focussed on the writings of Georg Lukács, Max Horkheimer, Theodor Adorno, Herbert Marcuse, and Jürgen Habermas, and he argued that all of these theorists relied on a self-contradictory distinction between "the natural" and "the social," with each simultaneously denying the existence of any unmediated and objective natural world accessible to human experience and knowledge, but also appealing to a kind of naturalism or objectivism when using a concept of Nature when explaining technology and its historical development. Social and critical theorists rely on a conception of Nature as a material substratum underlying society, and ultimately leave it to scientists to give the objective facts about that substratum in terms of instrumental reason and technical know-how, while denying that those facts can be known in asocial or ahistorical terms. This leaves a contradiction within social and critical theory, which grants the social and critical theorist an Archimedean point from the vantage point of which they are able to say something about something that they are supposedly unable to say anything about—Nature as an objective and material substratum—while simultaneously denying this right to natural scientists and all other natural philosophies. Nature is supposedly a social construct, with its historical and social conditions and contingencies—a cultural artefact based on human relations—while simultaneously also constituting the material foundation for human economies and technologies upon which society is constructed. Vogel explained that this contradiction arises from the assumption that Nature exists as "the extra-social realm" accessed through technological activity but understood in cultural terms. He considered all the methodological problems of social and critical theory to arise from the failure of social and critical theorists to clarify adequately their concept of Nature and how it relates to the development of their critiques of the technological society.[473] However, via an appeal to historical and sociological studies of science and technology, Vogel argued that these problems can be resolved in social and critical theory by taking the idea seriously that Nature is socially constructed.

As Vogel pointed out, "Western Marxism" continues to accept that the natural sciences disclose facts about the natural world, in accordance with a positivistic account of the scientific method, and this tradition reserves its criticisms for the social meaning and uses of the scientific method and its results. Even though he is quite correct, he failed to pay any attention to post-war developments in science studies that have shown positivism to be deeply flawed.[474] By the time Habermas' early works had been published, such as *Knowledge and Human Interests* in 1968, positivistic philosophy of science had even been criticised by positivists! Marxism has yet to catch up to these developments in the philosophy of science. Vogel also made the claim that a re-evaluation by critical theorists of the concept of Nature and how we understand natural science will provide insights relevant to environmental debates, as well as the social studies of science and technology. While I agree with Vogel that it is both timely and important that social and critical

---

[472] Vogel (1996)
[473] *Ibid* 2
[474] Vogel's social and critical theory of science largely relied on Kuhn (1962), Latour and Woolgar (1979), and Rouse (1987), but he also made reference to Knorr-Cetina and Mulkay (1983); Latour (1987); Haraway (1989), among others, and he argued that social and critical theorists have largely taken for granted a positivist philosophy of science.

theorists take Vogel's criticisms seriously and rethink how Nature and natural science are understood, I take issue with Vogel's social constructionism, which I largely take to be based on a series of misunderstandings and over-simplifications of natural science and its interpretation. My view is that Vogel's social constructionism fails to resolve anything at all and that we need a much deeper critical analysis of the concept of Nature in critical and social theory. Once this has been done then it is possible to develop an adequate critical and social theory of natural science.

Vogel considered social and critical theory to have been developed from within "the tradition of Western Marxism," which itself had been developed in opposition to the "orthodox tradition"—assuming these "traditions" really exist and are not a construction of Vogel's. These "traditions" boil down to a shared association with the natural philosophy of Engels, by assuming the validity of the natural sciences and of certain standard accounts of their methods. The "orthodox tradition" since Engels has tended to interpret the results of natural science as informing and validating dialectical materialism. Orthodox Marxists avoided applying Marxist analytical categories and concepts to the natural sciences, except to show how the sciences have been distorted or perverted by capitalism and bourgeois society. It is on this point that "Western Marxism" breaks with the "orthodox tradition." The key development in "Western Marxism" was a critical rejection of the assertion that Marxism is a science, akin to the natural sciences, with a clearly understood methodology. This involved the critical analysis of the social origin and historical development of the sciences, along with the rejection of *scientism* and the uncritical faith in science as an emancipatory and progressive force. Western Marxism represents science as being ideological in both form and content—a social construction. Hence an analysis of the class-structure is required to understand science as a social phenomenon and its results as social products. It is only on this premise that Western Marxism is able to mount a critique of the notions of scientific rationality, objectivity, and value-neutrality (or "value freedom"). Science is criticised for being unconsciously rooted in social reality, without having any means to become conscious of its own rootedness—the contingency of science cannot be an object for scientific investigation. Natural science is a historical and social phenomenon that expresses and reinforces the prevalent cultural norms, ideals, and culture of society. Thus, science tends towards being supportive of the *status quo* and "the ruling ideas" or "the ruling class," by representing human society as based on a natural order.

Vogel affirmed Alfred Schmidt's assessment of Lukács treatment of Nature in *History and Class Consciousness* as dissolving Nature "both in form and in content, into the social forms of its appropriation."[475] Following Schmidt, Vogel rejected Lukács' distinction between natural science and Marxist theory. According to Vogel, the former is concerned with the material substratum of society, whereas the latter is a concerned with society and social change. Yet, Vogel admits that this distinction is central to Western Marxist critiques of "orthodox Marxism" (i.e. Engels etc.) Natural science is restricted to studying the natural world and providing technological means of material manipulation and utilising natural forces, but any use of the methods of the natural science as a model for social theory is subjected to criticism and challenge. While this maintains "orthodox" Marxist materialism regarding how the natural world is understood, it rejects Engels' attempts to develop a dialectics of Nature into a model of social evolution and thereby unify Marxism as applicable to both the study of Nature and society. Lukács placed a limit on the applicability of both the methods of natural science and Marxism. This formed the basis of what Vogel termed as *the misapplication thesis*, which holds that the methods of the natural sciences are misapplied if they are taken to be the model for the

---
[475] Schmidt (1971: 96)

methods of social theory. Vogel's objection to this distinction is that it allows the methods and results of the natural sciences to continue unquestioned and challenged by Marxist theory, which can only deal with the social influences upon and social consequences of the results and methods of the natural sciences. His concern was that the misapplication thesis largely leaves the scientific conception of nature untouched and outside the realm of analysis, apart from showing how social influences can distort or pervert scientific methods. Vogel criticised Lukács' conception of the misapplication thesis because it could not be consistently applied, given that natural science is a social activity.

As Vogel noted, the misapplication thesis was untenable for the critical theories developed by Horkheimer and Adorno, both of whom mounted radical critiques of the methods of the natural science in relation to industrialization, war, and capitalism. This line of thinking led to "the domination of Nature" thesis. In Horkheimer and Adorno's writings, and later Marcuse's, "the domination of Nature" is inextricably related to the domination of human beings by other human beings. As a result, these writers rejected any notion that natural science has any hegemony as the means to study the natural world. Vogel argued that Marcuse's efforts to explicate an alternative conception of Nature led to a contradiction wherein nature is conceived as being "what it really is," thus promising the possibility of developing a "New Science" and "New Technology" that can rediscover Nature and leave it as-it-is, while on the other hand treating Nature as based on human relations and historical categories. Vogel claimed that Adorno's later attempts to develop the concept of the "non-identical" as a sign of Nature merely exacerbated the problem. He also argued that Habermas' attempts to reconcile the misapplication thesis with Critical Theory merely resorted to a neo-Kantian dualism between "the social" and "the natural," thereby, maintaining the distinction rather than dissolving it. Indeed, Habermas' early writings rely on a noumenal concept of "nature-in-itself," which is not developed consistently at all, while in his later work references to Nature simply disappear. Habermas basically side-stepped this problem, according to Vogel, by returning to Lukács' distinction between "the social" and "the natural," returning the focus to subject-object relations, and effectively banning Nature from philosophical discussion, while leaving the discovery of the facts of Nature to the natural sciences. However, Vogel criticised "the domination of Nature" thesis because it presupposes some Nature that is being dominated (thereby appropriated, transformed, exploited, or harmed), but this presupposition entails that it is possible to know of such a Nature, and this raises the question of by which epistemological standard (other than a kind of Kantian intuitionism) can such a concept be formulated. This suggests that "the domination of Nature" thesis requires an alternate conception of Nature, but that alternate conception is not made explicit or justified. It becomes an unspeakable Other—and never spoken of again.

Vogel identified this underlying conception of Nature within the Frankfurt School tradition (Horkheimer, Adorno, and Marcuse) as stemming from romanticism: "Nature-as-Other," whatever that means. The characterisation of science in terms of "the domination of Nature" thesis presupposed that the act of domination (manipulation, appropriation, exploitation, transformation) violates the otherness of Nature. This implies the existence of Nature, as being a realm beyond human experience and understanding—a noumenal thing-in-itself. Vogel also considered this conception of Nature as "Nature-as-Other" to have been apparent in the later writings of Martin Heidegger and the Deep Ecology movement. Vogel asked, how can we explain how we can come *to know* "Nature-as-Other"? Vogel argued that this implies a kind of naturalism (or intuitionism) that contradicts the claims regarding the historical and social character of human knowledge, and allows an Archimedean point not allowed to other philosophies or the natural sciences. By doing this, argued Vogel, Critical Theory is inherently inconsistent—or self-contradictory, to put a finer point on it—by allowing a supposedly unknowable and ungraspable Nature to be known and grasped as "Nature-as-Other." This leads to a naturalistic fallacy, according to Vogel. And that will not do.

Vogel's these inherent contradictions in Critical Theory as inherited from Marx's assumptions about the natural sciences and Nature. The difficulties and contradictions inherent to Critical Theory are the consequences of a tension within Marxism, from which Critical Theory was developed, between the Hegelian characterization of the social character of knowledge (wherein the subject influences the object known) and the commitment to materialism, by identifying a material substrate to social action, which makes social action possible and conditions it. Thus, argued Vogel, if we conceptualize this material substrate as Nature, on one hand we are compelled to consider the knowledge of Nature as historical and social in character, being the products of science as a social activity, but, on the other hand, this problematizes the kind of materialism upon which Marxism conceptually relies. It also problematizes the basis upon which the Western Marxist critique of objectivism and scientism is based. However, rather than explore the possibility of a refined or new understanding of the dialectical relation between the social and the natural, a possibility that Vogel does not seem even to consider, he simply chooses sides in favor of the Hegelian characterization of knowledge as active, and, in doing so, he argues for a social construction theory of Nature and natural science. Vogel states,

"By insisting on the significance of what was the great insight of classical German idealism—that knowledge, if it is to be possible at all, must be *active* and understood as involved from the very start with the object known—such a view asks us to "deconstruct" supposedly natural and familiar phenomena into the hidden social processes by which they were produced. It sees the relation of human subjects to the nature they inhabit as an active and world-changing one and wants therefore to take seriously the idea of nature as a 'social category' or more precisely as something *socially constructed*." (Vogel, 1996: 5)

Vogel affirmed the Hegelian view that the subject is historically and socially situated with regards to the natural sciences, and, by placing the emphasis on the conditioned, contingent, and dynamic character of how we understand the world, he rejected any other interpretation apart from the social construction of Nature. This does have elements of a compelling argument. Despite Popper's insistence, historicism cannot be readily dismissed. The ahistorical, static account of Nature offered by Popper *et al.* is based upon concealed social conditions and human relations, which have become reified ('congealed' labor, as Lukács termed it). An important role of Critical Theory is to show these conditions and show how representations of "the natural" are based upon socially organized activities and the products of labor, thereby exposing the ideology that underwrites the natural sciences and technology. Critical Theory also shows how that ideology conceals the social relations and conditions upon which those activities depend. Vogel argued that "alienation from Nature" should be understood as a social product that serves the *status quo*. He argued that we should understand that the *Umwelt* (the world that surrounds us) as the product of social activity and labor;[476] hence he argued that Critical Theory should reject naturalism and come to understand Nature *normatively* as a social construct. His whole thesis pivots on the claim that by adequately equating a Critical Theory of Nature with a Critical Theory of Society, while placing "the natural" under the category of "the social," naturalism and objectivism will be shown to be social constructs

---

[476] Vogel (1996: 7)

open to deconstruction, and human thinking will be liberated from the naturalistic fallacy. As he put it,

"To say that nature is 'constructed' then is in a sense simply a way of saying that an appeal to nature is always nothing more than an appeal to *us* and our own discursive process of justification." (Vogel, 1996: 10)

In as far as Vogel's thesis is limited to the claim that all acts of reference (in virtue of being human acts) to Nature are socially constructed then it is an uncontentious thesis, in my view. By pointing at something and speaking its name, we use language—a social construct—which is a human activity, with all its cultural interpretations, representations, and meanings. Sense and meaning are also socially constructed. However, my argument is that this is necessary but insufficient for any social and critical theory of natural science. Vogel's theory does not provide an adequate account of the role of technology in natural science. My argument is that we need to understand the insights of social constructionism and then move beyond it if we hope to develop an adequate understanding of Nature and natural science in terms of social and critical theory of technology. We need to move towards a social and critical theory of natural science—via historical and sociological accounts of natural science—by developing a theory of the dialectics of technology in relation to the ongoing historical and social construction of conceptions and representations of Nature, through labor, and in relation to the development of the technological infrastructure of science and society. In this way, we can understand the nature of the objective world revealed to us through technological activity. This has been my underlying motive in this book. Vogel (along with Western Marxists and social constructionists in general) assume an instrumentalist account of technology—'as a means to an end'—alongside an anthropological account—'as a human activity'—and that "the social" can be clearly understood and act as an adequate explanatory concept. "The social" has become the new Archimedean point—the new metaphysics. Yet, social theorists must recognize that, for any adequate social and critical theory, there must be some recognition that the concept of "the social" is itself historically conditioned and contingent—itself a social construct—and cannot be taken seriously as an objective ahistorical or absolute fact or truth about the world. The "us" and "our own discursive process of justification" are very much at stake, ambiguous, and subject to change and controversy. The meaning of "human" and "the social" are contingent and up for grabs. To assert any particular conception of "the social" as the category under which all other events and phenomena can be placed, is an activity based on a wider attempt to advocate a particular form of society and its associated political and economic institutions. It turns "society" into a noumenal "thing-in-itself" and subject to the same "Kantian intuition" for which Vogel denounced Marxists and critical theorists.

Vogel's own theory of environmental values is overly anthropocentric due to his appeal (naturalistic appeal, in my view) to humans as being the only determiner of value, even though he concedes that human beings can value non-human beings. Given his constructionist stance, in my view, he has 'jumped the gun' here by assuming, from the outset, that the meaning and truth of *being human* is something that is fixed and clearly identifiable. But, by what standard could we define what a human being is? Surely that would be a social construct too (given that "biology" and "physiology" are social constructs, according to Vogel). What is a human being? Surely this must be at stake. To treat it as an ahistorical and objective social construct leads us only to paradox and self-contradiction. As well as this self-contradiction, Vogel's social constructionism also makes and depends on essentialist and naturalist claims about values being a property of human intellectual and normative thinking, based on the assertion that only human beings are *known* to do this. Thus Vogel takes certain facts from the natural sciences as objective 'givens,' while simultaneously claiming that all scientific facts are social constructs. But, what standard of knowledge is being applied here? Vogel presents an *a priori* as if it were an empirical fact. This claim alone completely undermines his thesis. After all is said and done, if we were able to discover such empirical facts about human beings then why can't we discover empirical facts about non-human beings as well? The reliance on *a priori* definitions of the nature of "human beings" and "the social" is the self-contradiction at the heart of social construction theory.

Of course, as I have endeavored to explain in this book, one cannot hope to understand sciences and technologies without understanding their social contexts and relations within a culture. The "nature" revealed by science is a technological product because of the culture that developed technology as a means of disclosure. Once a device or technique is implemented as "the most efficient means" available, the technological imperative to use "the most efficient means" imposes a demand upon human beings to use them, and this demand is imposed by them on other human beings engaged in the same disciplined and specialized activity. This demand gives technology its social contexts of scientific justification and discovery. Metaphysical claims that "the most efficient mechanism" is "a natural mechanism" conceals the role of the technological imperative in the social construction of science. Indeed, the "most efficient means" are social outcomes of decisions and agreements about what can be said to be the well-understood and necessary means to achieve well-understood and necessary ends, in accordance with the already agreed upon objective, rational, and logical trajectories of scientific research and development in relation to its historical background and cultural resources. Human goals and practices are selected and structured in accordance with the necessities, possibilities, and limits of the human imagination, once disciplined through technological activity and seduced by its outcomes, but even this is constrained and shaped by the technological framework. The technological framework shapes how we think and act. Karl Manheim argued that, if taken to its extreme as a dominant factor, the assertion of technical rationality as the best approach to decision-making can suppress, distort, or damage the human capacity to use our intelligence and insight to critically evaluate the desirability of postulated ends and the proposed means to achieve them.[477] It is a choice that affects future choices. As I have endeavored to explain, this way of thinking has suppressed alternative conceptions of Nature and approaches to interacting with and living within the natural world. It reduces Nature to only those forms amenable to technological manipulation and control, as a resource; the success of which becomes the measure by which theory is tested and put into practice, as an objective pursuit with its technological consequences and their explanation. To declare that Nature is nothing other than a social construct simply presupposes a fixed representation of the human and the social, and, as a result, surrenders to the technological imperative and the rationality of the technological society to a historical and contingent image of modern man. My argument is that if we aim to get beyond this reductionism and move towards a deeper understanding of Nature, and hence also of science and ourselves, we need to understand the dialectics of technology and how it relates to how we live on Earth and relate to the natural world.

---

[477] Manheim (1951: 55)

The Ecotechnological Society:

Where can we go from here? Lewis Mumford advocated an "organic ideology" for society that called for the rational and harmonious convergence of human practical concerns and the need for fostering sustainable relations between all aspects of life.[478] Once we think about life in its full manifestation in terms of an organic whole—complex and comprised of interacting living beings—technology becomes subordinate to life, and the technological imperative subordinate to the instinctive imperative to sustain and enhance life. Subordinating the continuing refinement and integration of technology within an organic whole subordinates the technological imperative to the development of harmonious relations with all life on Earth. This ideology promises greater opportunities for human flourishing to become developed organically with the whole. How can we understand this ideology? Elsewhere I agreed with Mumford that the challenge facing us—something which is even more imperative today than during the time of Mumford's writing—is how to use science and technology to transform the development of the technological society in such a way as to place it at the service of bettering life for all, rather than the privileged few, and creating a more sustainable and democratic form of the life-world, which optimizes the chance for as many people as possible to live a good life in sustainable harmony with the natural world.[479] This involves *ecologizing* the technological society. If we assume that human beings are social beings with an instinctive imperative to sustain their existence, but uncertain about how to do that successfully in a complex and changing world, then ecologizing the technological society requires democratizing it to create the political superstructure for the technological construction of a sustainable society.[480]

However, as Mumford argued, a mature civilization—as part of the maturation process—must come to understand the rational limits of its technical abilities, which are imposed "by the very nature of the elements with which we work."[481] It is through such an understanding, once we have developed the technological society into a democratic and ecological society, that we will be able to develop the forces and relations of production into a sustainable state of equilibrium. We need to move beyond the exploitation of Nature and the naïve faith in our ability to construct a better artificial world through the domination of Nature, and, instead, we should democratically work at integrating our technologies within the natural world in order to choose and develop sustainable and rational ways of life in harmony with each other and other beings. This requires non-technical visions and shared values upon which we can explore through public communication our vision of a good life and build it into the sustainable development of our communities as members of a planetary ecology and an organic whole. Of course, this will involve the technical development of environmentally sustainable methods of constructing and organizing cities, transportation, industry, agriculture, and mining, as well as conserving the natural world through ecological interventions and restoration. Clearly technology will remain central to the human condition, but the ecologization and democratization of the technological society will involve a greater depth of understanding of the possibilities and benefits of cooperating and harmonizing with natural processes rather than competing with or replacing them. When practiced within a democratic society, it will help prepare, forewarn, and adapt us

---

[478] Mumford (1963: 426)
[479] Rogers (2006)
[480] Rogers (2008)
[481] *Op cit.* 429

to the capriciousness of Nature, as well as our own limitations, by optimizing our capacity to adapt and change in response to changes in the world.[482]

Thomas Hughes followed and developed Mumford's call for an integration of technics with the organic, and argued that we need to integrate our concerns about the conservation of the natural world, and the preservation of the health, integrity, and diversity of the lived-environment within the human built-world, in order to integrate ecology and technology to construct an *ecotechnological society* (Hughes, 2004).[483] When architects take into account the local climate and landscape when designing buildings, they are creating ecotechnological cities; when engineers and planners take into account the natural flow of rivers, streams, and prevailing winds when designing irrigation systems, they are constructing ecotechnological systems. Hughes argued that this architectural movement found its roots in the American puritans' idea of creating a pastoral New Jerusalem in the new world. Citing Perry Miller, Thomas Jefferson, Robert Beverly, J.A. Etzler, and Ralph Waldo Emerson, Hughes described how the early American settlers proposed a poetic vision of using science and technology to confront the wilderness and create pastoral communities that would nurture genuine virtue and return human beings to a state of grace (Hughes, 2004: 16-43). This idea was that through the prudent and careful construction of mills, canals, and steam engines, human beings could build a new country of interconnected, but self-sufficient communities, within green valleys, filled with orchards, farms, and the sound of songbirds. Agricultural combines would reap, sow, and plough, converting the wilderness into a bountiful garden landscape, free of the squalor and satanic mills of Europe, to bring about a new Eden within which all human beings would be equal, free, and happy. But, as history testifies, things did not quite work out that way. As Hughes put it,

"Technologically empowered, we have reason to doubt our values and competence as creators of the human-built world and as stewards of the remaining natural world. Slums in inner cities, ugly strip malls in the suburbs, hastily and cheaply built housing, polluted air and water, the loss of ecologically nurturing regions, and the likely threat of global warming give evidence of our failure to take responsibility for creating and maintaining aesthetically pleasing and ecologically sustainable environments." (Hughes, 2004: 153)

Instead of building ecologically sustainable environments, human beings have developed the technological society as a substitution for Nature, having appropriated the natural world as instrumentally available as a material substratum of resources and mechanisms that available through technological activity for discovery and manipulation, or represented as something dreadful and capricious which can destroy all our efforts without warning.[484] The ideal of a pastoral and sustainable society was eclipsed by the perceived need to industrialize and urbanize the world, which ushered in a detachment from the natural world and its exploitation. This attitude of exploitation (and fear) is brought very much into question when human activities, such as deforestation and agriculture, result in pollution, cause environmental disasters, or intensify natural events into disasters (via soil erosion). We should learn to recognize the extent that how we participate in the world irreversibly changes the world in ways that we simply cannot foresee—thereby endangering ourselves and generating greater uncertainty and vulnerability. The challenge facing us today is how we can develop our technologies and sciences in order to live harmoniously within the natural world. Today's urban planners and developers often use technologies that ignore or replace natural terrain rather than

---

[482] Rogers (2006)
[483] See also Rogers (2006, 2008); Bookchin (1996, 2005); Spirn (1984); Miller (1956).
[484] Rogers (2006)

interact and adapt to it. This can destroy natural areas that previously acted like natural buffers and defenses against flooding, wind, erosion, etc. Thoughtlessly destroying these areas increases the chance and intensity of natural disasters, such as the erosion of wetlands and offshore sand spits were major factors in the devastation of New Orleans by Hurricane Katrina in 2005.[485] As Hughes argued, the degradation of the natural world and the human spirit has been of our own making, and, therefore, we should not blame God or gods for the natural disasters that are caused and intensified as a result of our own ignorant and thoughtless actions. It is all too easy for us to blame someone else, such as capitalists, industrialists, developers, architects, engineers, bureaucrats, scientists, or politicians, but too many of us want cheap industrial and agricultural products, our own cars, low taxes, and we welcome and elect those that promise these things, regardless of their wider consequences for society and the natural world. Yet there are also numerous ways that we can participate in the decision-making process, or at least make some kind of protest regarding our exclusion, but most of us decline to do this and leave it to others. We should take responsibility and participate more in the design, construction, and development of the human-built world.

In order to do this in a way that is ecotechnological, while still remaining within the device paradigm and seeking technological solutions to ecological problems, human beings need to become more technically literate and informed about ecological science, along with all the other natural and social sciences. This literacy would be a limiting condition for democratic participation because it is only by satisfying this condition that we can consciously, rationally, and purposefully use technical knowledge to shape our society into a democratic and ecotechnological society. The ecotechnological society is as artificial as the technological society—in fact, it is an extension of it—following on from its paleotechnic and neotechnic phases, as Mumford termed them. The ecotechnological society would consider the urban landscape as being as much (if not more) the natural landscape for human beings than it would consider wilderness to be. Just as Mumford argued that the neotechnological society improves on the wastefulness and polluting aspects of the paleotechnological society, Hughes argued that the ecotechnological society improves on the consumptiveness and destructiveness of the neotechnological society. It is the movement from paleotechnics to neotechnics, and from neotechnics to ecotechnics that promises to avoid resource depletion, pollution, mass extinctions, and ecological degradation and collapse, and thereby maintain ecological diversity and health by extending the carrying capacity of the Earth and the upper limit to human population. However, the ecotechnological society continues the human effort to pacify existence for the benefit of human beings—and thereby continues to apply instrumental reason to the task of dominating Nature—but does so in a manner that leads to sustainable and renewable ecotechnological structures and relations, wherein human society becomes a sustainable and harmonious social ecology that is integrated into the non-human ecology of the planet Earth.

How should we proceed? In order to maximize adaptability and sustainability, both of which require diversity and pluralism, the ecotechnological society must embrace democratization as a prerequisite if it is to avoid totalitarianism and maximize its evolutionary potential.[486] Otherwise, not only would an undemocratic ecotechnological society be elitist (and probably fascist) in its structure and development, wherein all decisions were made by a scientific elite and dissent would be irrational and intolerable (and probably criminal), but it would repress and undermine its own creative potential and adaptability to unforeseeable events in an open-ended, changing, and complex world

---

[485] Rogers (2006)
[486] *Ibid*

by limiting the decision-making process to only a small group of people. As such, the ecotechnological society must grow out of a participatory democracy—understood as having more in common with community-based anarchism than representative liberal democracy—through communication and cooperation, whereby people are to learn from each other's experiences and mistakes while trying different solutions to shared problems, improve their understanding of their problems and how they are shared by others, and develop or change their practices as a result. A democratic society learns how to articulate and shape shared visions of a good life and the kind of pluralistic society wherein such visions could be realized in practice in the community, alongside alternative visions put into practice in other communities.[487]

By maximizing inclusion in the decision-making processes, a democratic society maximizes the available stock of creativity, experience, skills, imagination, and values, rather than relying on the presumed omniscience and infallibility of an elite (whether economic, political, religious, bureaucratic, technical, or scientific), and thereby increases the creative potential and adaptability of that society. In comparison, an elitist society is both compelled towards totalitarianism—thereby promoting the myth of the infallibility and necessity of its ruling class—which leads to rigid and inflexible thinking that can become obsolete in a changing world. It negates the worth of the vast majority of its population, except as instruments of the ruling class, and thereby reduces its creative potential and adaptability by putting all its eggs in one basket, so to speak, and gambling on "the wisdom" of the elite, and hoping that the world does not change in unpredictable and uncontrollable ways. Any totalitarian society is overspecialized, and, like any overspecialized species, runs the risk of its own extinction if its environment changes in ways to which it cannot adapt. A democratic society always maintains a healthy internal freedom, thereby allowing various responses to any change and the possibility of creative experimentation to occur at the local level, with different communities trying different possibilities. Maximizing inclusion in a free and democratic society, wherein not only is society governed with the consent of the governed, but also people are afforded the greatest chance of discovering ecological solutions to shared problems for themselves through democratic participation. In this sense, democracy becomes synonymous with a conscious and social evolution towards an ecotechnological society, wherein how to live and share life on Earth promises to be discovered through democratic participation, communication, and visioning, and adapting to the natural world.[488]

The ecotechnological society would be a sustainable democratic society because it would liberate human potential and creativity through community-based communication and practicality, wherein those who make decisions are those who suffer their consequences. Thus the ecotechnological society would transform human existence from being based on a parasitic/predatory relation with Nature (as well as each other) to being based on a symbiotic/cooperative relation with Nature (and each other). All life on Earth would flourish as a result of this transformation, which would be nothing less than the evolution of *Homo sapiens sapiens* into *Homo sapiens ecologicus*. Ultimately, this would become a physiological evolution, as well as an evolution of human consciousness, as our descendants are transformed by life on Earth, in all its aspects, into an ecological being integrated into the organic whole. The ecotechnological society would lead to a steady-state (zero growth) economy and sustainability at a planetary level, while improving the quality of human life through conscious and democratic participation. It would oppose and negate the capitalist economic imperative, while mediating and modifying the technological imperative in accordance with a definition of "efficiency" in terms of a

---

[487] Rogers (2008)
[488] See also Odum (1971).

concept of *integrated self-sufficiency*. "Economics" and "ecology" would be brought together and there would be a rediscovery of how they share the same etymological stem in the word *oikos* meaning "home." The Earth would no longer be seen as a composite of mechanisms and resources available for economic and technological exploitation, but, instead, *as our home*—the home of our spirit and the physical foundation of our being, shared with all life on Earth. Without doubt, this would be a radical change in how human beings think about ecology, as well as how we live on Earth. The pernicious and damaging effects of technology are a consequence of our failure to integrate technology within the organic whole. Pollution, resource depletion, and environmental degradation have been the result. Along with the destruction of organic, ecological communities, human relations have suffered along with those other beings in the natural world, as the organic whole degrades. This failure of ecological integration is the reason human beings have failed throughout history to create a sustainable civilization. When civilization is based on separation and exploitation of Nature, it is inevitably based also on the separation and exploitation of human beings as well, and inevitably disintegrates through ecological and social collapse. The mechanical world view presupposes separation and exploitation and any civilization based on it is also doomed to collapse.

Mechanical realism has resulted in technological activity (skilled labor) being represented as the primary mode of interaction upon which all the activities of natural science are performed and interpreted by scientists. Nature is represented as a mechanistic material substratum and an abstract of rules that govern how mechanisms work under specific conditions, and nothing else. Any understanding of how Nature works is itself proposed in terms of the knowledge of how to make into something that acts like the natural phenomenon in question, and analogously relating it to the scientific worldview to show how the natural phenomenon in question works as a part of "the world-system." Scientific experience is inherently a product of interventional technological activity, separating and exploiting, and the objective realm is confined to the technological framework itself. Knowledge takes the form of the interpretation of operational functions and numerical values determined through measurement and the mathematical projection a machine over the natural phenomenon. The test of knowledge is taken to be identical with measures of enhanced power and control. The success of science is measured in terms of its value for the further development of the technological society and the technological framework of science. How can we think differently about the relationship between science and Nature? How can we bring together science and the idea of Nature as an organic whole without representing ecology as a machine?

If we think of Nature as Other—the non-human and unknown—we need to understand the relation between the Other and ourselves. In the ancient world, this would have been accepted as mysterious and unknowable, perhaps as terrifying or divine, but in the modern world we are tempted immediately to relate the Other to our modes of technological activity. We make an "it" of the Other and explore how we encounter it. How can we think of the Other without technologizing it? The philosophy of physics proposed by physicist Ilya Prigogine can help us make sense of this question. If we take a step back from the act of mathematical projection of thermodynamics over the world implicit to his philosophy of science and instead apply his insights directly to the technological society, we can see that the relation between Nature-as-Other and the technological framework matches closely Prigogine's idea of *a dissipative structure*. He defined a dissipative structure as an open system maintained by extracting matter and energy from Nature; thereby reducing Nature to being a reserve of energy and matter (wherein matter and energy are interchangeable. The dissipative structure transforms energy and matter into heat, thereby destabilizing states of thermodynamic equilibrium in order to extract energy from Nature for work. Prigogine used the example of a whirlpool or vortex in water to explain a dissipative structure. Physically, the vortex does not exist; only the water (matter) and its motion (energy) exist. The vortex is a structured movement that exists only because it extracts energy from the water. It dissipates the energy of the water as the movement of the water is structured. The vortex is itself a dissipative structure in water. Representing the vortex as a pattern of forces is an abstract representation of the motion of matter and dissipated energy that is superimposed over the water and its movement. This act of projection is used to explains and makes sense of the experience of the vortex in terms of invisible causal relations, themselves visualized in terms of the abstract representation superimposed over the phenomenon of moving water, thereby allowing the further abstraction of these causal relations in terms of the more general form of a mechanism—centripetal force—which becomes the fundamental representation through which the phenomenon is explained and understood in terms of quantities and geometrical relations, differentials, vectors, and calculable variables and constants. Through the underpinning metaphysics of mechanical realism, this abstract representation is taken to be the objective reality of the vortex, while our experience of moving water is taken to be epiphenomenal or merely subjective. On this interpretation, the relation between the technological activity and Nature as system is entirely one of appropriation and dissipation, necessarily dissipating the natural world into heat as Nature is exploited, as the technological framework is structured, and ultimately dissipates the world system itself into a state of heat death. As an open-ended system, a dissipative structure inherently increases the entropy of the world-system to the point where its structure can no longer be sustained. Today we would talk about this in terms of anthropogenic climate change, global warming, and a runaway greenhouse effect—the result of industrial activity and pollution. A dissipative structure is inherently polluting, as the whole system moves from order to chaos, and least in the form of heat and energy unavailable for work. If the goal of the technological society is one of replacing Nature with itself, it dissipates Nature until it appropriates and consumes it, but by doing so decreases its ability to extract energy and matter from Nature, increasing the entropy of the world-system, until the technological society dissipates itself. If based on appropriation and exploitation, any technological civilization is doomed to collapse in on itself as it depletes and disorders the whole system within which it exists.

As the theory of the real, the scientific worldview represents mathematical projection itself as the ultimate objective reality (and all else is an illusion, subjective, or epiphenomenal), and tests it by transforming Nature into something amenable to appropriation and exploitation. It is this aspect of science—as the theory of the real—that makes it inherently reductive and destructive, especially when it is at its most revolutionary and creative stages. Moving beyond this way of thinking about Nature in terms of mechanical realism is central to the kind of Chaos and Non-Linear Systems Theory and the new philosophies of natural science proposed by James Gleick, Illya Prigogine and Isabella Stenglers, and the appeals to quantum wholeness proposed by David Bohm and Frijof Capra.[489] The shift between the old 'mechanistic' scientific paradigm to the new 'systems theory' paradigm appears (or promises) to be a new Scientific Revolution. But, if we take mechanical realism into account, we can see that this new paradigm is a continuation of the same metaphysical precepts, and the only distinction between the two paradigms rests on the shift from a linear (push-pull) concept of mechanism to a non-linear (reiterated feedback) concept of mechanism. While this shift radically transforms how we understand cause and effect, leading to a radical refinement of how we understand how natural change and persistence occur, Nature itself remains represented in technological terms and the test of any new representations or

---

[489] Gleick (1978); Prigogine and Stenglers (1984); Bohm (1980); and Capra (1975, 1982).

explanatory hypothesis is still informed and constrained by the technological framework of science. What we are witnessing is the construction of a new quantum machine-kind and associated set of fundamental representations—a new cybernetic stratum of the technological framework—rather than a new conception of science and technology.

The new quantum machine-kind (from which devices such as blackbody radiators, photovoltaic cells, and electron tunneling microscopes have been developed as members of the same machine-family) has its own set of fundamental representations and mathematical laws. This will generate new representations, but, rather than constituting "a new physics," quantum physics is a new stratum in the technological framework of experimental physics, and it is only by ignoring the technological framework within which scientific activity is situated that it can be considered to be a new paradigm at all. Hence the concepts of 'non-linearity' and 'self-organization' that underscore Chaos Theory or non-linear dynamics are refinements of the scientific worldview descended from mechanical realism, rather than a radical departure from it, even if it allows new speculative metaphysics and interpretations of how Nature "works" and the role of the human agent when making observations and intervening in the natural world. The unpredictability and complexity of the technological framework, itself extended via the science of ecology as a projection over the entire planet Earth and its varied forms of life, is mirrored by the new representations of Nature as inherently unpredictable and complex. Hence the long-term unpredictability of weather patterns can be presented as a confirmation of the projection of Chaos Theory over the whole atmosphere and surface of the Earth, thereby linking mathematical essentialism with ecology, meteorology, geology, and the other planetary sciences. Similarly, the uncertainty and indeterminacy of the technological framework, itself extended via quantum physics over Nature and all its possibilities, is mirrored in the discovery of the role of the human agent in the act of observation and making measurements in determining the outcomes of experiments. Hence the inability to predict with absolute certainty the measurement of physical variables can be presented as the confirmation of the Copenhagen Interpretation of Heisenberg's Uncertainty Principle that can itself be connected with Chaos Theory via Probability Theory and Number Theory, with all their associated fundamental representations (constructed from matrices, differential operators, vectors, and complex numbers) projected over all Nature. Thereby, "consciousness" is reintroduced into science and the object-subject distinction is itself bridged through scientific activity and mathematical projection, but it is done so as *an operator* capable of determining physical states or concrete particulars through the acts of observation and experimentation from within "a system."

The new refinement of the scientific worldview is one of stratification, non-linear complexity, non-reducibility, and probabilities. Once "the system" crosses certain thresholds new mechanisms, laws, and qualities emerge—a new order (stratum) emerges and is established as the new norm. In this respect, scientific inquiry is transformative and irreversible. It changes the object of inquiry in ways that are artificial, unknowable in advance, and uncontrollable, and, hence open-ended. Ironically, it is this unknowability and uncontrollability that situates scientific inquiry in the context of discovery and gives scientists a sense of objectivity and the participation of Nature in response to their interventions. It is the irresolvable tension and contradiction between attempts to control experimental interventions and the uncontrolled responses of the apparatus to those interventions that gives science its power within the context of discovery, which should be understood as a dialectical dynamic driven by technological innovation and novel representational techniques. Nature is not merely reduced to a collection of resources, but is also that which resists or frustrates our intentions and efforts. Nature frustrates our toils and labors. Machines fail and break down. Great ideas for inventions come to nothing. Boilers explode. Unsinkable ships sink. Nuclear reactors can leak or catch fire—even meltdown. Fail-safe systems fail. Nature resists its appropriation and reduction, and therein resides the site of struggle of human beings with Nature. Yet these 'failures' and 'problems' become the source of invention and innovation of new opportunities and solutions, each also bringing with them new problems and goals. Hence, as Prigogine observed, it is quite impossible to have complete knowledge of Nature. At most, human beings have a window on the human interaction with Nature through which we extrapolate the part we cannot see; and, hence, we are intimately and inextricably involved in how the Nature we describe reveals itself. The "hard sciences" like physics, chemistry, and biology reveal to us how beings respond to specific modes of intellectual inquiry, technological activity, and human intervention, and the ongoing nature of science tests its results by responding to these responses, and so on. It is this fundamental trajectory, most of all, which reveals the dialectics at the core of scientific activity. It is the ongoing construction of the technological framework that is incomplete. Nature is perpetually appropriated, reduced, and transformed by science and technology. Its abstraction into some mathematical or objective realm is itself a cultural product and result of specific modes of human thought and action. There is no way of reconciling the subjective and objective aspects of human existence when what is understood as the objective realm—against which the subjective is identified and given its own jurisdiction—is itself a product of specific modes of human intervention, imagination, and culture, yet is represented as independent of human thought and action. It is this irreconcilable dualism that has resulted in the irresolvable paradoxes and antimonies of the philosophy of science, which cannot be avoided or overcome until we recognize the dialectics of technology at the core of natural science.

 Once we take the dialectics of technology into account, we can see that science teaches us about how Nature responds to specific modes of human intervention. However, if Nature is understood under a unified concept as "non-human" then it transcends these interventions and cannot be reduced to its responses to them without doing violence to Nature. This violence cuts away aspects of Nature and discards them in order to select and appropriate only those aspects of Nature amenable to incorporation within the technological framework. This reductive aspect of scientific activity leaves us with a distortion of Nature as if it were the totality of Nature; this further distorts Nature by representing the cultural products of scientific activity as if these existed independently of scientific activity, thereby culturally positioning science as a special human activity in the context of discovery of the "non-human." It is this aspect of scientific activity that reveals the scientism (itself a form of religious fundamentalism) and totalitarianism inherent to the Scientific Revolution and how it has been bound-up with the construction of the technological society. This insight allows us to rethink both science and Nature. It does not mean that we should reject science—or, worse still, deny it—as the structure of the world has already changed through scientific activity. At this stage in history, to reject or deny science is an irrational and subjective individualism generated through cultural fragmentation and retardation, whereby individuals remain dominated by the device paradigm while rejecting or denying the scientific worldview and the technological framework. We cannot reject or deny science without failing to understand it and the technological society within which we exist. We cannot act like Ted Kaczynski and blow up the technological society. It will appropriate and incorporate the means of response (techniques of counter-terrorism and forensics) into its framework and further expand and develop its structure. The technological society cannot go back. It is irreversible. The facts produced by science *are the facts*. Science *explains* technological activity. What we need to do is incorporate science and technology into a broader and more comprehensive way of thinking about our existence and its meaning and purpose. We need to envision our goals and implement them into our practices through rethinking the scientific worldview and repurposing the technological society in accordance with these new goals. We need new and better imperatives.

Once we recognize that Nature—as revealed through scientific activity—is a response to our mode of inquiry, we must acknowledge that this Nature is mediated, appropriated, and reduced by technological activity. How we interact with Nature changes what we can learn or think about Nature. The facts and their explanation are product of technological activity—as well as discovered through resistance, things going wrong, and failures. When scientists measure global warming, they are not fabricating the measurements, as if they were making them up out of hot air, but are discovering what those measurements are from within the technological framework of science. This is a disciplined and technical activity, and objective within the framework. To disagree with their measurements of atmospheric temperature is to disagree with their methods and techniques—and the meaning of 'temperature'. The measurements produced by it are the facts, once we accept the methods of measurement. While their sense, interpretation, and explanation are a matter of social construction, dispute and controversy, the objectivity of their referent is established through the technological activity of making those measurements. This cannot be separated from the technological society which made those activities possible, and hence we are learning facts and gaining knowledge about the world as it responds to our interventions. In this sense, scientific measurements of climate change and global warming, and its interpretation as anthropogenic climate change, cannot be separated from the industrial society that generated the greenhouse gases in the first place, as well as making possible the techniques of the measurement. They are both emergent and developed within the same technological framework. Failures of models and predictions, and the need for their constant refinement, do not indicate flaws in the science—or some elaborate hoax—but, rather, reveal the incompleteness and ongoing construction of the technological framework.

It is the whole technological framework that is incomplete, unpredictable, and uncontrollable, as a dissipative structure, while it provides itself and its structures as the objective referents of science. Its representation as an *autopoiesis* wherein technology is a self-producing substitution for Nature, conceals how it is constructed through the appropriation of Nature in accordance with its own imperative the further develop the means to appropriate Nature. Yet its path of development is something that is not under control, as, while its successes and failures can be discovered, they are not determined by technology itself. For the realist, it is Nature that determines this path. The irony of the technological society is that it is a "self-organizing system" that is completely out of control while simultaneously attempting to place everything under control. Nature in the form of chaos, spontaneity, and irreversibility is antithesis at the core of the ongoing construction of the technological society, and hence it makes possible discoveries and brings new powers into the world by appropriating Nature. Yet, the Nature revealed to us is a substitution—a representation of Nature revealed, controlled, and constrained by technology—is placed before us as an act of the cultural concealment of our fear of uncertainty and vulnerability in the face of Nature beyond our control. It is this two-fold duality of the controllability and uncontrollability of Nature during the act of making that I term as the Fire of Hephaestus—it makes technology possible and shapes how it is done. Discovery and domination lie in its wake, alongside change and destruction, as new powers are brought into the world and wreak havoc. The Fire of Hephaestus is the power to transform the world through labor and material practice; it is the transformative power of technological activity itself—even though the origin of this power is mysterious and unknowable to us through technological activity. It transforms technology into an objective reality of its own making. To overcome uncertainty and mystery, thereby satisfying the psychological need for certainty and knowledge, epistemology conceals the Fire of Hephaestus behind technique and technical knowledge, abstracted into mechanisms of cognition and method supposedly based on the operations of logic and instrumental reason. This is justified by its results. Ontology is reduced to the abstraction of technology and its resources, and nothing else. Human experience is reduced to the results of operating on a complex of parallel or sequential mechanisms operating on matter in motion, with its justification given in terms of disciplined activity and its consequences. It frames and structures scientific experience to such an extent that we are simply unable to find any Archimedean point from which we can examine scientific experience in terms of scientific experience without descending into logical circularity and irrationalism. Science is a self-referential system of action and the interpretation of its results, given metaphysical sense as a means of discovering an objective reality outside of itself. Scientific theory is an explanatory abstraction of the *autopoiesis* of the technological framework that we represent in objective terms, as if it was nothing to do with us, and we 'just so happened' (as if by magic) to find it ready-made, as a self-evident means of producing an experience of the world in terms of matter and energy. The scientific worldview—no matter how refined and corrected in the light of further technological activity—is mediated and shaped by the very activity we are using to try to understand it. Hence the absence of any external point of reference from which we can scientifically observe ourselves engaged in scientific activity. Nature only enters into this self-contained worldview as material and force, both abstractions made operational as the mechanism through which force can act on matter.

Even enthusiasts of *Gaia* talk of feedback mechanisms. This is how entrenched mechanical realism has become in scientific thinking—even at the radical and revolutionary periphery of its post-Newtonian paradigms. Celebrated for it holism, James Lovelock's famous Gaia Hypothesis describes the planet Earth as a totalizing system—a non-linear dynamic machine, which like a steam engine governor—Lovelock's fundamental metaphor—readjusts itself to achieve a stable state of thermodynamic equilibrium at a planetary level, and modifies the chemical composition of the atmosphere and oceans via the activities of living organisms, such as bacteria and human beings, which in turn affect changes to the planet's atmosphere, oceans, and whole surface from the poles to the equator. Despite the portrayal of Gaia in popular science literature as a living organism or even a goddess (consciousness at a planetary level), within the terms of James Lovelock's original conception of Gaia, the hypothesis is concerned with identifying an objective feedback mechanism capable of causing atmospheric changes on a planetary level. The Gaia Hypothesis grew out of Lovelock's work with the Jet Propulsion Laboratory in Pasadena, California, to invent a technique and instrument capable of responding to these mechanisms in a way that would allow it to be represented as a means to detect life on Mars (or any other planet). This instrument measures the chemical composition of the atmosphere and representing those measurements in terms of entropic reductions (increases in the degree of order of the whole system) and deviations from states of thermal equilibrium, wherein the concept of life itself is reduced to a set of chemical and thermodynamic mechanisms capable of making measurable changes in the environment. Drawing its analogies from thermodynamic engines, thermostats, and autopilots, the Gaia Hypothesis projects cybernetics and non-linear dynamics over the whole planet, as a totality, to conceive of it as a complex machine capable of self-regulating its temperature, acidity, pressure, humidity, carbon dioxide level, etc., through the activities of living organisms and geological events, such as volcanoes and ice-ages. The norm is instability and disorder followed by periods of stability and order, what is termed as *punctuated homeostasis*, leading to distinct eras of climatic, geological, and ecological stability, whereby new species emerge as dominant in virtue of being best adapted to their environment. Unsustainable species, changes in the environment, and localized adaptation are the hallmarks of a punctuated evolution, whereby instead of continuous evolution, a new ecology grows out of the chaos of climate change, mass extinction, and the collapse of the old ecology, until balance is restored, albeit temporary. These are the lessons of the fossil record and punctuated equilibrium theory. One can readily apply these lessons to human history and the rise and fall of civilizations, systems of knowledge, and cultural practices and beliefs.

However, Lovelock warns us that we should not imagine that this is an anthropocentric machine, because "…if we transgress in our pollutions and forest clearance, Gaia can move to a new stable state and one that's no longer comfortable for us." (Lovelock, 1979: 359) We are one species of organism among billions, yet our actions have consequences that change the planet, both intentionally and in ways we cannot predict. Agriculture, deforestation, mining, and industry have changed the planet, arguably resulting in an improved world for human beings, but with responses from the planet that we could not foresee. The planet adjusts to a new equilibrium point. The consequences of these adjustments—in terms of atmospheric composition and temperature, along with ocean composition and temperature, might seem negligible initially but over centuries of intensification and expansion of human activity could result in a period of accelerating climate change. Gaia's readjustments to industrial pollution (especially carbon dioxide and other greenhouse gases) could lead to a runaway greenhouse effect, resulting in atmospheric temperature rises to such a high level that mass extinctions, including human extinction, would be the result. Climate change can also result in melting polar caps, ocean level rises, through extreme changes in atmospheric content, temperature and pressure, which can lead to catastrophic consequences for all species, including human beings, which are mostly populated around coasts and in cities. Like the inside of a pressure cooker, the planet will heat up until it no longer can sustain the very same organism that caused the change in the first place. Fever in response to a bacterial infection would also be an apt analogy. Over millions of years, the planet will reach a new equilibrium point, at which, the atmosphere will stabilize as a hot planet, like the planet Venus, or perhaps flip into an ice age through a correcting non-linear mechanism. These new conditions could afford the conditions for any surviving organisms to multiply and flourish—the punctuated evolution of a new ecology, which in turn changes the atmosphere, oceans, and land surface of the planet.

How can we reconcile the technological society and Gaia? What has emerged from our understanding of the incompleteness and heterogeneity of the technological framework—its dissipative structure—is an awareness of the need for a precautionary principle and democratic deliberation about the directions of the development of the technological society into an ecotechnological society, if we seek a sustainable existence for future generations.[490] Once we take sustainability requirements into account, any genuinely democratic society, itself driven by an imperative to maximize inclusion of everyone in deliberations and decision-making regarding the development of society, must also be an ecological society if it is to include future generations, or at least not exclude them from consideration by imposing irreversible decisions on them. The ecotechnological society is driven by an ecological imperative, wherein maximizing sustainability and the integration of organized human activity with the environment define the meaning of "efficiency" in ecological terms. This view still results in the reduction of the ecology of the Earth to a world system understood entirely in terms of

---
[490] Rogers (2008)

technological activity—and thereby remaining consistent with the technological framework from which it emerges. All attempts to mitigate and adapt to the dissipation caused by these technological interventions can only be technological solutions once the Earth is so reduced. Of course, any technological solution is doomed to generate instability, in a real sense, as it dissipates the world-system further, as it creates new possibilities and problems requiring the postulation of new technological solutions, and so on. This explains why much of the debate about the human destruction of the planet's ecology has become reduced to a debate about the best means to preserve the human species through technological innovation. This debate has been further abstracted through the myth of the invisible hand of the free market, allowing the faith in technological innovation to be represented in terms of the faith in the enduring human imperative to accumulate capital as the means of survival. State interventions (i.e. regulation and taxation) are themselves represented in terms of technological solutions. These myths provide the core axioms of a device paradigm for any and every possible problem and its future solution. The notion that we require a radical change in how human beings live on Earth and share it with other beings has become eclipsed by the domination of instrumental reason and anthropocentrism—subjectively embodied in the individual as fear, greed, and egoism. The capitalist economic imperative and the worship of Man are one and the same, having fed the ego and greed since ancient times, and bound up with human self-glorification in response to fear and vulnerability. Yet, instead of satisfying the ego by demonstrating its own cleverness and power to itself, through the acquisition of money and capital, the ego is further seduced by the Fire of Hephaestus, driven by the desire for more and more technological power and control, satisfied only momentarily within the device paradigm through consumption and conformity to the demands of the technological imperative. It is relentless and will not stop until all of the natural world has been consumed and replaced with itself. Technology has become a false idol upon which Nature is sacrificed. Will Nature's wrath be our reward?

It is ironic that, despite the romanticism and impossibility of the 'back to nature' movement, advocates of Deep Ecology (i.e. Baird Callicot and Warwick Fox) often use the same mechanistic science they reject as anthropocentric to justify and explain concepts of biocentrism, interconnectedness, and an ecological worldview. They have adopted the same uncritical mechanical realism as have the traditional philosophers of science. Its ways of thinking remain within the same technological framework, even if the shift in focus and vision is one of building and repurposing the ecotechnological society away from anthropocentrism and towards biocentrism through using and reinterpreting the scientific worldview. The objective realm of Nature remains those aspects of Nature revealed through technological activity; its fundamental representations—tested through technological activity—are used to justify a particular conception of biocentrism that we are supposed to believe rejects the instrumentalism and anthropocentrism of science and technology. It is for this reason that the anthropologist Tim Ingold termed biocentrism as *anthropocircumferentialism*: a region of the world is circumscribed by scientifically informed human beings as being 'non-human nature,' which, somehow, human beings can know and remain outside of.[491] This is a continuation of the same Cartesian conception of objectivity that is supposed to be rejected by Deep Ecology as an obstacle to ecological thinking. Taking this into account, despite the imperative to develop sustainable and harmonious relations between human beings and the other beings of the natural world, the ecotechnological society continues to treat Nature—the ecology of the Earth—as something instrumental for human survival and flourishing. Yet the notion of radical change holds together deeper concerns about how human life and society have been organized and structured in relation to the natural world, which has resulted in a

---

[491] Green College, Oxford, 1996

split within the environmentalist movement along the lines of deep and shallow interpretations of ecology. Hence Arne Naess wrote, "[Shallow] ecologically responsible policies are concerned only in part with pollution and resource depletion. There are deeper concerns which touch upon principles of diversity, complexity, autonomy, decentralization, symbiosis, egalitarianism, and classlessness."[492]

This requires considerable rethinking and self-reflection on our part. As a result, the future of the ecotechnological society, along with human existence, self-consciousness, and freedom would be one of *self-realization*, as Naess termed it, developed from Abraham Maslow's idea of self-realization in his *hierarchy of human needs*.[493] For Naess, the apex of human evolutionary potential is one of a valuable, meaningful, and ecologically sustainable existence on a shared planet. Philosophical and spiritual articulation of the meaning and value of existence is an essential core to any hope of intimacy with the natural world and self-realization. The ecotechnological society would be a fundamental transformation of the Earth into a stable and harmonious whole, wherein all beings can flourish—and this is *necessary* for human beings to have any hope of survival as a species. It is the self-realization of humanity in ecological integration and unity with all life on Earth. This does not imply the egoism or narcissism (neither of which are in any shortage) of anthropocentrism, but instead affords the real opportunity for human consciousness to transcend its anthropocentric trap (which should be considered as a form of egoism or narcissism in operation at a species level) and explore consciousness of a world that both transcends and conditions human existence and its meaning. The master-slave dialectic of Man v. Nature has not only generated suffering for the slave class (which includes animals as well as classes of human beings), but it has also eroded and negated any real possibility of a sustainable existence on Earth for the human species, as well as many other species, even if it has provided short-term benefits to the master class throughout history. The master-slave dialectic of Man v. Nature is ultimately cancerous and destructive to all life as it reduces the conception of life to the instrumental—and thereby consumes and destroys it, and by doing so, turns society inwardly upon itself. Avoiding this dialectic requires a conception of the Human-Nature relationship as based on unity rather than division, while respecting that human beings are both within and a part of Nature. The respect and protection of something outside of the ecotechnological society is essential for human activities to escape anthropocentrism and instrumentalism, and develop a genuine biocentrism and sense of unity and interconnection—as if our lives depended on it, because they surely do. The preservation of self-realization is an ecological imperative to integrate and sustain humanity alongside the ecology of the Earth from which we came and gain our meaning and value as living beings. The preservation of wilderness is essential for human self-realization because it preserves the connection with our origin and something outside of ourselves, older than us, which transcends us and re-connects us with who we are as self-conscious beings on this planet. Wilderness is essential for human freedom to be even possible, once we recognize the ecological conditions for human self-realization in unity and oneness with Nature.

What would resist the ecotechnological society becoming totalitarian? Naess termed this as 'the danger of ecologism.' The democratization of the technological society and its transformation to an ecotechnological society would prevent the new ecological paradigm and imperative from being oppressive and totalitarian. The construction of the ecotechnological society is an act of human self-realization and therefore an act of self-discovery, but made possible through unity and oneness with the Earth as the site of

---

[492] Naess (1973)
[493] Naess (1989)

human existence, meaning, and value. Democracy provides the crucible for humanity's self-discovery because it incorporates diversity and plurality within the communicable deliberation of societal development in relation to a reality that does not confirm to human intentionality and expectations. Human actions have unforeseeable consequences and we are fallible. The self-realization of a biocentric ecotechnological society would be the outcome of democratization in a genuinely open and pragmatic society, as a means of identifying and solving shared problems and realizing shared values, while aiming towards achieving maximum inclusion for the benefit and well-being of as many diverse communities and individuals as is possible, while recognizing the value of living sustainably on a shared planet as a condition of human existence. Human beings exist as a democratic society within an ecological community that responds to our actions in irreversible ways that we cannot predict with absolute certainty. Democracy is the best means to navigate this uncertainty, while also providing a precautionary principle regarding changing the conditions that support our existence. Having to persuade others should be the means by which this cautionary principle is exercised in democratic deliberation. Human interventions into already existing ecological relations between beings have negative consequences for (or presents obstacles to) human well-being and sustainability, as well as for other species, and the incorporation of science and technology into democratic and ecotechnological development is a prerequisite for making good, sustainable, and long-lasting decisions that benefit everyone. The ecological imperative arises as the outcome of a maximally inclusive participatory democracy concerned with stability and long-term solutions to the ecological problems of living on Earth and satisfying human needs, without destroying or consuming the ecology upon which human existence depends. Hence, we should not limit our understanding of the ecology of the Earth to being simply the means to reproduce human existence and satisfy human needs, but, instead, see it as the place wherein human beings can discover the meaning and value of existing at all. This involves respecting Nature as something outside of ourselves that is intrinsically valuable in its own right, while also providing the conditions and contexts of our own existence. Human beings self-realize through democratic participation, ecological awareness, and a respect for Nature as intrinsically valuable. Rather than being merely the material organization of systems by which human life is sustained and human needs are satisfied, the ecology of the Earth is the place wherein human enlightenment and liberation become possible, as situated, temporal, living, and spiritual beings that find lasting meaning, truth, and value from their unity with evolving consciousness and unified complexity of creation itself.

Democracy is the means to discover and realize shared visions of the ecotechnological society as a sustainable planet-wide society and how it could be further developed through the articulation and examination of those visions in terms that would be meaningful to everyone, even if they did not initially share those visions and the values they presuppose. This is not simply a matter of conforming to majority-rule, but is an ongoing process by which we can learn and discover how to live and cooperate with each other. In as much as democracy forms the basis for rational communication and deliberation between members of different communities and cultures, it also respects their differences, as it allows us to learn from each other. The 'veil of ignorance' demands that we consult others. Nature, as something outside of ourselves and our society, teaches us the conditions for human solidarity and our self-realization as an ecotechnological society. Egalitarianism, cooperation, and freedom must be fundamental principles of the aesthetics and logistics of the development of the ecotechnological society if self-realization and flourishing are possible for all beings on Earth (Rogers, 2008). Why is self-realization important for our future? Life on Earth exhibits such interconnected complexity that it has taken thousands of years for human beings to begin to become aware of these interconnections and the effects of disturbances in them. If we hope to gain knowledge of these interconnections, we cannot base our knowledge of Nature reductively on how Nature responds to our interventions—our disturbances—which is the only knowledge we gain from the experimental sciences. Instead, we must observe how animals and plants interconnect with each other, and their environment in general, *when left to their own devices in their natural environment*. That must be a given for any natural science. Thus biology needs to 'return' to natural history and phenomenology, but with the benefit of hindsight. Ecology should be based on observations of how plants and animals self-realize within a web of interconnections, which also includes us, while also maintaining awareness of the intimacy of consciousness, experience, and the world. Therefore, how each and every being self-realizes becomes the fundamental question for a proper ecological science, which must admit the aesthetic and ontological dimensions of life. The lived-world of self-realized meanings, experiences, and consciousness itself, would be the world explored by ecology, which would embrace the spiritual and philosophical aspects of consciousness as constitutive of ecological meaning and truth, as revealed through the unfolding otherness of the world. This new science promises to teach us how to self-realize and live freely.

We need to move beyond mechanistic interpretations. Ecology should be a *Wissenschaft*—or an *Ecosophy*, as Naess termed it—which embraces the historical, anthropological, sociological, philosophical, and psychological aspects of knowledge of science, as well as knowledge of the interconnections between organisms within shared natural environments, over and above technical knowledge. It is a broader and more comprehensive conception of science as providing and testing meaning and value, as well as truth and knowledge. It recognizes the social aspects to cognition and language, while also recognizing that meaning and truth arises in contexts of discovery as well as contexts of justification. This requires bringing knowledge and meaning together through the qualitative evaluation and interpretation of experience in a way that identifies positivistic interpretations of the empirical as a special case but seeks more general and comprehensive truth, meaning, and value. This new science is based on intimacy with the phenomenon and self-reflection on the context through which that intimacy is achieved. The qualia of experience and their hermeneutic (historical) interpretation and phenomenological (experiential) meanings are the fundamental elements of our knowledge, as cultural and subjective elements, but it through intimacy rather than abstraction that the qualia emerge from experience. The new science cannot be based on mechanisms and abstract units of matter and energy, but instead based on meaning and value arise through the process of self-realization. Any *Wissenschaft* must provide a methodology for ordering (categorizing) and evaluating (testing) interpretations of qualia, as well as facts and explanations of those facts, but this methodology cannot be assumed to be universal without being subjected to universally inclusive democratic processes of persuasion and enrollment. It is deeply personal and based on a holistic approach whereby this methodology is subjectively developed throughout a whole life in relation to being human within a society on a shared planet. Yet its meanings and values arise in relation to something outside and other to ourselves that we conceive as pre-existing and transcending human existence. Our choices have consequences that we cannot foresee and we need to discover them. We cannot remove ourselves from this holistic approach and therefore our choices and values are intimately implicated in the reality we discover through our self-realization as humanity.

Intimacy with Nature involves more than understanding Nature in terms of concepts, categories, and unfolding web of the meaning and value of an organic whole. It involves experiencing Nature *as Other*. The lived-world activity of the observer and the situated act of observation itself are integral to experiencing the otherness of Nature ("the worlding of the world" or what the Taoists call "the Tao"), but the otherness of Nature is experienced as something outside of the human subject—it transcends the observer and is the context of the act of observation—yet phenomena cannot also be understood as objective without transforming both observed and observer. Intimate experience of natural world reveals its intrinsic value that is independent from its usefulness to us, even though that value would not exist without us and we cannot understand intrinsic value except in relation to us. We discover Nature's value and objectivity through embodied engagement with it as something worthy of respect that is bigger than ourselves as subjects. Natural phenomena should be understood as emergent within a unified whole, which is inextricably bound to embodied consciousness of the whole and the emergent phenomena encountered through intimate experience, and it is this unity that gives everything within it intrinsic value once we situate ourselves within it. Ecological knowledge must be based on the phenomenological interpretation of self-consciousness in intimate experience of the natural world through the process of the self-realization of the meaning and value of being human within an ecology that is other than human. Knowledge of aspects of the natural world and their interrelation would take the form of qualitative reflections determined through reasoning and experience; the test of this knowledge would be measured in terms of the degree of liberation and enlightenment it would bring. Such knowledge is inherently good and would shape the way human beings explore the world and interact with other beings within it to discover the self-realization of being human in relation to the non-human. It would guide the evolution of new consciousness—an ascension of consciousness to higher levels of thinking and awareness in intimate relation to the otherness of natural world—and draw us towards its conception as transcendental and unified as Nature. Meaning and intimacy would form the primary bases for self-realization within the natural world in relation to living within the world and understanding one's consciousness of being alive within a world from which one comes. The science of ecology would be no longer premised on a confrontation with Nature, or its objectification, but by embracing the otherness of Nature alongside understanding Nature in terms of intimate experiences and lived relations within the place that we call home, and thereby found the basis for how we could understand ourselves. This calls for philosophy, spirituality, and care for existence as a whole to become the fundamental pursuits of ecological science and the evolution of human consciousness into an ecological consciousness. All peoples—all beings—would be *indigenous* to the Earth and have intrinsic value.

Back to Nature?

How does mechanistic science help us understand Nature as Other? It doesn't. Despite all our pretentions and conceits, we remain prisoners of the cave, gazing at the shadows on the wall, unable to understand, control, or even see the fire. We can only glimpse its penumbras on the cave wall and these shadows are what we call "Nature." Our object of study is our own skill at puppetry. This has resulted in an instrumentalist/exploitative relation between human beings and the natural world, as Nature is chopped up and dragged into the cave by the shadow puppeteers; some of it used to make the puppets and the rest thrown onto the fire and burned. Neither the puppeteers nor prisoners learn anything about the outside of the cave. Nature-as-Other is the outside of the cave. Trapped inside the cave, human beings do not encounter otherness at all, as they reduce everything to the art of shadow puppetry and their skill at predicting the procession of shadows on the cave wall. The cave is the totality of human existence—yet we are only dimly aware of the cave and reduce its walls to the boundary of the knowable and real. The technological society fills the cave, as technology itself becomes the boundary between inside and outside the cave, and Nature is reduced to technical relations (measures and mechanisms) between matter and energy—the flickering flames of the fire. This casts the shadows on which the scientific world view is built within the cave and therein finds its objective reality in the art of shadow puppetry—*our skill at making*. So, are we trapped without hope of freedom? How do we leave the cave? How do we go outside of ourselves, go out into the light of the Sun, and discover a greater reality and find our freedom in relation to the Other—*that which is not of our making*?

Nature—as experienced by human beings—is not merely comprised of the sum total of the relationships between entities, nor the sum total of entities, be they animals, plants, micro-organisms, forests, plains, tundra, oceans, mountains, the atmosphere, moons and planets, or stars and distant galaxies. The ecology of the planet Earth is comprised of the total interconnected whole of all such relationships—an organic whole—but it is also comprised of our consciousness of the whole and its value and meaning for us. Currently, mechanistic ecological science gives us only a glimpse—a shadow—of those aspects of the organic whole amenable to technological activity, either in terms of measurement or manipulation, but it cannot give us any insight into our conscious discovery of the meaning and value of that organic whole. How are we able to do this? To answer that question, we need to move beyond only viewing ecology in instrumentalist terms of how it is necessary for our life or how it benefits us—in terms of its function or mechanism for us. We need to recognize Nature-as-Other—something not of our making or imagination, something bigger than and beyond us—ineffable and transcendent. We need to recognize the extent that Nature transcends and pre-exists human existence, while simultaneously providing the site and conditions for human existence, freedom and consciousness to emerge and self-realize through experience. Clearly the argument here advocates a philosophical idealism that celebrates the Otherness of Nature as an ideal and value, while simultaneously embraces a philosophical realism by treating Nature with respect *as real*. This places aesthetics at the existential heart of the human relation with Nature, thereby conditioning the human relation with Nature to be one of intimacy—*a primordial intimacy* in a realist sense as being objectively fundamental to the satisfaction of human instincts and survival—based on experience, wonder, and awe at its intrinsic value and otherness in an idealist sense. What is needed is a revival of the sacredness of Nature and an awareness that stewardship entails obligation, restraint, care, and mindfulness—as natural virtues that are fundamental to the discovery of the reality, value, and truth of our being. However, far from advocating a 'return' to primitivism, if this is understood in terms of a primitive hunter-gatherer society or a romantic vision of Man alone in Nature, I am proposing that the way forward towards a rational, free, and sustainable society is through the ongoing emergence of conscious, spiritual, and philosophical enlightenment of the human relation with Nature.

The concept of Nature-as-Other is one that celebrates Nature as being transcendental—as sacred and beyond the human—from which we originate and within which we dwell, while also acknowledging it as unknowable and mysterious due to its otherness. This immediately puts us in an aesthetic and existential relation with Nature-as-Other through our primordial experience of Nature by being within it and of it. It demands humility from us while placing us in the context of the discovery of ourselves. The concept of Nature-as-Other also allows us to conceive of our encounters with Nature as unified and transcendental, within which we can find our identity and meaning as self-realized beings, while we develop and extend our ways of life through practical and intellectual activities along the lines of adaptation and stewardship. This places concepts of care and love at the heart of both philosophy and science, hence I can talk of spirit and spirituality, without descending into spiritualism or superstition. Our fundamental relation with Nature should be a spiritual one—which means one in which we discover the Spirit of the World through intimate experience, alongside practical and intellectual activity. It affords the opportunity for the development of an ecological imperative in relation to the natural world, and correspondingly to the becoming of our being in terms of reflecting on the meaning, value, and purpose of human existence within a unified and transcendent organic whole. Consequently, the natural and social sciences, alongside philosophy and spirituality would be brought together, and instead of imitating positivistic interpretations of the mechanistic sciences and their methods, would concern themselves with studying and better understanding the aesthetic and existential conditions for human self-realization within Nature and developing deeper conscious relations of the ecology of the organic whole within which we exist and act. In the political realm, decentralization and autonomy, democratization, diversity and pluralism, would all be understood through the ecological manifestations of self-realization within a greater organic whole, alongside a deeper exploration of the ideal and the existential nature of the human spirit as an ecological and conscious being situated on a shared planet that we call "home."

    Political action must necessarily become ecological in its focus—otherwise it would stand in opposition to self-realization, and therefore oppressive and harmful. The political aspects of democratic participation relate natural philosophy to social ecology—itself a political philosophy—in order to help guide the further construction and development of the ecotechnological society as technological manifestation of the self-realization of humanity as a sustainable and flourishing species. This may well prove to be an impossible task perpetually projected onto the asymptote of the End of History, as in many respects we are still stuck in the paleotechnic phase with our reliance on fossil fuels and polluting technologies, warfare, over-consumption and wastefulness, overfishing, deforestation, and pesticides and soil nutrient depletion in agriculture. Perpetual struggle until we can endure no more may well be our lot in life—along with our inevitable death and extinction—but if we take consciousness into account, along with the realization that this involves consciousness of our own existence and that this existence is of value to us, we should embrace the fact that we have freedom of action and can think and choose differently. If the current human trajectory is heading towards mass extinctions and planetary catastrophe, as both contemporary scientific evidence and common sense show us, we can recognize that, as a result of our consciousness of this trajectory, along with the freedom of consciousness to think and act differently, we can overcome the pathological anthropocentrism that has resulted in a psychotic (and fearful) human relation to all other forms of life, as threats or things of use, which is at the heart of human destructiveness and the current ecological crisis we face. In this respect, we can see that the artificial divisions between the sciences (both natural and social) and philosophy must be dissolved—as Edmund Husserl implored in his book *The Crisis of European Sciences and Transcendental Phenomenology*.[494] The new science of ecology would accept that understanding self-realization was a common goal to all life and intellectual activity, thereby bringing science and philosophy together into a single *Wissenschaft* capable of providing us with an intelligible and unified worldview within which we could better understand our place within existence, how to sustain it and realize its meaning and value through practical and intellectual activity, and how to share that existence with other beings. The unification of this world view would itself result from our shared dwelling on Earth and bring the Anthropocene to an end as humanity self-realizes itself as the

---

[494] First published in German in 1936 and in English by Martinus Nijhoff Publishers in 1954.

ecotechnological society given spiritual and philosophical meaning and truth from within an organic whole.

Naess termed this as *an ecophilosophical approach*, with all of its normative aspects, rather than as something derived from the science of ecology as a mechanistic science. *Ecosophy* is based on a scientific and philosophical understanding of harmony and equilibrium—itself based on the facts of mechanistic ecological science, linguistics, and psychology—but is directed towards discovering and articulating the norms and values of the self-realization of human relations, meaning, and value, as a holistic approach from within an organic whole. This involves reflexive descriptions of experience and reflection on *the quality of existence*; itself admitting phenomenological and narrative descriptions of how self-realization has been achieved or thwarted in lived-world contexts—lifeworlds—rather than restricting itself to the prediction and explanation of the persistence or change of measurable and operational quantities through technological activity. It involves including human beings as agents of thinking and cognition that intervene into and change the world through our activities, but it also involves recognizing that our existence is something that is both for us and meaningful to us. It involves reflection of the kind of being which can have meaningful experiences and values. In this sense, as a relational and holistic approach built on an ecophilosophical approach, Ecosophy is inherently self-diagnostic and therapeutic as a dialogic and dialectical method of articulation and self-reflection about one's relations with other beings and the natural world, interpreted through the lens of holistic and systems thinking as well as humanistic psychology and linguistics, as the basis for human wellbeing and social ecology. By focusing on human relations with Nature, applying both gestalt psychology and systems thinking to our lifeworlds, Ecosophy promises to provide us with a philosophical self-evaluation and critical worldview inspired by our relations with each other and the ecology of the Earth. It is biocentric in virtue of respecting the intrinsic value of natural beings—and hence of all existence—while also respecting the human need to understand and find meaning, value, and purpose within the world through intimate experience, alongside practical and intellectual activity. It also recognizes the error of totalizing and reductive worldviews generated through scientism (especially ecologism), due to the ongoing, complex, open-ended, and incomplete processes of life itself. As Naess observed, and as I've argued through this book, mechanistic science cannot provide us with principles of action, norms, meanings, or values, simply because these cannot be reduced to quantified and operational variables or mechanisms. They involve choices and self-reflections. They are outside of the remit of mechanistic science.

Naess observed that in terms of all the great philosophies of the past two millennia, whether Platonism, Aristotelianism, Buddhism, Confucianism, or Stoicism, the technologism of the current era would be deemed wanting. As Naess put it (p.87), "No matter which of the great philosophies one considers to be valid, our current role would be evaluated negatively." In the great philosophies, the good life is not equated with the pursuit of power nor mindless consumption, and human values and norms were not reduced to market relations, modes of production, and levels of consumption (beyond those necessary to stay alive). While all these great philosophies had an ascetic core, they were concerned with the quality, meaning, and character of human life, which defined their understanding of wealth and abundance, rather than as the acquisition of the means to improve one's "standard of living" in terms of economic growth. Ecological consciousness in modern society has tended to be limited to a focus on "life-style choices" made in terms of patterns of consumption and modes of production. As such, it has tended to become focused on materialism and the ecological preconditions for maintaining an environment suitable for human existence—thus it has remained anthropocentric and shallow. It also remains quite focused on the individual and their consumer choices rather than the ways that human beings can organize themselves to change the current system. Critical of this way of thinking, Naess argued (p. 91) that human consciousness needs to change for the ecological movement to become affirmative of a new way of life rather than merely damning of the old. This change in consciousness must transcend ideology and 'stern political lines,' and, instead, transform the ecological movement into a "renewing and joy creating movement." While levels of consumption need to be reduced to avoid resource depletion and pollution, as well as reducing waste, the thoughtless use of natural resources, and inequality, it is essential that the ecological movement is based on raising awareness and mutual support to transform human consciousness and behavior in accordance with the need for coordinated efforts to transform society from a technological society to an ecological one.

The technological society tends towards the centralization of ever-increasing production and consumption as its measure of success. Regardless of whether it is socialist or capitalist, the technological society is totalitarian in this respect. Technique has become autonomous. Naess argued that to counter this tendency requires that we put technique to the cultural test and consider its implementation from the standpoint of its impact on the whole of society. Technical advances must be subordinated to wider cultural norms and values before they can be considered as advances at all. In this sense, it is foolish to think that there is a technological solution to the current ecological crisis. Instead, what is needed is a radical reevaluation of human needs and the ecological impact of the means to satisfy them. This involves a change of consciousness in order to transform human beings into the solution to the ecological crisis. The ecotechnological development of society involves not only technological advancement, in the limited and positivistic sense of the term, and the implementation of ecologically 'soft technologies,' but it involves a fundamental transformation in human consciousness and culture. This involves a radical democratization of society along the lines of the cautionary principle, maximum inclusion and diversity, decentralization and the empowerment of local communities.[495] This converges with Naess' notion of self-realization as a way of showing how individuals can identify and satisfy their needs, and realize their potential, through coordinated activities and communication with others, as well as learning the interconnectivity between all things. Apathy and egoism are the enemies of our minds, if self-realization is our goal. Although it is the ultimate goal and guiding norm, self-realization is an ongoing process, rather than the attainment of a perfect state of being, that guides human thought and action towards the connection of the individual to a higher Self, which is understood as distinct from the ego. Therefore Nature is not something against which we must struggle, or to which we must adapt, but is something of intrinsic value and the site wherein human beings can discover this higher Self. For this purpose, Naess considered gestalt psychology to be a useful tool for individuals to understand their own existence in relation to meaningful experiences of Nature in relation to the question of what it means to live a good life.

Hence, the task facing us involves not only countering and reversing the ecologically destructive aspects of human technological activity, which will ultimately lead to either our own extinction as a species or our exile into completely artificial environments, but involves developing ecological wisdom (*sophia oikos*) and, as such, would be a continuance, synthesis, and culmination of philosophy, spirituality, religion, art, and science, bringing together all of these aspects into a mode of inquiry directed primarily towards our conscious enlightenment within the natural world as sustainable and integrated ecological beings. This involves moving beyond the anthropocentrism,

---

[495] Rogers (2008)

which has dominated philosophy since the Scientific Revolution (despite its removal of Man from the center of the Universe) and the Enlightenment (with its inherent anthropocentrism and individualism), and towards what has been termed as Deep Ecology. While there are many philosophical problems with Deep Ecology and its concept of biocentrism[496], along with its use of speculative interpretations of the results of the mechanistic sciences, its fundamental insight is that human beings are not separate from the ecology of the Earth, and therefore our existence has ecological consequences that cannot be ignored if we not only hope to continue to survive, but instead flourish and realize our potential or inner nature to its fullest degree by adopting a biocentric approach. Nature is something we need to respect as our home and source of meaning and value, and, combined with our intimate experiences of the natural world, opens us up to ecophilosophical thinking and provides us with intuitive insights that we need to affirm and incorporate into our Ecosophy. It is not a doctrine or set of categorical imperatives, but rather an ongoing and revisable intellectual process directed towards living a meaningful and valuable life on Earth. It involves doing away with ideas of Nature as something that should be preserved only because it is of instrumental value to human beings, as something separate and outside of us that only has value either as a place of natural resources or somewhere offering entertainment and spectacle, or simply as the environmental means to our continued existence. The anthropocentric approach and the egoism it entails prevents us from discovering our potential and nature to the fullest degree because it prematurely anticipates and curtails our reality.

Ecosophy involves doing away with our idea of Nature as something which our primary relation with is one of survival or fear, as something to dominate and use. Instead, we conceive of Nature as the place of our being-in-the-world—our home—with ecology being the science of home stewardship, as 'eco-' in Greek meant 'household' or 'home'. Ecosophy brings ecology and economics together. As Native American Luther Standing Bear (1971) put it,

"We did not think of the great open plains, the beautiful rolling hills, and winding streams with tangled growth, as "wild." Only to the white man was Nature a "wilderness" and only to him was the land "infested" with "wild" animals and "savage" people. To us it was tame. Earth was bountiful and we were surrounded with the blessings of the Great Mystery. Not until the hairy man from the east came and with brutal frenzy heaped injustices upon us and the families we loved was it "wild" for us. When the very animals of the forest began fleeing from his approach, then it was that for us the "wild west" began."

Ecosophy offers a deeper ontological consciousness of the intrinsic value of Nature. Through intimate experience of wilderness and the Otherness of Nature, while also requiring pragmatic and intimate relationships between beings within the natural world, we can discover a deeper reality and truth about the world. Wilderness is not a place to be feared, but a place of self-discovery and self-realization, providing we encounter it in the spirit of openness and humility. Such reflections can be found in the writings of Baruch Spinoza, Robinson Jeffers, Arne Naess, George Santayana, Aldo Leopold, Johann Wolfgang de Goethe, Henry David Thoreau, Mahatma Gandhi, John Muir, Murray Bookchin, Val Plumwood, Vandana Shiva, Charlene Spretnak, Carolyn Merchant, and Wangari Muta Maathai, and many others. We can learn about the spiritual value of wilderness from the Buddha, Francis of Assisi, Jesus of Nazareth, and Lao Tzu too. Far

---

[496] A concept the anthropologist Tim Ingold termed as *anthropocircumferentialism*, given that human beings circumscribe a region of existence and identify it as "the biological" and place themselves outside of it, looking in, as if looking at it in a petri dish.

from utopian, these diverse voices from out of the wilderness all call on us to shed the view of Nature as a site of struggle and toil—within which labor is the primary mode of interaction—and replace it with a new vision wherein the Earth is understood afresh as an Eden wherein we belong.[497] The wilderness is a place of discovery and revelation of the truth, beauty, and value of life itself—as it is in wilderness that we discover the wildness, the other that is both without and within ourselves—and with it the truth, freedom, humility, and openness of our consciousness of Nature and our reality. Wilderness teaches us a sacred sense of the Earth as our home and as a condition for our own being, and as such is the teacher of spiritual liberation from egoism; it is the source of our sense of the interconnectivity and unity of all existence, including our own and countless other species all situated in interconnected lifeworlds on a shared planet.

By calling for "the sacred sense of the Earth," I am not advocating any return to so-called natural ('primitive') religion or any kind of Nature-worship (both of which are anthropomorphizations), as if Nature was a god or goddess. I'm not suggesting we become the hunter gatherers of a post-industrial society—living in a science fiction of enlightened nomads dancing at the End of History. Instead, I advocate a form of phenomenological primitivism regarding our experiences of Nature, as a place wherein we exist and experience Nature-as-Other—beyond us, yet within us—whereby it draws our consciousness away from individual egoism and the anthropocentric worldview and towards self-realization of human being as not only existing within the whole of Nature but also because of Nature. The human journey towards self-realization is a manifestation of Nature. We carry our history and culture with us on this journey, but they are not burdens we must shed—they are our stepping stones. They are expressions of our lineage and evolution—our 10,000 plateaus. They are our guides on this journey, and they have shaped our paths and drawn our maps up to this point in our journey. Language mediates our experiences, gives them sense and referents, and makes them meaningful for us as beings who are born and learn how to speak and think about the world as social beings within a culture with its history. Yet there comes a point in this journey after which we must proceed without guide nor map—without the certainty of language-given sense and reference—beyond the 'givens' of language and culture. They are inadequate to the task of the cognition and articulation of the intimate experience and consciousness of the otherness of Nature, which is unique alongside its interconnectivity and unity within experience—this aspect of experience cannot be conveyed in public and shared terms. Nature is beyond culture, history, and language; within and without us all, pre-subjective and not entirely within our cognitive grasp, it eludes description. Beyond that point where language is adequate to the task, the journey must continue in solitude and silence, as thereafter we truly enter into the wilderness, within and without each of us, and reconnect to intimate experience of Nature free from our cultural projections of representations and categories. The wilderness within and without is "the clearing space," to put it in Heideggerian terms. By going beyond the culturally 'given' senses of the meaning of "Nature," by placing on epoché (a suspension of judgement) on our systems of representation and conceptual frameworks, we have the possibility of reconnecting to Nature as a referent through the intimate experience of being in the natural world—as the phenomenological experience of an active and embodied subject within a world that precedes and transcends us.[498]

---

[497] Many cultures have myths about a state of primordial paradise on Earth. The Sumerians had *Dilum*; the Iranians the Garden of *Yima*; the Egyptians *Tep Zepi*; and the Greeks had their *Golden Age*.
[498] Merleau-Ponty (1996; 1987)

This avoids the pathetic fallacy, as John Ruskin termed it.[499] Intimate experience is not a projection of emotions over Nature and the experience of our sensibilities, as Ruskin considered the romantic poets to have done, just as it is not a projection of mechanical realism and empiricism over Nature and experience. The pathetic fallacy is to anthropomorphize Nature by projecting subjective experience over Nature by describing Nature as cruel, brooding, or fearsome, for example. Marcuse committed this fallacy when he attempt to reconstruct Nature as a subject—and thereby free and conscious, and an object of moral concern. The intimate experience of Nature—as phenomenological experience—is of Nature beyond the categories of subject and object, and hence beyond the declarative use of language. The referent of language is experience itself, rather than something 'outside' of experience, as the embodied subject is situated within that experience as a being for whom it is meaningful as an experience, and also is active in having that experience. What can we make of the experience of Nature-as-Other in relation to the intimate experience of Nature? I think that anyone who has walked 20 miles along a mountainside trail on a hot summer day, or spent some time in the forests of this world—or, heaven forbid, found themselves marooned or lost at sea in a dingy or clinging on to a piece of driftwood, awaiting first sight of a shark's dorsal fin break the surface of the waters, or finding themselves tossed under waves, unable to breathe—all will testify that Nature is distinctly *other*. Not only in the sense that "mountainside trails" "forests," and "seas," exist independently of the embodied subject, even though these words have clearly objective referents, but in terms of the undeniable sense of reality (of which I am a part) being beyond me, while including me, and which does not conform to my intentions or thoughts, while my intentions and thoughts are part of it. It is the experience of reality as being larger than me and happening irrespective of my expectations and worldview, while I am a part of it and come from it. This reality is simultaneously transcendent and imminent in the experience of being *of* the world.

The Otherness of Nature draws its otherness from within us—revealing the open emptiness and silence within, void of ego and identity, the inner core of Being, beyond sense and reference—beyond language and culture—and beyond our conscious grasp and control. Otherness is not 'out there' as a thing-in-itself, but is emergent from our experience of Nature. Our attempts to express Nature beyond-control-and-cognition—the sense that we cannot grasp it in its entirety as a whole or totality—are attempts to go beyond the limits of language—thereby entering into poetical use and metaphor. Yet, our thought and action is drawn towards grasping it in its totality as a matter of survival and understanding who we are. It acts on us—if we are open to it. From Nature come our instincts and drives, our reality and power, as we surrender to the otherness without us, all around us, and within us, as it changes us by naturalizing us. It is revealed in Nature's constraints and interventions, which shape us physically and psychologically in how we live and understand our alethic modalities (our necessities, possibilities, and limits), but it also shapes our thinking about reality and our place within it. Openness to the Otherness of Nature draws our free consciousness from within ourselves—from the primordial openness and silence that lies beyond ego and language— and beyond our self-conscious cognition and categorization of ourselves as human beings. It comes from the place wherein there is no division between the human and non-human—between self and other. The Otherness of Nature remains ineffable and unknowable to us, just as our own otherness within is ineffable and unknowable to us. We can only speak of it in metaphors. Unspeakable—of which Wittgenstein implored us not to speak, and of which Lao Tzu riddled us with the Tao—but from within which we find the openness and silence that transforms us if we surrender to it. Beyond the limits of consciousness is also beyond

---

[499] Ruskin coined this term in *Modern Painters* ('Of the Pathetic Fallacy", vol. iii, pt. 4), first published in 1856.

language and representation, beyond science and quantification, and emanates from its potential the possibility of being 'here and now'. Poetically, we can say that it is the wellspring of the becoming of being—that which Taoists called the Tao, and early Christians called the Kingdom of God—within and without, all around and nowhere. Yet, these words do not describe it—they miss the mark completely, as they are substitutes, metaphors, as there is no "it." Otherness is not a thing in the world that exists 'outside' of experience. There is no objective referent "it" outside of our experience. Otherness is a quality of being that is made possible only because consciousness, imagination, reason, and logic—and language—make "it" thinkable as something for us as beings, while human consciousness and experience should be viewed as emergent properties of our being-in-the-world, with language, culture, and history as intrinsic elements in the possibilities and actualities of that experience, as an embodied being that is 'thrown' into the world. Hence, we can term the Otherness of Nature as its Spirit.

Profound shifts in our consciousness of the interconnectivity and unity of the Earth—as a place of intrinsic value—involve changes how we think about and interact with natural beings, including each other. This is just as important within the built-environment of the technological society as it is from within the wilderness. The positivistic reduction of Nature to its instrumental value or functionality not only involves a conscious separation of Nature from the human subject, but it also involves a separation of the human 'self' from the human body. Instrumentalism is an obstacle to self-realization because it rejects the intrinsic value of all existence and leads to an impoverished and reductive view of life itself—the technologization of Nature and the body reduces it to matter organized to satisfy the ambitions of the ego. In contrast, self-realization involves deeper relations with language and culture, as even these cease to be mere means to an end, but condition and shape our expressions of the intrinsic value of our being within a world that simultaneously is our home and transcends us as something other, something "non-human," which is not of human making and is not under our control. It involves the realization that being human is also something that is not entirely of human making and under our control—and transcends the ego. Nature makes self-realization possible by surrendering to its Spirit as something greater than ourselves to which we belong; as it is from the intimate experience of going 'back to Nature'—beyond language, history, and culture—that we learn how to live within an interconnected organic whole. When we 'return' to the built environment and social being, we can recognize that, as much as "Nature" is socially constructed, "the social" is naturally constructed when we live in an intimate relation with the ecology of the natural world, which shapes our practices and experiences. Nature shapes us. Yet, how we understand and interact with the ecology of the natural world is shaped by language, history, and culture. We shape Nature. We are bound together. Pragmatic and lived relations within the natural world involve recognizing the sacredness of existence itself, which is a spiritual recognition that represents Nature accordingly as something of intrinsic value of which we are a part and within which we exist, and which draws from us our being-in-the-world, and provides us with all we need to exist and self-realize. Should we recognize Nature in this way, it becomes something that we learn from by letting it intervene upon us and 'put us to the test' on a journey of self-discovery, while we learn how to experience Nature by leaving it unconstrained and something not of our making. This involves an openness towards Nature-as-Other, which brings us into solidarity with the natural world. By respecting Nature and celebrating "the sacred sense of the Earth," we share that sense with each other and all other living beings—moral existence involves solidarity with all life—and by doing so self-realize our own potential and value as both a species and as individuals, as each is a part of the potential and value of the whole Earth. Discovering the goodness of the Earth as our home is how we escape the cave and develop natural virtue through recognizing the intrinsic value of the ecology of the Earth and all existence, and integrating our behavior into that ecology to respect and preserve it.

Of course, this view of Nature-as-sacred runs counter to the dominant ideology of Western civilization wherein the technological society is premised on human dominance over Nature. The natural world is a collection of forces and untapped resources, and nothing else. Yet the values and meanings underlying this ideology did not come from science and technology themselves, even if they are the most advanced expression of it. This ideology is premised on the anthropocentric image of Man and Nature. Perhaps instinctive and itself natural from the point of view of an organism seeking to survive and reproduce in an often seemingly hostile and hazardous world, it finds its cultural origins in ancient mythology and parables, such as the biblical story of Man's fall from grace and expulsion from Eden into a world of suffering, struggle, and toil. It is an expression of the dissatisfaction with the natural world, and the aspiration of human beings to make a better world—one that better suits human aspirations and expectations—and will overcome our vulnerabilities and fears. The walls and fences of civilization are built to keep Nature out, while its substitution is constructed within. This dissatisfaction with Nature also was apparent in the toils of the ancient and medieval alike, be they engineers, inventors, and alchemists, along with the Renaissance mechanists and the revolutionaries of the Scientific Revolution, and other aspirant and Faustian improvers of the world from ancient cities of the Sumerians to the modern day industrial revolution and post-industrial Futurists. The anthropocentrism of the Enlightenment and Humanism, ironically if we take into account its secular rationalism, find its cultural origin in *the Book of Genesis* and the belief in the fundamental superiority of Man over all other living beings as derived from the divine image, alone granted reason, freedom, and consciousness by God. It is this self-image of anthropocentric superiority that underlies that view of Nature as being something of use to Man:

Genesis 1:26 "…and let them have dominion over the fish of the sea, and over the fowl of the air, and over the cattle, and over all the earth, and over every creeping thing that creepeth over the earth."

While the ecotechnological society is clearly an improvement over both the neotechnic and paleotechnic phases, and something we should be moving towards constructing if we hope to survive on this planet as a species, we need to move beyond the ancient aspiration to dominate and pacify Nature. We need to view other beings (including other human beings) as beings with intrinsic value rather than instrumental value so we can learn how to share this planet with other beings, and discover the value of each other, and thereby ourselves. This transformation cannot be undertaken as a radical departure from the human domination of Nature unless there is also an evolution of human consciousness to embrace mindfully the interconnected value of all living beings, for their own sake as co-inhabitants of the planet Earth, rather than merely viewing all things as technological objects understood only in terms of the usefulness for something else. It is only through an evolution of consciousness—itself another plateau in a journey of ten thousand plateaus—that arises the possibility of enlightened, harmonious, and sustainable being-in-the-world.

In his last interview, Heidegger quipped "only a god can save us now."[500] Regardless of what he meant by this cryptic remark, it is clear to me that this god would not be the god Hephaestus. Technology cannot save us. While we consider the world as

---

[500] *Der Spiegel*, Nr. 23 (1976): 193-219

something instrumental, even treating ecology in terms of its function and use for us, we maintain a shallow and consumptive relation with the world and our own existence. Without any concept or vision of the good, or the purpose of our existence beyond satisfying our desires and appetites, we remain in an impoverished and meaningless relation to life and the world. Science and technology are complicit in the consumption of Nature by normalizing instrumental reason and the technologization of the natural world. It is the reduction of Nature to those aspects amenable to quantification, manipulation, and control that place it at the disposal of consumption and commodification, and this results in the negation of ecological, philosophical, and spiritual thinking. It leaves us crawling on our bellies and eating dust all of our lives. The technological appropriation and rapacious consumption of the natural world is a violation of the Earth and is destructive of the sources and resources of life itself. Putting aside that in *Old Testament* terms, Man's worship of technology—the worship of the things made by Man—is the worship of false idols; while we talk of the utility of the ecology of the world—with rainforests being "the lungs of the planet" and wetlands being "buffers"—and think in instrumentalist terms, we maintain a technological relation with Nature as if it is something that only exists for us in a world of our making. While this is clearly an improvement on pathological ecocide, it is also a reproduction of a shallow attitude regarding the natural world. It reinforces the device paradigm and societal gamble that we can "save the planet" through technology and acts of consumption by inventing and buying hybrid cars, energy saving devices, recycling, installing solar power cells, growing and eating organic food, and by using biodegradable laundry detergent. While all of these are valuable technological innovations for the construction of the ecotechnological society, "green consumerism" is merely an extension and continuation of instrumentalism (exploited by capitalism) and the totalitarian outlook that Nature is something that is for us and nothing else. The Earth cannot be saved by "market forces"; it can only be appropriated, transformed, and consumed by them. "Green consumerism" is simply a way of saving capitalism and a high level of mass consumption and profit. Regardless of its method of acquisition, all money is debt—a promise to pay its exchange value in goods and services—a debt that is ultimately extracted from Nature. We need to change our patterns of production and consumption at a deeper level by reducing and simplifying both in accordance with an ecological imperative. We need to transform the technological society into the ecotechnological society while deepening our ecological thinking and conscious awareness of the unity and interconnectivity of life on Earth. We need to transcend our egoism and anthropocentrism to become good stewards and sustainably integrated into the organic whole. Green consumerism and shallow environmentalism, while preferable to the rapacious cancer of unchecked mass consumerism and the exploitation of the natural world, are little more than the capitalist co-option of ecotechnological thinking to transform it into another source of profit by supplying the means for "environmentally conscious" consumers to continue their patterns of consumption while salving their own consciences. It simply delays the inevitable.

There is no formula or method to get us out of this shallow, anthropocentric and destructive relation with Nature. There is no technological solution—we cannot invent a technique or instrument that will save us from ourselves. We face an existential crisis. Artificiality cannot save us at all. It enslaves us to its imperatives and discipline. Even if we speak of the rights of future generations, this means nothing if their lives would be a continuation of the same—for the sake of future generations. If we hope to live meaningful lives within the natural world without destroying it, we must learn from Nature by surrendering to it, adapting to it, through intimacy and developing a sense of place: of belonging through the integration of our patterns of production and consumption with the ecology of the world, and through the conscious awareness of our unity and oneness with the world. This involves a profound paradigm shift which would result in changing our material practices and level of consumption, as it would involve radical changes to agriculture and manufacture while constructing the ecotechnological society. It is essential that the ecotechnological society is built on a shared vision of how human beings should live on Earth. The problem facing us is a political problem of how to come to this shared vision—it is not a technological problem. It is a problem of inclusion in the process by which the future is envisioned—whose world is it?

Already we have sufficient technological development for human beings to live in self-sufficient ecocities, fed by organic permaculture without polluting and degrading the environment, while reforesting the world, restoring the plains and savanna, protecting and replenishing the oceans, using technology to reduce waste and emissions to zero, controlling population, and conserving the remaining wilderness. While these ecocities would be advanced technologically speaking, as fully neotechnic and capable of satisfying all the material and social needs of their denizens, they would also be based on a shared vision of self-sufficiency and how sustainable human life necessarily involves living a simpler life, ideally based on gardening and philosophical spiritual enlightenment, love, compassion, and humility, within egalitarian and democratic societies, and being 'in touch' with Nature, within and without. The people of the ecotechnological society would reject the "device paradigm" as wasteful and the imperative to innovate for its own sake as irrational. They would reject the insatiable avarice of profiteering and the desire for power over others as pathological and harmful. Ecological sustainability requires a concept of technological sufficiency—as a limit—along with a vision of how to satisfy human needs and live a good life without fantasies of limitless economic growth alongside a perpetually growing population engaged in unrestrained production and consumption. Growing our own food, being good stewards of the Earth, and becoming ecologically conscious of the means of our own existence and the consequences of our interventions are all only part of this new paradigm shift. It requires that we learn the discipline of self-restraint and the cautionary principle.

The conservation of wilderness and the protection of fragile ecologies, as well as reforestation and repairing the damage caused by the old paradigm, although vital and necessary, are also important parts of this paradigm shift. We must also go 'back to Nature' within our thinking and being—within ourselves—as well as outwardly towards the otherness without us. This requires solitude, reflection, and intimacy with the natural world. This does not necessarily mean that we should all live as hunter gatherers or as hermits in caves, even though these ways of life should always be possible for human beings. The ecological imperative demands natural virtue from us, but we do not need to wander into the wild to survive by means of primitive technologies and our own wits—the anarcho-primitivist's utopia—á la Robinson Crusoe or even like the biblical Cain. Although less destructive by far, this way of life would be merely a repetition of the same, albeit on a lesser, more primitive scale. It also does not mean that we are destined to create a new Eden and the ersatz Parousia of some futuristic new society, as that may well be a Faustian bargain we might regret by realizing the nightmares of dystopian science fiction. Going back to Nature means that we must radically change our thinking and being to open ourselves to the Otherness of Nature, by being receptive to that otherness that is something sacred and miraculous that is already within us and to which we must surrender to learn how to live better—*by becoming natural again*. It is a revolution in the sense of turning full circle. The whole world is both within and without us. We must go 'back to Nature' by reconnecting with our intimate experience of the natural world and becoming conscious of the Earth as our home and the place where we learn how to live through intimacy and adaptation to natural being. By becoming conscious of our ecological interdependency, we thereby learn to share the world with other beings as we develop 'higher consciousness' and surrender to the Otherness of Nature. We must learn its Spirit—the wildness of Nature—within and without ourselves. By going 'back to Nature', we return to ourselves and become one with the world.

REFERENCES:

Adorno, T., 1994, *The Stars Down to Earth, and Other Essays on the Irrational in Culture* (Crook, ed., London: Routledge)
Althusser, L., 1971, *Lenin and Philosophy* (New York: Monthly Review Press)
Appfel-Marglin, F., and Marglin, S., eds., 1996, *Decolonising Knowledge: From Development to Dialogue* (Oxford: Clarendon Press)
Arendt, H., 1958, *The Human Condition* (University of Chicago Press)
———, 1959, *The Origins of Totalitarianism* (New York: Meridian Books)
———, 1963, *Eichmann in Jerusalem* (New York: Viking Press)
Aristotle, 1984, *The Complete Works*, 2 volumes (Barnes, ed., trans., Princeton University Press)
Aronowitz, S., et al., (eds.), 1996, *Technoscience and Cyberculture* (Routledge)
Authier, M., 1995, "Archimedes: The Scientists' Canon." in *A History of Scientific Thought* (Serres, ed., Blackwell Publications)
Bachelard, G., 1985, *The New Scientific Spirit* (Goldhammer, trans., Beacon Press: Boston)
———, 2002, *The Formation of the Scientific Mind* (McAllester Jones, trans., Clinamen: Bolton)
Bacon, F., 1989, *New Atlantis and the Great Insaturation* (Harlan Davidson).
———, 2000, *New Organon* (Jardine and Silverthorne, eds., Cambridge University Press)
Bailes, K., 1978, *Technology and Society Under Lenin and Stalin* (Princeton University Press)
Bakunin, M., 1970, *God and the State* (Dover Publications)
———, 1990, *Statism and Anarchy* (Shatz, ed., Cambridge University Press)
Barber, B., 1984, *Strong Democracy: Participatory Politics for a New Age* (University of California Press)
Barnes, B., 1974, *Scientific Knowledge and Sociological Theory* (Routledge)
Barrow, J.D., and Tipler, F.J., 1989, *The Anthropic Cosmological Principle* (Oxford University Press)
Baudrillard, J., 1981, *For a Critique of the Political Economy of the Sign* (Saint Louis: Telos Press)
Bäuerle, C.B., et al., 1995, "Field Dependence of the Magnetization of adsorbed 3He Films at ultra low Temperatures." *J. Low Temp. Phys.* 101: 457.
———, 1996a, "Laboratory Simulation of Cosmic String Formation in the Early Universe using Superfluid 3He." *Nature* 382: 332.
———, 1996b, "Simulated Cosmic Strings in a "Big Bang" in Superfluid 3He at 160muK." *Czech. Journal of Physics* 46: 5.
———, 1996c, "Systematic Study of 3He Absorbed on Graphite by NMR Techniques." *Czech. Journal of Physics* 46: 399.
———, 1996d, "3He/Graphite Commensurate Bilayer Films in the Antiferromagnetic Regime." *Czech. Journal of Physics* 46: 401.
———, 1996e, "Magnetic Field Dependence of the Nuclear Magnetisation of 3He Films Absorbed on Graphite in the Ferromagnetic Regime." *Czech. Journal of Physics* 46: 403.
———, 1996f, "The New Grenoble 100mK Refrigerator." *Czech. Journal of Physics* 46: 279.
———, 1996g, "L'univers dans une eprouvette?" *La Recherche* 291, 26
———, 1996h, "El Universo en una probeta." *Mundo Cientedfico*, 174.
———, 1997, "On the Temperature Scale and the Heat Capacity of Superfluid 3He-B in the 100mK range." *Phys. Rev. B.* 57, 14381.
———, 1998, "Superfluid 3He Simulation of Cosmic String Creation in the Early Universe." *J. Low Temp. Phys.* 110: 13.
Bellis, P., 1979, *Marxism and the USSR: The Theory of Proletarian Dictatorship and the Marxist Analysis of Soviet Society* (Atlantic Highlands, NJ: Humanities Press)
Bennett, J.A., 1986, "The Mechanics' Philosophy and the Mechanical Philosophy" in *History of Science*, pp. 1-28.
———, 1989, "A viol of water or a wedge of glass." in Gooding et al., eds., 1989, pp. 105-112.
Bhaskar, R., 1975, *A Realist Theory of Science* (Leeds Books)
———, 1986, *Scientific Realism and Human Emancipation* (London: Verso)
———, 1989, *The Possibility of Naturalism* (Harvester Wheatsheaf)
———, 1993, *Dialectic: The Pulse of Freedom* (Routledge)
Biehl, J., 1998, *The Politics of Social Ecology: Libertarian Municipalism* (New York: Black Rose)

Bijker, W., *et al.*, 1987, *The Social Construction of Large Technological Systems* (Cambridge Mass: MIT Press)
Billig, D, 1987, *Arguing and Thinking* (Cambridge University Press)
Bimber, B., 1994, "Three Faces of Technological Determinism", Smith & Marx, eds., 1994: 80-100
Black, M., 1962, *Models and Metaphors* (New York: Cornell University Press)
Blackledge, P., and Kirkpatrick, G., 2002, *Historical Materialism and Social Evolution* (Basingstoke: Palgrave Macmillan)
Bloor, D., 1976, *Knowledge and Social Imagery* (London: Routledge)
Boas, F., 1911, *The Mind of Primitive Man* (New York: Macmillan)
Bobik, J., 1963, "Matter and Individuation," in *The Concept of Matter in Greek and Medieval Philosophy* (Mullin, ed., University of Notre Dame Press), pp. 281-98.
Bohm, D., 1980, *Wholeness and the Implicate Order* (Boston: Routledge and Kegan Paul)
Bookchin, M., 1971, *Post-scarcity Anarchism* (New York: Black Rose)
——————, 1974, *Our Synthetic Environment* (New York: Harper Colophon)
——————, 1981, *Toward an Ecological Society* (Montreal: Black Rose)
——————, 1996, *The Philosophy of Social Ecology: Essays on Dialectical Naturalism* (New York: Black Rose)
——————, 2005, *The Ecology of Freedom* (Oakland, CA: AK Press)
Borgmann, A., 1984, *Technology and the Character of Contemporary Life: A Philosophical Inquiry* (University of Chicago Press)
——————, 1990, "Communities of Celebration: Technology and Public Life" in Ferré, ed., *Research in Philosophy and Technology*, vol. 10 (Greenwich: JAI Press), pp. 315-45
——————, 2000, "Reply to My Critics," in Higgs, Light, & Strong, eds., 2000, pp. 341-70.
Boyd, R., 1984, "The Current State of Scientific Realism," in *Scientific Realism* (Leplin, ed., Berkeley: University of California Press), pp.41-82
Boyle, R., 1965, *The Origin of Forms and Qualities According to the Corpuscular Philosophy*, vol. 3 of *The Works of the Honourable Robert Boyle* (Birch, ed., 6 vols. Hildesheim: Georg Olms)
Bradley, D.I., *et al.*, 1995a, "Potential Dark Matter Detector? The Detection of Low Energy Neutrons by Superfluid 3He." *Phys. Rev. Lett.* 75: 1887.
——————, 1995b, "A Highly Sensitive Nuclear Recoil Detector Based on Superfluid 3He-B." *J. Low Temp. Phys.* 101: 9.
Bréhier, É, 1968, "The Creation of the Eternal Truths in Descartes' System." in *Descartes: A Collection of Critical Essays* (Doney, ed., University of Notre Dame Press), pp.192-208.
Brett, G., 1939, "Byzantine Watermill" in *Antiquity*, xiii, pp. 354-6
Bruno, G., 1998, *Cause, Principle and Unity: And Essays on Magic* (Cambridge Texts in the History of Philosophy, University of Cambridge)
Bryant, L., 1919, *Six Red Months in Russia* (London: Heinemann)
Bukharin, N., 1921, "Era of Great Works," in *The Communist Index*, January 27, 1921, p. 9
——————, 1925, *Historical Materialism: A System of Sociology* (International Publishers)
——————, 1929, *Imperialism and World Economy* (International Publsihers)
——————, 1931, *Science at the Crossroads: Papers Presented to the International Congress of the History of Science and Technology Held in London from June 29th to July 3rd, 1931 by the delegates of the U.S.S.R* (London: Frank Cass and Co.)
Bunkov, Yu, M., *et al.*, 1992a, "Resonant observation of the Landau field in superfluid 3He-B by NMR." *Phys. Rev. Lett.* 68: 600.
——————, 1992b, "A new NMR mode in the Landau field in superfluid 3He-B." *J. Low. Temp. Phys.* 89: 27.
——————, 1992c, "Persistent spin precession in 3He-B in the regime of Vanishing Quasiparticle Density." *Phys. Rev. Lett.* 69: 3092
——————, 1994, "Persistent spin precession in superfluid 3He." *Physica B* 194-196: 827.
——————, 1995a, "The Ultimate Performance of a Superfluid 3He-B Detector.", *Superconductivity and Particle Detectors*, World Scientific Publishing.
——————, 1995b, "NMR in Superfluid 3He at Very Low Temperatures." *J. Low Temp. Phys.* 101: 123

———————, 1996, "Texture Dependence of the Persistent NMR Signal in Superfluid 3He-B." *Czech. Journal of Physics* 46: 233

Burke, J.G., ed., 1983, *The Uses of Science in the Age of Newton* (Berkeley)

Burke, L., *et al.,* eds., 1980, *Alice Through the Microscope* (London; Virago)

Burstall, A.F., 1963, *A History of Mechanical Engineering* (London: Faber & Faber)

Burtt, E.A., 1954, *The Metaphysical Foundations of Modern Science* (New York: Doubleday)

Calian, F., 2010, "Some Modern Controversies on the Historiography of Alchemy," in *Annual of Medieval Studies at CEU*, 2010: 186.

Cantor, G., 1989, "The Rhetoric of Experiment", in Gooding, *et al.*, 1989, pp. 159-80.

Capra, F., 1978, *The Tao of Physics: An Exploration of the Parallels Between Modern Physics and Eastern Mysticism* (Shambhala Publications)

———————, 1982, *The Turning Point: Science, Society, and the Rising Culture* (Bantam Books)

Carnap, R., 1995, *The Unity of Science* (Black, trans., Thoemmes Press)

Carr, E.H., 1966, *The Bolshevik Revolution* 1917-1923, *Volume* 1 (Baltimore: Penguin Books)

Cartelon, H., 1975, "Does Aristotle have a Mechanics?" in Barnes, *et al.*, (eds.), *Articles on Aristotle*: 1: *Science* (London: Duckworth)

Cartwright, N., 1983, *How the Laws of Physics Lie* (Oxford: Clarendon Press)

———————, 1999, *The Dappled World: A Study of the Boundaries of Science* (Cambridge University Press)

Case, P. F., 2009, (Rosicrucian Order of the Golden Dawn)

Cervantes Saavedra, M., de, 1998, *Don Quixote de la Mancha* (Riley, ed., Jarvis, trans., Oxford University Press)

Chaloupka, W., 1992, *Knowing Nukes: The Politics and Culture of the Atom* (University of Minnesota Press)

Chatly, H., 1942, "The Development of Mechanisms in Ancient China" in *Engineering*, cliii, p. 145

Chakravartty, A., 2010, *A Metaphysics for Scientific Realism: Knowing the Unobservable* (Cambridge University Press)

Chomsky, N., *Media Control: The Spectacular Achievements of Propaganda* (New York: Seven Stories Press, 2002)

Clagett, M., 1959a, in Kearney, ed., 1964, pp. 40-4

———————, ed., 1959b, *The Science of Mechanics in the Middle Ages* (Madison: University of Wisconsin Press)

———————, 1978, *Archimedes in the Middle Ages*, 3 volumes (The American Philosophical Society)

Clagett, M., and Moody, A., (eds.) 1952, *The Medieval Science of Weights* (Madison: University of Wisconsin Press)

Clark, D.M, 1982, *Descartes' Philosophy of Science* (Pennsylvania State University Press)

Cohen, J.W., 1955, "Technology and Philosophy", *Colorado Quarterly* 3, no. 4 (spring), pp. 409-20

Cohen, R.S., ed., 1997, *Experimental Metaphysics: Quantum Mechanical Studies for Abner Shimony*, vol.1 (Kluwer Academic Publishers)

Collins, H.M., 1985, *Changing Order: Replication and Induction in Scientific Practice* (London: Sage Publications)

Cook, M., 2001, "Divine Artifice and Natural Mechanism: Robert Boyle's Mechanical Philosophy of Nature" in *Osiris*, 2nd series, 16, pp. 133-50

Copenhauer, B., 1990, "Natural Magic, Hermetism, and Occultism in Early Modern Science" in Lindberg and Westman, 1990, pp. 261-301

Coulson, J., Whitfield, D.H., and Preston, A., eds., 2003, *Keeping Things Whole* (Chicago, IL: Great Books Foundation)

Cousins, D.J., *et al.*, 1994, "T-3 Temperature Dependence and a Length Scale for the Thermal Boundary Resistance between Saturated Dilute 3He-4He Solution and Sintered Silver." Phys. Rev. Lett. 73: 2583

———————, 1995a, "A geometry dependent thermal resistance between a saturated dilute 3He/4He solution and sintered silver powder." *J. Low Temp. Phys.* 101: 259.

———————, 1995b, "Andreev Reflection of a Beam of Ballistic Quasiparticle Excitations Incident on a Static B-A Phase Interface in Superfluid 3He." *J. Low Temp. Phys.* 101: 293

———————, 1996a, "Concentration Dependence of the Thermal Boundary Resistance between a Dilute 3He-4He Solutions in Sintered Silver Powder." *Czech. Journal of Physics* 46: 155.

——————, 1996b, "Andreev Reflection from a B-A Interface in Superfluid 3He." *Czech. Journal of Physics* 46: 249.

——————, 1996c, "Probing the A-B Phase Interface in Superfluid 3He by Andreev Reflection of a Quasiparticle Beam." *Phys. Rev. Lett.* 77: 5245

——————, 1997, "Dynamics with a non-newtonian gas: the force on a body moving through a beam of excitations in superfluid 3He." *Phys. Rev. Lett.* 79: 2285

——————, 1999, "Persistent coherent spin precession in superfluid 3He-B excited by off-resonant excitation." *Phys. Rev. Lett.* 82: 4484

Cunningham, F., 1987, *Democratic Theory and Socialism* (Cambridge University Press)

Crease, R.P., 1993, *The Play of Nature: Experimentation as Performance* (Bloomington: Indiana University Press)

Crosland, M., ed., 1975, *The Emergence of Science in Western Europe* (Macmillan Press)

Curley, E.M., 1984, "Descartes on the Creation of the Eternal Truths", *Philosophical Review*, 93, pp. 569-97.

Dalton, L., and Park, K., 1998, *Wonders and the Order of Nature* (New York: Zone Books)

Dampier, W.C., 1938, "From Aristotle to Galileo", in Needham and Pagel, eds., 1975, *Background to Modern Science* (New York: Arno Press).

Danziger, K., 1990, *Constructing the Subject* (Cambridge University Press)

Darwin, C., 1859, *On the Origin of Species by Means of Natural Selection* (John Murray)

——————, 1871, *The Descent of Man in Relation to Sex* (John Murray)

Davis, R.W. 1980, *The Socialist Offensive: The Collectivisation of Soviet Agriculture* 1929-1930 (Basingstoke: Palgrave Macmillan)

Day, R.B., Beiner, R., and Masciulli, J., eds., 1988, *Democratic Theory and Technological Society* (Armonic, N.Y.: Sharpe)

Dear, P., 1988, *Mersenne and the Learning of the Schools* (Cornell University Press)

Deleuze, G., and Guattari, F., 1994, *What is Philosophy?* (Burchell and Tomlinson, trans., London-New York: Verso)

Derrida, J., 1967, *Of Grammatology* (Spivak (trans.), Baltimore and London: John Hopkins University Press)

de Santillana, G., 1961, *The Origins of Scientific Thought* (London: Weidenfield and Nicolson)

Descartes, R., 1961, *Rules for the Direction of the Mind* (Lafleur, trans., Indianapolis: Library of the Liberal Arts)

——————, 1966-76, *Oeuvres de Descartes* (Adam and Tannery, eds., Paris: Vrin/C.R.N.S)

——————, 1968, *Discourse on Method and The Meditations* (Sutcliffe, trans., Penguin Books)

——————, 1983, *Principles of Philosophy* (Miller and Miller, trans., Dordrecht: Reidel)

de Solla, D.J., 1959, "Ancient Greek Computer; with Biographical Sketch" *Scientific American* (June)

DeVries, K., 1992, *Medieval Military Technology* (Broadview Press)

Dickson, D., 1974, *Alternative Technology and the Politics of Technological Change* (Collins)

——————, 1984, *The New Politics of Science* (University of Chicago Press)

Dobbs, B.J.T., 2002, *The Janus Face of Genius: The Role of Alchemy in Newton's Thought* (Cambridge University Press)

Drake, S. (ed., trans.) 1975, *The Discoveries and Opinions of Galileo* (Garden City NY.: Doubleday Anchor)

——————, 1979, *Galileo At Work* (University of Chicago Press)

Drake, S., and Drabkin, I.E., 1969, *Mechanics in the Sixteenth Century Italy* (University of Wisconsin Press)

Dugas, R., 1955, *A History of Mechanics* (Maddox, trans., Routledge and Keegan Paul)

Duhem, P., 1969, *To Save the Phenomena: An Essay on the Idea of Physical Theory from Plato to Galileo.* (Doland and Chaninah (trans.), University of Chicago Press)

Dunne, J., 1993, *Back to the Rough Ground: "Phronesis" and "Techne" in Modern Philosophy and in Aristotle* (Indiana: University of Notre Dame Press)

Dunne, J.W., 1948, *An Experiment With Time* (London: Faber & Faber)

Dupuy, J-P., 2000, *The Mechanization of the Mind: On the Origins of Cognitive Science* (DeBevoise, trans., Princeton University Press)

Durbin, P.T., 1972, "Technology and Values: A Philosophical Perspective." in *Technology and Culture* 13, no. 4, October, pp. 556-76.

—————, 1978, "Towards a Social Philosophy of Technology" in *Research in Philosophy and Technology*, pp. 67-97.
Easlea, B., 1983, *Fathering the Unthinkable: Masculinity, Scientists, and the Nuclear Arms Race*, (London: Pluto Press)
—————, 1986, "The Masculine Image of Science with Special Reference to Physics: How Much Does Gender Really Matter?" in Harding (1986), pp.132-58.
Easterbrook, G., 1996, *A Moment on the Earth: The Coming Age of Environmental Optimism* (London: Penguin)
Eisenstein, E.L., 1979, *The Printing Press as an Agent for Change: Communications and Cultural Transformations in Early-Modern Europe*, vol. 2 (Cambridge University Press)
Ellul, J., 1964, *The Technological Society* (Wilkinson, trans., New York: Knopf)
—————, 1965, *Propaganda: The Formation of Men's Attitudes* (Vintage Books)
—————, 1972, *The Political Illusion* (Vintage Books)
—————, 1972b, *The Politics of God and the Politics of Man* (Grand Rapids, Mich.: Eerdmans)
—————, 1976, *The Ethics of Freedom* (Grand Rapids, Mich.: Eerdmans)
—————, 1990, *The Technological Bluff* (Bromiley, trans., Grand Rapids, Mich.: Eerdmans)
—————, 1991, *Anarchy and Christianity* (Bromiley trans., Grand Rapids, Mich.: Eerdmans)
Elzinga, A., 1972, *On a Research Program in Early Modern Physics* (New York: Humanities Press)
Emerson, R.W., 1860, *English Traits* (Boston)
Emerton, N.E., 1984, *The Scientific Re-Interpretation of Form* (Cornell University Press)
Engels, F., 1947, *Anti-Duhring* (Burns, trans., London: Progress Publishers)
—————, 1934/1883, *Dialectics of Nature* (Dutt, trans., London: Progress Publishers)
Enrico, M.P., et al., 1993, "Direct Observation of the Andreev Reflection of a Beam of Excitations in Superfluid 3He-B.", *Phys. Rev. Lett.* 70: 1846
—————, 1994a, "Temperature dependence of the nuclear spin-lattice relaxation time in copper metal to below 10 mK." *Phys. Rev. B* 49, 6339.
—————, 1994b, "A proposed new nuclear cooling refrigerator." *Physica B* 194-196: 47.
—————, 1994c, "Modelling the Damping Force Exerted on a Macroscopic Object Moving through Superfluid 3He-B in the Ballistic Regime." *Physica B* 194-196: 787.
—————, 1994d, "Quasiparticle beams in superfluid 3He-B at very low temperatures." *Physica B* 194-196: 789.
—————, 1994e, "Direct measurement of the Andreev reflection of a beam of excitations in superfluid 3He-B." *Physica B* 194-196: 791
—————, 1994f, "Pumping 3He in a dilute 3He-4He solution by a magnetic field." *Physica B* 194-196: 837.
—————, "Diffuse scattering model of the thermal damping on a wire moving through superfluid 3He-B at very low temperatures." *J. Low Temp. Phys.* 98: 81
Erasmus, D., 1925, *In Praise of Folly* (Chicago: P. Covici)
Euclid, 1952, *The Thirteen Books of The Elements*, in *Great Books of the Western World* (Heath, trans., University of Chicago Press)
Feenberg, A., 1991, *Critical Theory of Technology* (New York: Oxford University Press)
Feibleman, J.K., 1982, *Technology and Reality* (London: Martin Nijhoff Publishers)
Fermi, L., and Bernadini, G., 1961, *Galileo and the Scientific Revolution* (New York: Basic Books)
Feyerabend, P.K., 1975, *Against Method: Outline of an Anarchistic Theory of Knowledge* (London: New Left Books)
Findlay Hendry, R., 1995, "Realism and Progress: Why Scientists Should Be Realists." in *Philosophy and Technology* (Fellows, ed., Cambridge University Press) pp. 53-72.
Fisher, S.N., et al., 1989, "Beyond the two-fluid model: Transition from linear behaviour to velocity-independent force on a moving object in 3He-B.", *Phys. Rev. Lett.* 64: 2566
—————, 1990a, "Anomalous dynamic behaviour of a gas of quasiparticles in superfluid 3He-B." *Physica B* 165 & 166: 651
—————, 1990b, "The Force on a Wire Moving through Superfluid 3He-B." *Physica B* 165 & 166: 683
—————, 1991a, "A microscopic calculation of the force on a moving wire in superfluid 3He-B." *J. Low Temp. Phys.* 83: 225
—————, 1991b, "Distortion of Superfluid 3He-B as a Function of Magnetic Field and the First-Order Transition to 3He-A at the T=0 Limit." *Phys. Rev. Lett.* 67: 1270

———————, 1991c, "Exponential boundary resistance between superfluid 3He-A and silver sinter at temperatures down to 230muK." *Europhysics Letters* 16: 385
———————, 1991d, "Exponential, rather than power-law, temperature dependence of the damping of a vibrating wire resonator in 3He-A at low temperatures." *Phys. Rev. Lett.* 67: 3788
———————, 1992a, "The superfluid energy gap in 3He-B as a function of magnetic field in the low temperature limit." *Physica B* 178: 336
———————, 1992b, "Andreev exclusion of bulk state quasiparticles from a surface in 3He-A at the low temperature limit: implications for transport properties." *Physica B* 178: 304
———————, 1992c, "Blackbody Source and Detector of Ballistic Quasiparticles in 3He-B; Emission Angle from a Wire Moving at Supercritical Velocity." *Phys. Rev. Lett.* 69: 1073
———————, 1992d, "Ballistic quasiparticle beam experiments in superfluid 3He-B." *J. Low. Temp. Phys.* 89: 477
———————, 1992e, "The one-dimensional excitation gas in superfluid 3He-A at very low temperatures." *J. Low Temp. Phys.* 89: 481
———————, 1994, "The A-Phase at very low temperatures, a one-dimensional excitation gas." *Physica B* 194-196: 793.
———————, 1995, "A Pumping Experiment in Dilute 3He-4He Solutions at Millikelvin Temperatures." *J. Low Temp. Phys.* 100: 241
Fitzpatrick, S., 1994, *The Russian Revolution* (Oxford University Press)
Fodor, J.A., 1994, *The Elm and the Expert* (Cambridge, MA: MIT Press)
Forbes, R.J., 1955, *Studies in Ancient Technology* (9 volumes, Leiden: Brill)
Foster, M., 1988, "Sir Kenelm Digby (1603-1665) as Man of Religion and Thinker" in *Downside Review*, 106, pp. 101-25.
Foucault, M., 1980, *Knowledge/Power: Selected Interviews and Other Writings, 1972-1977* (Gordon, trans., ed., New York: Pantheon Books)
———————, 1994, *The Order of Things: An Archaeology of the Human Sciences* (New York: Vintage Books)
Fournier D'Albe, E.E., 1926, *Hephaestus or The Soul of the Machine* (London: Keegan-Paul)
Franklin, A., 1986, *The Neglect of Experiment* (Cambridge University Press)
———————, 1990, *Experiment, Right or Wrong?* (Cambridge University Press)
Fraasen, B.C., van, 1980, *The Scientific Image* (Oxford: Clarendon Press)
French, S., 2014, *The Structure of the World: Metaphysics and Representation* (Oxford University Press)
Friedman, M., 1962, *Capitalism and Freedom* (University of Chicago Press)
Frost, R.L., 1992, "Mechanical Dreams in Twentieth-Century France" in Winner, ed., 1992, pp. 51-77
Fuller, S., 1993, *Philosophy, Rhetoric & the End of Knowledge* (University of Wisconsin Press)
Funkenstein, A., 1986, *Theology and the Scientific Imagination from the Middle Ages to the Seventeenth Century* (Princeton University Press)
Gadamer, H-G., 1966, *Philosophical Hermeneutics* (University of California Press)
Galeano, E., 1997, *The Open Veins of Latin America* (Monthly Review Press)
Galen, 1947, *On the Natural Faculties* (Brock, A.J., trans., London: Heinemann,)
Galileo Galilei., 1914, *Dialogues Concerning Two New Sciences* (Crew and de Salvio, trans., Evanston: Northwestern University Press)
———————, 1960, *On Motion and Mechanics* (Drabkin and Drake, trans., Madison: University of Wisconsin Press)
Galison, P., 1987, *How Experiments End* (Chicago: Chicago University Press)
———————, 1997, *Image and Logic: A Material Culture of Microphysics* (Chicago University Press)
Galison, P., and Stump, D., (eds.), 1996, *The Disunity of Science: Boundaries, Contexts and Power* (Stanford University Press)
Garber, D., 1978, "Science and Certainty in Descartes." in *Descartes: Critical and Interpretive Essays* (Hooker, ed., John Hopkins University Press) pp. 114-51
———————, 1992, *Descartes' Metaphysical Physics* (University of Chicago Press)
Gaukroger, S., 1976, "Bachelard and the problem of epistemological analysis" in *Studies in the History and Philosophy of Science*, **7**, pp.189-224
Gille, B., 1964, *The Renaissance Engineers* (London: Lund Humphries)
Gleick, J., 1978, *Chaos: The Making of a New Science* (New York: Viking)
Goldman, E., 1923, *My Disillusionment in Russia* (New York: Doubleday, Page & Company)

Gooding, D., 1990, *Experiment and the Making of Meaning* (Dordrecht: Kluwer Academic Publishers)

———, 1992, "Putting human agency back into observation" in Pickering, ed. (1992), pp. 65-112.

———, 1996, "Creative Rationality: Towards an Abductive Model of Scientific Change" in *Philosophica* 58, pp. 73-102.

Gooding, D., Pinch, T., Schaffer, S., (eds.), 1989, *The Uses of Experiment: Studies in the Natural Sciences* (Cambridge University Press)

Gooding, J., *Socialism in Russia: Lenin and his Legacy*, 1890-1991 (Basingstoke: Palgrave Macmillan, 2002)

Goodman, N., 1978, *Ways of Worldmaking* (Indianapolis: Hackett)

Gordo-Lopez, A.J., and Parker, I., (eds.), 1999, *Cyberpsychology* (London: Macmillan Press)

Gorokhov, V., "Politics, Progress, and Engineering: Technical Professionals in Russia" in Winner, L., ed., 1992, pp. 175-86

Georgescu-Roegen, N., 1971, *Entropy Law and the Economic Process* (Cambridge: Harvard University Press)

Gouldner, A.W., 1982, *The Two Marxisms* (New York: Seabury)

Grafton, A., 1988, "The Availability of Ancient Works", in Schmitt, Skinner, and Kessler (eds.), *The Cambridge History of Renaissance Philosophy* (Cambridge University Press), pp. 767-91.

Graham, L., 1998, *What Have We Learned About Science and Technology from the Russia Experience?* (Stanford, CA: Stanford University Press)

Grant, E., 1978, "Aristotelianism and the Longevity of the Medieval World View." in *History of Science*, 16, pp. 93-106.

———, 1981, *Much Ado About Nothing: Theories of Space and Vacuum from the Middle Ages to the Scientific Revolution* (Cambridge University Press)

———, 1987, "Ways to Interpret the Terms 'Aristotelean' and 'Aristoteleanism' in Medieval and Renaissance Natural Philosophy", *History of Science*, 25, pp. 333-58.

Greenberg, D., 1967, *The Politics of Pure Science* (New York: New American Library)

Griffin, S., 1978, *Woman and Nature: The Roaring Inside Her* (New York: Harper & Row)

Gulluzzi, P., 1987, "The Career of a Technologist" in *Leonardo da Vinci: Engineer and Architect*, (Gulluzzi, ed., Montreal: Montreal Museum of Fine Arts), pp. 41-109.

Guthrie, W.K.C, 1971, *The Sophists* (Cambridge University Press)

Habermas, J., 1970, *Toward a Rational Society* (Boston: Beacon Press)

———, 1973, *Theory and Practice* (Viertel, trans., Boston: Beacon Press)

———, 1979, *Communication and the Evolution of Society* (Boston: Beacon Press

———, 1987, *Knowledge and Human Interests* (Shapiro, trans., Cambridge: Polity Press)

Hacking, I., 1982, "Language, Truth, and Reason." in *Rationality and Relativism* (Lukes and Hollis eds., Oxford: Blackwell)

———, 1983, *Representing and Intervening: Introductory Topics in the Philosophy of Science* (Cambridge University Press)

———, 1992, "'Style' for Historians and Philosophers." *Studies in History and Philosophy of Science*, 23, pp. 1-20.

———, 2000, *The Social Construction of What?* (Harvard University Press)

Hackmann, W. D., 1978, *Eighteenth Century Measuring Devices* (Oxford University Press)

———, 1986, "Scientific Instruments: models of brass and aids to discovery", in Gooding et al., eds., 1986, pp. 31-66.

Hackett, J., ed., 1997, *Roger Bacon and the Sciences* (Leiden: Brill)

Hall, A.R., 1972, "Science, Technology, and Utopia." in Mathias, ed., 1972, pp.33-53

Hanks, J.M., 1984, *Jacques Ellul: A Comprehensive Bibliography* (Greenwich, Con: JAI Press)

Haraway, D., 1989, *Primate Visions: Gender, Race, and Nature in the World of Modern Science* (New York: Routledge)

———, 1991, *Simians, Cyborgs, and Women: The Re-Invention of Nature* (London: Free Association Press)

Harding, J., ed., 1986, *Perspectives on Gender and Science* (London: Falmer Press)

Harré, R., 1972, *The Philosophies of Science: An Introductory Survey* (Oxford University Press)

———, 1986, *Varieties of Realism* (Oxford: Basil Blackwood)

Harré, R., and Madden, E.H., 1977, *Causal Powers* (Oxford: Blackwell)

Harrison, P., 1987, *The Greening of Africa: Breaking Through the Battle for Land and Food* (London: Paladin Grafton Books)

Hatfield, G., 1989, "Reason, Nature, and God in Descartes," *Science in Context*, 3, no. 1, pp. 175-201.

―――――――, 1990, "Metaphysics and the New Science." in Lindberg & Westman, eds., 1990, pp. 93-166.

Haugeland, J., 1985, *Artificial Intelligence: The Very Idea* (Cambridge, MA: MIT Press)

Hayek, F.A., *The Road to Serfdom* (New York: Routledge, 2001)

Heath, T.L., ed., n.d., *The Works of Archimedes with the Method of Archimedes* (New York: Dover Publications)

―――――――, 1949, *Mathematics in Aristotle* (Oxford University Press)

Heathcote, N. H. de V. (1950) "Guericke's sulphur globe," *Annals of Science*, 6: 293-305

Heelan, P., 1983, *Space-Perception and the Philosophy of Science* (Berkeley: University of California Press)

Hegel, G.W.F., 1977, *Phenomenology of Spirit* (Miller, trans., Oxford University Press)

Heidegger, M., 1939, "On the Essence and Concept of *Phusis* in Aristotle's *Physics* Bk.I" in McNeill ed., *Martin Heidegger: Pathmarks* (Sheenan, trans. Cambridge University Press, 1998), pp. 183-230

―――――――, 1962, *Being and Time* (Macquarie and Robinson, trans., New York: Harper & Row)

―――――――, 1977, *The Question Concerning Technology and Other Essays* (Lovitt, trans., Harper Torchbooks)

―――――――, 1977a, "The Question Concerning Technology", in Heidegger (1977, pp.3-35).

―――――――, 1977b, "The Age of the World Picture", in Heidegger (1977, pp.115-54).

―――――――, 1977c, "Science and Reflection", in Heidegger (1977, pp.155-82).

―――――――, 1999, *Basic Writings* (Farrel Krell (ed.), London: Routledge)

―――――――, 1999a, "Building, Dwelling, Thinking", in Heidegger (1999, pp. 343-364).

―――――――, 1999b, "Modern Science, Metaphysics, and Mathematics", in Heidegger (1999, pp. 267-306).

―――――――, 1999c, "What is Metaphysics?" in Heidegger (1999, pp.82-96).

―――――――, 1999d, "Letter on Humanism", in Heidegger (1999, pp. 217-265).

Heilbroner, R.L., 1967, "Do Machines Make History?" first published in *Technology and Culture*, July 1967, pp. 335-45; reproduced in Smith and Marx, eds., 1994, pp. 53-65

Heisenberg, W., 1971, *Physics and Philosophy* (New York: Harper & Row)

Hempel, C., 1966, *Philosophy of Natural Science* (Pearson)

Herman, E.S., and Chomsky, N., 1988, *Manufacturing Consent: The Political Economy of the Mass Media* (New York: Pantheon Books)

Herzen, A., *From the Other Shore and The Russian People and Socialism* (Budberg, trans., London, 1956)

Hickman, L.A., 1990, *John Dewey's Pragmatic Technology* (Bloomington: Indiana University Press)

―――――――, 1997, "Populism and the Cult of the Expert" in Winner, ed., 1997, pp. 91-103

Higgs, E., Light, A., and Strong, D., eds., 2000, *Technology and the Good Life?* (University of Chicago Press)

Hill, D., 1984, *A History of Engineering in Classical and Medieval Times* (London: Croom Helm)

Hobbes, T., 1962, *Elements of Philosophy: The First Section, Concerning Body* (1655), vol. I, in *The English Works of Thomas Hobbes of Malemsbury*, Molesworth, ed., II vols, Aalen: Scientia.

Hodges, H., 1970, *Technology in the Ancient World* (Penguin Press)

Hooke, R., 2003, *Micrographia: Or Some Physiological Descriptions of Minute Bodies Made by Magnifying Glass, with Observations and Inquiries Thereupon* (Courier)

Hooykaas, R., 1972, "Beekman, Isaac" in volume I the *Dictionary of Scientific Bibliography* (Gillispie ed., 16 volumes, New York: Scribner), p. 566

Horkheimer, M., 1974, *Critique of Instrumental Reason: Lectures and Essays since the End of World War II* (O'Connell et al., trans., New York: Continuum)

Horkheimer, M., and Adorno, T., 2002, *Dialectic of Enlightenment* (Stanford University Press)

Hughes, T.P., 1983, *Networks of Power* (Baltimore: John Hopkins Press)

―――――――, 2004, *Human-Built World: How To Think About Technology and Culture* (University of Chicago Press)

Hume, D., 1965, *The Essential Works of David Hume*, (Ralph Cohen, ed., New York: Bantam Books)

Hunter and Davis, eds., 2000, *The Works of Robert Boyle* (London: Pickering & Chatto)

Husserl, E, 1960, *Cartesian Meditations* (Cairns, trans., The Hague: Martinus Nijhoff)

———, 1970, *The Crisis of European Sciences* (Carr, trans., Northwestern University Press)

Hutten, E.H., 1958, *The Language of Modern Physics* (London)

Huxley, A., 1932, *Brave New World* (London: Chatto & Windus)

Idel, M., *Jewish Magical and Mystical Traditions on the Artificial Anthropoid* (Albany: State University of New York Press, 1990)

Igov, M., 2005, "Physics as a Technical Phenomenon" *Khimiya*, Vol. 15, Issue 2

Ihde, D., 1979, *Technics and Praxis: A Philosophy of Technology* (Boston: Reidel)

———, 1983, *Existential Technics* (Albany: State University of New York Press)

———, 1991, *Instrumental Realism: The Interface Between Philosophy of Science and Philosophy of Technology* (Bloomington: Indiana University Press)

Ince, M., 1986, *The Politics of British Science* (Brighton: Wheatsheaf)

James, R., 2013, *Why Cows Need Names: And More Secrets of Amish Farms* (Kent State University)

Jefferson, T., 1998, *Notes on the State of Virginia* (Penguin Classics)

Jennings, F, 1976, *The Invasion of America: Indians, Colonialism, and the Cant of Conquest* (Norton)

Johnson, F.R., 1940, in Kearney, ed., 1964, pp. 44-50.

Jonas, H., 1974, *Philosophical Essays: From Ancient Creed to Technological Man* (Englewood Cliffs, N.J: Prentice-Hall)

———, 1984, *The Imperative of Responsibility: In Search of an Ethics for the Technological Age*. (Chicago University Press)

Joravsky, D., 1961, *Soviet Marxism and Natural Science 1917-1932* (New York, NY: Columbia University Press)

———, *The Lysenko Affair* (University of Chicago Press, 1970)

Kant, I., 1964, *Critique of Pure Reason* (London: Macmillan)

———, 1995, *Opus postumum* (Forster, ed., Rosen & Forster, trans., Cambridge University Press)

Kapp, E., 1877, *Philosophie der Technik*, pp. 44-5, quoted and trans. Mitcham, ed., 1994, pp. 23-4.

Kargon, R.H., 1966, *Atomism in England from Hariot to Newton* (Oxford)

Kaufman, T.D., 1993, "Astronomy, Technology, Humanism, and Art at the Entry of Rudolf II into Vienna, 1577" in *The Mastery of Nature: Aspects of Art, Science, and Humanism in the Renaissance* (Princetown University Press) pp.136-50.

Kearney, H.F., ed., 1964, *Origins of the Scientific Revolution* (London: Longmans)

Keller, A.G., 1964, *A Theatre of Machines* (London: Chapman & Hall)

———, 1971, "Archimedean Hydrostatic theorems and Salvage Operations in 16th-century Venice" in *Technology and Culture*, pp. 612-17.

——— 1975, "Mathematics, Mechanics, and Experimental Machines in Northern Italy in the Sixteenth Century" in Crosland (ed.), pp.15-34.

Kellner, D., *Herbert Marcuse and the Crisis of Marxism* (London: Macmillan, 1984)

Kemp, M., 1981, *Leonardo Da Vinci: The Marvellous Works of Man and Nature* (Harvard University Press)

Knorr-Cetina, K., 1981, *The Manufacture Of Knowledge* (Pergamon Press)

Knorr-Cetina, K., and Mulkay, M., eds., 1983, *Science Observed* (London: Sage)

Koyré, A., 1992, *Metaphysics and Measurement* (Maddison, trans., Gordon and Breach)

Krieger, M., 1991, *Doing Physics: How Physicists Take Hold of the World* (Bloomington: Indiana University Press)

Krige, J., 1989, "Why did Britain join CERN?" in Gooding, D., *et al.*, (eds.), 1989, pp.385-405

Kuhn, T.S., 1957, *The Copernician Revolution* (Harvard University Press)

———, 1962, *The Structure of Scientific Revolutions* (Chicago University Press)

———, 1970, "Postscript" in *The Structure of Scientific Revolutions* (second edition, Chicago University Press)

———, 1977, *The Essential Tension* (Chicago University Press)

Ladyman, J., and Ross, D., 2009, *Everything Must Go: Metaphysics Naturalized* (Oxford University Press)

Laird, W.R., 1986, "The Scope of Renaissance Mechanics", OSIRIS, 2nd series, 2: pp. 43-68.

———, 1991, "Archimedes among the Humanists", *ISIS*, 82: 629-38.

Landes, D., 1959, *Bankers & Pashas: International Finance and Economic Imperialism in Egypt* (Cambridge, Mass.: Harvard University Press)

Landels, J.G., 1978, *Engineering in the Ancient World* (University of California Press)

Latour, B., 1987, *Science in Action* (Milton Keynes: Open University Press)
————, 1990, "The Force and Reason of Experiment," in Le Grand (ed.), 1990, pp.49-80.
————, 1991, *We Have Never Been Modern* (Harvester Wheatsheaf)
Latour, B., and Woolgar, S., 1979, *Laboratory Life* (Sage Publications)
Lecourt, D., 1977, *Marxism and Epistemology: Bachelard, Canguilheim, Foucault* (Brewster, trans., London: New Left Books)
Leiss, W., 1972, *The Domination of Nature* (New York: George Braziller)
Lenin, V.I., *Collected Works* (Moscow: Progress Publishers, 1972)
Lennon, T.M., 1993, *The Battle of the Gods and Giants: The Legacies of Descartes and Gassendi, 1655-1715* (Princeton University Press)
Levidow, L., and Young, B., (eds.), 1981, *Science, Technology, and the Labour Process: Marxist Studies*. Vol. 1 (London: CSE)
Lewin, M., 1975a, *Lenin's Last Struggle* (London: Pluto Press)
————, 1975b, *Political Undercurrents in Soviet Economic Debates: From Bukharin to the Modern Reformers* (London, 1975)
Liebman, M., 1975, *Leninism under Lenin* (London: Jonathon Cape)
Lindberg, D.C., (ed.), 1978, *Science in the Middle Ages* (Chicago University Press)
Lindberg, D.C., and Westman, R.S., 1990, *Reappraisals of the Scientific Revolution* (University of Cambridge Press)
Long, P.O., 1997, "Power, Patronage, and the Authorship of Ars: From Mechanical Know-how to Mechanical Knowledge in the Last Scribal Age." *ISIS*, 88, pp. 1-41.
Lovejoy, A.O., 1936, *The Great Chain of Being: A Study of the History of an Idea* (Harvard University Press)
Lovekin, D., 1991, *Technique, Discourse, and Consciousness: An Introduction to the Philosophy of Jacque Ellul* (Bethlehem, Pa: Lehigh University Press)
Lovelock, J., 1979, *Gaia: A New Look at Life on Earth* (Oxford University Press)
————, 1988, *The Ages of Gaia: a Biography of Our Living Earth* (New York: Norton)
Lovins, A.B., 1979, *Soft Energy Paths: Toward a Durable Peace* (New York: Harper & Row)
Lukács, G., 1967, *History and Class Consciousness* (Livingston, trans., London: Merlin Press)
————, 1978, *The Ontology of Social Being: 3. Labour* (London: Merlin Press)
Luxemburg, R., 1918, *The Russian Revolution* (London)
Lyotard, J-F., 1991, *The Postmodern Condition: A Report on Knowledge* (Bennington and Massumi (trans.), Manchester University Press)
Machamer. P., 1998, "Galileo's machines, his mathematics, and his experiments" in *The Cambridge Companion to Galileo* (Machamer, ed., Cambridge University Press)
MacIntyre, A., 1984, *After Virtue: A Study in Moral Theory*, Indiana: University of Notre Dame Press.
Mackenzie, D., 1984, "Marx and the Machine," *Technology and Culture*, 25, July 1984, pp. 473-502
Mamdani, M., 1972, *The Myth of Population Control* (New York: Monthly Review Press)
Mandel, D., 1984, *The Petrograd Workers and the Soviet Seizure of Power* (Basingstoke: Palgrave Macmillan),
Mandelbaum, M., 1964, *Philosophy, Science, and Sense Perception: Historical and Critical Studies* (Baltimore: John Hopkins University Press)
Manheim, K., 1936, *Ideology and Utopia* (Routledge and Kegan-Paul)
————, 1951, *Man and Society in an Age of Reconstruction* (New York: Harcourt & Brace)
Marcus, G.E., (ed.), 1986, *Technoscientific Imaginaries*, University of Chicago Press.
Mathias, P., 1972, "Who Unbound Prometheus? Science and Technical Change, 1600-1800" in Mathia (ed.), 1972, pp. 54-80.
————, (ed.), 1972, *Science and Society 1600- 1900* (Cambridge University Press)
Marchant, C., 1994, *Ecology* (NJ: Humanities Press)
Marcuse, H., 1958, *Soviet Marxism* (New York, NY: Columbia University Press)
————, 1971, *An Essay on Liberation* (Boston: Beacon Press)
————, 1974, *Eros and Civilization: A Philosophical Inquiry into Freud* (Boston: Beacon Press)
————, 1978, *The Aesthetic Dimension* (Boston: Beacon Press)
————, 1989, *Counterrevolution and Revolt* (Boston: Beacon Press)
————, 1991, *One-Dimensional Man: Studies in the Ideology of Advanced Industrial Society*, (Boston: Beacon Press)

―――――, 1999, *Reason and Revolution: Hegel and the Rise of Social Theory* (Humanity Books)
―――――, 2001a, *Technology, War, and Fascism* (Kellner, ed., New York: Routledge)
―――――, 2001b, *Towards a Critical Theory of Society* (Kellner, ed., New York: Routledge)
―――――, 2001c, *The New Left and the 1960s* (Kellner, ed., New York: Routledge)
Marx, K., 1904, *A Contribution Towards the Critique of Political Economy* (New York: International Publishers)
―――――, 1938, *Critique of the Gotha Program* (New York: International Publishers)
―――――, 1967, *Capital, vol. 1, A Critical Analysis of Capitalist Production* (Engels ed., More and Aveling, trans., New York: International Publishers)
―――――, 1970, *The Economic and Philosophical Manuscripts of 1844* (Lawrence & Wishart)
―――――, 1975, *The Holy Family* (New York, International Publishers)
―――――, 1992, *Poverty of Philosophy* (New York: International Publishers)
―――――, 1993, *Grundrisse: Foundations of the Critique of Political Economy* (Penguin)
―――――, 2011, *The German Ideology* (Martino Fine Books)
Masterman, M., 1965, "The Nature of a Paradigm," in Lakatos, I., and Musgrave, A., *Criticism and the Growth of Knowledge: Proceedings of the 1965 International Colloquium in the Philosophy of Science 4* (3 ed., Cambridge: Cambridge University Press, 1970), pp. 59–90
Maxwell, N., 1984, *From Knowledge to Wisdom* (Oxford University Press)
―――――, 1998, *The Comprehensibility of the Universe* (Oxford University Press)
―――――, 2009, "Replies," in Leemon McHendry, ed., *From Knowledge to Wisdom: Studies in the Thought of Nicholas Maxwell* (Frankfurt: Ontos Verlag, 2009)
McClintock, P.V.E., et al., 1992, *Matter at Low Temperatures* (Glasgow: Blackie)
McMullin, E., (ed.), 1967, *Galileo: Man of Science* (New York: Basic Books)
Meadows, A.J., 1972, *Science and Controversy: A Biography of Sir Norman Lockyer* (London: Macmillan Press)
Merleau-Ponty, M., 1999, *Phenomenology of Perception* (Smith, trans., London: Routledge)
―――――, 1987, *The Visible and the Invisible* (Lefort, ed., Lingis, trans., Northwestern University Press)
Merton, R.K., 1938, "Science, Technology, and Society in Seventeenth Century England" *Osiris*, Vol. IV, pt. 2, pp. 360–632
―――――, 1968, *Social Theory and Social Structure* (Free Press)
―――――, 1979, *The Sociology of Science: Theoretical and Empirical Investigations*
Mesthene, E., 1970, *Technological Change* (New York: Signet)
Miller, A., 1978, *A Planet to Choose: Value Studies in Political Ecology* (New York: Pilgrim Press)
Miller, P., ed., 1956, *Errand into the Wilderness* (Cambridge, Mass.: Belknap Press)
Mintz, S.I., 1969, *The Hunting of the Leviathan: Seventeenth Century Reactions to the Materialism and Moral Philosophy of Thomas Hobbes* (Cambridge University Press)
Mitcham, C., 1994, *Thinking Through Technology: The Path Between Engineering and Philosophy* (Chicago and London: University of Chicago Press)
Moran, B., 2005, *Distilling Knowledge: Alchemy, Chemistry, and the Scientific Revolution* (Harvard University Press)
More, L.T., 1941, "Boyle as Alchemist," *Journal of the History of Ideas* 2 (1): 61–76
More, T., 1975, *Utopia* (New York: Norton)
Morhard, K.,D., et al., 1995, "2D liquid 3He near solidification: a highly correlated Fermi Liquid." *J. Low Temp. Phys.* 101: 161
―――――, 1996, "A two-dimensional Fermi-Liquid in the highly correlated regime: the second layer of 3He adsorbed on graphite." *Phys. Rev. B.* 53: 2658
Morpurgo, G., 1972, *A Search for Quarks (a Modern Version of the Millikan Experiment): One Researcher's Personal Account* (Genoa: Mimeo)
Mueller, M, 1987, "Technology out of Control," *Critical Review*, 1.4, pp. 24-40
Mumford, L., 1963, *Technics and Civilization* (Harcourt Brace & Company)
Murdoch, J.E., and Sylla, E.D., 1978, "The Science of Motion", in Lindberg, ed., 1978, pp. 206-224
Myklebust, S., ed., 1997, *Technology and Democracy: Obstacles to Democratisation: Productivism and Technocracy* (Oslo: Centre for Technology and Culture)
Naess, A., 1973, "The Shallow and the Deep, Long-Range Ecology Movement: A Summary" *Inquiry* 16 (1973): 95-100

———, 1989, *Ecology, Community, and Lifestyle* (Rothenberg, trans., Cambridge University Press)
Nagel, E., 1961, *The Structure of Science: Problems in the Logic of Scientific Explanations* (New York: Harcourt, Brace and World)
Nagel, T., 1986, *The View from Nowhere* (Oxford University Press)
Naam, R., 2005, *More Than Human: Embracing the Promise of Biological Enhancement* (Broadway)
Naylor, R.H., 1989, "Galileo's Experimental Discourse.", in Gooding, *et al.*, eds., 1989, pp. 117-34.
Newman, W.R., 2004, *Promethean Ambitions* (University of Chicago Press)
———, 2007, "Newton and Alchemy," *The Chymistry of Isaac Newton Project*
Neurath, O., and Carnap, R., ed., 1955, *Foundations of the Unity of Science: Toward an International Encyclopedia of Unified Science* (University of Chicago Press)
Newton, I., 1962, *The Mathematical Principles of Natural Philosophy and The System of the World*, 2 vols. (Cajori, ed., trans., University of California Press)
Nietzsche, F.W., 1983, *The Birth of Tragedy Out of the Spirit of Music* (Tanner, ed., Whiteside, trans., Penguin Books)
Norris, C., 2000, *Quantum Theory and the Flight from Realism: Philosophical Responses to Quantum Mechanics* (Routledge)
Odum, H.I., 1971, *Environment, Power, and Society* (New York: Wiley)
Olsen, R.G., 2009, *Technology and Science in Ancient Civilizations* (ABC-CLIO)
Ortega y Gasset, J., 1993, *Revolt of the Masses* (New York: Norton)
Osler, M.J, 1994, *Divine Will and the Mechanical Philosophy: Gassendi and Descartes on Contingency and Necessity in the Created World* (Cambridge University Press)
Pacey, A., 1974, *The Maze of Ingenuity: Ideas and Idealism in the Development of Technology* (Allen Lane)
Pandit, G.L., and Dosch, H.G., 2014, *The Frontiers of Theory Development in Physics: A Methodological Study in its Dynamic Complexity* (Los Angeles, CA: Trébol Press)
Parsons, H., 1977, *Marx and Engels on Ecology* (Westpoint, Conn.: Greenwood Press)
Peirce, C.S., 1932-58, *The Collected Works of Charles Saunders Peirce*, 8 volumes (vols. 1-6, Hartshorne and Weiss, eds.; vols. 7-8, Burks, ed., Mass.: Cambridge)
———, 1955, "The Doctrine of Necessity Examined." in *Philosophical Writings of Peirce* (Buchler ed., Dover Publications: New York), pp.324-38.
Phillipson, S.L., *et al.*, 1998, "Mesoscopic behaviour of the neutral Fermi gas 3He confined in quantum wires." *Nature* 395: 578.
Pickett, G.R., *et al.*, 1994, "Superfluid 3He at very low temperatures: a very unusual excitation gas." *Physica B* 197: 390
Pickering, A., 1987, *Constructing Quarks: A Sociological History of Particle Physics* (Edinburgh University Press)
———, ed., 1992, *Science as Practice and Culture* (University of Chicago Press)
———, 1995, *The Mangle of Practice: Time, Agency, and Science* (University of Chicago Press)
Pines, S., 1996, "The Origin of the Tale of Salaman and Absal: A Possible Indian Influence," in
Pines, S., *Studies in the History of Arabic Philosophy* (Jerusalem: Magnes Press, 1996), pp.343-53
Plato, 1997, *The Complete Works* (Cooper, ed., Hackett Publishing)
Pliny, 1991, *Natural History* (Healy, trans., Penguin Classics)
Plutarch, 1961, *Lives* (Perrin, trans., Cambridge: Harvard University Press)
Polanyi, M., 1958, *Personal Knowledge* (Routledge & Keegan Paul)
———, 1946, *Science, Faith, and Society: A Searching Examination of the Meaning and Nature of Scientific Inquiry* (University of Chicago Press)
Popper, K.R., 1974, "Replies to Critics", in Schilpp (ed.), *The Philosophy of Karl Popper* (Open Court, La Salle, Ill.)
———, 1975, *The Logic of Scientific Discovery* (London: Hutchinson)
———, 1963, *Conjectures and Refutations: The Growth of Scientific Knowledge* (Routledge)
———, 1945, *The Open Society and its Enemies: Volume 1 The Spell of Plato, and Volume 2 The High Tide of Prophecy: Hegel, Marx, and the Aftermath* (Routledge)
Postman, N., 1992, *Technopoly: The Surrender of Culture to Technology* (New York: Vintage Books)
Price, B., ed., 2003, *Francis Bacon's The New Atlantis: New Interdisciplinary Essays* (Manchester University Press)
Prigogine, I., and Stengers, I., 1984, *Order Out of Chaos* (Bantam Books)

Principe, L., 2000, *The Aspiring Adept: Robert Boyle and his Alchemical Quest* (Princeton University Press)
Proudhon, J-P., 2011, *The Philosophy of Poverty* (Theophania Publishing)
Psillos, S., 1999, *Scientific Realism: How Science Tracks the Truth* (Routledge)
Putnam, H., 1973, "Explanation and Reference" in *Conceptual Change* (Pearce and Maynard (eds.), Dordrecht: Reidel), pp. 199-221)
————, 1975, "The meaning of 'meaning'" in *Language, Mind and Knowledge*, Gunderson (ed..), (University of Minnesota Press, pp. 131-193)
————, 1985, *Philosophical Papers: Volume 3, Realism and Reason* (Cambridge University Press)
Rautman, M.L., 2006, *Daily Life in the Byzantine Empire* (Greenwood Publishing Group)
Redondi, P., 1987, *Galileo Heretic* (Rosenthal, trans., Princeton University Press)
Reed, J., 1966, *Ten Days that Shook the World* (Penguin Books)
Rescher, N., 1969, "What is Value Change? A Framework for Research," in Baier, K., and Rescher, N., eds., *Values and the Future* (New York: The Free Press, 1969)
Rhys Williams, A., 1921, *Through the Russian Revolution* (New York: Bons and Liveright)
Richards, J., 1965, *The Philosophy of Rhetoric* (New York)
Ricoeur, P., 1987, *The Rule of Metaphor* (Czerny et al. (trans.), Routledge & Keegan-Paul)
Rogers, K.A., 1992, "Selection of Tau-Pair Events from the DELPHI 1991 data." (Lancaster University)
————, 1994, "Measurements of Tau-lepton Polarisation and Forward-Backward Asymmetry as Tests of Electroweak Theory" (Lancaster University/CERN)
————, 2005, *On the Metaphysics of Experimental Physics* (Basingstoke & New York: Palgrave Macmillan)
————, 2006, *Modern Science and the Capriciousness of Nature* (Basingstoke & New York: Palgrave Macmillan)
————, 2008, *Participatory Democracy, Science and Technology* (Basingstoke & New York: Palgrave Macmillan)
————, 2009, "Metaphysics and Methodology: Aim-Oriented Empiricism" in Leemon McHendry, ed., *From Knowledge to Wisdom: Studies in the Thought of Nicholas Maxwell* (Frankfurt: Ontos Verlag, 2009).
————, 2011, *Debunking Glenn Beck: How to Save America from Media Pundits and Propagandists* (Santa Barbara, CA: Praeger)
————, 2012, *Occupy Media! Propaganda and the Free Press* (Los Angeles, CA: Trébol Press)
————, 2014, "Education and the Ninth Amendment" in Byczkiewicz, V., ed., *Democracy and Education: Collected Perspectives* (Los Angeles, CA: Trébol Press)
————, 2015, "Plato and the Banishment of the Artist" in Reeve, H., ed., *Ymedaca* (Yorkshire Sculpture Park)
————, 2016, *Propaganda and the Free Press* (Los Angeles, CA: Trébol Press)
Roll-Hansen, N., 2005, *The Lysenko Effect: The Politics of Science* (New York: Humanity Books)
Rorty, R., 1981, *The Mirror of Nature* (Princeton University Press)
Rose, P.L., 1975, *The Italian Renaissance of Mechanics* (Geneva: Droz)
Rose, P.L., and Drake, S., 1971, "The Pseudo-Aristotelian *Questions of Mechanics* in Renaissance Culture" in *Studies in the Renaissance*, 18, pp.65-104
Rossi, B., 1968, *Francis Bacon: From Magic to Science* (Rabinovich, S., trans., London: Routledge & Keegan Paul)
Rossi, P., 1970, *Philosophy, Technology, and the Arts in the Early Modern Era* (Attanasio (trans.), Nelson (ed.), New York: Harper & Row)
Rostovtzeveff, M., 1941, *Social and Economic History of the Hellenistic World* (Oxford)
Rouse, J., 1987, *Knowledge and Power* (Ithaca, NY: Cornel University Press)
Rousseau, J-J., 1987, *The Basic Political Writings* (Indianapolis: Hackett)
Rubidge, B., 1990, "Descartes' *Meditations* and Devotional Meditations," *Journal of the History of Ideas*, 51, pp. 22-49.
Russel, B., 1962, *The Scientific Outlook* (Norton & Company)
Sambursky, S., 1987, *The Physical World of the Ancient Greeks* (Dugut (trans.), London: Routledge & Keegan-Paul)
Sargent, R-M., 1995, *The Diffident Naturalist: Robert Boyle and the Philosophy of Experiment* (University of Chicago Press)

Sarles, H., 1985, *Language and Human Nature: Toward a Grammar of Interaction and Discourse* (University of Minnesota Press)

Sayer, K.M., 1976, *Cybernetics and the Philosophy of Mind* (International Library of Philosophy and Scientific Method, Routledge & Keegan Paul)

Schadewaldt, W., 1979, "The Concepts of *Nature* and *Technique* according to the Greeks." in *Research in Philosophy and Technology* 2 (Caroll, Mitcham, and Mackey, trans.), pp. 159-71

Schaffer, S., 1986, "Glass works: Newton's prisms and the uses of experiment." in Gooding *et al.* (eds.), 1986, pp. 67-104.

Schaffer, S., *et al.* (eds.), 1991, *William Whewell: A Composite Portrait* (Cambridge University Press: Cambridge)

Schmidt, A., 1971, *The Concept of Nature in Marx* (Fowkes, trans., London: New Left Review)

Schmitt, C.B., 1975, "Science in the Italian Universities in the Sixteenth and Early Seventeenth Centuries", in Crosland (ed.), 1975, pp. 33-56.

————, 1983, *Aristotle and the Renaissance* (Harvard University Press)

Sclove, R.E., 1995, *Democracy and Technology* (New York: Guildford Press)

Seeger, R.J., 1966, *Galileo Galilei, His Life and His Works* (Oxford: Pergamon Press)

Seifer, M., 2001, *Wizard: The Life and Times of Nikola Tesla* (Citadel Press)

Settle, T., 1967, "Galileo's Use of Experiment as a Tool of Investigation" in *Galileo: Man of Science* (McMullin, ed., New York: Basic Books)

Shanin, T., 1983, *Late Marx and the Russian Road: Marx and the 'Peripheries of Capitalism'* (London)

Shapin, S., and Schaffer, S., 1985, *Leviathan and the Air-Pump: Hobbes, Boyle, and the Experimental Life* (Princeton University Press)

Sharp, L., 1973, "Walter Charleston's Early Life, 1620-1659, and the Relationship to Natural Philosophy in mid-Seventeenth Century England" in *Annals of Science*, 30, pp. 311-40.

Shaw, N.S., *et al.*, 1996, "Quasiparticle Molasses: the Giant Force on an Object Moving through a Beam of Thermal Excitations in Superfluid 3He." *Czech. Journal of Physics* 46: 247.

————, 1998a, "Persistent Precessing Texture-Trapped Domains in Low Temperature Superfluid 3He." *J. Low Temp. Phys.* 110: 57.

————, 1998b, "Ultra-low temperature magnetic properties of liquid 3He films." J. Low Temp. Phys. 110: 333.

Shea, W.R., 1972, *Galileo's Intellectual Revolution* (London)

————, 1991, *The Magic of Numbers and Motion: René Descartes' Scientific Career* (Mass: Science History Publications)

Silvard, R.L., 1987, *World Military and Social Expenditures* (Washington D.C.: World Priorities)

Simms, D.L., 1987, "Archimedes and the Invention of Gunpowder." in *Technology and Culture*, 28, pp. 67-79.

————, 1988, "Archimedes' Weapons of War and Leonardo" in *British Journal for the History of Science*, 21, pp. 195-210

Simon, H., 1981, *The Sciences of the Analytical* (Cambridge: MIT Press)

Singer, P., 2000, *Darwinism and the New Left: Politics, Evolution, and Cooperation* (Yale University Press)

Skolimowski, H., 1976, "The Ecology Movement Re-examined" *Ecologist* 6

Smith, M.R., and Marx, L., eds., *Does Technology Drive History? The Dilemma of Technological Determinism* (Cambridge, Mass: MIT Press, 1994)

Snow, C.P., 1964, *Two Cultures and a Second Look* (Cambridge University Press)

Solzhenitsyn, A., *The Gulag Archipelago: An Experiment in Literary Investigation* (Whitney, trans., London: Harper & Row, 1973)

Sophocles, 1993, "Ode to Man" in *Antigone* (New York: Dover Publications)

Sorabji, R., 1988, *Matter, Space, and Motion: Theories in Antiquity and Their Sequel* (Cornell University Press)

Sourbut, E., 1996, "Gynogenesis: A Lesbian Appropriation of Reproductive Technologies" in Lykke, N., and Braidotti, R., eds., *Between Monkeys, Goddesses, and Cyborgs: Feminist Confrontations with Science, Medicine, and Cyberspace* (London: Zed Books, 1996), pp.227-41.

Spirn, A., 1984, *The Granite Garden: Urban Nature and Human Design* (New York: Basic Books)

Spragen, Jr., T.A., 1973, *The Politics of Motion: The World of Thomas Hobbes* (University of Kentucky Press)

Stalin, J. V., 1938, *Dialectical and Historical Materialism* (International Publishers)

Standing Bear, L., 1971, *Touch the Earth* (McLuhan, ed., New York: Simon and Schuster)
Stiegler, B., 1998, *Technics and Time, 1: The Fault of Epimetheus* (Beardsworth and Collins, trans., Stanford University Press)
Strong, E.W., 1966, *Procedures and Metaphysics: A Study in the philosophy of mathematical-physical science in the sixteenth and seventeenth centuries* (Georg Ohms)
Suchman, L., 1987, *Plans and Situated Actions* (Cambridge University Press, 1987)
Sutela, P., 1991, *Economic Thought and Economic Reform in the Soviet Union* (Cambridge University Press)
Suter, R., 1969, "The Scientific Work of Alessandro Piccolomini" in *ISIS*, 60, pp. 210-222.
Svenson, L.S., 1972, *The Ethereal Aether: A History of the Michelson-Morley-Miller Aether-drift Experiments, 1880-1930* (University of Texas)
Temple, R., 2000, *The Crystal Sun* (Century Publications)
Thomas, P., 1994, *Alien Politics: Marxist State Theory Revisited* (New York: Routledge)
Thorndike, L., 1934, *History of Magic and Experimental Science* (New York: Columbia University Press)
Torrance, R.M., (ed.), 1999, *Encompassing Nature: Nature and Culture from Ancient Times to the Modern World* (Washington: Counterpoint)
Trotsky, L., 1960, *Literature and Revolution* (University of Michigan)
Tuchel, K., 1982, "Friedrich Dessauer as Philosopher of Technology: Notes on his Dialogue with Jaspers and Heidegger" in *Research in Philosophy and Technology*, vol. 5, pp. 269-280 (Durbin, ed., Greenwich, CT: JAI Press, 1982)
Umber, D.Z., 2000, *Holding the Line: The Telephone in Old Order Mennonite and Amish Life* (John Hopkins University Press)
Usher, A.P., 1962, *A History of Mechanical Inventions* (Harvard University Press)
Van Doren, C., 1938, *Benjamin Franklin* (New York: Viking Press)
Van Helden, A., 1983, "The birth of the modern scientific instrument" in Burke, ed., 1983, pp. 49-84
Vogel, S., 1996, *Against Nature: The Concept of Nature in Critical Theory* (NY: State University of New York Press)
Wajcman, J., 1984, *Feminism Confronts Technology* (Cambridge: Polity)
Wallace, W.A., 1984, *Galileo and his Sources: The Heritage of the Collego Romano in Galileo's Science* (Princeton University Press)
Wallis, R., and Baran, S., 1990, *The Known World of Broadcast News* (London: Routledge)
Walsh, V., 1980, "Contraception: The growth of a technology." in Burke, eds., 1980, pp. 182-207
Ward, C., 1983, *Housing: An Anarchist Approach* (London: Freedom Press)
Wartofsky, M., 1997, "Technology, Power, and Truth," in Winner, ed., 1997, pp.15-34
Waterlow, S., 1982, *Nature, Change, and Agency, in Aristotle's Physics* (Oxford: Clarendon Press)
Watson, J.D., 1968, *The Double Helix* (New York: Mentor Books)
Weber, M., 1958, *The Protestant Ethic and the Spirit of Capitalism* (Parsons, trans., New York: Scribners)
————, 1972, "Science as a Vocation" in *From Max Weber* (in Gerth and Mills, ed., New York: Oxford University Press)
————, 1978, *Economy and Society* (2 vol., Berkeley, CA: University of California Press)
Weiner, N., 1965, *Cybernetics or Control and Communication in the Animal and the Machine* (MIT Press)
————, 1964, *God and Golem, Inc.* (Cambridge: MIT Press)
Westfall, R.S., 1962, "The Foundations of Newton's Philosophy of Nature" in *British Journal for the History of Science*, I: pp. 171-82.
————, 1971, *Force in Newton's Physics: The Science of Dynamics in the Seventeenth Century* (New York: American Elsevier)
————, 1983, "Robert Hooke, mechanical technology, and scientific investigation." in Burke (ed.) 1983, pp. 85-110.
Westman, R.S., 1975, "Three Responses to Copernican Theory: Johannes Praetorius, Tycho Brahe, and Michael Maestlin." in *The Copernican Achievement* (Westman, ed., University of California Press), pp. 285-345.
White, K.D., 1984, *Greek and Roman Technology* (Thames and Hudson)
White, L., 1962, *Medieval Technology and Social Change* (Oxford University Press)
————, 1967, "The Historical Roots of Our Ecological Crisis" in *Science* 155, March

Whitehead, A.N., 1997, *Science and the Modern World* (Free Press)

—————, 1964, *The Concept of Nature* (Cambridge University Press)

Whittaker, E.T., 1989, *A History of the Theories of Aether & Electricity: The Classical Theories/The Modern Theories* (2 Volumes, Dover Publications)

Wightman, W.P.D., 1962, *Science and the Renaissance*, vol.1 (Edinburgh: Oliver and Boyd)

Winner, L., 1977, *Autonomous Technology: Technics-out-of-Control as a Theme in Political Thought.* (Cambridge: MIT Press)

—————, 1886, *The Whale and the Reactor* (University of Chicago Press)

—————, ed., 1992, *Democracy in a Technological Society* (Kluwer Academic Publishers)

Woddis, J., 1972, *New Theories of Revolution* (New York: *International Publishers*)

Wolpert, L., 1992, *The Unnatural Nature of Science* (London: Faber and Faber)

Wolters, G., and Salmon, W.C., (eds.), 1994, *Logic, Language, and the Structure of Scientific Theories: Proceedings of the Carnap-Reichenbach Centennial, University of Konstanz, 21-24 May 1991* (University of Pittsburgh Press)

Woodruff, P., 1993, *Thucydides on Justice, Power and Human Nature* (Indianapolis: Hackett)

Yates, F., 1964, *Giordano Bruno and the Hermetic Tradition* (Chicago University Press)

Yearly, S., 1988, *Science, Technology, and Social Change* (London: Unwin Hyman)

Yoder, J.G., 1988, *Unrolling Time: Christiaan Huygens and the Mathematization of Nature* (Cambridge University Press)

Zaleski, E., *et al.*, 1969, *Science Policy in the USSR* (Pairs: OECD Publications)

Zamyatin, Y., 1924, *We* (New York: E.P. Dutton)